Biochemistry of Wounded Plant Tissues

Biochemistry of Wounded Plant Tissues

Editor Günter Kahl

Walter de Gruyter · Berlin · New York 1978

Editor

Prof. Dr. Günter Kahl
Fachbereich Biologie/Botanik
Johann Wolfgang Goethe-Universität
Siesmayerstraße 70
D - 6000 Frankfurt am Main, Germany

With 178 Figures

Cip-Kurztitelaufnahme der Deutschen Bibliothek

Biochemistry of wounded plant tissues/ed. Günter Kahl. –
1. Aufl. – Berlin, New York: de Gruyter, 1978.
ISBN 3-11-006801-X

NE: Kahl, Günter [Hrsg.]

Library of Congress Cataloging in Publication Data

Main entry under title:

Biochemistry of wounded plant tissues.

Includes index.
1. Plant cells and tissues. 2. Plants- Wounds and injuries. I. Kahl, Günter, 1936 –
QK725.B57 581.2 78-14468
ISBN 3-11-006801-X

 Printing: Mercedes-Druck, Berlin. – Binding: Lüderitz & Bauer, Buchgewerbe GmbH, Berlin. – Printed in Germany

PREFACE

Over a hundred years ago - at a time when classical plant physiology, then dominated by Julius Sachs and Wilhelm Pfeffer, had yet to come to the fore - there arose the idea of studying parts of plants rather than whole organisms. An understanding of the parts was deemed indispensable for understanding the whole. For this reason pieces of various tissues were excised from intact plant organs and examined critically under the microscope, by then already developed. The events triggered by excision were easily reproducible, and stimulated the imagination and enthusiasm of Hugo von Mohl (then in Tübingen, Germany), leading him to remark in 1849: "Even the smallest part of a living tissue, when isolated from the organism, reacts as does the organism itself: it does its utmost to survive. To study the part is to recognize the whole." ("Selbst das kleinste Stück an lebendem Gewebe reagiert, vom Organismus isoliert, wie dieser selbst: es setzet alles dran zu überleben. Studiret man das Stück, erkennet man das Ganze.")

The century since the death of von Mohl has seen the fulfillment of his prophecy: the tissue slice system - and particularly the storage tissue slice system - has now been widely accepted as a model system in plant physiology and plant biochemistry, principally because of the versatility of its application within the whole spectrum of problems in plant biology.

Excision of such tissues brings about cellular events which lead to a de-differentiation of differentiated storage cells. The CYTOLOGY of cells of excised tissues may then answer general questions of plant biology such as, for example, the transformations of cellular membranes, the assembly of microtubuli, drastic changes in nuclear and nucleolar structure, the genesis of various organelles, or the formation of cellulose walls and their modifications. MORPHOGENETIC PROCESSES in excised tissues resemble those occurring in the "normal" life cycle of the whole plant, i.e. the formation of new tissues or tissues with new capabilities and the organisation of various organs. In fact, this area was the cradle of plant tissue culture; at a time when the isolation and culture of single cells were a mere dream to physiologists, small pieces of plant storage tissue were kept alive over extended periods of time. The METABOLIC PROCESSES of excised tissues aroused deep interest, and their thorough analysis brought many insights into cell metabolism and its regulation. The metabolic basis for the transition of a cell from its dormant to its activated state can be studied in detail, aiding our understanding of breaking dormancy in the plant kingdom in general. The fascinating problem of the interplay between nucleus, cytoplasm and the various organelles with processes involving various aspects of nucleic acids and proteins and their regulation by hormones and other parameters has been attacked successfully using excised tissues. These systems communicate intensively with their environment - as is documented by the active uptake of solutes - as do their different parts with one another (through membranes and membrane-dependent conductive processes). The WOUND-HEALING PROCESS as a whole leads to the effective closure of wound sites, guarantees the survival of the threatened organism and convincingly bears witness to the capability of the plant to resist challenges from the environment.

Most PLANT PATHOGENS penetrate into host tissues through wounded areas. The interrelationship between such pathogens of viruses, bacteria, fungi, higher plants or even lower animals, and the recipient cells could be illuminated by plant storage tissue model systems at a cellular and molecular level.

After 50 years of scientific struggle TUMOR INDUCTION in higher plants is now open to a molecular approach. The inciting Agrobacterium tumefaciens bears a large plasmid, part of which is transferred from its vehicle to the host cell by a sex-factor-like activity. This transposon-like element is very likely integrated into the cell genome and uncontrolledly transcribed. The latter system is on a molecular level strikingly similar to the animal and probably also to the human tumor system. In contrast to the latter, however, the wounding of recipient tissues is a necessary prerequisite for the attachment of the bacterium at specific sites of the host tissue as well as the transfer of the oncogenes and probably their integration. Again, the storage tissue slice system offers a unique possibility to study tumor induction and also adds to our limited knowledge in this important area of molecular biology.

The contributors to this volume have striven to offer a complete review of all aspects of wound-healing in plant storage organs. It is hoped that the book will not only be informative, but will stimulate thinking on many topics which once seemed settled, bringing about new activity and kindling enthusiasm in this important field of biology.

Frankfurt am Main December 1977 Günter Kahl

CONTENTS

LIST OF CONTRIBUTORS

ASAHI,TADASHI
Laboratory of Biochemistry,Department of Agricultural Chemistry,
Nagoya University,Chikusa,Nagoya,Japan

BARCKHAUSEN,RALF
Department of Biology,University of Frankfurt/Main,Siesmayerstr.70
6000 Frankfurt/Main,GFR

BENVENISTE,PIERRE
Institut de Botanique,Laboratoire de Biochimie Végétale,28 Rue Goethe,
67083 Strasbourg Cedex,France

BYRNE,HENRY
Department of Biology,E.L.B.A.,Carleton University Ottawa,Ontario
K1S 5B6,Canada

CHERRY,JOE H.
Purdue University,Agricultural Experiment Station,Department of
Horticulture,Lafayette,Indiana 47907,USA

CHRISPEELS,MAARTEN,J.
Department of Biology,University of California,La Jolla,California
92037,USA

DAVIES,DAVID D.
University of East Anglia,School of Biological Sciences,University
Plain,Norwich NOR 88C,England

DIZENGREMEL,PIERRE
Université Pierre et Marie Curie,Laboratoire de Biologie Végétale IV,
12,Rue Cuvier,75005 Paris,France

GALLIARD,TERENCE
Agricultural Research Council,Food Research Institute,Colney Lane,
Norwich NR4 7UA,England

KADER,JEAN-CLAUDE
Laboratoire de Physiologie Végétale Appliquée,Physiologie cellulaire,
12,Rue Cuvier,Paris 5°,France

KAHL,GÜNTER
Department of Biology,University of Frankfurt/Main,Siesmayerstr.70,
6000 Frankfurt/Main,GFR

KOLATTUKUDY,P.E.
Department of Agricultural Chemistry,Washington State University, Pullman,Washington 99163,USA

KUĆ,JOE
Department of Plant Pathology,University of Kentucky,Lexington, Kentucky 40506,USA

LANCE,CLAUDE
Université Pierre et Marie Curie,Laboratoire de Biologie Végétale IV, 12,Rue Cuvier,75005 Paris,France

LATIES,GEORGE G.
Biology Department,Molecular Biology Institute,University of California,Los Angeles,California 90024,USA

LISKER,NORBERTO
Department of Plant Pathology,University of Kentucky,Lexington, Kentucky 40506,USA

MAZLIAK,PAUL
Laboratoire de Physiologie Végétale Appliquée,Physiologie cellulaire,12,Rue Cuvier,Paris 5°,France

ŌBA,KAZUKO
Laboratory of Biochemistry,Department of Agricultural Chemistry, Faculty of Agriculture,Nagoya University,Chikusa,Nagoya,Japan

PRATT,HARLAN K.
Department of Vegetable Crops,Agricultural Experiment Station,University of California,Davis,California 95616,USA

RHODES,MICHAEL
Food Research Institute,Agricultural Research Council,Colney Lane, Norwich NR4 7UA,England

ROSENSTOCK,GÜNTER
Department of Biology,University of Frankfurt/Main,Siesmayerstr.70
6000 Frankfurt/Main,GFR

SADAVA,DAVID
Joint Science Department,The Claremont Colleges,Claremont,California
91711,USA

SETTERFIELD,GEORGE
Department of Biology,E.L.B.A.,Carleton University Ottawa,Ontario
K1S 5B6,Canada

SPARKUHL,JOACHIM
Department of Microbiology and Parasitology,University of Toronto,
Toronto,Ontario,Canada M5S 1A1

URITANI,IKUZO
Laboratory of Biochemistry,Faculty of Agriculture,Nagoya University,
Nagoya,Chikusa 464,Japan

VAN STEVENINCK,ROBERT F.M.
University of Queensland,Department of Botany,St.Lucia,Brisbane,
Queensland 4067,Australia

WOOLTORTON L.S.C.
Food Research Institute,Agricultural Research Council,Colney Lane,
Norwich NR4 7UA,England

YANG,SHANG FA
Department of Vegetable Crops,Agricultural Experiment Station,University of California,Davis,California 95616,USA

Ultrastructural Changes in Wounded Plant Storage Tissue Cells

Ralf Barckhausen

INTRODUCTION

Plants are subject to various environmental conditions or factors, the action of which may entail a partial or even complete destruction of the plants' organism. If plants are not lethally damaged, however, they are able to regenerate destroyed tissue or to protect themselves against parasitic attack and loss of water through a process of wound-healing quite similar to animals. Usually the differentiated and hence specialized cells beneath a wound-surface, uncapable of cell division, are dedifferentiated, loose their original functions, regain mitotic activity and ultimately form meristematic tissues. This meristem or wound-cambium in the course of a new differentiation phase replaces the damaged tissue functionally or produces a layer of boundary tissue in the wounded region such as a periderm, thus sealing the wound (Fig. 1a,b,c,d). Wound-healing as a cytophysiological process then seems best suited for investigations of the developmental capabilities of plant cells.

For such investigations the storage tissue of plants has often been used, i.e. potato tubers, rhizomes of the Jerusalem artichoke or roots of the carrot (33, 34). These storage tissues altogether have the advantage of being composed of more or less homogeneous, parenchymatic cells and of being available in large quantities all through the year. Wounded plant tissues often form callus mainly in the presence of exogenously added plant growth substances (IAA; 2,4 D; kinetin or gibberellic acid). Such a callus consists of less specialized, mitotically active cells (Fig. 1c) and represents the starting material for tissue cultures which play an important role

Biochemistry of Wounded Plant Tissues

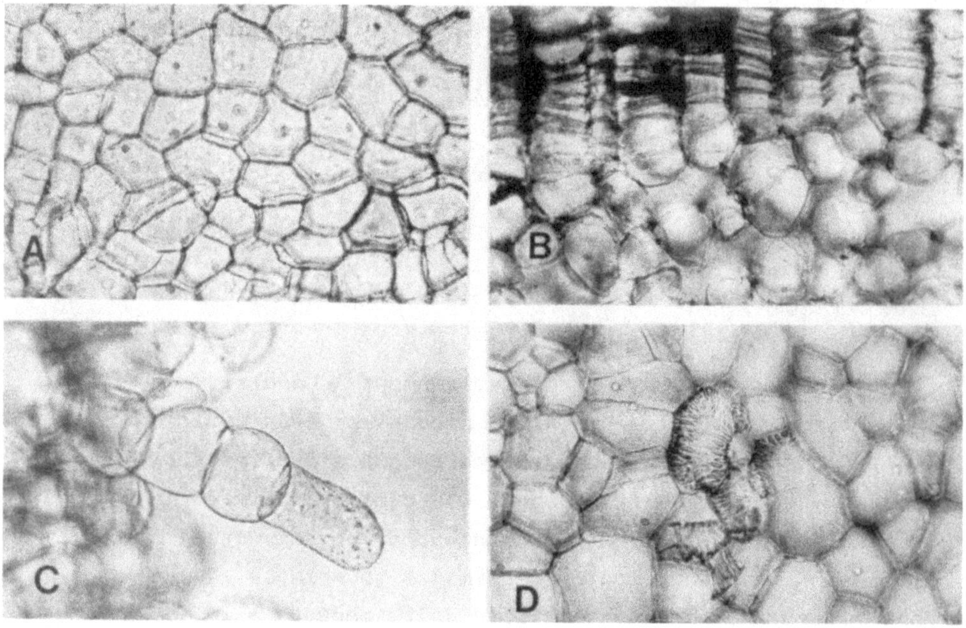

Figure 1. Developmental tendencies of wound-healing comfrey tissue.
A. Prototype.
B. Within 72 hours of wound-healing the injured tissue has formed a wound periderm.
C. Callus-like proliferations at the wound surface.
D. Regeneration of tracheidal elements in injured tissue.

in developmental physiology. The carrot system has become famous in this respect (43, 60).

Physiological investigations of the developmental tendencies in wounded plant tissues first of all require exact anatomical and morphological studies on the level of light and electron microscopy. For experimental work wounding of storage tissues is simply brought about by cutting it into small discs. These discs are placed into containers with high moisture or in aqueous solutions and left there for wound-healing. For electronmicroscopic examinations small pieces of tissue are fixed in O_sO_4, glutaraldehyde followed by O_sO_4, chromate/O_sO_4 or other substances. Then the pieces are embedded into plastics (e.g. epoxy resin) and cut into ultrathin sections. Finally the sections are treated with uranylacetate and/or lead-citrate and examined under the electron microscope.

LIGHTMICROSCOPIC CYTOLOGY

Ultrathin sections prepared for electronmicroscopy usually contain only a very small part of the relatively large cells of wound-healing storage tissue. Furthermore, it is impossible to follow changes in living cells of wound-healing tissues with the electron microscope. For these reasons it is necessary at the beginning to give a short light-microscopic view of the changes in injured tissues.

In many plant storage tissues the wound reaction is not linked to any meristematic activities of cells near the wound (33). In this case the cell walls beneath the wound surface become lignified only (e.g. Helianthus tuberosus). In Raphanus sativus a few irregular cell divisions and the development of callus occur side by side together with lignification. Wounded potato tissue finally develops a wound periderm.

All three types of wound reactions have in common that immediately after wounding the nucleus migrates to the cell wall which is adjacent to the wound or it moves to the middle of the cell, if mitotic division follows (3, 9, 25). At the same time, usually within 24 hours after the injury, the area of the nucleus increases by about 50% and that of the nucleolus by about 135% in Helianthus tuberosus (17). The corresponding values for the potato system are 168% and 300% respectively (35, 57). Afterwards the area of the nucleus and nucleolus decreases again. These are the only changes which can be observed with the light microscope in Helianthus tuberosus (17). In the case of potato and comfrey tissues the first cell divisions can be seen from the second up to the fifth cell layer beneath the wound surface within 24 hours after injury. Within

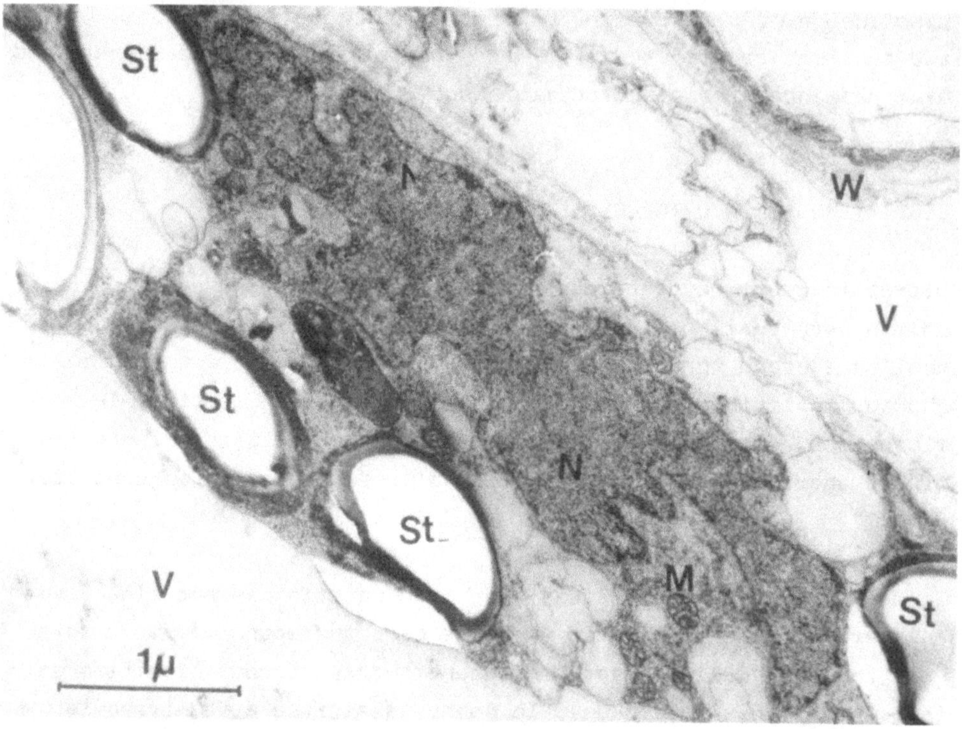

Figure 2. Ultrathin section of a cell from intact tuber tissue of Solanum tuberosum. The strongly lobed nucleus (N) is surrounded by numerous amyloplasts containing starch grains (St) and mitochondria (M). The cytoplasm of these highly vacuolated cells (V=vacuole) is restricted to the protoplasmic bag of the nucleus and to a thin cytoplasmic layer at the cell wall (W).

few days the mitotically active cells develop a solid wound periderm. The cells which directly border the wound, however, do not divide. They suberize and soon die. During the development of a wound periderm the starch of potato tuber cells is almost completely used up within the actively dividing cells. Developmental tendencies of wound healing in plant storage tissues which can be observed with the light microscope can be readily demonstrated with the comfrey (Scorzonera hispanica) as an example (Fig. 1a, b, c, d).

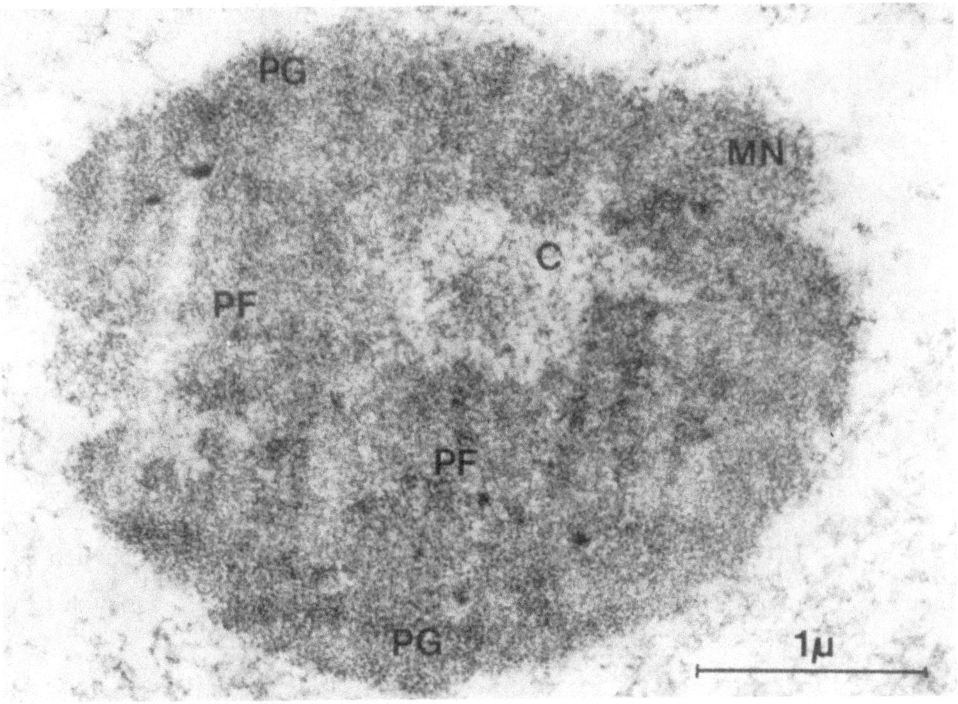

Figure 3. Nucleolus in cells of intact tuber tissue of Solanum tuberosum with a small peripheral layer of pars granulosa (PG). The nucleolus consists mainly of pars fibrosa (PF). C=chromatin, MN= micronucleolus.

ELECTRONMICROSCOPIC CYTOLOGY OF INTACT STORAGE TISSUE

The electron micrographs of the cells of intact storage tissue are generally very poor in structures (3, 5, 17, 25, 68). A giant vacuole nearly completely fills the interior of the cell. The cytoplasm contains only few organelles and covers the cell wall as a thin layer. Cytoplasmic strands are hardly to be found (Fig. 2). The nucleus is carried by a protoplasmic bag and usually located near the cell wall. Most frequently it is flattened and lens-shaped. Most organelles of the cell can be found in the area around the nucleus. This may be

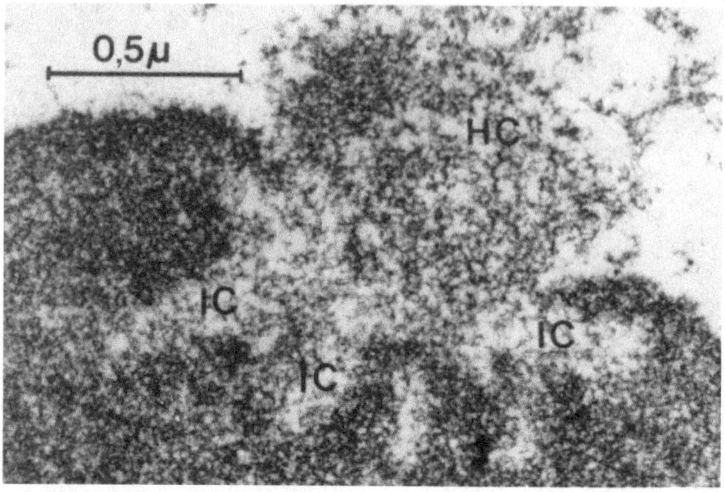

Figure 4. The surface of the nucleolus in cells of intact potato tuber tissue is associated with heterochromatin (HC), which is connected to the intranucleolar chromatin (IC).

especially well observed in the cells of the potato tuber. Here the nucleus is surrounded by numerous amyloplasts (Fig. 2). In the potato as well as in the Jerusalem artichoke system the surface of the nucleus is strongly lobed (3, 5; Fig. 2). The chromatin of plant storage tissue, e.g. eu- and heterochromatin, is more or less uniformely dispersed over the area of the nucleus. The nuclear membrane resembles the usual structure with its double unit membrane including many pores (3, 5, 17, 30). Each nucleus contains one or more nucleoli which have nearly identical structures in all storage tissues so far examined. They consist of a large central fibrillar part, the pars fibrosa, and a much smaller granular part, the pars granulosa. In quiescent cells the pars granulosa may be absent. This is especially true for dormant tuber tissue of Helianthus tuberosus (17, 29, 46). Within the pars fibrosa, to a lesser extent also in the pars granulosa, more or less electron dense regions may exist.

They are still fibrillar and are structurally equivalent to the intranucleolar chromatin (Fig. 3). In Helianthus tuberosus, Daucus carota and Solanum tuberosum the nucleolus of the interphase nucleus is always associated with heterochromatin, which quite frequently is linked to the intranucleolar chromatin (29, 30, 38, 46; Fig. 4). Aside of these elements of the nucleolus so far described, the pars fibrosa, pars granulosa and nucleolar chromatin, nucleolar vacuoles can also be demonstrated in quiescent cells of the potato tuber. They are relatively small and contain electron transparent material, which consists of few granules and fibrils (5, 6). Generally nucleolar vacuoles in nucleoli of resting tuber tissues are rarely to be found (29, 30, 46).

The nucleolus of potato tuber cells is always associated with a spherical body of 0,8 µm in diameter. According to Lafontaine (37) this body is called "micronucleolus". It contains densely packed fibrils and granules. The granules are definitely bigger than the ribosomal precursors of the pars granulosa (5; Fig. 3).

The plastids in quiescent cells of storage tissues are either leucoplasts or chromoplasts. After illumination of the tissue chloroplasts may also appear. In intact carrot tissues the plastids contain ribosomes and either starch or starch together with carotene (26). The plastids of the Jerusalem artichoke are characterised by numerous tubules and cisternas, the membranes of which can be connected to a relatively large membrane-bounded irregular body. Its content is more electron-dense than the material of the surrounding plastid stroma (63). This membrane-bounded "intraplastid body" is also characteristic for the plastids of the potato tuber (5, 41; Fig. 5). It contains much protein, which is used up for synthesis of the thylakoids when chloroplast development takes place (2, 21, 22, 26, 59). For this reason the "intraplastid body" also bears the denotation "prethylacoidal centre" (26). The plastids or amyloplasts respectively of potato tuber tissue also harbor a large amount of starch, phytoferritin, lipid droplets (plastoglobuli) and tubules similar to the widely known prolamellar body (Fig. 5,6).

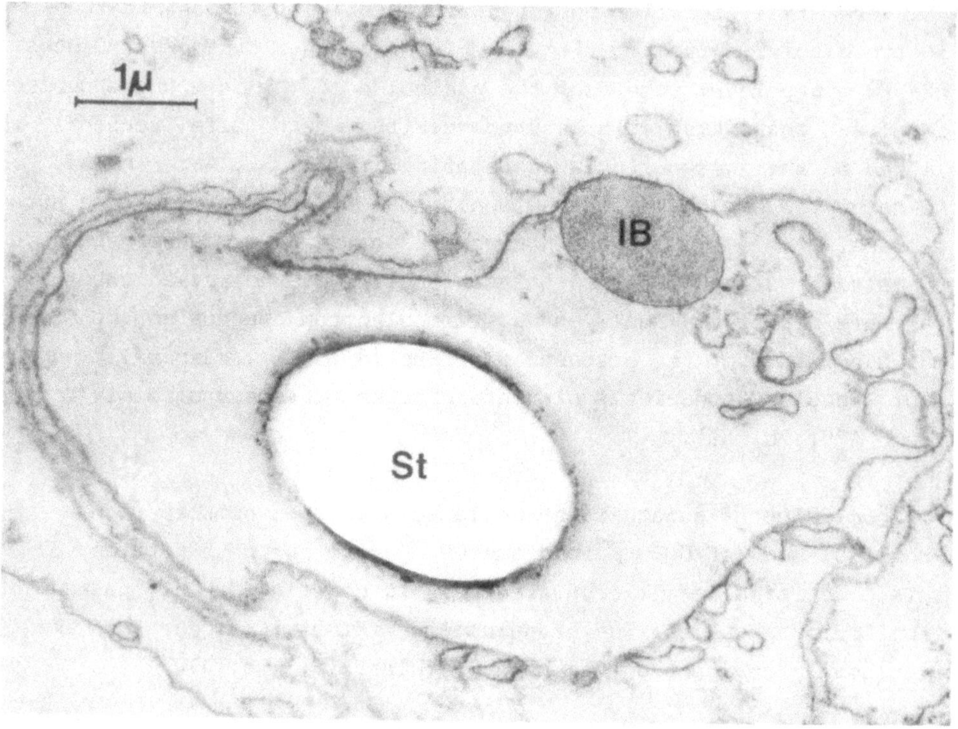

Figure 5. Amyloplast in cells of intact potato tuber tissue with starch grain (St) and intraplastid body (IB).

In tubers of the Jerusalem artichoke Tulett, Bagshaw and Yeoman (63) found compact clusters of plastids near the nucleus and solitary plastids scattered in the other cytoplasmic regions. These clustered plastids consist of a head and a tail. The heads are arranged towards the outside of such a cluster and the tails are intertwined in its centre. Sometimes the tails are intermingled with ER and microtubules. The function, formation and fate of these plastid clusters is not yet definitely known.

The mitochondria of the intact plant storage tissues are usually shaped more or less spherically. Their internal structure shows tubular foldings of the inner membrane, characteristic for plant

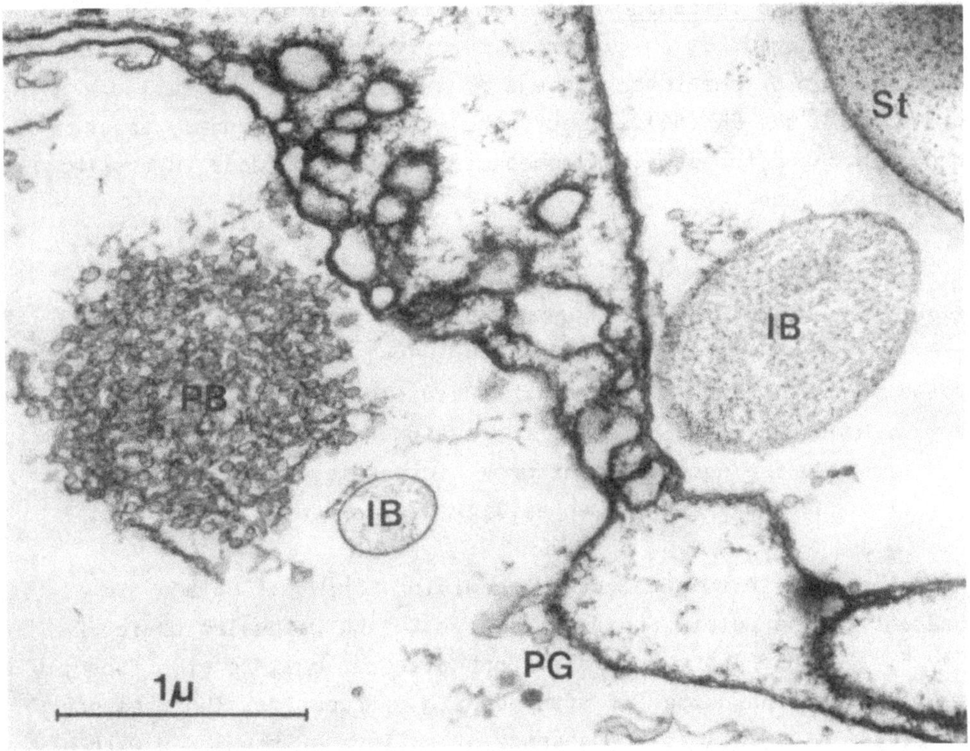

Figure 6. Prolamellar body (PB) in an amyloplast of intact white potato tuber tissue. St=starch grain, IB=intraplastid body, PG= plastoglobuli.

mitochondria (15, 25, 64). Bagshaw, Brown and Yeoman (4) found mitochondria with different profiles such as ring-, band-, and bell-shape. However, these results could not be confirmed by Fowke and Setterfield (17), who had also examined the Jerusalem artichoke system.

In intact tissues numerous mitochondria frequently are aggregated to big clusters in the cytoplasm, which are located mainly around the nucleus. Number and inner structure of the mitochondria does not allow a definite conclusion as to their metabolic activity.

The endoplasmic reticulum of all objects so far investigated is poorly developed. It only carries few ribosomes. In the Jerusalem artichoke 4% of the total content of the ribosomes per cell are associated with the ER (56). "Free" ribosomes are also rarely to be found in the cytoplasm. Polysomes usually do not occur in resting storage tissues.

Dictyosomes are either not present or only rarely to be seen. One exception was reported by Bagshaw (3), who in contrast to Fowke and Setterfield (17) found a relatively well developed Golgi apparatus in the storage tissue of the Jerusalem artichoke. This, however, cannot be interpreted as indication for high metabolic activity. In this case the occurrence of many dictyosomes could rather be explained as a relict from an earlier developmental stage (68).

Microbodies in form of crystal containing bodies ("CCB's") are present in the potato tuber as well as in the Jerusalem artichoke (3, 5, 17, 68). The number of microbodies per cell is significantly smaller than the number of mitochondria or plastids. The presence of crystals as well as the absence of contact of the CCB's with the ER suggest a low metabolic activity of these organelles (19).

Microtubules can only be observed near the cell wall. They have been found very rarely in storage tissues, however.

Lipid droplets ("spherosomes") are present in the cells of potato tuber tissue in a similar frequency as the microbodies. Characteristic elements of the cytoplasm of some cells of the potato tuber tissue are large storage protein crystals. They are not membrane-bounded and measure from 2 to 25 µm in size (5, 23).

There are no remarkable features of the cell wall except that the plasmodesmata in cells of Helianthus tuberosus as well as Solanum tuberosum are extended to bigger cavities in the region of the middle lamella and primary wall (3, 5).

The electronmicroscopic investigations of cells of intact storage tissues in different plants then clearly show that these tissues mainly consist of resting cells. Their metabolic activity is very low as judged from the appearance of the cellular organelles. Especially characteristic features of these cells are poorly developed structures for protein synthesis. The nucleoli contain no or few ribosomal precursors, i.e. a small pars granulosa only. The number of ER cisternae and ribosomes is strongly reduced. The number and structure of dictyosomes, microbodies and plastids also point to a low metabolic activity in general.

ELECTRONMICROSCOPIC CYTOLOGY OF WOUND-HEALING STORAGE TISSUE

It has already been mentioned that injury of plant storage tissue causes either lignification of cell walls only or mitotic activity and wound periderm formation. In both cases the ultrastructural changes will certainly be different. Therefore both types of wound reactions will be described separately.

WOUND-HEALING WITHOUT ANY MITOTIC ACTIVITY

Most of our knowledge of this process originates from cytological examinations that were related to the behaviour of tissue explants under the influence of grwoth substances. Electronmicroscopical investigations of wound-healing tissues without any cell divisions have been performed especially by Israel and Steward (25), Fowke and Setterfield (17), Jordan and Chapman (29, 30), and Rose, Setterfield and Fowke (46) in the Jerusalem artichoke and the carrot. Accordingly the ultrasturctural changes in the course of wound-healing exhibit the following characteristics:

The most typical changes occur during the first 24 hours after injury. In the case of Helianthus tuberosus the nucleus and nucleolus increase significantly within this period. The total mass of cytoplasm remains constant, however, and it does not change even within 72hours after wounding (17). Jordan and Chapman (30) were able to demonstrate that in carrot slices together with an enlargement of the nucleus the number of nuclear pores increases from about 1000 to some 3400 within 24 hours. At the same time the flattened nucleus changes to an irregular, extensively lobed form. The nucleus of Helianthus cells reaches its maximal size within 24 hours after wounding and slowly decreases afterwards.

The distribution of eu- and heterochromatin regions changes characteristically. The nuclei of intact tissue contain larger heterochromatic regions, which after wounding dissolve into smaller parts and disperse evenly throughout the whole area of the nucleus (29).

In the course of wound-healing the nucleolus undergoes characteristic changes. Three hours after wounding the heterochromatin, which is associated with the surface of the nucleolus, seems to migrate into and hence is peripherally embedded in the nucleolus. Within 24 hours parts of this special heterochromatin (L-zone; 29, 30) or even its total mass (46) is evenly distributed within the fibrillar regions of the nucleolus. The intranucleolar chromatin in turn is restricted to the pars fibrosa.

The rapid growth of the nucleolus signals its high metabolic activity, which after the first 7,5 hours following wounding reaches its maximum as judged from labelling of nucleolar material with ^{3}H-uridine in Helianthus tuberosus tissue (46). After another five hours the incorporation of ^{3}H-uridine decreases significantly. The increase of the nucleoli is mainly caused by an enormous growth of the pars granulosa. As already mentioned, nucleoli of intact tuber tissue of Helianthus tuberosus possess no pars granulosa (17, 29, 46). Within 3 hours after injury, however, the formation of ribosomal precursors begins (29) and after 24 hours a large part of the nucleolus consists exclusively of pars granulosa (comp. Fig. 8). The pars granulosa is mainly located in the periphery of the nucleolus, but may well extend into the pars fibrosa. Both parts of the nucleolus are always clearly distinguishable from one another.

In the carrot (30) as well as Jerusalem artichoke (17, 46) nucleolus vacuoles develop in the course of wound-healing. They occur in the pars granulosa as well as in the pars fibrosa (comp. Fig. 8, 9, 10). In Helianthus the first nucleolar vacuoles appear after 7,5 hours and after 25 hours 66% of all nucleoli contain vacuoles of different size (46). 25 hours after excision of the tissue the largest vacuoles can be observed. This fact is of interest inasmuch as the metabolic activity in the nucleolus begins to decrease already, i.e. ^{3}H-uridine labelling is markedly slowed down. Also both contraction and reformation of vacuoles in Helianthus nucleoli occur (46). The function of the nucleolus vacuoles will be discussed in detail.

The plastids and mitochondria do not change their shape and internal structure in wound-healing storage tissues of Helianthus and Daucus.

The ER is strongly build up in the course of wound-healing. This is especially true for the rough ER. At the same time the number of ribosomes increases rapidly. This is true for the membrane-bound ribosomes as well as for the free ribosomes. In agreement with this findings is the intensive production of ribosomal precursors in the nucleoli. In contrast to these results Sparkuhl, Gare and Setterfield (56) found the rRNA content in the wounded tissues to be decreased by 28% within two hours after injury. During the subsequent 48 hours of wound-healing no dramatic change in rRNA content takes place.

These contradictory results may be explained as follows. Most of the ribosomes of intact tissues are dissociated into their subunits (56). For this reason hardly any ribosomes can be seen in the electron microscope. But within four hours after injury most of these subunits aggregate to complete ribosomes (56). Then they can easily be observed in the electron microscope. Since the rRNA content in wound-healing tissue remains constant (except for the first two hours) and since an intensive rRNA synthesis in the nucleolus occurs at the same time, a high turnover of ribosomes in the cells seems to take place (45, 56). Together with the rapid aggregation of subunits into complete ribosomes immediately after excision of intact tissue a dramatic formation of polysomes ensues. After four hours a large amount of ribosomes is engaged in polysomal aggregates only. After 24 hours about 68% of all ribosomes present in wound-healing tissues of Helianthus occur as polysomes. Only after 48 hours the number of polysomes decreases again (56). Concerning the formation of polysomes similar results have been obtained with other tissues by Leaver and Key (39), Cherry (14) and Kahl (31).

All other organelles in wound-healing tissues without any mitotic activity such as dictyosomes, microbodies and microtubules do not show any significant change. Taken altogether these cytological observations suggest an increase in protein synthesis in all these injured storage tissues.

WOUND-HEALING TOGETHER WITH MITOTIC ACTIVITY

The wound-induced ultrastructural changes to be described in this chapter only refer to potato tuber tissue. The wound reaction of other plant storage tissues, which are also induced to perform cell divisions has not yet been examined. In cells which regain their mitotic activity after wounding and finally develop a wound periderm the following changes can be observed. Within 12 hours after wounding the cytoplasmic content of the cell increases rapidly. The cytoplasmic layer adherent to the cell wall becomes more voluminous and numerous cytoplasmic strands pass through the vacuole. The starch content decreases and within 48 to 96 hours the starch disappears from the cell. At the same time the number of most organelles increases considerably.

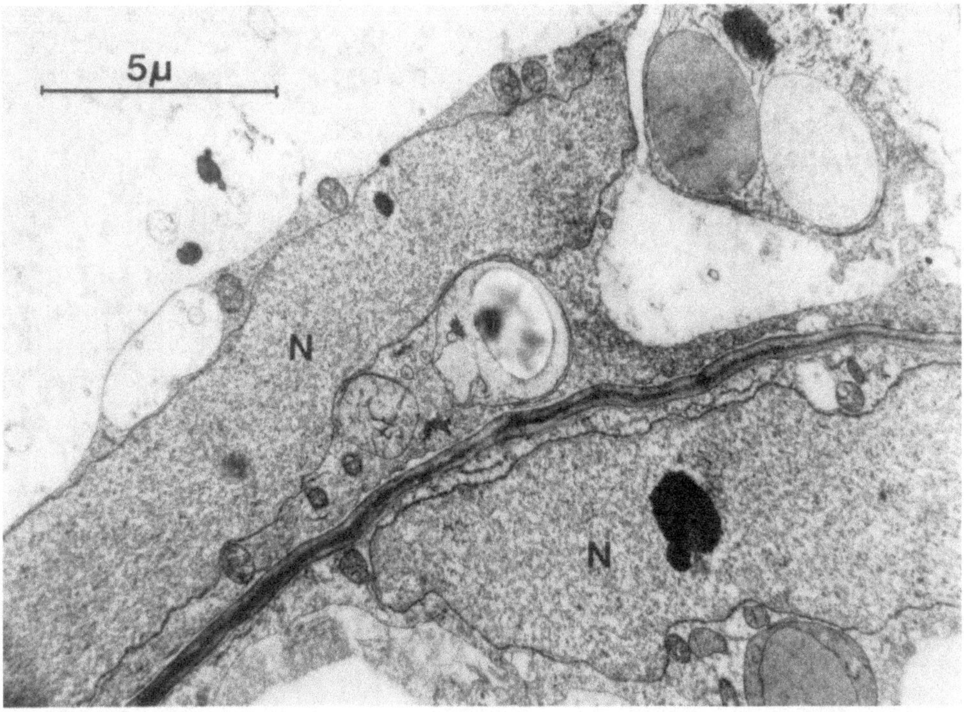

Figure 7. Nuclei (N) with more or less regular, elliptic shape 192 hours after injury (Solanum tuberosum).

The <u>nuclei</u> do not undergo any significant changes within the first 48 to 96 hours. They still possess deep invaginations which contain many organelles (Fig. 15) in contrast to the intact tuber tissue. Only 192 hours after injury the nuclei assume a more or less regular, elliptical shape (Fig. 7), which is characteristic for tissues with mitotic activity. No changes occur in the distribution of the eu- and heterochromatin. The nucleolar pores in wound-healing tissues contain central grana. Therefore a transfer of RNA from the nucleoplasm to the cytoplasm seems to be likely (18, 67).

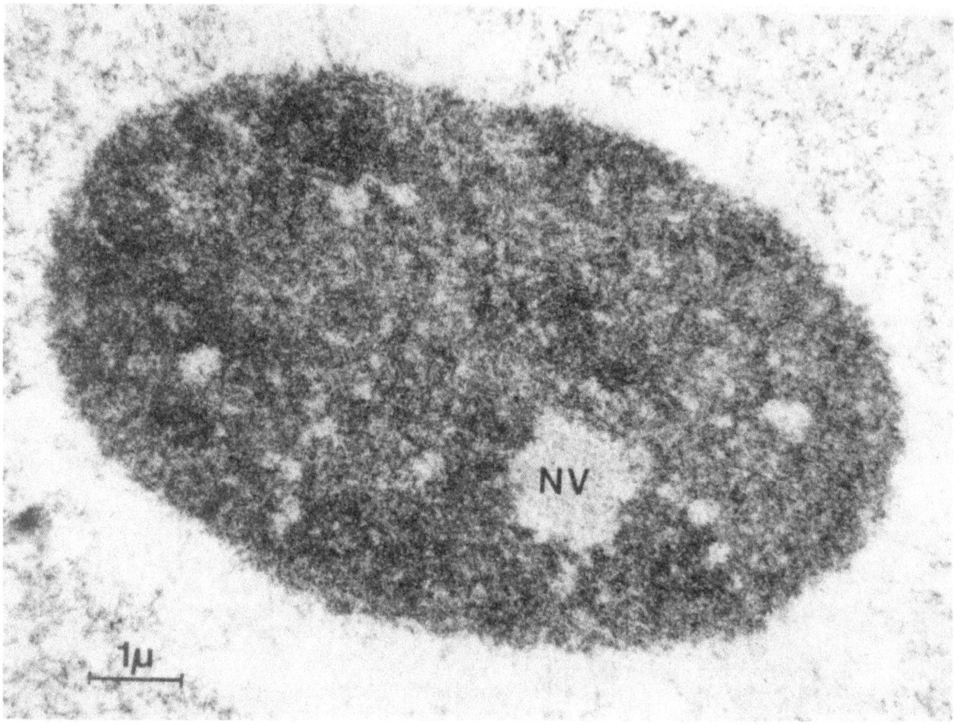

Figure 8. Nucleolus in cells of wound-healing potato tuber tissue 12 hours after injury. The nucleolus consists mainly of pars granulosa and contains a small vacuole (NV).

The changes of the nucleoli mostly correspond to the conditions in Helianthus and Daucus tissues, which has been described before. So the pars granulosa increases considerably within 12 hours after excision. At the same time smaller vacuoles have developed in most nucleoli (Fig. 8). After 24 hours the nucleolus mainly consists of pars granulosa. The number of vacuoles per nucleolus definitely decreases but instead their size enormously increases. 48 hours after wounding many nucleoli contain giant vacuoles (6). The remaining nucleolar material, pars granulosa, pars fibrosa and chro-

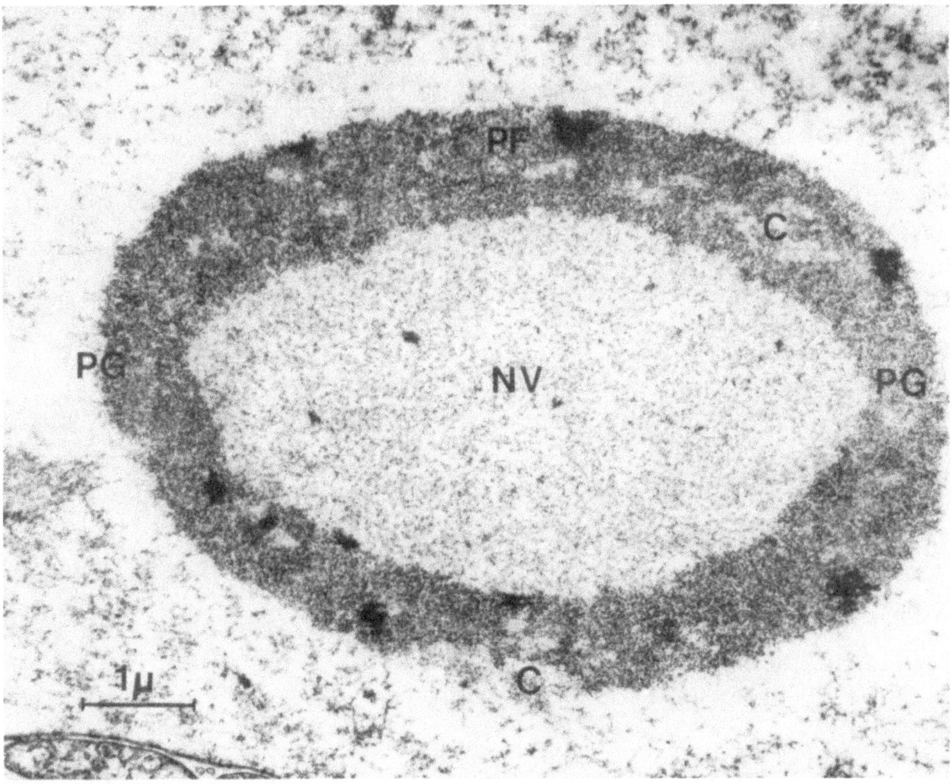

Figure 9. Nucleolus in cells of wound-healing potato tuber tissue with a giant vacuole (NV) 48 hours after injury. Within the vacuole only few granules and fibrils are located. PG=pars granulosa, PF=pars fibrosa, C=chromatin.

matin, is restricted to a small area around the vacuole (Fig. 9). The content of the vacuole consists of few granules and fibrils (Fig. 9). Contraction and reformation of such giant vacuoles could not be observed with the light microscope. After 96 to 192 hours of wound-healing no giant vacuoles are to be found. The size of the vacuoles decreases significantly at the end of the wound reaction (Fig. 10). Nevertheless the pars granulosa still occupies a large part of the nucleolus (Fig. 10). In wound-healing potato tuber tissue nucleoli without any giant vacuoles are also common. In this

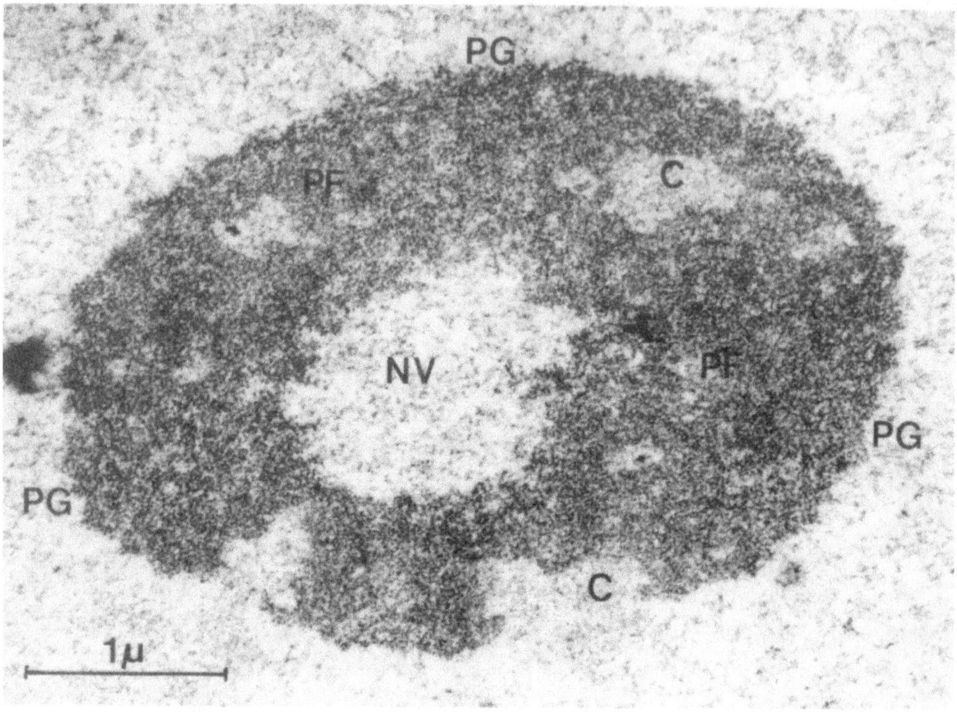

Figure 10. 192 hours after injury of potato tuber tissue the nucleolus vacuole (NV) has decreased in size but the pars granulosa (PG) still comprises most of the nucleolar mass. C=chromatin, PF=pars fibrosa.

case, however, the nucleoli always contain many smaller vacuoles (5).

The intranucleolar chromatin is not restricted to the pars fibrosa in contrast to Helianthus tuberosus (29). The chromatin occupies all parts of the nucleolus except for the vacuole (Fig. 10). Special changes of the intranucleolar chromatin could not be found in potato tuber cells.

In the course of wound-healing the micronucleolus frequently accompany the nucleolus and increases in size only for a short time, reaching its maximum after 12 hours (5).

The investigations of wounded storage tissue of Helianthus tuberosus and Solanum tuberosum both show a rapid enlargement of the pars granulosa immediately after injury, a fact, which is characteristic both for tissues with and without mitotic activity during the wound reaction. The nucleolus is known to be the site of rRNA synthesis in the cell. The rRNA synthesis begins in the pars fibrosa. The ribosomal precursors, which are already combined with proteins, accumulate in the pars granulosa before leaving the nucleus (20). For this reason a well developed pars granulosa always is an indicator for an active production of ribosomes. Numerous investigations with other objects confirm these results (12, 13, 24, 27, 37, 44).

The reason for the development and function of the nucleolus vacuoles are difficult to interpret, although they occur in many tissues concomitant with a high rate of rRNA synthesis (6, 8, 12, 13, 27, 28, 30, 36, 46). The experiments of Rose, Setterfield and Fowke (46), however, clearly indicate that the maximum of vacuole formation coincides with a decrease in rRNA synthesis. Similar results have been obtained with potato tuber slices. Here the giant vacuoles appear 48 hours after injury (6). At this time the total volume of the nucleolus has already decreased (57) and the formation of polysomes and protein synthesis are slowed down considerably (31, 32). Furthermore, in Zea mays nucleolar vacuoles are formed immediately after the onset of germination, although the rRNA synthesis did not yet start (8). At the same time the nucleolus decreases in size and the pars granulosa becomes smaller. These results clearly show that vacuoles are to be found, if the rRNA synthesis either has not yet started or has already reached its maximum. Consequently a low rate of rRNA synthesis leads to a rapid decrease in the ribosomal precursor pool in the nucleolus, especially in the case of a high demand of ribosomes in the cytoplasm. This may be the ultimate reason for the formation of vacuoles. If the rRNA synthesis again proceeds at a high rate, then the vacuoles disappear (8).

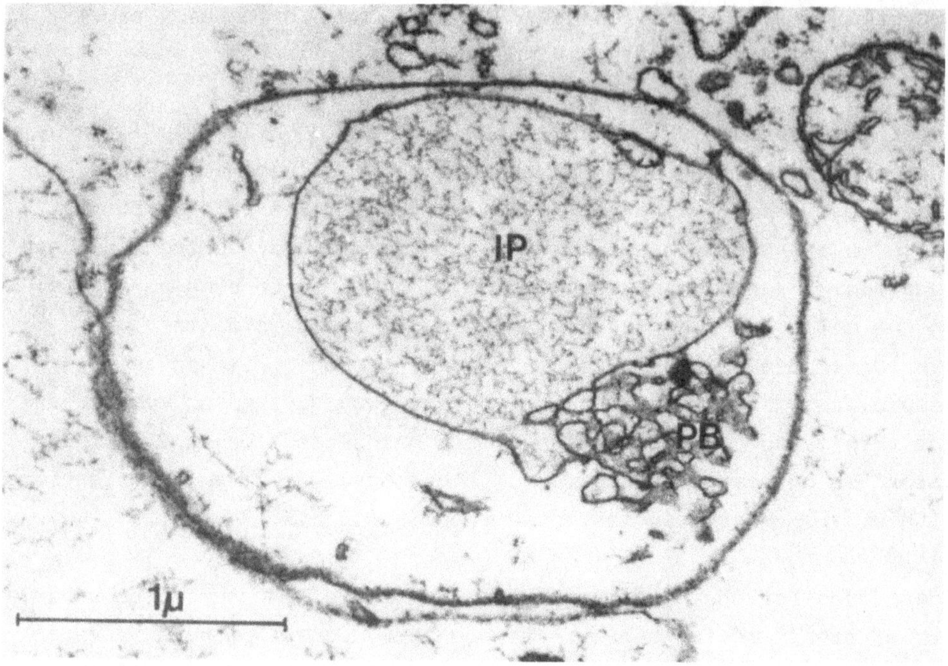

Figure 11. Section of a plastid 24 hours after injury. A large intraplastid body (IP), which originates from the prolamellar body (PB), has developed (Solanum tuberosum).

In starch storing plastids of the potato tuber the injury causes somewhat different changes. First of all a rapid breakdown of starch in the plastids starts, which goes together with a shrinkage of the plastids. Desintegration of amyloplasts respectively their double membranes could not be found in the course of wound-healing. Concomitantly with the starch degradation the intraplastid bodies increase in size. This can already be observed 24 hours after injury (Fig. 11). After 48 to 96 hours almost no starch is visible in the region of the wound periderm and after 192 hours one or more big intraplastid bodies replace the starch grains in all larger plastids (Fig. 12). The intraplastid bodies are in contact with the prolamellar bodies. The membranes of the intraplastid bodies in fact originate from the latter (Fig. 11). Furthermore, the intraplastid bodies

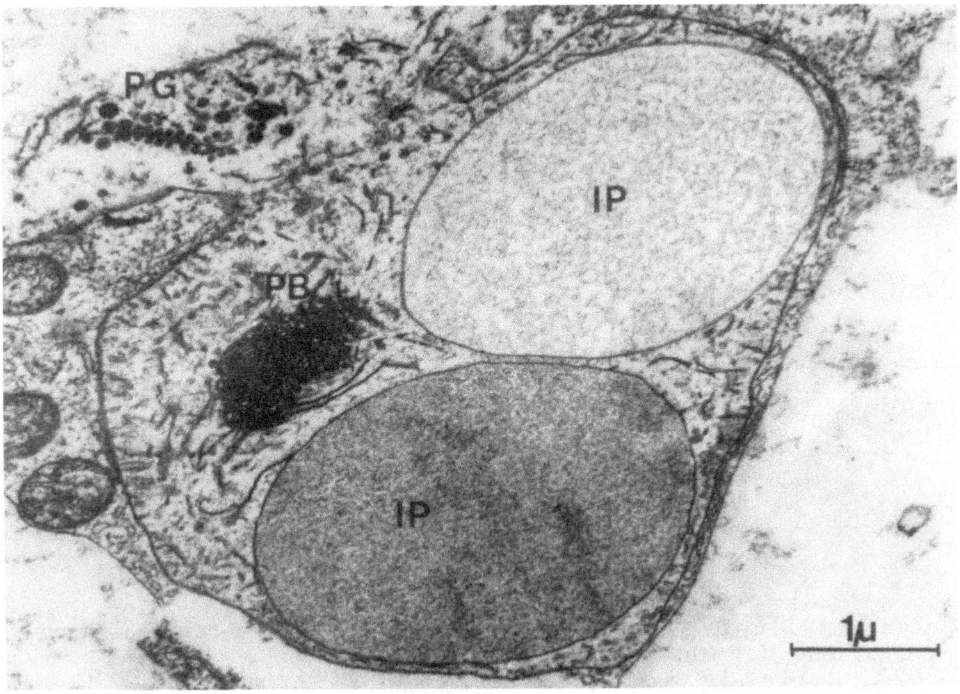

Figure 12. Plastid in a cell of wound-healing potato tuber tissue 192 hours after injury. It contains two large intraplastid bodies (IP) with different proteinaceous content. Furthermore a prolamellar body (PB) and clustered plastoglobuli (PG) are to be seen.

of the same plastid may either contain various amounts of the same material or also different material (Fig. 12). Most certainly this material represents protein (2, 21, 22). This means, that in the course of wound-healing the plastid-function changes from starch to protein storage. The number of plastoglobuli also increases and often form clusters within the plastids (Fig. 12).

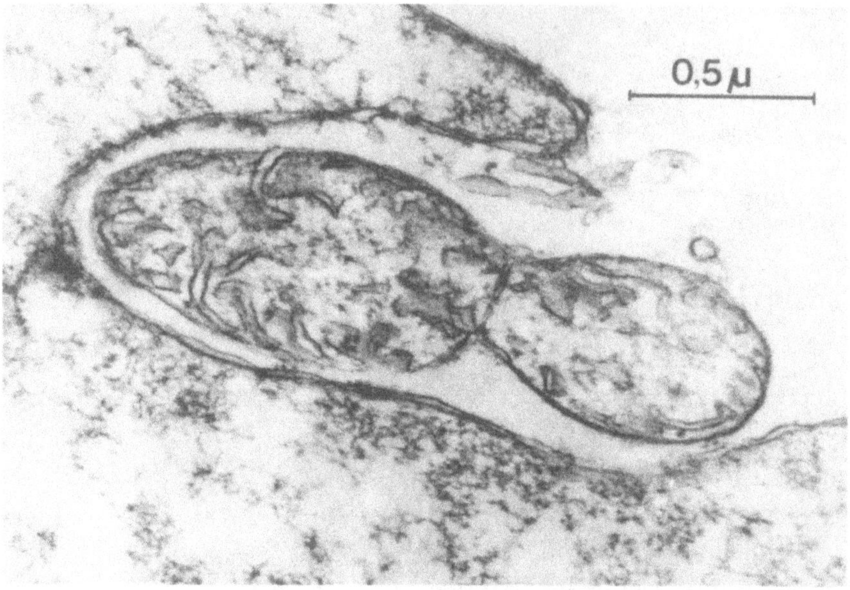

Figure 13. Dividing mitochondrium in cells of wound-healing potato tuber tissue (48 hours after injury).

Besides the large protein storing leucoplasts numerous proplastids develop in wound-healing potato tuber tissue. These proplastids have a diameter of about one μm. They are variable in form and may also occur as long wormshaped structures in the cytoplasm (Fig. 14). The proplastids multiply either by division or by budding of larger leucoplasts ("amyloplasts"). The process of wound-healing gives rise to a third type of plastids, which is differentiated into a "head" and a "tail". These plastids form clusters near the nucleus. Their "tails" are intertwined in the center of the cluster and are intermingled with ER. The surface of the cluster is build up by the "heads" of these plastids. Identical plastid clusters were found by Tulett, Bagshaw and Yeoman (63) in quiescent tuber cells of the Jerusalem artichoke.

The development of plastids in the course of wound-healing definitely

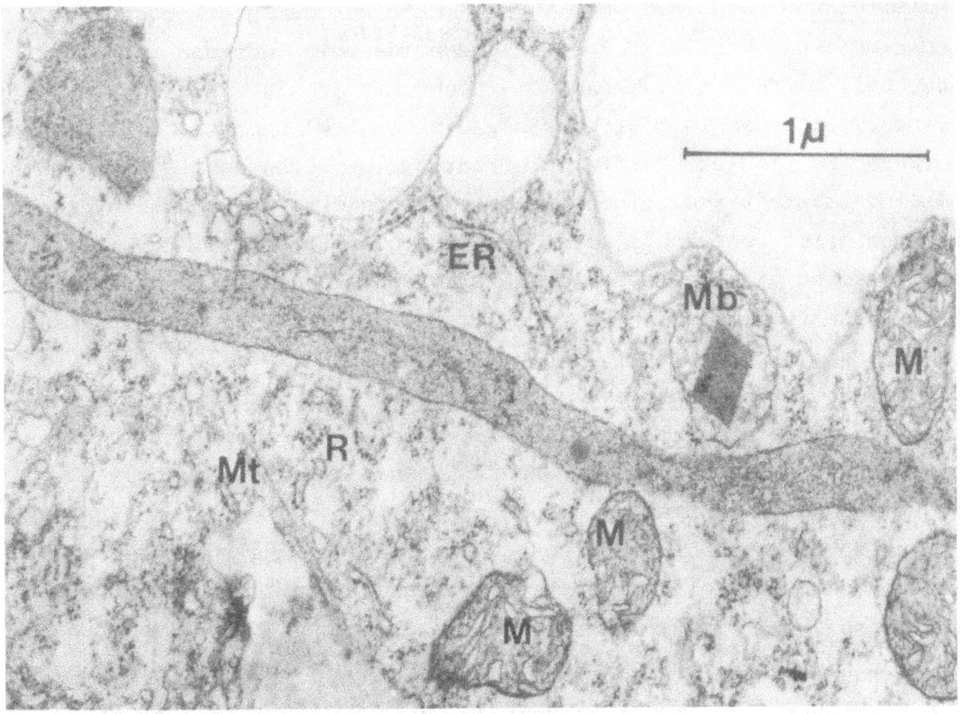

Figure 14. Wormshaped plastid in cells of wound-healing potato tuber tissue (48 hours after injury). The cytoplasm is rich in ribosomes (R), mitochondria (M), rough endoplasmic reticulum (ER), microtubules (Mt) and microbodies (Mb) containing crystals.

shows that the tissue becomes meristematic. This is especially indicated by the development of proplastids. Furthermore leucoplasts with intraplastid bodies are often to be found in actively dividing cells (41, 42, 58). The investigations of Israel and Steward (26), Stetler and Laetsch (59), Boasson, Laetsch and Price (10), Ames and Pivorum (2) and Henry (21) revealed that etioplasts are structurally similar to the plastids formed in wound-healing potato tuber tissue. From all these data it may be concluded that amyloplasts change to etioplasts in injured potato tuber tissue.

The mitochondria do not show any definite change in the course of wound-healing (Fig. 14). They only considerably increase in number (40, 65). Different stages of divisions are mainly visible 24 to 48 hours after excision (Fig. 13). Verleur (64) and Verleur and co-workers (65) isolated different fractions of mitochondria from wound-healing potato tissue. These fractions were distinguishable by their sedimentation behaviour, physiology and ultrastructure. Even thorough examinations of a whole series of electron micrographs could not confirm these findings.

The volume of the endoplasmic reticulum and the number of membrane-bound and free ribosomes rapidly increase within the first 24 hours after injury. In contrast to intact tissue numerous polysomes attached to the membranes of the ER as well as freely distributed in the cytoplasm are now to be seen (31; Fig. 14). The increase of ER is accompanied by differentiation processes into two types. One of these two types represents the cisternal, rough ER (Fig. 14), which is also typical for wound-healing tissues of Helianthus and Daucus. The other type consists of tubular, smooth ER (Fig. 15). It is significant, that the tubular, smooth ER accumulates in special cytoplasmic regions. In the course of this process nuclear invaginations and cytoplasmic strands develop, which only contain smooth ER tubules and no other organelles (Fig. 15). 48 hours after injury the accumulation of smooth ER tubules reaches its maximum. After this time the accumulation of this ER type decreases and disappears completely after 192 hours.

The formation of two different ER types in wound-healing potato tissue may occur for the following physiological reasons. The activation of protein synthesis after injury (32) requires a rapid building up of rough ER, free ribosomes and polysomes (31). It is assumed, that the formation of smooth, tubular ER is functionally related to the suberization of the cell wall (7). In the course of wound-healing suberin and monomolecular wax lamellae are deposited onto the wall alternately (5, 7, 16, 53, 54, 55, 66; Fig. 17, 18). This process

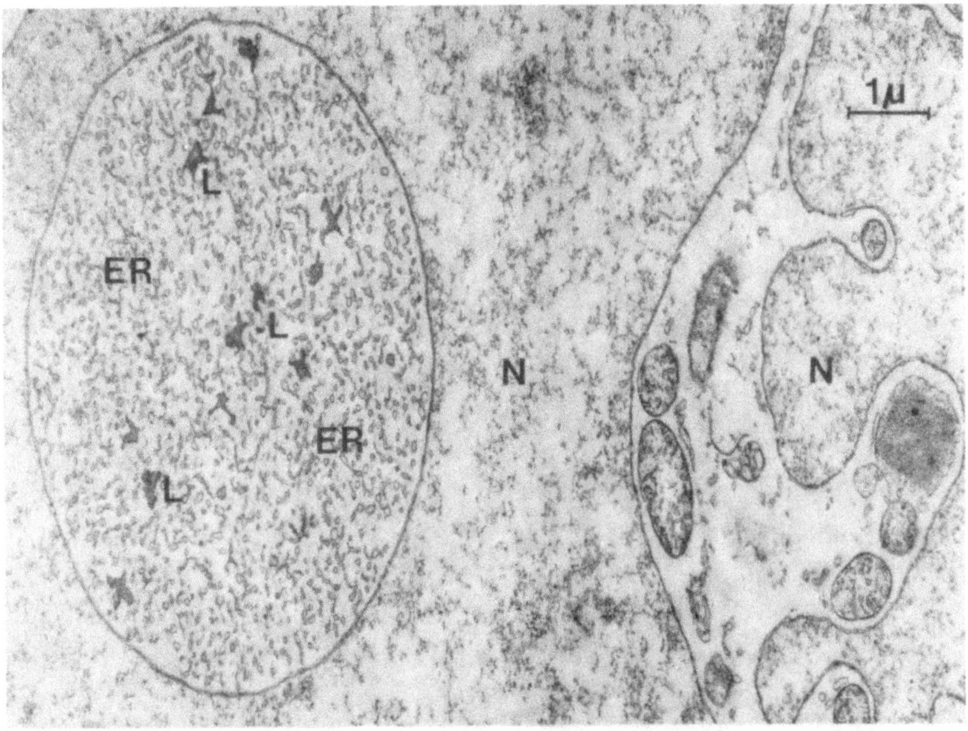

Figure 15. Strongly lobed nucleus (N), which contains a large cavity filled with tubular smooth ER, synthezising lipids (L). Wound-healing potato tuber tissue 48 hours after injury.

reaches its maximum 48 hours after excision together with the maximal accumulation of smooth ER. These ER elements usually contain lipids (Fig. 15). Probably they represent products of the suberin and/or wax synthesis. Lipid synthesis by the tubular, smooth ER is also assumed by numerous other investigators (1, 47, 48, 49, 50, 51, 52). The separation of lipid synthesizing ER elements from other organelles within nuclear invaginations and cytoplasmic strands cannot be explained satisfactorily. On the other hand there exist feed-back mechanisms in metabolism between fatty acid synthesis in the cytoplasm and the citric acid cycle within the mitochondria. Through the separation of tubular smooth ER and mitochondria a mu-

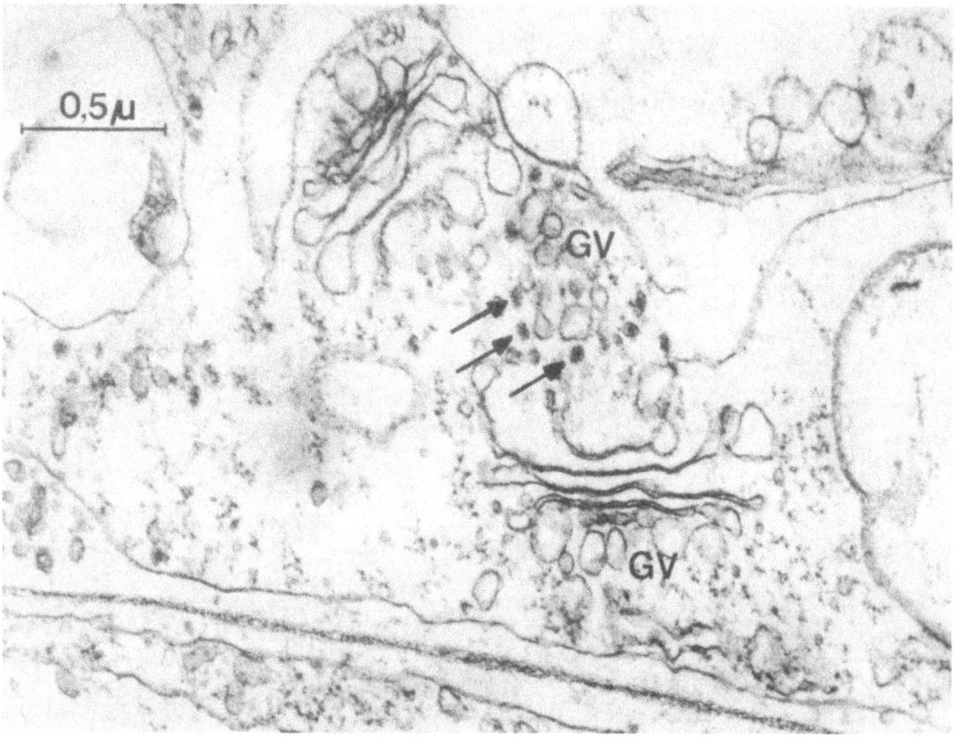

Figure 16. Clustered dictyosomes with "coated vesicles" (arrows) and Golgi vesicles (GV). Wound-healing potato tuber tissue 48 hours after injury.

tual inhibition of some of the metabolic pathways of these organelles can be reduced.

The amount of dictyosomes strongly increases immediately after injury. These organelles can be found without any difficulties after 12 hours of wound-healing. 24 hours after excision of the tissue a well-developed Golgi apparatus occur in all cells (Fig. 16). The development of numerous Golgi vesicles is an indication for a high activity of dictyosomes. Mitotic activity is intimately correlated with cell wall formation and therefore a high production of Golgi vesicles may be expected. Number and activity of dictyosomes remain almost unchanged after 192 hours. Dictyosomes, oftenly clustered, are frequently associated with "coated vesicles" (Fig. 16). The occurrence of many microtubules in wound-healing potato tuber tissue (Fig. 14) is correlated to cell wall formation and mitotic activity.

Most microbodies containing crystals (CCB's) do not change their structure during wound-healing (Fig. 14). There are only few microbodies with crystal fragments, which indicate degradative or synthetic processes of crystals.

Lipid droplets or spherosomes only rarely are constituents of wound-healing tissue, except those which originate from tubular ER. In comparison to intact tissue the number of spherosomes decreases. The large storage protein crystals, which lie "freely" in the cytoplasm of some cells of intact tuber tissue, disappear within 12 hours after injury. It is concluded that this storage protein is used up for metabolic processes.

It should be stressed that the wound-induced modification of original and newly formed cell walls are different from each other. Already 24 hours after excision the original cell walls immediately beneath the wound surface are covered by suberin layers. With increasing distance from the wound surface itself the suberization process is retarded. For this reasons cell walls can be observed, which are suberized on one side only (Fig. 17). The plasmodesms of these original walls are not closed by suberin. They still function as channels in the suberin layer (55).

The newly formed cell walls are already suberized before the formation of the secondary wall starts. Here the suberin layer is deposited directly onto the primary wall (Fig. 18). After the suberization the synthesis of these cell walls is completed by the addition of a cellulose layer (tertiary wall). Finally these cell walls altogether consist of middle lamellae, primary wall, suberin and a final tertiary cellulose layer (66; Fig. 18). The mechanism of the alternating excretion of suberin and wax is not known. However, some

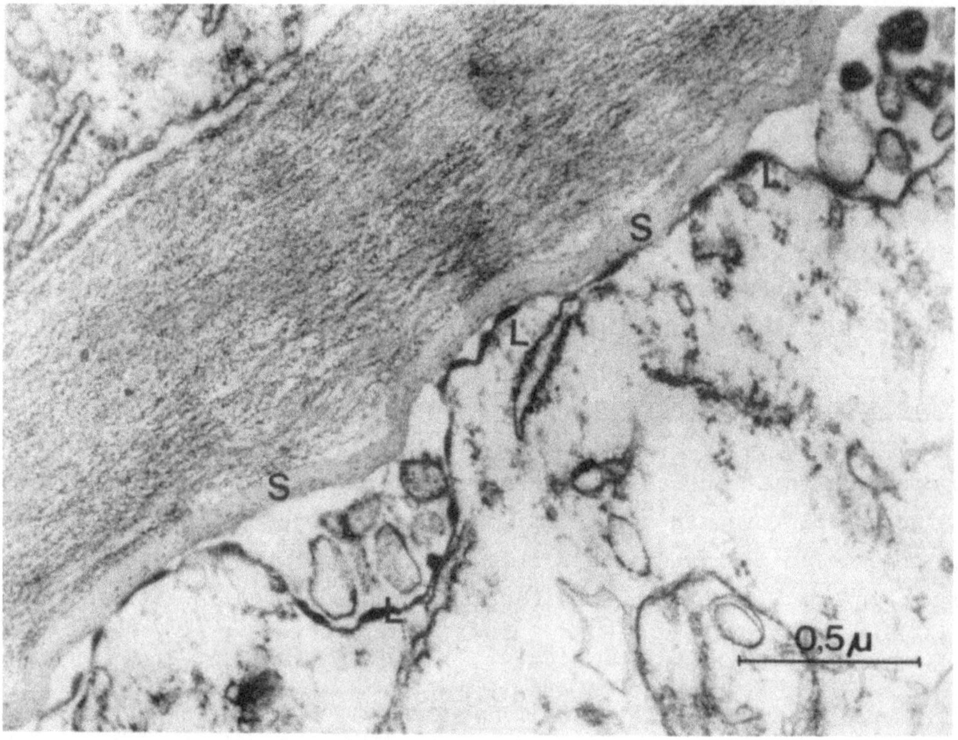

Figure 17. Cell wall coated by a suberin layer (S) on only one side. Lipids (L) are passing through the plasmalemma towards the cell wall (48 hours after injury, Solanum tuberosum).

data point to an accumulation and uniform distribution of suberin or wax at the plasmalemma with subsequent passage of this material through the membrane via the cell wall (Fig. 17).

Cells with suberized cell walls soon die. Degeneration of the cytoplasm already starts if only a thin suberin layer has been excreted. This rapid cell death probably prevents the formation of a final cellulose layer. At this point it should be stressed that not all wound periderm cells actually suberize. The periderm cells most remote from the wound surface remain as living phellogen with a typical cambial character (Fig. 19). Their cell walls are orientated in parallel to each other and the cells each contain a relatively large

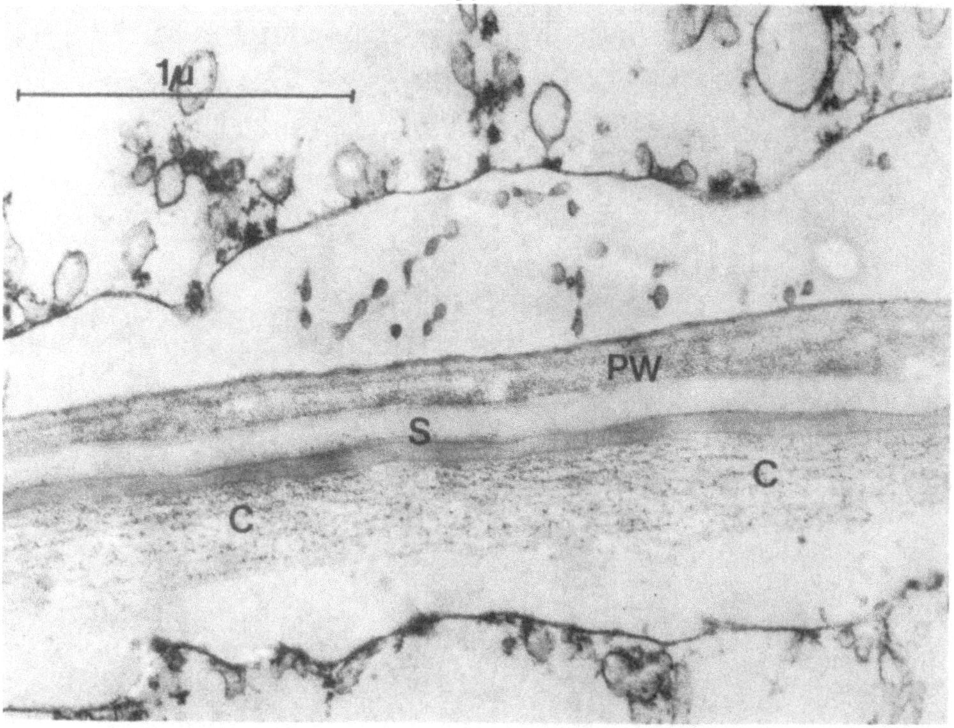

Figure 18. Newly formed cell wall, which is already suberized before the formation of the secondary wall has started. The suberin (S) is deposited directly onto the primary wall (PW). Then the formation of the cell wall is completed by the addition of a tertiary cellulose layer (C) onto the suberin layer (96 hours after injury, Solanum tuberosum).

central vacuole. In contrast to the original storage tissue cells these periderm cells are smaller and rich in cytoplasm. The ultrastructure of these surviving wound periderm cells is similar to the structure of the normal phellogen cells of the intact potato tuber tissue.

The cells immediately beneath the wound surface of potato tuber tissue are not induced to mitotic activity. However, they suberize within 24 hours and subsequently die. With the exception of the formation of a large pars granulosa in the nucleolus, the rough ER,

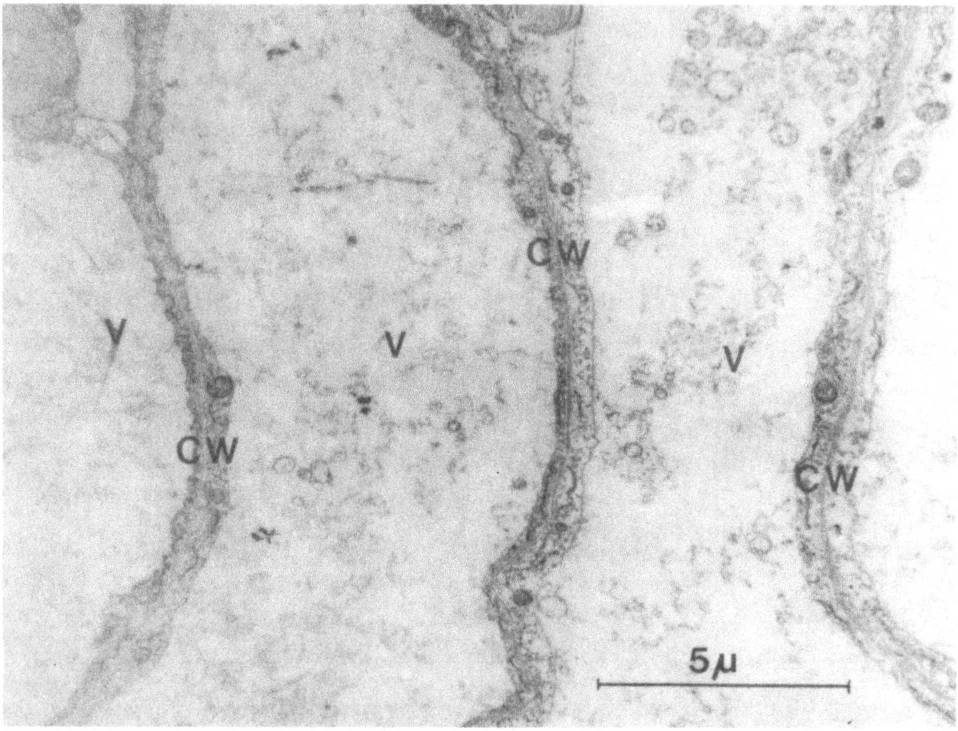

Figure 19. Part of the wound periderm of white potato tuber tissue. The cambial character of this tissue is demonstrated by parallelly orientated cell walls (CW), the relatively large cytoplasmic content together with many organelles and cytoplasmic strands passing through the vacuole (V). Wound-healing potato tuber tissue 192 hours after injury.

smooth tubular ER and polysomes no other changes could be detected. So the wound reaction of these cells resembles closely that of storage tissue cells of Helianthus tuberosus and Daucus carota.

CONCLUSION

The ultrastructural changes of wound-healing storage tissue of Helianthus tuberosus, Daucus carota and Solanum tuberosum mostly follow the same tendency in the first 24 hours after injury (Fig.

20, 21, 22), although the cytological response in the different tissues is quite different (induction of mitosis in white potato tuber tissue; chemical modification of cell walls only in Jerusalem artichoke and carrot tissue). Nucleus and nucleolus both considerably enlarge. The nucleolus forms an extensive pars granulosa as well as vacuoles. At the same time rough ER, free ribosomes and polysomes strongly increase (5, 6, 17, 25, 29, 30, 31, 32, 46, 56).

The only differences between Helianthus and Daucus on one hand and the potato system on the other in wound-healing within the first 24 hours is an increase in the amount of cytoplasm and dictyosomes in the potato tissue. In the later stages of wound-healing, i.e. the time after the first 24 hours no further significant changes are to be observed in Helianthus and Daucus tissue. Contrastly the potato tissue perform the first cell divisions after 24 hours.

Treatment of injured storage tissue of Helianthus and Daucus with plant growth substances (e.g. IAA; 2,4 D; kinetin) induces mitotic activity. The cell divisions in contrast to the potato tissue are not limited in time and finally lead to a callus tissue. Inspite of these differences it is of interest to compare the ultrastructural changes of wound-healing potato tissue with tissue of Helianthus and Daucus which has been activated by hormones to perform cell divisions.

In Daucus as well as in Helianthus the induction of mitotic activity first of all causes a rapid increase in cytoplasmic mass. Numerous cytoplasmic strands develop. The nucleus becomes spherical and migrates into the center of the cell. Here it is hold in position by many cytoplasmic strands. Afterwards division is prepared by these cells, which is terminated by the formation of the phragmoplast. The position of the latter is determined by cytoplasmic strands (3, 17, 25, 68). These results agree well with those of the potato system. Likewise the changes of the nucleolus show the same tendency in the Jerusalem artichoke as well as the potato. Thus in Helianthus tissue a rapid increase of the pars granulosa and the formation of nucleolus vacuoles can also be observed in the presence of 2,4 D (3, 46, 68).

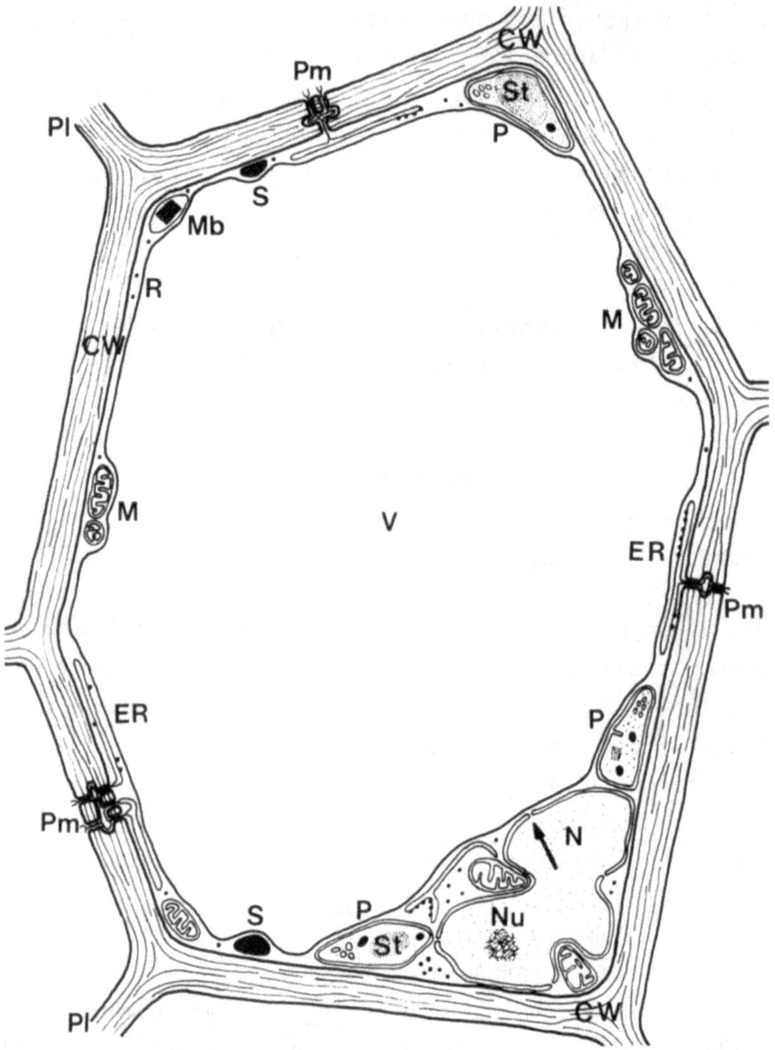

Figure 20. Diagram of a typical parenchymatic cell of intact plant storage tissue. The cell content mainly consists of a large vacuole (V) and a thin cytoplasmic layer on the cell wall (CW). The cytoplasm is poor in cell organelles, i.e. especially ribosomes (R), endoplasmic reticulum (ER) and dictyosomes (D). The plastids (P) contain starch (St). The nucleus (N) is situated near the cell wall and only possesses few nuclear pores (arrows). The nucleolus (Nu) is very small and build up by fibrils of the pars fibrosa. IP=intraplastid body, M=mitochondrium, Mb=microbody, Mt=microtubules, NV=nucleolus vacuole, Pl= plasmalemma, Pm=plasmodesma, Po=polysome, S=spherosome.

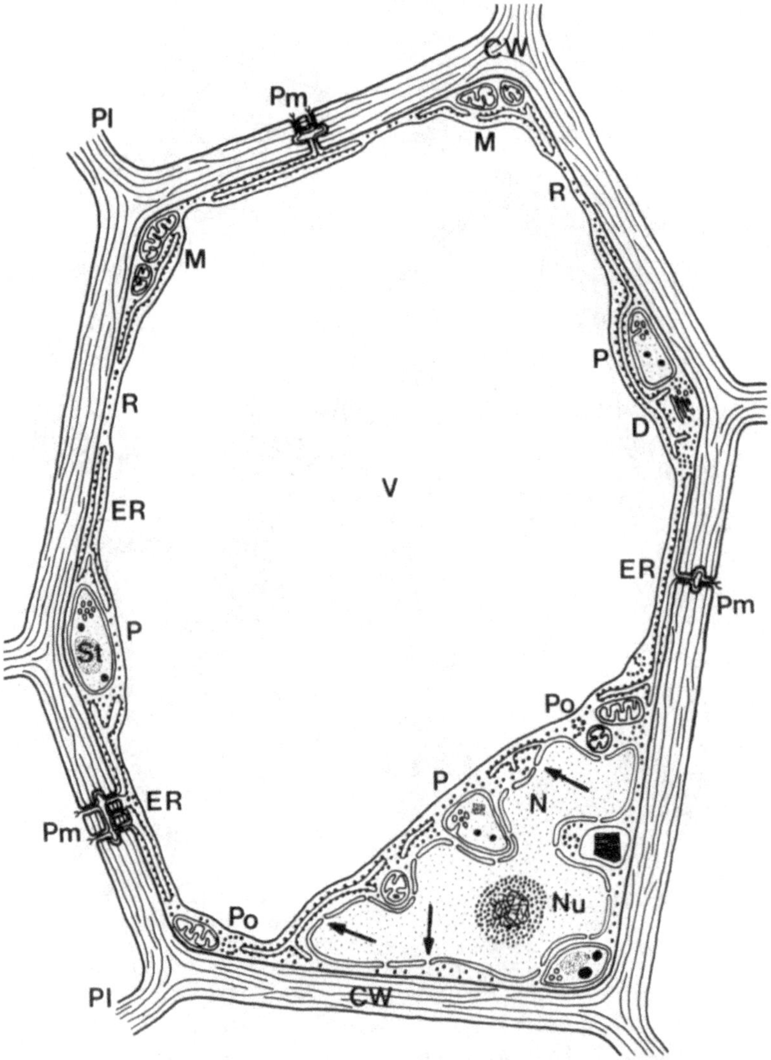

Figure 21. Diagram of a cell of injured storage tissue of Helianthus tuberosus or Daucus carota 24 hours after excision. Wound-healing of these storage tissues is not linked to any meristematic activity of cells near the wound surface. Their cell walls become lignified only. The ultrastructural changes in these cells are restricted to an increase in ribosomes, polysomes, rough ER membranes, nuclear pores and the size of the nucleus and nucleolus. The latter develops a typical pars granulosa. (Key to abbreviations see Figure 20.)

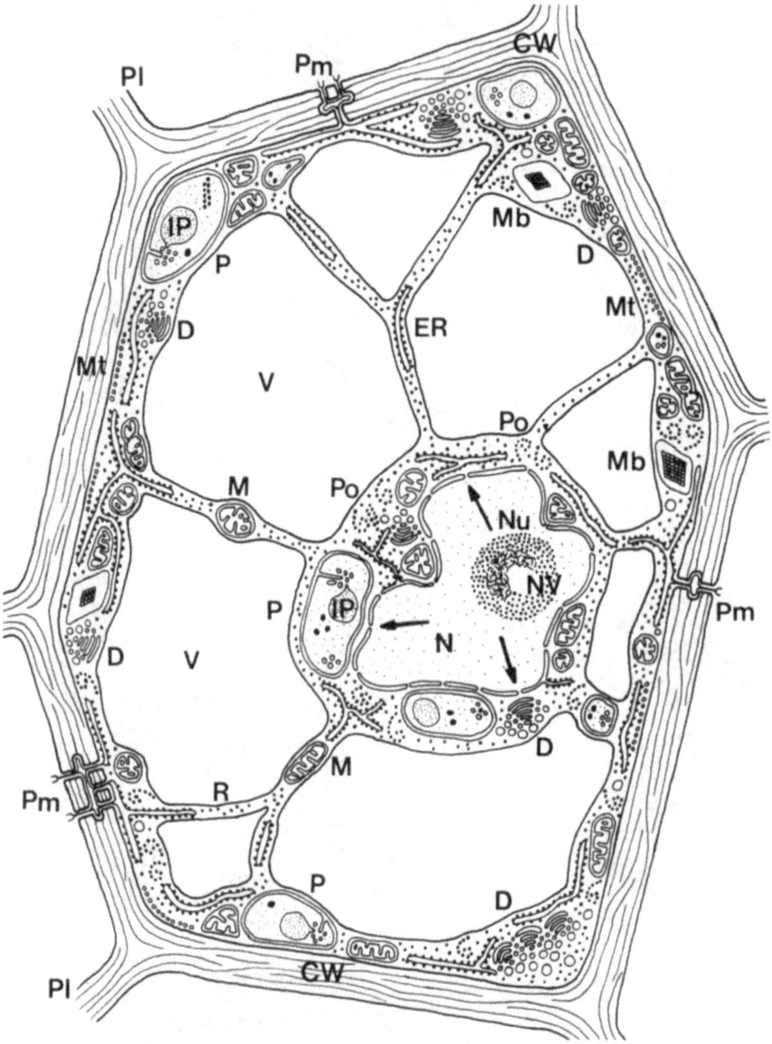

Figure 22. Diagram of a cell of injured storage tissue of Helianthus tuberosus or Daucus carota treated with plant hormones and of untreated storage tissue of Solanum tuberosum 24 hours after excision. The wound-reaction of cells of these tissues is always correlated with mitotic activity. The cytoplasmic content has strongly increased including most of the organelles such as ribosomes, polysomes, ER membranes, dictyosomes and mitochondria. Microtubules are common. The plastids now contain intraplastid bodies instead of starch. The nucleus is situated in the middle of the cell and will soon divide. It contains a large nucleolus with a well developed pars granulosa and nucleolus vacuole. Numerous nuclear pores are to be seen. (Key to abbreviations see Figure 20.)

The addition of 5-fluoro-uracil in Helianthus leads to contractile giant vacuoles in the nucleolus, which correspond to those of wound-healing potato tissue (6, 45). However, the nucleolus vacuoles in the potato tuber cells do not seem to be contractile.

The plastids in quiescent tissues of Solanum, Helianthus and Daucus are very different in structure (5, 26, 41, 63). In spite of that fact, they show the same developmental tendencies in wound-healing tissue respectively in injured tissue treated with plant hormones. Storage substances such as starch and phytoferritin disappear. Intra-plastid (prethylakoidal) bodies increase in size and number. If not already present in the original tissue, typical prolamellar bodies develop. Their membranes are connected with the prethylakoidal body. At the same time the plastoglobuli propagate. Often they form clusters within the plastid (Fig. 12). The plastids of wound-healing potato tissue as well as of actively dividing tissues of Helianthus and Daucus altogether develop to typical etioplasts, a tendency, which becomes most pronounced in illuminated tissues (26, 63). Furthermore the plastids of dividing cells of different storage tissues also increase in number.

Inspite of many investigations no definite changes in mitochondrial structure could be observed neither in the potato nor in the Jerusalem artichoke and carrot. However, mitochondria frequently undergo divisons in all these tissues.

Storage tissue exhibiting mitotic activity of those three plants altogether show a strong increase in rough ER, ribosomes and polysomes (3, 5, 25). However, there exists a significant difference between the potato and the carrot or Jerusalem artichoke: additionally to the rough ER smooth tubular ER for lipid synthesis is formed in the potato tissue.

In all storage tissues so far examined the number of dictyosomes strongly increases. These organelles are highly active in producing Golgi vesicles (3, 5, 17, 25, 68). Microtubules are also common in tissues of Solanum, Helianthus and Daucus, because they are involved in cell divisions similar to dictyosomes (Fig. 14).

In wounded potato tuber tissue the microbodies or CCB's remain unchanged (Fig. 14), whereas in Helianthus slices these organelles as well as the enclosed crystals are evidently degraded (3). Because of lack of information for all other organelles such as multivesicular bodies, vacuoles and spherosomes no comparison of these three objects is possible.

The characteristical changes in wound-healing tissues with or without application of growth substances are summarized in Fig. 20, 21 and 22. Interpretation of all these results allow the following conclusions. In Helianthus, Daucus as well as Solanum storage tissues a species-specific wound-reaction is evoked by mechanical injury. This wound-reaction for example may be restricted to lignification processes involving only few organelles (i.e. nucleus, nucleolus, ribosomes, ER). On the other hand the reaction may involve extensive changes within the cells, which go together with cell divisions and affect nearly all organelles. Notwithstanding the cytological differences the first ultrastructural changes which are to be seen are nearly identical, however. From this it can be deduced that in the case of lignification (Helianthus, Daucus) the wound-reaction does not surpass an initiating phase, i.e. the cells do not enter the division phase in contrast to the cells of potato tuber tissue.

If the self-regulation of the wound-healing process is "disturbed" by the application of plant hormones, then the wound-reaction in Helianthus as well as Daucus tissues does not stop at an early stage but continues towards cell division. Now in these tissues changes can be observed, which are similar to those in injured potato tubers. However, these changes run into an "endless" growing callus.

If the ultrastructual changes of wound-healing storage tissues of Helianthus, Daucus and Solanum are compared with those of leaves and normal roots, far reaching similarities are obvious. This is certified by investigations of Taylor and coworkers (61), Bowes (11), Bowes and Torrey (12) and Tucker (62). The wound-reaction of storage tissue can therefore be regarded as representative for higher plants and certainly be used as a model.

References

1 Amelunxen, F., Arbeiter, H.: Untersuchungen an den Drüsenhaaren von Cleome spinosa L.
Z. Pflanzenphysiol. 61, 73-80 (1969)

2 Ames, I.H., Pivorum, J.P.: A cytochemical investigation of a chloroplast inclusion.
Amer. J. Bot. 61, 794-797 (1974)

3 Bagshaw, V.: Changes in ultrastructure during the development of callus cells.
Ph. D. Thesis, University of Edinburgh (1969)

4 Bagshaw, V., Brown, R., Yeoman, M.M.: Changes in the mitochondrial complex accompanying callus growth.
Ann. Bot. 33, 35-44 (1969)

5 Barckhausen, R.: Feinstrukturelle Untersuchungen nach traumatogener Aktivierung ausdifferentierter Knollenparenchymzellen von Solanum tuberosum L.
Ph. D. Thesis, J.W. Goethe-Universität, Frankfurt/M (1973)

6 Barckhausen, R., Rosenstock, G.: Feinstrukturelle Untersuchungen zum Vorkommen von "Riesenvakuolen" in Nucleoli verletzter Parenchymzellen aus der Knolle von Solanum tuberosum L.
Biochem. Physiol. Pflanzen (BPP) 164, 327-337 (1973)

7 Barckhausen, R., Rosenstock, G.: Feinstrukturelle Untersuchungen zur traumatogenen Suberinisierung beim Knollenparenchym von Solanum tuberosum L.
Z. Pflanzenphysiol. 69, 193-203 (1973)

8 Barsy, T. de, Deltour, R., Bronchart, R.: Study of nucleolar vacuolation and RNA synthesis in embryonic root cells of Zea mays.
J. Cell. Sci. 16, 95-112 (1974)

9 Block, R.: Wound-healing in higher plants.
Bot. Rev. 7, 110-146 (1941)

10 Boasson, R., Laetsch, W.M., Price, I.: The etioplast-chloroplast transformation in tobacco: correlation of ultrastructure, replication and chlorophyll synthesis.
Amer. J. Bot. 59, 217-223 (1972)

11 Bowes, B.G.: Polar regeneration in excised roots of Taraxacum officinale Weber: a light and electron microscopic study. Ann. Bot. 40, 423-432 (1976)

12 Bowes, B.G., Torrey, J.G.: Ultrastructural changes in cells of pea root cortical explants cultured in vitro. Protoplasma 90, 99-118 (1976)

13 Chaly, N.M., Setterfield, G.: Organization of the nucleus, nucleolus, and protein-synthesizing apparatus in relation to cell development in roots of Pisum sativum. Can. J. Bot. 53, 200-218 (1975)

14 Cherry, J.H.: Regulation of invertase in washed sugar beet tissue. In: Biochemistry and physiology of plant growth substances (Eds. F. Wightman, G. Setterfield), 417-431. Runge Press, Ottawa (1968)

15 Ducet, G., Cotte-Martinon, M., Coulomb, P., Diano, M., Meunier, D.: Les mitochondries du tubercule de pomme de terre. Physiol. Vég. 8, 35-54 (1970)

16 Falk, H., El Hadidi, M.N.: Der Feinbau der Suberinschichten verkorkter Zellwände. Z. Naturforsch. 16b, 134-137 (1961)

17 Fowke, L., Setterfield, G.: Cytological responses in Jerusalem artichoke tuber slices during aging and subsequent auxin treatment. In: Biochemistry and physiology of plant growth substances (Eds. F. Wightman, G. Setterfield), 584-602. Runge Press, Ottawa (1968)

18 Franke, W.W., Scheer, U.: The ultrastructure of the nuclear envelope of amphibian oocytes: A reinvestigation. II. The immature oocyte and dynamic aspects. J. Ultrastruct. Res. 30, 317-327 (1970)

19 Frederick, S.E., Newcomb, E.H., Vigil, E.L., Wergin, W.P.: Finestructural characterization of plant microbodies. Planta 81, 229-252 (1968)

20 Hay, E.D.: Structure and function of the nucleolus in developing cells. In: The nucleus (Eds. A. Dalton, F. Haguenau), 2-79. Academic Press, New York (1968)

21 Henry, E.W.: Peroxidases in tobacco abscission zone tissue. III. Ultrastructural localization in thylakoids and membrane-bound bodies of chloroplasts. J. Ultrastruct. Res. 52, 289-299 (1975)

22 Hoefert, L.L., Esau, K.: Plastid inclusions in epidermal cells of Beta. Amer. J. Bot. 62, 36-40 (1975)

23 Hölzl, J., Bancher, E.: Über die Eiweißkristalle von Solanum tuberosum.
Österr. Bot. Z. 105, 385-407 (1958)

24 Hyde, B.B.: Changes in nucleolar ultrastructure associated with differentiation in the root tip.
J. Ultrastruct. Res. 18, 25-54 (1967)

25 Israel, H.W., Steward, F.C.: The fine structure of quiescent and growing carrot cells: its relation to growth induction.
Ann. Bot. 30, 63-79 (1966)

26 Israel, H.W., Steward, F.C.: The fine structure and development of plastids in cultured cells of Daucus carota.
Ann. Bot. 31, 1-18 (1967)

27 Johnson, J.M.: A study of nucleolar vacuoles in cultured tobacco cells using radioautography, actinomycin D, and electron microscopy.
J. Cell Biol. 43, 197-206 (1969)

28 Johnson, J.M., Jones, L.E.: Behavior of nucleoli and contracting nucleolar vacuoles in tobacco cells growing in microculture.
Amer. J. Bot. 54, 189-198 (1967)

29 Jordan, E.G. Chapman, J.M.: Ultrastructural changes in the nucleoli of Jerusalem artichoke (Helianthus tuberosus) tuber discs.
J. Exp. Bot. 22, 627-634 (1971)

30 Jordan, E.G., Chapman, J.M.: Nucleolar and nuclear envelope ultrastructure in relation to cell activity in discs of carrot root (Daucus carota).
J. Exp. Bot. 24, 197-209 (1973)

31 Kahl, G.: Synthesis of rRNA, tRNA and other RNA-species concomitant with polyribosome formation in aging potato tuber slices.
Z. Naturforsch. 26b, 1058-1064 (1971)

32 Kahl, G.: Activation of protein synthesis in aging potato tuber slices.
Z. Naturforsch. 26b, 1064-1067 (1971)

33 Kahl, G.: Genetic and metabolic regulation in differentiating plant storage tissue cells.
Bot. Rev. 39, 274-299 (1973)

34 Kahl, G.: Metabolism in plant storage tissue slices.
Bot. Rev. 40, 263-314 (1974)

35 Kahl, G., Wielgat, B.: Regulation of transcriptional activity in wounded potato tuber tissues.
Physiol. Vég. 14, 725-738 (1976)

36 Kohlenbach, H.W.: Nucleoli mit "pulsierenden Vakuolen" bei kultivierten Rhoeo-Epidermiszellen.
Z. Pflanzenphysiol. 58, 187-190 (1967)

37 Lafontaine, J.G.: Structural components of the nucleus in mitotic plant cells. In: Ultrastructure in biological systems. Vol. 3: The nucleus (Eds. A. J. Dalton, F. Haguenau), 151-196.
Academic Press, New York (1968)

38 Lafontaine, J.G., Lord, A.: A correlated light- and electron-microscope investigation of the structural evolution of the nucleolus during the cell cycle in plant meristem cells (Allium porrum).
J. Cell Sci. 16, 63-93 (1974)

39 Leaver, C.J., Key, J.L.: Polyribosome formation and RNA synthesis during aging of carrot root tissue.
Proc. nat. Acad. Sci. 57, 1338-1344 (1967)

40 Lee, S.G., Chasson, R.M.: Aging and mitochondrial development in potato tuber tissue.
Physiologia Plantarum 19, 199-206 (1966)

41 Marinos, N.G.: Multifunctional plastids in the meristematic region of potato tuber buds.
J. Ultrastruct. Res. 17, 91-113 (1967)

42 Newcomb, E.H.: Fine structure of protein-storing plastids in bean root tips.
J. Cell. Biol. 33, 143-164 (1967)

43 Reinert, J.: Über die Kontrolle der Morphogenese und die Induktion von Adventivembryonen an Gewebekulturen aus Karotten.
Planta 53, 318-333 (1959)

44 Rose, R.J.: Changes in nucleolar activity during the growth and development of the wheat coleoptile.
Protoplasma 79, 127-143 (1974)

45 Rose, R.J., Setterfield, G.: Cytological studies on the inhibition by 5-fluorouracil of ribosome synthesis and growth in Jerusalem artichoke tuber slices.
Planta 101, 210-230 (1971)

46 Rose, R.J., Setterfield, G., Fowke, L.C.: Activation of nucleoli in tuber slices and the function of nucleolar vacuoles.
Exp. Cell Res. 71, 1-16 (1972)

47 Schnepf, E.: Über den Feinbau von Öldrüsen. I. Die Drüsenhaare von Arctium lappa.
Protoplasma 67, 185-194 (1969)

48 Schnepf, E.: Über den Feinbau von Öldrüsen. II. Die Drüsenhaare in Calceolaria-Blüten.
Protoplasma 67, 195-203 (1969)

49 Schnepf, E.: Über den Feinbau von Öldrüsen. III. Die Ölgänge von Solidago canadensis und die Exkretschläuche von Arctium lappa.
Protoplasma 67, 205-212 (1969)

50 Schnepf, E.: Über den Feinbau von Öldrüsen. IV. Die Ölgänge von Umbelliferen: Heracleum spondylum und Dorema ammoniacum.
Protoplasma 67, 375-390 (1969)

51 Schnepf, E.: Tubuläres endoplasmatisches Retikulum in Drüsen mit lipophilen Ausscheidungen von Ficus, Ledum und Salvia.
Biochem. Physiol. Pflanzen 163, 113-125 (1972)

52 Schnepf, E., Nagl, W.: Über einige Strukturbesonderheiten der Suspensorzellen von Phaseolus vulgaris.
Protoplasma 69, 133-143 (1970)

53 Sitte, P.: Der Feinbau verkorkter Zellwände.
Mikroskopie 10, 178-200 (1955)

54 Sitte, P.: Zum Feinbau der Suberinschichten im Flaschenkork.
Protoplasma 54, 555-559 (1962)

55 Sitte, P.: Die Bedeutung der molekularen Lamellenbauweise von Korkzellwänden.
Biochem. Physiol. Pflanzen 168, 287-297 (1975)

56 Sparkuhl, J., Gare, R.L., Setterfield, G.: Metabolism of free and membrane-bound ribosomes during aging of Jerusalem artichoke tuber slices.
Planta 129, 97-104 (1976)

57 Sponholz, W.R., Rosenstock, G.: Über die Beziehung zwischen dem Alter von Speichergeweben und dessen Nucleinsäurespiegel vor und nach Verwundung.
Ber. Deutsch. Bot. Ges. 86, 485-497 (1974)

58 Srivastava, L.: On the fine structure of the cambium of Fraxinus americana L.
J. Cell Biol. 31, 79-93 (1966)

59 Stetler, D.A., Laetsch, W.M.: Chloroplast development in Nicotiana tabacum "Maryland Mammoth".
Amer. J. Bot., 56, 260-270 (1969)

60 Steward, F.C.: Growth and development of cultivated cells. III. Interpretations of the growth from free cell to carrot plants.
Amer. J. Bot. 45, 709-713 (1958)

61 Taylor, M.S., Wu, J.H., Daniel, R.S.: Fine structure of virus lesion periderm formation in Nicotiana glutinosa.
J. Ultrastruct. Res. 48, 173 (1974)

62 Tucker, S.C.: Wound regeneration in the lamina of magnoliaceous leaves.
Can. J. Bot. 53, 1352-1364 (1975)

63 Tulett, A.J., Bagshaw, V., Yeoman, M.M.: Arrangement and structure of plastids in dormant and cultured tissue from artichoke tubers.
Ann. Bot. 33, 217-226 (1969)

64 Verleur, J.D.: Observation on the induction of mitochondrial particles in potato tuber tissue after wounding.
Z. Pflanzenphysiol. 61, 299-309 (1969)

65 Verleur, J.D., Uritani, D.: Respiratory activity of the mitochondrial fractions isolated from healthy potato tubers and from potato tuber tissue incubated after cutting and infection with Ceratocystis fembriata.
Plant Physiol. 40, 1008-1012 (1965)

66 Wattendorff, J.: The formation of cork cells in the periderm of Acacia senegal Willd. and their ultrastructure during suberin deposition.
Z. Pflanzenphysiol. 72, 119-134 (1974)

67 Wunderlich, F.: The macronuclear envelope of Tetrahymena pyriformis GL in different physiological states.
II. Frequency of central granules within the pores.
Z. Zellforsch. Mikroskop. Anat. 101, 581 (1969)

68 Yeoman, M.M., Street, H.E.: General cytology of cultured cells.
In: Plant tissue and cell culture (Ed. H.E. Street), 121-160. Blackwell Scientific Publications, Oxford (1973)

Chemistry and Biochemistry of the Aliphatic Components of Suberin

P. E. Kolattukudy

INTRODUCTION

Suberization involved in wound-healing has been studied for over half a century mainly using cytochemical techniques. With the advent of modern analytical techniques such as chromatographic techniques and combined gas-chromatography and mass spectrometry, the composition of at least the aliphatic components of suberin of several species of plants has become known. Therefore, during the past several years we have begun to understand the chemical nature of suberin. This knowledge concerning the chemical composition has led to biochemical investigations which are beginning to reveal how the aliphatic components of suberin are synthesized and how suberization might be regulated. However, it must be pointed out at the outset that very little is known about the chemistry of the phenolic components of suberin. Therefore, I shall limit the discussions in this chapter to the aliphatic components of suberin.

There is some confusion regarding the definition of suberin. Since suberin is beginning to change its identity from an anatomical one to a chemical entity a more clear definition of what is meant by suberin is needed. Obviously the anatomical entity, suberin complex, is composed of a polymeric material and soluble waxes are associated with this layer. In this respect the suberin complex is analogous to the cuticle where the polymeric material called cutin is embedded in soluble waxes. With this analogy in mind, I propose to use the term suberin to refer to the polymeric material of the suberin complex. The extensive chemical investigation of barks is beyond the scope of this chapter although many compounds revealed by such studies most probably are constituents of the suberin contained in the bark samples. The results of such studies have been discussed elsewhere (1,2,3,4).

The work from the author's laboratory discussed in this Chapter was supported by a grant, BMS 74-09351, from the National Science Foundation.

Biochemistry of Wounded Plant Tissues

STRUCTURE OF SUBERIN

Methods of Structure Determination

Isolation. Since suberin is deposited in close association with other cellular components, particularly the cell wall, it is virtually impossible to isolate pure suberin. Therefore, only suberin-enriched preparations can be used for chemical studies. Methods for isolating suberin-enriched preparations generally involve removal of as much of the other cellular components as possible by physical, chemical and enzymological techniques. For example, to isolate suberin-enriched preparations from the natural skin of underground parts of plants such as roots and tubers, first physical techniques can be used to remove the peel. Mixtures of enzymes such as cellulase and pectinase are used to remove as much carbohydrate material as possible (5,6). The soluble lipids can be removed by thorough extraction with organic solvents such as $CHCl_3$. Repetition of these treatments is usually employed to prepare the suberin-enriched material. Preparation of suberin-enriched material from wound periderm is more difficult as the suberized layer is not as easily removed as the natural skin. However, treatment with cellulase and pectinase can be used to remove the bulk of the nonsuberized tissue and a procedure similar to that described has been used to prepare suberin-enriched preparations from wound periderm of potato tubers (7). Preparations enriched in suberized endodermis were also prepared using similar techniques (8). These techniques could result in varying degrees of losses of soluble lipids such as the waxes. The most convenient way to isolate the waxes which might be present in the suberin layers is to isolate them from the solvent extract of the suberin-enriched layer isolated after the enzymatic treatments indicated above.

Depolymerization. The methods thus far employed to depolymerize suberin involve cleavage of ester bonds, and therefore the available information concerning the polymer is restricted to the monomers which are involved in ester bonds. Three methods of depolymerization have been used: hydrolysis with alcoholic alkali (9,10,11,12,13), transesterification with methanol containing a catalyst such as BF_3 (5,6) or sodium methoxide (14), and hydrogenolysis with $LiAlH_4$ (Fig. 1). The first method which gives the free acids has the disadvantage of oxidative reactions occurring during hydrolysis, particularly in the presence of the phenols released during hydrolysis.

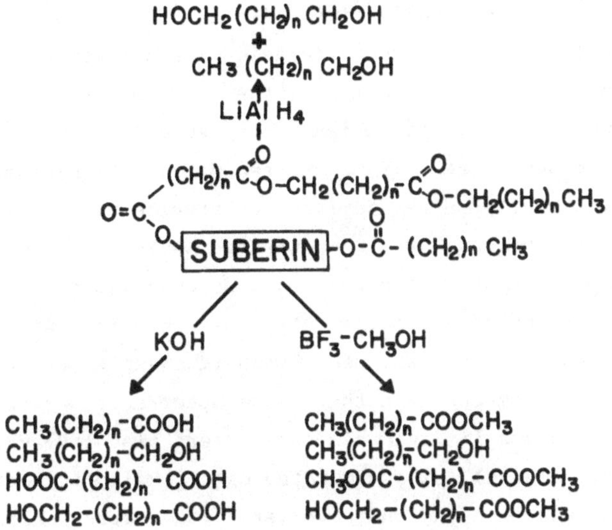

Figure 1. Depolymerization techniques used for the analysis of the aliphatic components of suberin.

The second method is probably the most suitable method for structural studies, while the third method has a disadvantage in that it converts both ω-hydroxy acids and dicarboxylic acids into indistinguishable diols. Examination of the phenolic components would also be aided by the hydrogenolysis technique as the reducing environment would minimize oxidative destruction of the phenolic materials. Since both ω-hydroxy acids and dicarboxylic acids are measured together as diols, higher sensitivity is achieved by this technique (7). Therefore, for quick determination of the presence of suberin and to determine whether a phytopolymer is of cutin-type or suberin-type, the hydrogenolysis technique is very convenient (15). On the other hand, for a detailed analysis of the composition of the monomers, a transesterification technique such as that using 14% BF_3 in methanol should be used. However, it should be born in mind that the prolonged treatment, which is usually necessary to get maximum depolymerization, could result in the addition of methanol at the double bonds of the monomers (6,7).

Structure Determination of Monomers

The monomers isolated from the depolymerization step can be readily resolved by thin-layer chromatography (6,9,10,12,13,16). For example, the transesterification technique usually gives rise to four classes of monomers; fatty acid methyl esters, dicarboxylic acid dimethyl esters, fatty alcohols, and ω-hydroxy acid methyl esters (6). Other more polar hydroxy acids are found only in much smaller quantities in the suberin of most plants. Each class can then be subjected to further resolution by gas-liquid chromatography and trimethylsilyl ether derivatives provide a convenient method for the analysis of the hydroxylated components. As indicated earlier in this chapter, combined gas-chromatography and mass spectrometry provides the most convenient method of structure determination. The fatty acids which constitute a universal component of all suberin samples are readily resolved by gas-liquid chromatography and their mass spectra are readily interpretable (17). The dicarboxylic acid dimethyl esters also give characteristic mass spectra (6,17,18,19). The saturated dicarboxylic acid dimethyl esters give a weak molecular ion and characteristic ions corresponding to M-31, M-73, M-92, M-105, and M-123, whereas the unsaturated counterparts give a fairly prominent molecular ion and strong diagnostic ions at M-32, M-50, M-64, and M-92 (Fig. 2). The mass spectra of trimethylsilyl derivatives of saturated ω-hydroxy acids show weaker molecular ions than those observed in the spectra of their unsaturated counterparts (6,18,20). Both types of ω-hydroxy acids show diagnostic ions corresponding to M-15, M-31, and M-47 (Fig. 3). A metastable ion representing the transition [M-15]→[M-47] is a very characteristic feature of the mass spectrum of the ω-hydroxy acid derivative.

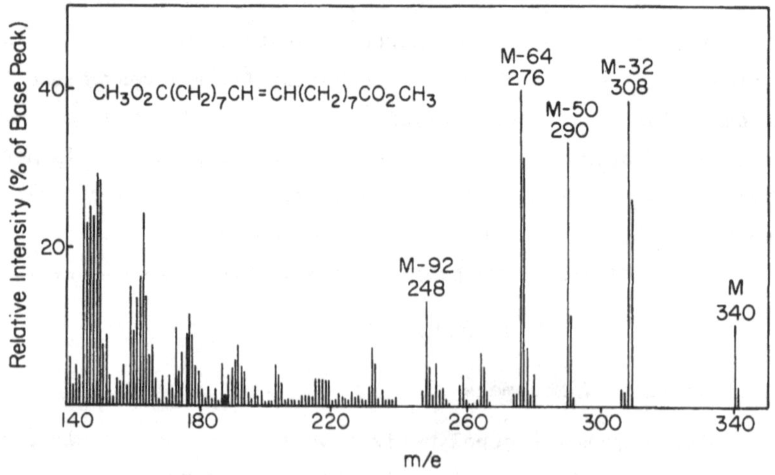

Figure 2. Mass spectrum of dimethyl octadecenedioate (6).

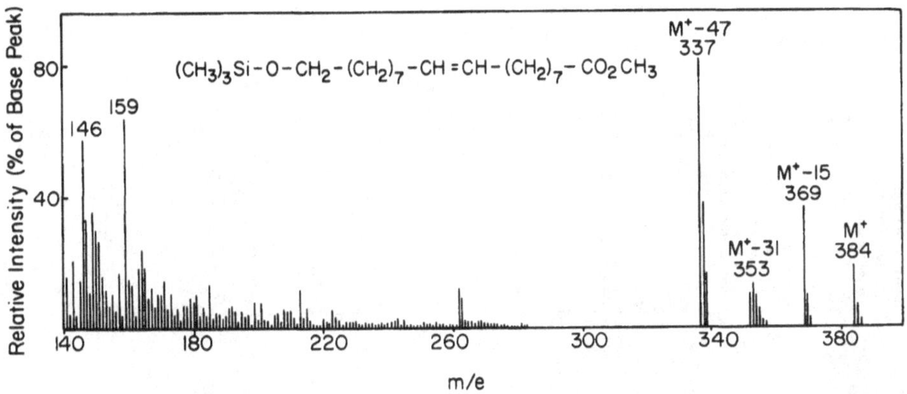

Figure 3. Mass spectrum of the trimethylsilyl ether of methyl 18-hydroxy-octadecenoate (6).

Hydrogenolysis of most suberin samples give rise to very long-chain alcohols and diols as dominant components (5). The mass spectra of trimethylsilyl ethers of alcohols give a dominant ion at M-15 and other characteristic ions in the lower mass region (21,22). The mass spectra of the trimethylsilyl ethers of unsaturated diols show a stronger molecular ions than those observed with saturated diols. Both types of diols show characteristic ions corresponding to M-15, M-31, M-90, M-15-90, and M-31-90 (Fig. 4). In both cases diagnostic doubly-charged ions corresponding to (M-30)/2 and their first isotope ions are observed (23,24). Identification of dihydroxypalmitic acid, epoxy-C_{18} acid and trihydroxy-C_{18} acid and the corresponding dicarboxylic acids which are present as minor components of most suberin samples can be identified also by combined gas-liquid chromatography and mass spectrometry and their spectra are discussed elsewhere (6,14,18,20).

The phenolic fraction obtained from the hydrogenolysate ($LiAlH_4$) or deuterolysate ($LiAlD_4$) of suberin also can be analyzed by combined gas-chromatography and mass spectrometry (25). Hydrogenolysates of the suberin samples from potato, sweet potato, turnip, rutabaga, carrot, and red beet gave dihydroconiferyl alcohol which could be analyzed as their acetates or trimethylsilyl ethers. The characteristic ions found in the mass spectra of these derivatives are indicated in Figure 5.

Composition of Suberin

Presently available information concerning the composition of suberin is

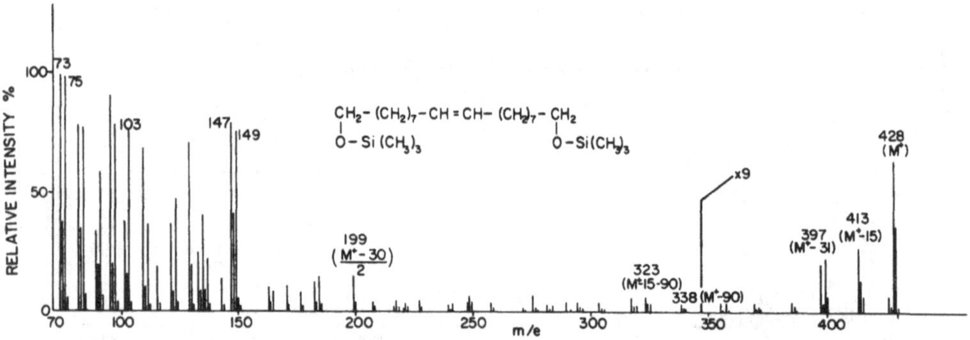

Figure 4. Mass spectrum of the trimethylsilyl ether of octadecene-1,18-diol (24).

derived mainly from studies on bark or the natural periderm of underground parts of plants such as roots and tubers. In the case of most suberin preparations, the commonly used depolymerization techniques, such as hydrogenolysis, transesterification, and hydrolysis, release only 20 to 50% (by weight) of the crude sample (5,6,9,10,16). Since suberin is tightly associated with cell wall (26,27,28,29,30,31), it cannot be isolated in pure form, and therefore the residual insoluble material left behind after the depolymerization techniques probably contains cell wall material. The intense color of such residues would indicate that they probably also contain phenolic materials and possibly aliphatic chains attached by bonds not susceptible to the depolymerization techniques. However, the composition of such insoluble materials has not been examined.

Phenolic materials also constitute a large portion (in many cases, two-thirds) of the material released by the depolymerization techniques (5,9,10, 15,25). A systematic investigation of these phenolic constituents is essential to our understanding of the composition of suberin, but none has been reported. However, ferulic acid has been identified as a covalently attached component of suberin in potato, sweet potato, turnip, rutabaga, carrot, and red beet (25). The residue recovered after partial alkaline hydrolysis of the suberin samples from the above plants contained ferulic acid, strongly suggesting that this acid is covalently attached to suberin. However, this esterified ferulic acid constitutes only a small (<1%) portion

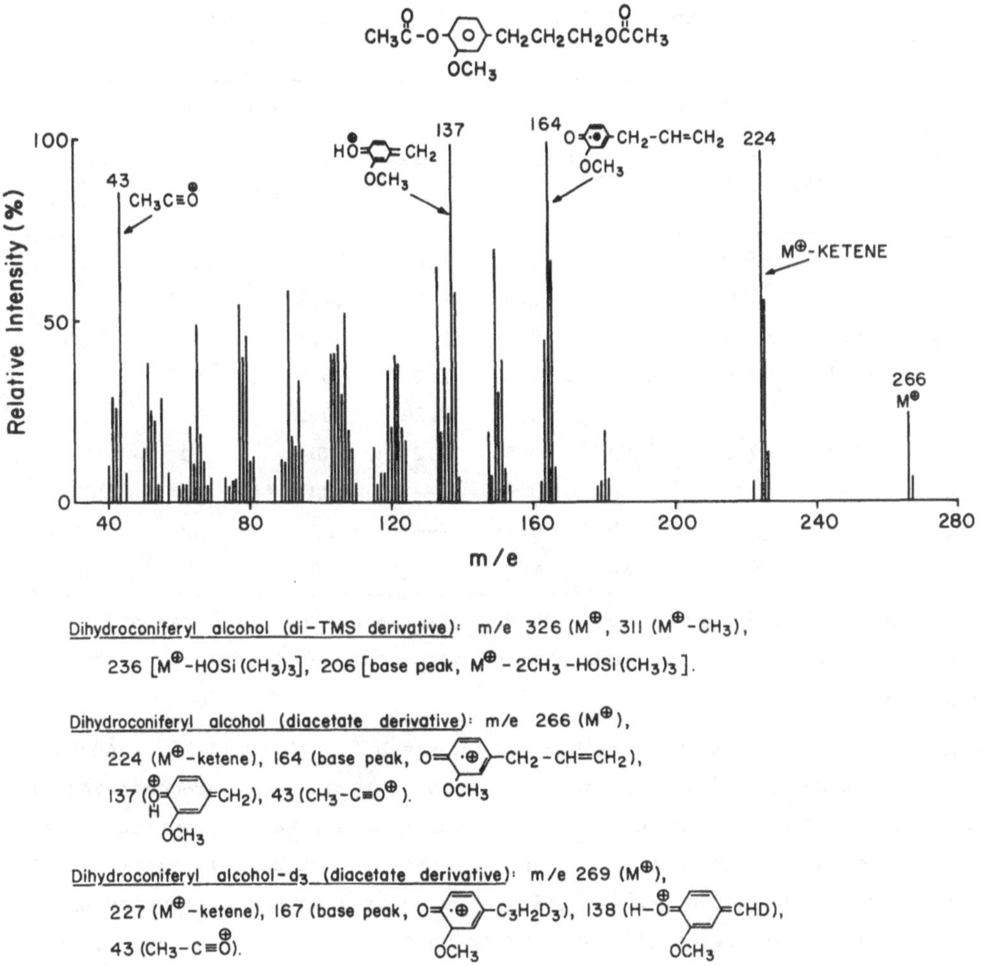

Figure 5. Mass spectrum of dihydroconiferyl alcohol diacetate (top) and the major ions in the mass spectrum of other derivatives of dihydroconiferyl alcohol (bottom). Data from reference 25.

of the suberin sample and much of the phenols contained in these samples remain unidentified.

The major aliphatic components of suberin, often constituting not more than 10% of the total crude skin (suberin-enriched) preparations from storage tissues, are ω-hydroxy acids and the corresponding dicarboxylic acids together with smaller amounts of very long ($>C_{20}$) acids and corresponding alcohols (5,6). In tree bark samples, the aliphatic components of suberin

Table I. The Percent Composition of ω-Hydroxy Acids and Dicarboxylic Acids of Suberin

Chain Length	ω-Hydroxy Acids				Dicarboxylic Acids			
	Ribes (50.1%)	Carrot (31%)	Rutabaga (36%)	Sweet Potato (36%)	Ribes (35.9%)	Carrot (24%)	Rutabaga (22%)	Sweet Potato (21%)
C_{16}	48.0	23.0	41.5	4.9	42.8	26.2	42.7	6.6
C_{18}	5.7	6.9	4.0	1.3	19.7	14.9	11.2	9.1
$C_{18:1}$	28.3	53.7	42.0	90.7	32.6	54.6	35.9	80.5
C_{20}	15.7	7.8	9.1	0.3	2.5	3.0	5.4	0.2
C_{22}	2.1	7.8	3.4	0.3	2.2	0.2	3.3	ND
C_{24}	0.2	0.3	ND	0.5	0.2	ND	D	ND
C_{26}	---	ND	ND	0.9	---	ND	ND	ND
C_{28}	---	ND	ND	0.7	---	ND	ND	ND
C_{30}	---	ND	ND	ND	---	ND	ND	ND

Data taken from references 5 and 10; D, detectable; ND, not detectable. The numbers in parentheses indicate the percentage of the total aliphatic monomers.

might be substantially higher, but seldom reach 50% (8,10,11,16). Among the bifunctional molecules of suberin, monounsaturated C_{18} predominates in many plants (Table I), although in some cases saturated C_{16} constitutes an equally dominant component (5,15) and in others, particularly in tree barks, saturated C_{22} dominates (9,10,11,16). The chain length distribution of dicarboxylic acids is usually similar to that of the ω-hydroxy acids, but in many cases the latter contains a higher proportion of longer chains than the former. For example, in potato suberin, dicarboxylic acids longer than C_{18} were not detected, whereas ω-hydroxy acid fraction contained aliphatic chains as long as C_{28} (6). In *Malus pumila* the major ω-hydroxy acid is C_{22}, whereas C_{16} and C_{18} are the major dicarboxylic acids (16). In suberin from most storage tissues, either monounsaturated C_{18} alone or together with saturated C_{16} dominate the bifunctional molecules (5). The fatty acids and fatty alcohols of suberin are characterized by the presence of very long ($>C_{18}$) aliphatic chains as the dominant components (Table II). The more polar acids such as those containing additional functional groups in the middle of the aliphatic chains are only minor components in most

Table II. The Percent Composition of Fatty Acids and Fatty Alcohols of Suberin

Chain Length	Fatty Acids				Fatty Alcohols			
	Ribes (0.3%)	Carrot (14%)	Rutabaga (4%)	Sweet Potato (9%)	Ribes (0.7%)	Carrot (3%)	Rutabaga (5%)	Sweet Potato (4%)
C_{16}	---	3.0	3.0	0.9	---	ND	ND	ND
C_{18}	---	9.4	1.8	1.1	37.1	9.4	28.4	68.3
$C_{18:1}$	---	2.3	3.2	3.9	---	ND	10.7	ND
C_{20}	47	17.6	71.4	0.9	33.9	37.7	38.8	15.5
C_{22}	47	39.8	16.4	3.3	29.0	52.8	22.4	6.1
C_{24}	---	23.3	3.9	4.7	D	D	D	3.0
C_{26}	---	0.6	ND	16.6	---	D	D	3.0
C_{28}	---	ND	ND	20.1	---	ND	ND	0.6
C_{30}	---	ND	ND	34.3	---	ND	ND	0.1

Data taken from references 5 and 10; D, detectable; ND, not detectable. The numbers in parentheses indicate the percentage of the total aliphatic monomers.

suberin samples, whereas such components are the dominant ones in cutin (32). However, suberin samples from some barks were reported to contain a large proportion of in-chain substituted polar acids (9). Since there is a possibility that bark samples, particularly those from young stems, might include cutin, care must be taken to isolate suberin samples without contamination from cutin. In any case, the suberin samples from underground parts of plants, including storage tissues, shown in Tables I and II, tapioca roots and *V. faba* roots as well as aerial roots of *Monstera deliciosa* Liebm., do not contain in-chain substituted polar acids as substantial components (33). However, most of them do contain such polar acids as minor components (5,6). Even these minor polar components are distinguished from cutin components in that the in-chain substituted dicarboxylic acids are nearly as significant as their ω-hydroxy counterparts. For example, 9,10-epoxyoctadecanedioic acid is almost as significant a component of potato suberin as is 18-hydroxy-9,10-epoxyoctadecanoic acid (6), whereas in cutin samples the former is at best a minor component when compared to the latter (32). Dihydroxy-C_{16} acid is a common minor component of suberin, and the positional isomer composition of this dihydroxy acid in the cutin of the aerial parts of a plant may or may not be identical to that found in the suberin of the underground part of the same plant. For example, 10,16-, 9,16- and 8,16-dihydroxypalmitic acid was found in potato leaf cutin in a ratio of 61:26:13, and the ratio was quite similar (67:23:10) in the dihydroxy-C_{16} acid of the suberin from the tubers (34). In *V. faba*, on the other hand, the cutin in aerial parts (stem, petiole and leaf) contained 10,16- and 9,16-dihydroxy-C_{16} acid in a ratio of about 17:3, whereas in the root suberin a much higher proportion of 9,16-dihydroxy acid was found, although this dihydroxy acid was only a minor component of suberin (33).

The composition of the aliphatic components of suberin, although probably not an adequate parameter for chemotaxonomy, does reveal chemotaxonomic relationships (5,10). For example, suberin composition of carrot and parsnip, two members of the Umbelliferae family, was similar; likewise rutabaga and turnip, both of which belong to the Cruciferae family, have fairly similar suberin composition (5). The similarities between the suberin composition of sweet potato and potato tuber are consistent with the close taxonomical relationship between Convolvulaceae and Solanaceae within the same order. Even though red beet (Chenapodiaceae) is not closely related to turnip (Cruciferae), the composition of the dicarboxylic acids and the ω-hydroxy acids of the suberin of these two plants showed surprising similarity.

The soluble waxy materials which might be an important component of the lamellar structure of the suberin complex (31) have not been comprehensively examined. The soluble waxes from tree bark samples reach as high as 25% of the crude samples (16). However, it appears likely that the composition of the waxes of suberin layers is similar to that of the better known cuticular waxes (86). For example, alkanes, which constitute one of the most common components of cuticular waxes, are also found in potato peel (12,35). Just as in the cuticular waxes, a strong predominance of alkanes with odd-number of carbon atoms was observed in the alkanes from potato peel. In the potato peel alkane fraction, $\underline{n}$-C_{25} is the major component with smaller amounts of $\underline{n}$-C_{23} and $\underline{n}$-C_{27}, whereas the major alkanes in the wound periderm are $\underline{n}$-C_{23} and $\underline{n}$-C_{21} (35). This difference is consistent with the findings that hydrocarbons of the cuticular lipids of plants contain longer carbon chains than those found in internal alkanes. The fatty alcohols and very long fatty acids recovered from the neutral lipid fraction of the soluble lipids of suberized tissue most probably are derived from wax esters (35). In addition to these, terpenes such as the pentacyclic triterpenes, betulinol and friedelin have also been found as bark constituents, suggesting that they might possibly be part of the wax fraction of suberin (4,11). Much further work is needed on the wax fraction associated with the suberin layers of underground parts of plants before generalizations can be drawn. However, from the fragmentary data presently available, it would appear that this wax is probably similar to cuticular waxes.

Intermolecular Structure of Suberin

Little is known about the intermolecular structure of suberin. The major aliphatic monomers of suberin namely dicarboxylic acids and ω-hydroxy acids appear to be located exclusively as components of the polymer and not in the soluble lipids (7). An attempt was made to determine whether the hydroxyl groups of the ω-hydroxy acids of suberin are free in the polymer. Methane sulphonyl chloride treatment of suberin followed by $LiAlH_4$ or $LiAlD_4$ treatment of the mesylated polymer resulted in no increase in fatty alcohols and no incorporation of deuterium at the ω-carbon was detected (36). These results showed that the ω-hydroxy acids are attached to polymer at both ends of the aliphatic chains. The total number of hydroxyl groups and carboxyl groups present in the aliphatic monomers clearly show that they themselves cannot form a polymer. The presence of large amounts of phenolic materials in suberin (5,9,10,16,25) suggests that the aliphatic components are esterified to the phenolics. An insoluble core, rich in phenolic material, remains after the depolymerization techniques, which cleave ester bonds. Therefore it appears that this core is held together by other types of bonds, probably similar to those found in lignin type material (12). On this basis the model shown in Figure 6 was proposed for suberin (32). The phenolic materials are probably attached to the cell wall components in a manner similar to that suggested for lignin attachment to carbohydrates (37,38) and the hydroxyl groups in the phenolic matrix are esterified to the aliphatic components. Such cross linkages and esterification of the phenolic materials would result in water proofing of the suberized wall. In addition soluble waxes which are present in association with these layers would also help in water-proofing the

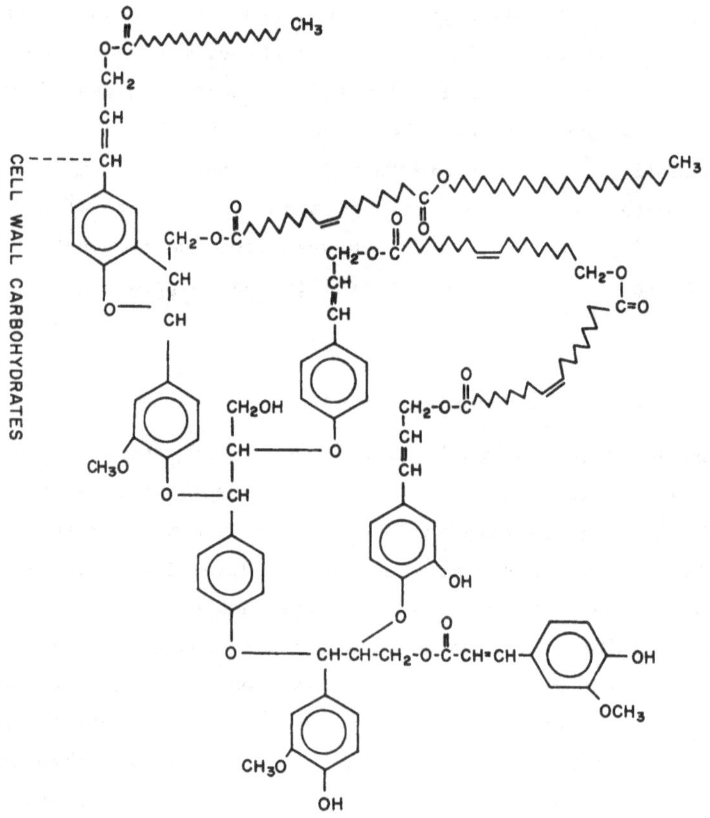

Figure 6. A tentative model proposed for the structure of suberin (32).

suberized walls. As the bark lignin usually contains less methoxyl groups than the internal lignin (39), the phenolic core of suberin shown in Figure 6 is indicated as having less number of methoxyl groups than that usually indicated in lignin models. The suberin model proposed above would suggest that disruption of synthesis of the phenolic core or the synthesis of the aliphatic components would prevent the formation of the normal suberin layer and thus result in a failure to water-proof the wound. All experimental evidence thus far available on the biosynthesis of the phenolic components discussed elsewhere in this volume and the aliphatic components (discussed in a following section in this chapter) strongly suggest that during wound healing the phenolic components are formed prior to the formation of the aliphatic components. This sequence of events is also consistent with the suberin model proposed above. However, this model is only a tentative hypothesis based on fragmentary evidence and it is intended to be only a working model which needs to be tested.

Suberization During Wound Healing of Aerial Parts of Plants

Several authors have concluded from histological studies that wounded leaves, stems, and fruits produce suberin (40,41,42) but others have suggested that cutin, the natural protective polymer of such tissues is resynthesized (43,44). In fact such wound healing tissues have been used by what was called biosynthetic studies on cutin (44). Such conflicting views persisted mainly because the chemical nature of the polymer deposited during wound healing was unknown. Some recent experimental evidence helps to resolve this conflict and establishes suberization as a general process involved in wound healing.

The structure and composition of the aliphatic monomers of the polymeric material deposited during wound-healing of tomato fruit, bean pods, and Jade leaves were examined (15). After removing the cuticle-containing layer of tissue, the wounds were healed for 14 days and the resulting surface layer was excised, lyophilized, solvent-extracted, and depolymerized by hydrogenolysis with $LiAlH_4$ or transesterified with BF_3 in methanol. The products obtained by the chemical depolymerization were subjected to thin layer chromatography and combined gas chromatography and mass spectrometry. The major aliphatic components isolated from the hydrogenolysate of the wound polymer produced by tomato fruit were hexadecane-1,16-diol and octadec-9-ene-1,18-diol, which were shown to be derived from a 1:1 mixture of ω-hydroxy and dicarboxylic acids of the appropriate chain length by $LiAlH_4$ reduction. Also identified in the wound polymer were long chain ($>C_{20}$) fatty acids and alcohols (Table 3). This monomer composition is typical of suberin polymers and is in sharp contrast with that of the cutin of tomato fruit which contains dihydroxy C_{16} acid as the major aliphatic component (Fig. 7). The hydrogenolysis of the wound material from bean pods gave octadecene-1,18-diol as the major aliphatic component, and smaller amounts of hexadecane-1,16-diol and long chain alcohols. Similar treatment of the normal cuticular tissue of these pods gave hexadecane triol, as well as C_{16} and C_{18} alcohols. Hydrogenolysis of wound material from the Jade leaves gave octadecene-1,18-diol, C_{16} and C_{22} diols, as well as alcohols

Table III. Percent Composition of the Major Aliphatic Components of Suberin Deposited in the Wound Periderm of Tomato Fruits

Chain Length	Fatty Acids	Fatty Alcohols	ω-Hydroxy Acids	Decarboxylic Acids
C_{16}	50.5	D	37.8	52.7
C_{18}	14.5	16.6	D	8.6
$C_{18:U}$	21.8	3.1	22.0	35.8
C_{20}	6.7	33.7	11.6	0.8
C_{22}	6.5	46.6	25.4	2.0
C_{24}	---	---	3.3	---

Data from reference 15.

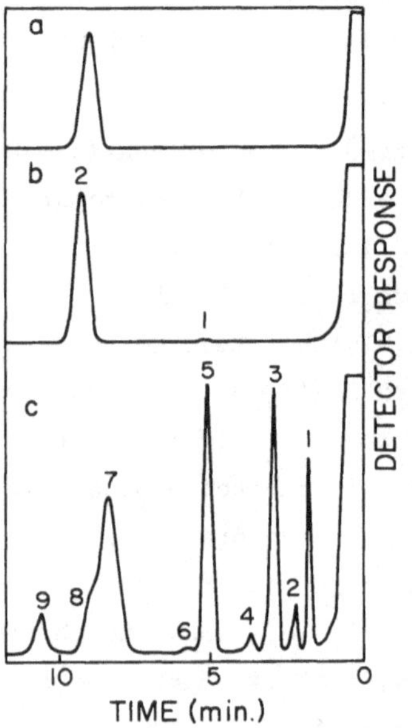

Figure 7. Gas-liquid chromatograms of trimethylsilyl ethers of the hydrogenolysate of normal tomato peel (a), control tomato tissue (b), and wound-healed tomato tissue (c). In c: 1, hexadecanol; 2, heptadecanol; 3, a mixture saturated and unsaturated C_{18} alcohol; 4, nonadecanol; 5 (and b1), hexadecane-1,16-diol; 6, eicosanol; 7, octadecene-1,18-diol; 8 (and b2), hexadecanetriol; 9, docosanol. Data from reference 15.

from C_{16} to C_{26}, whereas similar treatment of the cutin-containing tissue from these leaves gave C_{16} triol as the major aliphatic component. Thus the major aliphatic monomers of the polymeric material deposited during the wound-healing of bean pods and Jade leaves are very similar to those of suberin, although the natural protective polymer of these tissues is cutin. From these results, it was concluded that suberization is a fundamental process involved in wound-healing in plants, irrespective of the chemical nature of the natural protective polymer of the tissue.

BIOSYNTHESIS OF SUBERIN

Virtually all of the information presently available on the biosynthesis of suberin is derived from studies on potato tubers. Since the composition of potato suberin suggested that the monounsaturated C_{18} components, which constitute major aliphatic components of suberin in this species, could be derived from oleic acid, initial biosynthetic studies were done with [1-^{14}C]oleic acid. Expecting that the expanding young tubers would synthesize suberin most rapidly, such young tubers were coated with an emulsion of [1-^{14}C]oleic acid and incubated (45). The insoluble material remaining after a thorough extraction of the soluble lipids from the periderm, was labeled suggesting that the exogenous [1-^{14}C]oleic acid was incorporated into suberin. Hydrogenolysis of the insoluble material followed by analysis of the products by thin-layer and gas chromatography revealed that octadec-9-ene-1,18-diol was one of the major labeled components present in the hydrogenolysate of the labeled periderm derived from [1-^{14}C]oleic acid. Thus, exogenous oleic acid was incorporated into 18-hydroxyoleic acid and/or the corresponding dicarboxylic acid of suberin in young potato tubers. However, the intact tuber is not a convenient system for detailed biochemical studies on suberization, and it is virtually impossible to get uniform potato tubers throughout the year.

Since the chemical composition of the suberin formed at the wound periderm is identical to that of the suberin of the natural skin of potato tubers (7), wound-healing tissue disks might be a suitable system for biosynthetic studies. Therefore, a time-course of development of suberin in wound-healing potatoes was determined. Since ω-hydroxyoleic acid and the corresponding dicarboxylic acid are the major components of potato suberin, the hydrogenolysis technique was used to convert both of these acids into octadecene-1,18-diol and the amount of this diol was used as a measure of suberin. With this technique the time-course of deposition of suberin in tissue cylinders was found to be identical to that in tissue disks (35). After about 3 days of aging the formation of aliphatic components of suberin

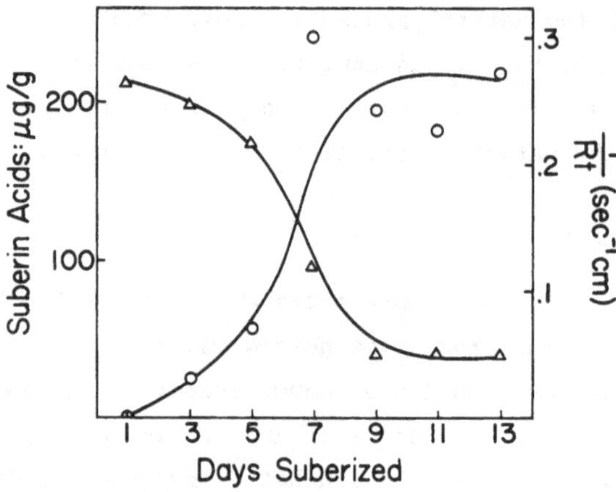

Figure 8. Time course of suberization in potato tuber tissue and development of diffusion resistance of tissue surface to water vapor. o—o, octadecenediol as a measure of suberin; Δ—Δ, diffusion resistance; data from reference 7.

started, and suberization became quite rapid between the 5th and 6th day and leveld off after 7 days (Fig. 8). With the deposition of the aliphatic components of suberin, the diffusion resistance of the tissue surface to water vapor increased and as the deposition of suberin ceased the diffusion resistance also leveled off (Fig. 8). This clear correlation between the formation of the aliphatic components and development of diffusion resistance provided us with a simple and rapid method for monitoring suberization in tissue slices during biosynthetic studies.

Incorporation of Labeled Precursors into Suberin

In 4-day aged potato disks, [1-^{14}C]oleic acid was incorporated into an insoluble residue, which, upon exhaustive hydrogenolysis with $LiAlH_4$, released labeled chloroform-soluble products (35). Radio thin-layer and gas-chromatographic analyses showed that the label was contained exclusively in octadecenol and octadecene-1,18-diol. Chemical degradation of the labeled diol with OsO_4 and periodate followed by chromatographic analyses of the products showed that virtually all of the label contained in the diol was in octadec-9-ene-1,18-diol, the product expected from the major components of suberin, ω-hydroxyoleic acid and the corresponding dicarboxylic acid. As expected, the other characteristic components of potato suberin, namely the very long-chain acids and alcohols, were not labeled. In order to examine the biosynthesis of these components in suberizing potato disks, [1-^{14}C]acetate was used as the precursor which was incorporated into both alkanol and alkane diol fractions obtained by $LiAlH_4$ treatment of suberin. Radio gas-liquid chromatographic analyses of these two fractions showed that substantial amounts of ^{14}C contained in the alkanol fraction were present in the aliphatic chains longer than C_{20} (Fig. 9a). In the diol fraction the major labeled component was octadec-9-ene-1,18-diol, but C_{16}, C_{18}, C_{20}, C_{22}, and C_{24} alkane-α,ω-diols were also labeled (Fig. 9b).

Soluble lipids such as those found in cuticular waxes might be present in the suberized walls. If such is the case, the soluble lipids derived from [1-^{14}C]acetate in the suberizing potato disks might contain labeled wax components. A detailed chromatographic examination of the soluble lipids showed the presence of labeled very long acids (Table IV) and labeled C_{22} and C_{24} alcohols. These labeled components were present in the neutral lipid fraction. In addition, alkanes were also labeled and the distribution of label among the alkanes was consistent with the composition of n-alkanes found in the wound periderm of this tissue; C_{21} and C_{23} were the major components with lesser amounts of C_{19} and C_{25} (Fig. 10).

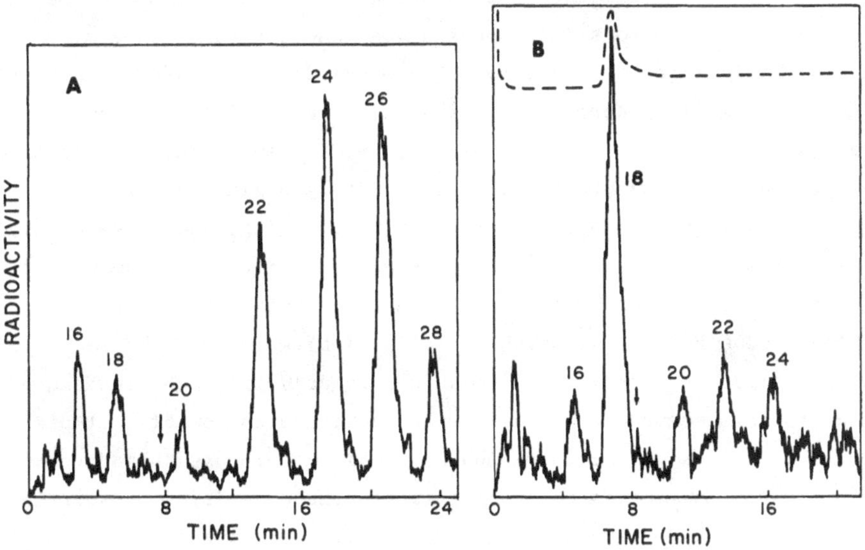

Figure 9. Radio gas-liquid chromatogram of acetates of alcohols (A) and diols (B) isolated from the hydrogenolysate of suberin from 6-day aged potato disks which had been incubated with [1-^{14}C]acetate for 6 hr. The number on each peak represents chain length. Data from reference 35.

Table IV. Distribution of Radioactivity in Fatty Acid Methyl Esters Obtained from the Total Soluble Lipids Derived from [1-^{14}C]Acetate in Suberizing Potato Disks

Days Aged	Radioactivity						
	C_{16} s+u[1]	C_{18} s+u	C_{20}	C_{22}	C_{24}	C_{26}	C_{28}
	% *total*						
0	35	50	6	9	ND[2]	ND	ND
1	29	40	12	11	8	ND	ND
2	35	47	9	5	5	ND	ND
3	37	35	10	11	7	ND	ND
4	36	31	10	14	12	ND	ND
5	33	37	6	9	9	7	ND
6	34	34	6	8	10	4	4
7	41	28	7	11	9	3	3
8	39	27	7	10	6	4	8

[1]s, saturated; u, unsaturated. [2]ND, not detected. Data from reference 35.

To determine whether the incorporation of the exogenous labeled precursors into suberin monomers reflected the suberization process, time-course of incorporation of precursors into suberin was examined (35). The amounts of [1-^{14}C]oleic acid incorporated into ω-hydroxyoleic acid and/or dicarboxylic acid (measured as diol) were 0, 1.5, 2.5, 0.8, and 0.3% of the applied substrate in tissues suberized for 0-, 2-, 4-, 6-, and 8-days, respectively. Similarly, incorporation of [1-^{14}C]acetate into suberin was low up to the 3rd day of aging, quite rapid the next 4 days of aging, and subsequently the rate decreased (Fig. 11). As the tissue aged, increasing amounts of [1-^{14}C]acetate were incorporated into longer aliphatic chains in the insoluble residue as well as soluble lipids (Tables IV and V). Incorporation of exogenous 16-hydroxy[G-^{3}H]hexadecanoic acid into suberin also reflected the time-course of suberization indicated above (61). These changes in the rates of incorporation of exogenous precursors into characteristic suberin components clearly reflected the deposition of suberin as measured by both the amount of octadecene diol and the development of diffusion resistance of the tissue surface to water vapor. Thus, the wound-healing potato disks provide a convenient experimental system for biosynthetic studies on suberin.

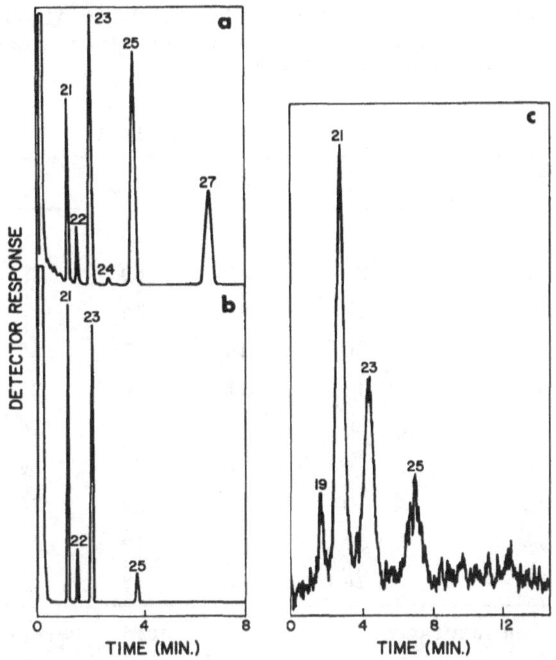

Figure 10. Gas-liquid chromatogram of n-alkanes isolated from (A) potato tuber skin and (B) wound periderm. C, Radio gas-liquid chromatogram of alkanes derived from [1-^{14}C]acetate in 5-day aged potato disks. Number on each peak represents chain length. Data from reference 35.

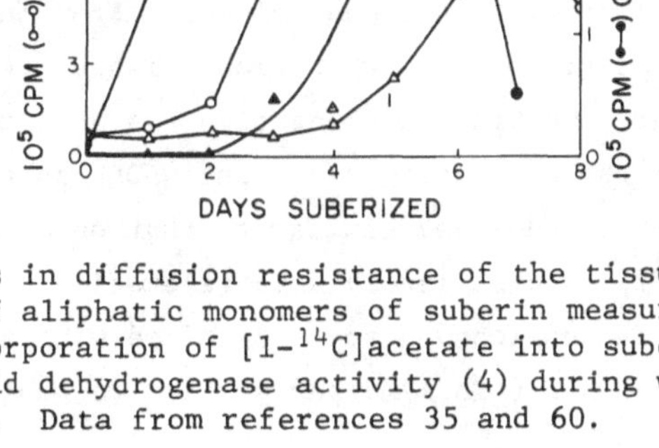

Figure 11. Changes in diffusion resistance of the tissue surface to water vapor (1), level of aliphatic monomers of suberin measured as octadecene-1,18-diol (2), incorporation of [1-^{14}C]acetate into suberin (3), and ω-hydroxy fatty acid dehydrogenase activity (4) during wound healing of potato tuber disks. Data from references 35 and 60.

Table V. Distribution of Radioactivity in the Fatty Alcohols Isolated from the Hydrogenolysate of the Insoluble Residue Obtained from Suberizing Potato Disks Incubated with [1-^{14}C]Acetate

Days Aged	Radioactivity						
	C_{16}	C_{18} s+u[1]	C_{20}	C_{22}	C_{24}	C_{26}	C_{28}
			% total				
0	ND[2]	ND	ND	ND	ND	ND	ND
1	20	32	18	14	17	ND	ND
2	27	30	14	12	15	3	ND
3	13	12	11	26	24	17	ND
4	11	9	8	27	26	17	4
5	6	6	4	30	39	36	10
6	10	9	6	18	24	22	11
7	14	12	6	19	21	15	12
8	26	14	12	18	11	9	12

[1]s, saturated; u, unsaturated. [2]ND, not detected. Data from reference 35.

Pathways Involved in the Biosynthesis of Suberin

From the chemical composition of suberin and the biosynthetic studies discussed above, the pathway indicated in Figure 12 was proposed for the biosynthesis of the major aliphatic components of suberin. One of the major reactions involved in the biosynthesis of suberin is chain elongation which results in the formation of the very long aliphatic chains. Even though chain elongation has not been demonstrated in cell-free preparations from suberizing tissues, elongation has been detected even after one day of aging (46), and the results discussed in a previous section clearly demonstrate chain elongation in tissue slices. From the results obtained with cell-free preparations from other plant tissues (47,48), it would appear probable that a membranous elongation system catalyzes elongation with malonyl-CoA as the elongating agent and NADPH as the reducing agent. Presumably an acyl-CoA reductase similar to that observed in other plant systems (49,50,51,52) is also involved in generating the fatty alcohols found in suberin. These alcohols are esterified with acyl-CoA and/or phospholipids as the acyl doner in a manner similar to that observed in other plants (53).

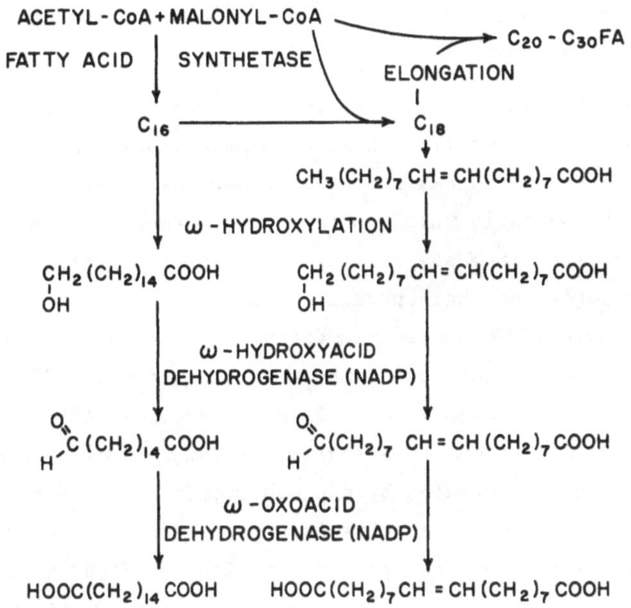

Figure 12. A probable pathway for the biosynthesis of the major aliphatic monomers of suberin.

Since ω-hydroxy acids and dicarboxylic acids are major aliphatic components of suberin, ω-hydroxylation and the oxidation of ω-hydroxy acids to the dicarboxylic acids would be key reactions involved in suberin biosynthesis. ω-Hydroxylation has not been demonstrated in suberizing plant tissues and indeed in any higher plant. Recently, however, a microsomal preparation from the shoots of germinating *V. faba* was shown to catalyze ω-hydroxylation of fatty acids (54). This system requires NADPH and O_2 and probably involves cytochrome P_{450}. A similar system is most probably involved in suberizing tissues and the endoplasmic reticulum observed to lie parallel to the cell wall in wound-healing tissues (27) is probably involved in both ω-hydroxylation and chain elongation. The ω-hydroxy acid is then converted into the corresponding dicarboxylic acid (as discussed in the succeeding section) by two dehydrogenases each with a preference for NADP. How the monomers are incorporated into the polymer is unknown. It would appear probable that activation of the carboxyl group is involved in the esterification of the monomers. Even though it is possible that the reactions involved in the synthesis of the monomers use either a thioester or another acyl derivative as the substrate so that the acyl group can then be transferred to the polymer, no experimental evidence concerning this possibility is available.

Biosynthesis of the small amounts of in-chain substituted polar acids found in suberin has not been studied in suberizing tissues. However, information concerning the biosynthesis of such compounds is available from biosynthetic studies on cutin which contains large amounts of such monomers. Incorporation of exogenous specifically labeled precursors into such polar acids strongly suggested that epoxidation of ω-hydroxyoleic acid gives rise to 18-hydroxy-9,10-epoxystearic which undergoes hydration of the oxirane function to generate the trihydroxy stearic acid (55). These two steps were also demonstrated in cell-free preparations (56,57). The first step catalyzed by a particulate preparation from *Spinacea oleracea* involves cytochrome P_{450} as demonstrated by a photoreversible inhibition of the enzymatic reaction by CO and in this case activation of the carboxyl group seven methylenes away from the epoxidation site was found to be required. Epoxide hydration step, also catalyzed by 3000*g* particles from young apple fruits did not require any cofactors and the enzymatic reaction resulted in a *trans* hydration giving rise to a *threo* product from the *cis* epoxide. This stereochemistry of the epoxide hydrase is consistent with the diastereoisomer composition of the trihydroxystearic acid found in cutin as well as suberin. Hydroxylation of ω-hydroxypalmitic acid at C-10, C-9, and/or C-8 is most probably involved in the biosynthesis of the dihydroxy C_{16} acid found in suberin. Such a hydroxylation has been demonstrated in cell-free preparations from a cutin synthesizing tissue (58). Recently it was found that this hydroxylation is catalyzed by a microsomal preparation from germinating *V. faba* and with this system the involvement of cytochrome P_{450} was demonstrated by a photoreversible inhibition by CO (59). These polar acids undergo further oxidation of the primary hydroxyl groups in suberizing tissues to generate the corresponding dicarboxylic acids which appear to be present in suberin in much greater extent than in cutin.

Enzymes Involved in Suberin Biosynthesis

Conversion of ω-hydroxy acids to corresponding dicarboxylic acids is a characteristic reaction involved in suberization. This conversion has been demonstrated in cell-free preparations from suberizing potato disks.

With 16-hydroxy[G-^{3}H]hexadecanoic acid as the test substrate, the oxidation to dicarboxylic acid required NADP or NAD with a slight preference for the former (60). This enzymatic activity was located mainly in the 105,000*g* supernatant and showed a pH optimum near 9.5. A typical Michaelis-Menton substrate saturation pattern was observed with an apparent Km of 5×10^{-5} M for 16-hydroxyhexadecanoic acid. The oxidation of the ω-hydroxy acid was inhibited by thiol reagents such as p-chloromercuribenzoate, N-ethyl maleimide and iodoacetamide. That the conversion of ω-hydroxyhexadecanoic to the corresponding dicarboxylic acid involved an ω-oxo acid intermediate was suggested by the observation that 16-oxo[G-^{3}H]hexadecanoic acid was rapidly converted into the corresponding dicarboxylic acid by the same cell-free preparation with an absolute specificity for NADP (Table VI).

Table VI. Cofactor Specificity of the Dehydrogenases Involved in Suberization in Potato Disks

Cofactors	Rate[a] nmole/min/mg protein
Soluble Supernatant	
No cofactor	0.10
NAD (1 mM)	0.38
NADP (1 mM)	0.47
NAD (0.5 mM) + NADP (0.5 mM)	0.51
ω-Hydroxy Fatty Acid Dehydrogenase (from DEAE)	
No cofactor	0.07
NAD (1 mM)	0.48
NADP (1 mM)	3.25
NAD (0.5 mM) + NADP (0.5 mM)	3.45
ω-Oxo Fatty Acid Dehydrogenase (from Sepharose 6B)	
No cofactor	2.5
NAD (1 mM)	1.2
NADP (1 mM)	9.1
NAD (0.5 mM) + NADP (0.5 mM)	6.3

[a]With soluble supernatant the formation of C_{16} dicarboxylic acid from ω-hydroxy C_{16} acid was measured because the crude preparation contained both ω-hydroxy acid dehydrogenase and ω-oxo acid dehydrogenase. With the DEAE-purified ω-hydroxy acid dehydrogenase ω-oxo C_{16} acid formation was measured (60,61).

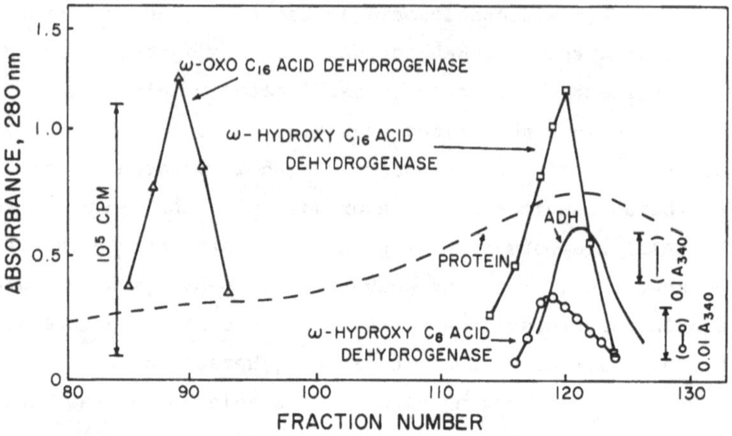

Figure 13. Sepharose 6B gel filtration of a 30-60% $(NH_4)_2SO_4$ fraction obtained from an acetone powder extract of potato disks suberized for six days. Data taken from reference 61.

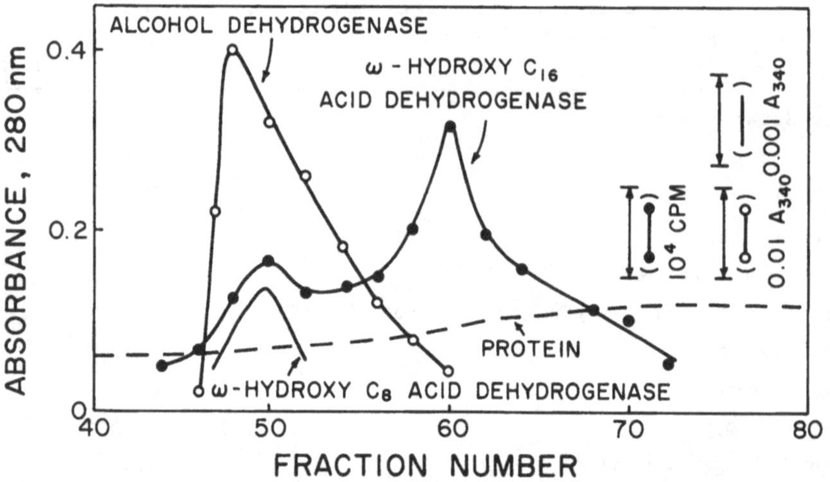

Figure 14. DEAE-Cellulose chromatography of the protein fraction (shown in Fig. 13) containing ω-hydroxy fatty acid dehydrogenase activity. Data from reference 61.

Gel filtration of the 105,000*g* supernatant, prepared from suberizing potato disks, using a Sepharose 6B column revealed that the ω-hydroxy acid dehydrogenase activity was clearly separated from the ω-oxo acid dehydrogenase activity (Fig. 13); the former was much smaller than the latter. The ω-hydroxy acid dehydrogenase activity, however, was not resolved from alcohol dehydrogenase. Since there was a possibility that alcohol dehydrogenase itself might catalyze oxidation of 16-hydroxypalmitic acid, further protein fractionation was attempted. Anion exchange chromatography on DEAE-Sephadex resolved ω-hydroxy acid dehydrogenase from alcohol dehydrogenase (Fig. 14). Once resolved from alcohol dehydrogenase, the ω-hydroxy acid dehydrogenase showed a high degree of preference for NADP. Most probably the oxidation of ω-hydroxy acid observed with NAD in the crude cell-free preparation is due to alcohol dehydrogenase. Further purification of the ω-hydroxy acid dehydrogenase was accomplished by affinity chromatography with an NADP-agarose column. The bulk of the protein was not bound to the column and the ω-hydroxy acid dehydrogenase which was bound to the column could be eluted with a salt gradient (61). Electrophoresis of this purified dehydrogenase showed that all of the enzymatic activity was located in one band (Fig. 15), and this band was by far the most intense protein band.

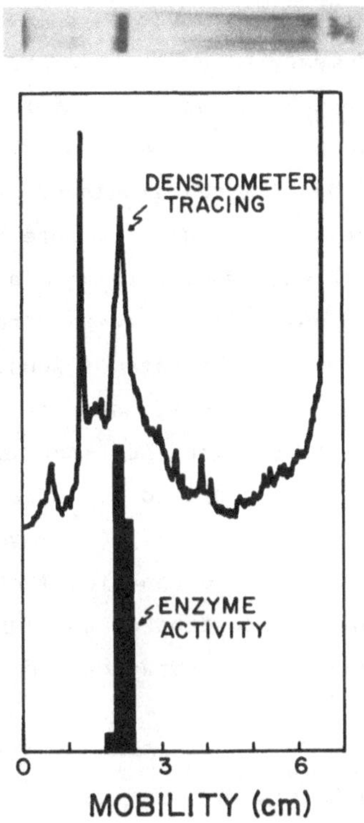

Figure 15. Polyacrylamide gel electrophoresis of an ω-hydroxy fatty acid dehydrogenase preparation obtained from suberizing potato disks using NADP-agarose affinity chromatography. Top, gel stained for dehydrogenase; middle, coomassie blue-stained gel scan; bottom, enzyme activity measured using labeled ω-hydroxyhexadecanoic acid. Data from reference 61.

REGULATION OF SUBERIN SYNTHESIS

Environmental and Physiological Factors which Affect Suberization

The effects of environmental factors such as temperature, relative humidity and oxygen concentration as well as physiological factors such as age, on wound-healing in storage tissues, particularly potatoes, have been studied since the latter part of the last century and this work is summarized in earlier reviews (62,63). Numerous investigators have examined the effects of such factors on suberization and wound periderm formation using histological techniques and the results of such studies have been summarized recently (64). Apparently due to the problems arising from the nonquantitative nature of the histological techniques of measuring suberization and the variability of the tissues, conflicting conclusions have been drawn from such studies. A comprehensive study on the effects of environmental factors on suberization, as measured by the recently developed gas chromatographic technique (7), has not been done. However certain tentative conclusions can be drawn from the information presently available from the cytological studies as well as our limited amount of studies using the gas chromatographic measurement of suberin. The rate of suberization in potatoes is highly dependent on temperature with a Q_{10} of 2 to 3 in the range of 5°C to 25°C (64,65,66). Higher storage temperature prior to wounding has an adverse effect on the rate of wound healing (67,68). High relative humidity, preferably above 90%, has a favorable effect on suberization (64,65,69,70). Oxygen is required for suberization but the minimal oxygen content of the atmosphere required to sustain maximal rates of suberization remain disputed (64,69,71,72). However, it appears clear that oxygen content higher than 20% has no beneficial effect (33,64). Concensus is that the rate and probably the extent of wound healing are highest in the young tubers and decrease with maturation and further with storage (64,67,68,73).

Changes in Enzyme Activity Levels

Incorporation of exogenous precursors into suberin did not begin until after about 3 days of aging (35). This lag period would indicate that the enzymes involved in the synthesis of suberin were induced as a result of wounding. Since ω-hydroxy acid dehydrogenase is the only enzyme which has been demonstrated in cell-free preparations, we used this enzyme activity as a model for suberin synthesizing enzymes and measured the changes in the levels of its activity during wound-healing (60,61). Extracts from fresh tissue showed little ω-hydroxy acid dehydrogenase activity, but conversion of ω-hydroxypalmitic acid to the corresponding dicarboxylic acid was measurable in one-day aged tissue (Fig. 11). This activity reached its maximal levels after four days of aging and decreased thereafter, the sharpest decline occurring after 6 days of aging. This time-course of development is quite consistent with the time-course of suberization.

Since the oxidation of ω-hydroxy acid to the corresponding dicarboxylic acid involves two enzymes, changes in the two activities during the suberization process were examined with 16-hydroxy[G-^{3}H]hexadecanoic acid and 16-oxo[G-^{3}H]hexadecanoic acid as the substrates. The ω-hydroxy acid dehydrogenase activity showed the changes indicated above, but little change in ω-oxo acid dehydrogenase activity was detected (61). These results suggested that the enzyme catalyzing the first and the rate-limiting step in the conversion of the ω-hydroxy acid to the dicarboxylic acid is the one which is regulated to bring about suberization.

The time-course of development of ω-hydroxy acid dehydrogenase activity during the aging process is complicated by the possibility that alcohol dehydrogenase might catalyze a slow oxidation of ω-hydroxyhexadecanoic acid to the ω-oxo acid. Thus, the significant amount of ω-hydroxy acid dehydrogenase activity observed in the extracts of one- and two-day aged tissue might not be due to actual ω-hydroxy acid dehydrogenase. In order to examine this possibility, acetone powder extracts prepared from fresh and 5-day aged tissue were fractionated by ammonium sulphate precipitation followed by gel filtration on a Sepharose 6B column. The protein fraction which contained both alcohol dehydrogenase and ω-hydroxy acid dehydrogenase was assayed for NAD dependent and NADP dependent oxidation of ω-hydroxy hexadecanoic acid. The NADP dependent oxidation of the ω-hydroxy acid by protein fraction from the 5-day aged tissue was about 16 times as fast as the rates obtained with similar protein fractions from the fresh tissue. On the other hand, NAD dependent oxidation of the ω-hydroxy acid was only 50% higher in the 5-day aged tissue when compared to the fresh tissue (61). This difference in the NAD dependent rate was similar to the difference in the alcohol dehydrogenase activity of the two tissues. Thus it is probable that the rise in alcohol dehydrogenase activity which occurred during the initial two days of aging is responsible for the increased oxidation of the ω-hydroxy C_{16} acid observed in crude extracts prepared from the tissue aged for one or two days. The NADP-dependent ω-hydroxy acid dehydrogenase which is most probably involved in suberization is induced only after 3 days.

Evidence for Induction of ω-Hydroxy Acid Dehydrogenase

Since the cells involved in suberization at the cut surface do not synthesize suberin in the intace tissue, it is apparent that the wounding triggered this process possibly by inducing the synthesis of enzymes necessary for suberization. To determine whether the increase in the activity of ω-hydroxy acid dehydrogenase and suberization require protein synthesis, the effects of cycloheximide and actinomycin D on these two processes were examined. Application of cycloheximide after one, two or three days of aging, virtually prevented suberization, whereas application of cycloheximide after 4 and 5 days of aging had little effect on suberization (60). These results suggest that protein synthesis is required in the suberization process and that the proteins required in this process are synthesized before the fourth day after wounding.

To determine the approximate time when the gene transcription, involved in the formation of the aliphatic components of suberin, takes place, the effect of actinomycin D treatment on suberization was determined (60). Maximum inhibition of suberization occurred when the disks were treated with this inhibitor of nucleic acid synthesis three days after cutting and little inhibition of suberization occurred when similar treatment was done four days after cutting (Fig. 16). Therefore, it appears probable that the transcriptional processes directly related to the biosynthesis of the aliphatic components of suberin take place between the third and the fourth day after wounding. However, significant inhibition of suberization occurred also when the disks were treated after 1 and 2 days of aging. Therefore, it is possible that the nucleic acid synthesis which takes place during this early stage plays a direct or indirect role in suberization. On the other hand, the possibility that actinomycin D applied after 1 and 2 days of aging persisted in the tissue even after the third day cannot be ruled out.

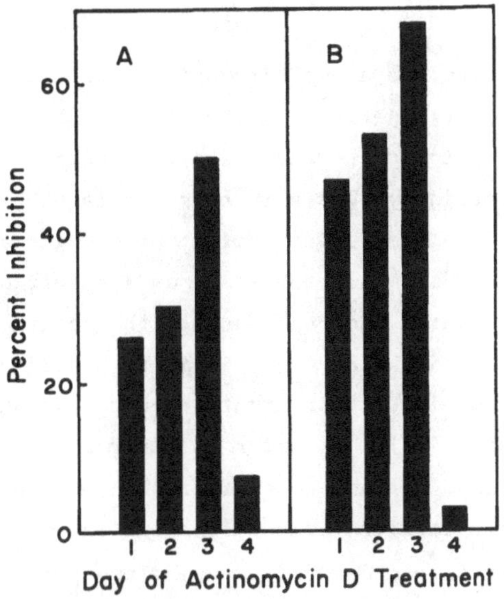

Figure 16. Effect of actinomycin D on the formation of aliphatic components of suberin (A) and the development of ω-hydroxyfatty acid dehydrogenase activity (B) in suberizing potato tuber disks. Data from reference 60.

The changes in ω-hydroxy acid dehydrogenase activity during wound-healing suggested that the most rapid rate of appearance of this activity occurred between the third and fourth day after wounding. From the effect of actinomycin D treatment on suberization discussed above, it would appear probable that the transcriptional processes involved in the synthesis of this enzyme occur between the third and fourth day after cutting. In fact, maximum inhibition of the appearance of ω-hydroxy acid dehydrogenase activity occurred when potato disks were treated with actinomycin D three days after wounding (Fig. 16). Similar treatment after 4 days of aging resulted in little decrease in the level of this enzyme. Actinomycin D treatment prior to the third day of aging did result in substantial inhibition of appearance of ω-hydroxy acid dehydrogenase, possibly due to reasons similar to those discussed above in connection with cycloheximide treatments. As discussed in a succeeding section, treatment of the disks with water (washing) also inhibits suberization and the appearance of ω-hydroxy acid dehydrogenase activity. The simultaneous inhibition of suberin formation and ω-hydroxy acid dehydrogenase activity by actinomycin D and water strongly suggests that this enzyme is involved in suberization and that gene transcription induced by wounding is involved in suberization.

Suberization Inducing Factor

It appears probable, from the above discussion, that wounding results in suberization by somehow triggering, in the cells in the immediate vicinity of the wound, transcription and translation of a segment of DNA which does not participate in protein synthesis in these cells when they are in the intact tuber. These localized effects might be brought about by some chemical(s) produced at (or near) the wound as a result of the injury. Recently it was found that a thorough wash of the potato disks immediately after preparation of the disks severely inhibits suberization, suggesting that a suberization inducing factor might be removed from the surface by washing the disks (Table VII). In order to determine whether such a factor is generated at the wound surface soon after the wounding process, potato disks were washed after various periods of aging and the effect of this treatment on suberization was determined using two parameters namely, the diffusion resistance of the tissue surface to water vapor and octadecene-1,18-diol generated by $LiAlH_4$ treatment. The results of such a study showed that thorough washing of the disks soon after preparation or after one, two or three days of aging resulted in the most severe inhibition of suberization (Table VIII). Washing of the disks after 4 days of aging had very little influence on suberization. Similarly, the appearance of

Table VII. Effect of Washing of Potato Disks on the Time-course of Development of Suberin and Diffusion Resistance of the Tissue Surface to Water Vapor

Time After Cutting (days)	Suberin[a] μg C_{18} diol/g dry wt		Diffusion Resistance[a] $sec.cm^{-1}$	
	Washed[b]	Unwashed	Washed[b]	Unwashed
2	---	---	2.6	3.6
4	22	155	2.9	6.7
6	147	370	3.7	16.5
8	128	327	7.4	24.0
10	119	320	7.1	26.0

[a]Expressed as μg octadecene-1,18-diol obtained per g dry weight of the tissue by $LiAlH_4$ treatment (7); diffusion resistance was measured as before (7). [b]Tissue disks were washed for 3 hr immediately after cutting.

Table VIII. Effect of Washing on Suberization of Potato Disks

Day Washed[a]	Diffusion Resistance[b] sec. cm^{-1}	Suberin[b] µg C_{18} diol/g dry wt
0	11	165
1	13	80
2	10	110
3	17	75
4	23	105
5	29	220
6	29	250
Unwashed	29	260

[a]Washed for 3 hr in water after the number of days of suberization indicated. [b]Diffusion resistance and the amount of suberin as diol were measured as described in reference 7 after 8 days of suberization. Data taken from reference 74.

ω-hydroxy acid dehydrogenase activity also was inhibited by washing of the disks. For example thorough washing of disks within four days after cutting inhibited the appearance of ω-hydroxy acid dehydrogenase activity by about 50% (measured at 5 days). Thus, it appears that washing of the disks anytime prior to the occurrence of the transcription and translation processes, involved in suberization, severely inhibits suberization. Therefore, the presence of the chemical(s), which is removed by washing, is required for the enzyme induction process. It is possible that a series of reactions which take place at the surface results in the formation of the suberization inducing factor(s) which diffuses into the cells near the surface and thus affects the suberization process.

Measurement of suberization by diffusion resistance always agreed with the results obtained by octadecene diol measurement except in the case of disks washed during the early period of aging (74). Thorough washing of the disks immediately after cutting resulted in severe inhibition of development diffusion resistance of the tissue surface to water vapor but only a moderate inhibition of the formation of the aliphatic components of suberin (Table VIII). This difference might indicate that the inhibition of synthesis of the phenolic components of suberin, which are formed earlier in the wound-healing process, results in the disruption of the suberization process. Even when the aliphatic components are synthesized by the enzymes induced during the latter part of wound-healing (3-4 days after cutting), these components cannot be esterified to the phenolic matrix if the synthesis of this matrix had been inhibited by treatments during the early periods of wound-healing. The water proofing of the phenolic matrix by the aliphatic chains is probably responsible for the development of diffusion resistance.

Involvement of Abscisic Acid (ABA) in Suberization

The inhibition of suberization brought about by the thorough washing of the disks was partially restored by the addition of ABA. This suberization promoting effect of ABA was concentration dependent, optimum level being near 10^{-4} M (74). This observation raised the possibility that washing inhibited suberization by removal of ABA from the disks. In order to examine this possibility, the wash was examined for free and bound ABA by a high pressure liquid chromatographic technique. ABA was in fact detected in the wash and the identity of ABA isolated from the wash by high pressure liquid chromatography was confirmed by mass spectrometry (74). The level of ABA which could be removed by washing increased in a linear fashion at least up to 6 hr. Since the level of bound ABA present in the wash was very low compared to that of free ABA, it appears unlikely that the increase in ABA was due to release of bound ABA unless the increase represented release of ABA from a polymer bound form which might be present at the cut surface. However, this possibility appears unlikely because mild alkali (pH 10) treatment of the tissue surface did not release much ABA. Therefore, if ABA is released from a bound form at the surface, the linkage involved must be stable to the mild alkali treatment used in this study and the release might be catalyzed by hydrolytic enzymes. *De novo* synthesis of ABA cannot be ruled out as a possible source as no experiments have been performed to test this possibility.

The results discussed above provided suggestive evidence that ABA is involved in the induction of suberization. If such is the case, ABA might also trigger suberization in tissue culture of potato. In fact, addition of ABA to the medium resulted in suberization of potato tuber tissue culture of suberin as measured by generation of octadecene diol by $LiAlH_4$

treatment (Table IX). Identity of the aliphatic suberin component was confirmed by mass spectrometry. Thus, the capacity of ABA to induce suberization was clearly demonstrated in tissue culture. The molecular events involved in the induction of suberization by ABA are unknown. It is also not known whether removal of ABA is the sole reason for the inhibition of suberization by a thorough washing of the potato disks. Attempts to reverse the inhibition of suberization caused by the thorough washing of potato disks showed that ABA appeared to reverse the inhibition only in disks washed immediately or one day after cutting, the most severe inhibition observed in disks which were washed two or three days after cutting was not reversed by ABA. Furthermore measurable amounts of ABA could not be washed off the tissue 2 days after cutting. Therefore it appears that the biochemical events triggered by ABA during the early part of wound healing produce a chemical(s) by the third day of wound-healing and this chemical is the true inducer of the enzymes involved in the biosynthesis of the aliphatic components of suberin. The nature of this chemical is unknown but it appears to be located near the surface so that it can be washed out. From the results thus far obtained the sequence of events involved in suberization is tentatively postulated to be as shown in Figure 17. Further studies on these important aspects of suberization are likely to provide meaningful answers concerning the molecular biology of the suberization process.

Table IX. Effect of ABA on Suberization in Potato Tissue Culture

Day	Suberin μg C_{18} diol/mg dry tissue
3	0.32
4	0.38
5	0.40
6	0.44
8	0.60
Control after 10 days	0.12

The low amount of octadecene diol obtained from control with no addition of ABA most probably represents the low amount of suberin present in the starting tissue. 10^{-4} M ABA was used in all cases (59).

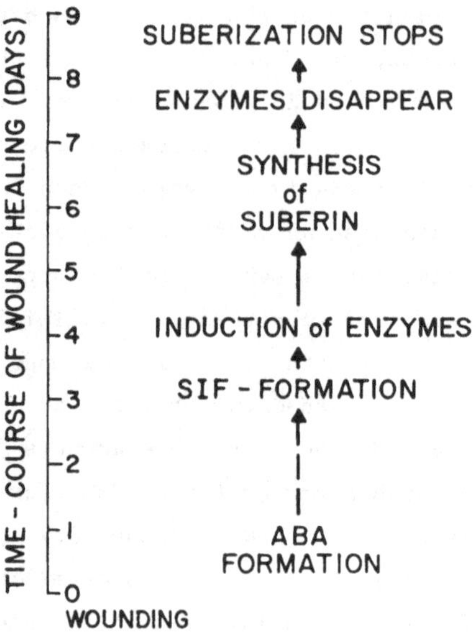

Figure 17. A tentative scheme showing the time-course of the processes involved in suberization in potato disks. SIF, a hypothetical suberin inducing factor.

ENZYMATIC DEGRADATION OF SUBERIN

There is a possibility that microorganisms which attack underground parts of plants excrete enzyme(s) which depolymerizes suberin. In fact one such organism, *Streptomyces scabies*, a pathogen which infects potatoes, grows quite well on powdered potato suberin preparation as the sole source of carbon. The search for enzymes which depolymerize suberin is hindered by the fact that we do not have a reliable method for measuring suberin hydrolyzing activity. However, using hydrolysis of radioactive cutin as a measure of the depolymerizing activity we purified polyester hydrolyzing enzymes from the extracellular fluid of the suberin grown *S. scabies* with Sephadex G-100 gel filtration and ion exchange chromatography on QAE-Sephadex and SP-Sephadex (Fig. 18). A nonspecific esterase and polyester hydrolyzing enzymes are obtained by this procedure (75). The nonspecific esterase was measured by the hydrolysis of a model substrate p-nitrophenyl palmitate while the polymer hydrolyzing activity was measured by the hydrolysis of another model substrate p-nitrophenylbutyrate. The latter enzyme does not catalyze hydrolysis of p-nitrophenyl palmitate. The patterns of fractionation of enzymes obtained by this procedure were similar

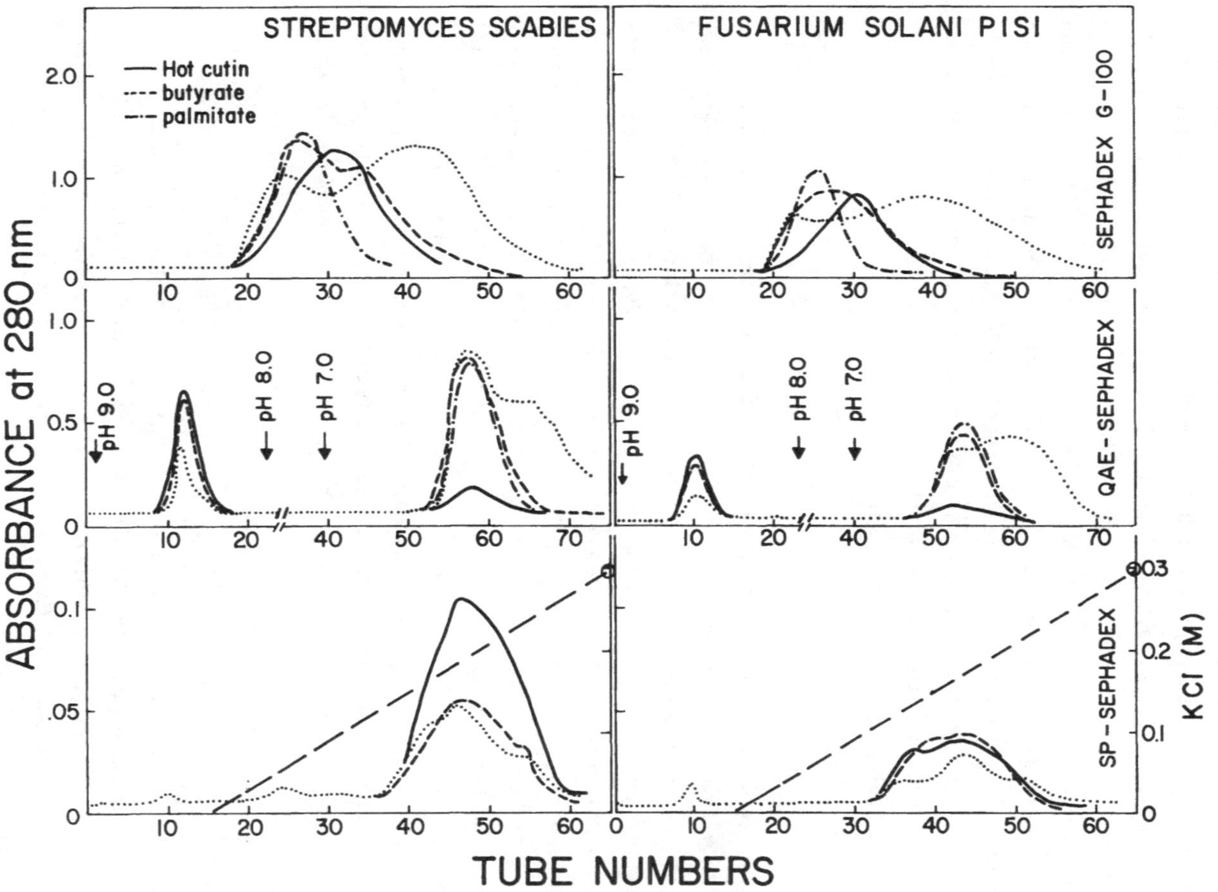

Figure 18. Purification of cutin hydrolyzing (and suberin hydrolyzing) enzymes from the extracellular fluid of two microorganisms grown on suberin as the sole source of carbon. The steps used were similar to those described before (76). The dotted lines indicate absorbance at 280 nm. Data from reference 75.

to those obtained with the extracellular fluid of another fungus, *Fusarium solani pisi* grown on suberin. The latter fungus also grows on cutin and produces extracellular enzymes which have been fractionated by a procedure similar to that indicated above (76,77). Even though the purification of the polymer hydrolyzing enzymes from the extracellular fluid of suberin-grown fungi was done using cutin hydrolysis as an assay, analysis of products using combined gas-liquid chromatography and mass spectrometry showed that the protein fractions, which catalyzed hydrolysis of labeled cutin also catalyzed release of the aliphatic components of potato suberin. Thus methods are available to examine whether suberin hydrolyzing enzymes are involved in host-pathogen interaction.

CONCLUSION

The knowledge of the chemistry of suberin obtained in recent years allowed us to probe into the biochemistry of suberization process. As a result we are beginning to understand the biochemical pathways involved in the synthesis of the aliphatic components of suberin. Even though the importance of suberization in wound healing has been recognized for a long period of time we are only beginning to get a glimpse at the biochemical events which trigger suberization and regulate this process. Much of these events are yet to be defined at a molecular level and a clearer understanding of this process should help us to devise means to enhance suberization in agriculturally important plant tissues. Recent results concerning suberin hydrolyzing enzymes set the stage for a systematic attack on this problem so that we may understand the role suberin plays in host-pathogen interaction.

REFERENCES

1. Ribas, I., Seoane, E.: The chemistry of cork. VII. Phellonic and phellolic acids. An. R. Soc. Esp. Fis. Quim. 47B:61-66 (1951).
2. Jensen, W., Tinnis, W.: Suberin. IX. Identification of unsaturated acids in suberin of *Quercus suber* and *Betula verrucosa*. Paperi Puu 39:261-268 (1957).
3. Gonzalez-Gonzalez, A., Sola, J., Inglesias-Martin, F., Rivas-Paris, V., Ribas-Marques, I.: Chemistry of cork. Study of the acids. Quim. Ind. (Bilbao) 15:139-146 (1969).

4. Martin, J. T., Juniper, B. E.: The cuticles of plants. St. Martin's Press, Inc., New York. 1970.

5. Kolattukudy, P. E., Kronman, K., Poulose, A. J.: Determination of structure and composition of suberin from the roots of carrot, parsnip, rutabaga, turnip, red beet, and sweet potato by combined gas-liquid chromatography and mass spectrometry. Plant Physiol. 55:567-573 (1975).

6. Kolattukudy, P. E., Agrawal, V. P.: Structure and composition of aliphatic constituents of potato tuber skin (suberin). Lipids 9:682-691 (1974).

7. Kolattukudy, P. E., Dean, B. B.: Structure, gas chromatographic measurement, and function of suberin synthesized by potato tuber tissue slices. Plant Physiol. 54:116-121 (1974).

8. Robards, A. W., Payne, H. L., Gunning, B. E. S.: Isolation of the endodermis using wall-degrading enzymes. Cytobiologie 13:85-92 (1976).

9. Holloway, P. J.: The composition of suberin from the corks of *Quercus suber* L. and *Betula pendula* Roth. Chem. Phys. Lipids 9:158-170 (1972).

10. Holloway, P. J.: The suberin composition of the cork layers from some *Ribes* species. Chem. Phys. Lipids 9:171-179 (1972).

11. Jensen, W.: A study of the outer bark of birch (*Betula verrucosa*) and cork oak (*Quercus suber*) by scanning electron microscopy. An. R. Soc. Esp. Fis. Quim. 68:871-878 (1972).

12. Brieskorn, C. H., Binnemann, P. H.: Chemische zusammensetzung des suberins der kartoffelschale. Z. Lebensm. Unters. Forsch. 154: 213-222 (1974).

13. Brieskorn, C. H., Binnemann, P. H.: Carbonsauren und alkanole des cutinse und suberins von *Solanum tuberosum*. Phytochemistry 14:1363-1367 (1975).

14. Holloway, P. J., Deas, A. H. B.: Epoxyoctadecanoic acids in plant cutins and suberins. Phytochemistry 12:1721-1735 (1973).

15. Dean, B. B., Kolattukudy, P. E.: Synthesis of suberin during wound healing in jade leaves, tomato fruit, and bean pods. Plant Physiol. 58:411-416 (1976).

16. Holloway, P. J., Baker, E. A., Martin, J. T.: The chemistry of plant cutins and suberins. An. Quim. 68:905-916 (1972).

17. Ryhage, R., Stenhagen, E.: *In* Mass Spectrometry of Organic Ions, F. W. McLafferty, ed., Academic Press, New York, 399 (1963).

18. Eglinton, G., Hunneman, D. H.: Gas chromatographic-mass spectrometric studies of long-chain hydroxy acids. I. The constituent cutin acids of apple cuticle. Phytochemistry 7:313-322 (1968).

19. Adams, K. R., Bonnett, R.: Long chain α,ω-dicarboxylic acids from the spores of *Equisetum telmateia* and *E. arvense*. Phytochemistry 10:1885-1890 (1971).

20. Eglinton, G., Hunneman, D. H., McCormick, A.: Gas-chromatographic-mass spectrometric studies of long chain hydroxy acids. III. The mass spectra of the trimethylsilyl ethers of aliphatic hydroxy acid methyl esters. A facile method of double bond location. Org. Mass. Spectrom. 1:593-611 (1968).

21. Kolattukudy, P. E., Walton, T. J.: The biochemistry of plant cuticular lipids. Prog. Chem. Fats Other Lipids 13:121-175 (1973).

22. Sharkey, A. G., Friedel, R. A., Langer, S. H.: Mass spectra of trimethylsilyl derivatives. Anal. Chem. 29:770-776 (1957).

23. McCloskey, J. A., Stillwell, R. N., Lawson, A. M.: Use of deuterium-labeled trimethylsilyl derivatives in mass spectrometry. Anal. Chem. 40:233-236 (1968).

24. Walton, T. J., Kolattukudy, P. E.: Determination of the structures of cutin monomers by a novel depolymerization procedure and combined gas chromatography and mass spectrometry. Biochemistry 11:1885-1897 (1972).

25. Riley, R., Kolattukudy, P. E.: Evidence for covalently attached p-coumaric acid and ferulic acid in cutins and suberins. Plant Physiol. 56:650-654 (1975).

26. Sitte, P.: Der feinbau verkorkter zellwande. Mikroskopie 10:178-200 (1955).

27. Wattendorff, J.: Ultrahistochemical reactions of the suberized cell walls in *Acorus*, *Accacia*, and *Larix*. Z. Pflanzenphysiol. 73: 214-225 (1974).

28. Dean, B. B., Kolattukudy, P. E., Davis, R. W.: Chemical composition and ultrastructure of suberin from hollow heart tissue of potato tubers (*Solanum tuberosum*). Plant Physiol. 59: in press (1977).

29. Falk, V. H., El-Hadidi, M. N.: Der feinbau der suberinschichten verkorkter zellwande. Z. Naturforsch. 166:134-137 (1961).

30. Parameswaran, N., Kruse, J., Liese, W.: Aufbau und feinstruktur von periderm und lenticellen der fichtenrinde. Z. Pflanzenphysiol. 77:212-221 (1976).

31. Sitte, P.: Die bedeutung der molekularen lamellenbauweise von korkzellwanden. Biochem. Physiol. Pflanzen. 168:287-297 (1975).

32. Kolattukudy, P. E.: *In* Recent Advances in Phytochemistry, Vol. 11, F. Loewus and V. C. Runeckles, eds., Plenum Press, New York, 185-247 (1977).

33. Dean, B. B., Kolattukudy, P. E., unpublished results.

34. Kolattukudy, P. E.: *In* Recent Advances in the Chemistry and Biochemistry of Plant Lipids, T. Galliard and E. I. Mercer, eds., Academic Press, New York, 203-246 (1975).

35. Dean, B. B., Kolattukudy, P. E.: Biochemistry of suberization: incorporation of [1-^{14}C]-oleic acid and [1-^{14}C]-acetate into the aliphatic components of suberin in potato tuber discs (*Solanum tuberosum*). Plant Physiol. 59: in press (1977).

36. Riley, R., Kolattukudy, P. E.: unpublished results.

37. Freudenberg, K.: *In* Constitution and biosynthesis of lignen, K. Freudenberg and A. C. Neish, eds., Springer-Verlag, Inc., New York, 47-122 (1968).

38. Lai, Y. Z., Sarkanen, K. V.: *In* Lignins: Occurance, formation structure and reactions, K. V. Sarkanen and C. H. Ludwig, eds., John Wiley and Sons, Inc., New York, 165-240 (1971).

39. Swan, E. P.: A study of western red cedar bark lignen. Pulp Paper Mag. Can. 67:T456-T460 (1966).

40. Priestly, J. H., Woffenden, L. M.: Physiological studies in plant anatomy. Part V. Causal factors in cork formation. New Phytol. 21:252-268 (1932).

41. Schneider-Orelli, O.: Versuche uber Wundung und Wundverschluss an Pflanzeorganen. Centr. Bakt. 2 Abt. 30, 420-429 (1911).

42. Wachter, W.: Wundverschluss bei *Hippuris vulgarus* L. Beih. Zum. Bot. Centr. 18:1 (1905).

43. Blackman, F., Frost, G., Matthae, L. C.: On the reaction of leaves to traumatic stimulation. Ann. Bot. 15:533-546 (1901).

44. Bredemeijer, G., Heinen, W.: Cutin synthesis in plants. I. Free fatty acid movement during cutin synthesis in injured *Gasteria verrucosa* leaves. Acta Bot. Neerl. 17:15-25 (1968).

45. Kolattukudy, P. E., Brown, L.: unpublished results.

46. Willemot, C., Stumpf, P. K.: Fat metabolism in higher plants. XXXIII. Development of fatty acid synthetase during the "aging" of storage tissue slices. Can. J. Bot. 45:570-584 (1967).

47. Kolattukudy, P. E., Buckner, J. S.: Chain elongation of fatty acids by cell-free extracts of epidermis from pea leaves (*Pisum sativum*). Biochem. Biophys. Res. Commun. 46:801-807 (1972).

48. Macey, M. J. K., Stumpf, P. K.: Fat metabolism in higher plants. XXXVI: Long chain fatty acid synthesis in germinating peas. Plant Physiol. 43:1637-1647 (1968).

49. Khan, A. A. and Kolattukudy, P. E.: A microsomal fatty acid synthetase coupled to acyl-CoA reductase in *Euglena gracilis*. Arch. Biochem. Biophys. 158:411-420 (1973).

50. Khan, A. A. and Kolattukudy, P. E.: Solubilization of fatty acid synthetase, acyl-CoA reductase, and fatty acyl-CoA alcohol transacylase from the microsomes of *Euglena gracilis*. Arch. Biochem. Biophys. 170:400-408 (1975).

51. Kolattukudy, P. E.: Reduction of fatty acids to alcohols by cell-free preparations of *Euglena gracilis*. Biochemistry 9:1095-1102 (1970).

52. Kolattukudy, P. E.: Enzymatic synthesis of fatty alcohols in *Brassica oleracea*. Arch. Biochem. Biophys. 142:701-709 (1971).

53. Kolattukudy, P. E.: Mechanisms of synthesis of waxy esters in broccoli (*Brassica oleracea*). Biochemistry 6:2705-2717 (1967).

54. Soliday, C. L., Kolattukudy, P. E.: Biosynthesis of cutin: ω-Hydroxylation of fatty acids by a microsomal preparation from germinating *Vicia faba*. Plant Physiol. 59: in press (1977).

55. Kolattukudy, P. E., Walton, T. J., Kushwaha, R. P. S.: Biosynthesis of the C_{18} family of cutin acids: ω-Hydroxyoleic acid, ω-hydroxy-9,10-epoxy acid, 9,10,18-trihydroxystearic acid, and their Δ^{12} unsaturated analogs. Biochemistry 12:4488-4498 (1973).

56. Croteau, R., Kolattukudy, P. E.: Biosynthesis of hydroxy fatty acid polymers. Enzymatic epoxidation of 18-hydroxyoleic acid to 18-hydroxy-*cis*-9,10-epoxystearic acid by a particulate preparation from spinach (*Spinacia oleracea*). Arch. Biochem. Biophys. 170:61-72 (1975).

57. Croteau R., Kolattukudy, P. E.: Biosynthesis of hydroxy fatty acid polymers. Enzymatic hydration of 18-hydroxy-*cis*-9,10-epoxystearic acid to *threo*-9,10,18-trihydroxystearic acid by a particulate preparation from apple (*Malus pumila*). Arch. Biochem. Biophys. 170: 73-81 (1975).

58. Walton, T. J., Kolattukudy, P. E.: Enzymatic conversion of 16-hydroxypalmitic acid into 10,16-dihydroxypalmitic acid in *Vicia faba* epidermal extracts. Biochem. Biophys. Res. Commun. 46:16-21 (1972).

59. Soliday, C. L., Kolattukudy, P. E.: unpublished results.

60. Agrawal, V. P., Kolattukudy, P. E.: Biochemistry of suberization. ω-Hydroxyacid oxidation in enzyme preparations from suberizing potato tuber disks. Plant Physiol. 59: in press (1977).

61. Agrawal, V. P., Kolattukudy, P. E.: unpublished results.

62. Shapovalov, M., Edson, H. A.: Wound-cork formation in the potato in relation to seed piece decay. Phytopathology 9:483-496 (1919).

63. Werner, H. O.: Wound healing in potatoes (Triumph variety) as influenced by type of injury, nature of initial exposure, and storage conditions. Res. Bull. Neb. Agric. Exp. Stn. 102 (1938).

64. Wigginton, M. J.: Effects of temperature, oxygen tension and relative humidity on the wound-healing process in the potato tuber. Potato Res. 17:200-214 (1974).

65. Priestly, J. H., Woffenden, L. M.: The healing of wounds in potato tubers and their propagation of cut sets. Ann. Appl. Biol. 10: 96-115 (1923).

66. Radatz, W.: Die wundkorkbildung der kartoffelknolle in abhangigkeit von lagerungsbedingungen. Landbforsch. Volkenrobe 17:153-158 (1967).

67. Kny, L.: Uber die bildung des wundperiderms an knollen in ihrer abhangigkeit von ausseren einflussen. Ber. Dt. Bot. Ges. 7:154-168 (1889).

68. Lange, H., Rosenstock, G.: Kausalanalytsche untersuchungen zum alterungsvorgang bei der wundkompensation von speicherorganen unter besonderes berucksichtigung des nahrstoffaktors. Phytopath. Z. 52:188-201 (1965).

69. Lange, H., Rosenstock, G., Kahl, G.: Induktions bedingungen der suberinsynthese und zellproliferation. Planta 90:109-118 (1970).

70. Artschwager, E.: Wound-periderm formation in the potato as affected by temperature and humidity. J. Agric. Res. 35:995-1000 (1927).

71. Brauner, L., Brauner, M., Hasman, M.: The relations between water-intake and oxybiosis in living plant tissues. Rev. Fac. Sci. Univ. Istanbul B5:266-309 (1940).

72. Lipton, W. J.: Some effects of low-oxygen atmospheres on potato tubers. Am. Potato J. 44:292-299 (1967).

73. Lange, H., Rosenstock, G.: Uber alterungserscheinungen bei traumatisch induzierter meristematischer aktivitat. Phytopath. Z. 51:136-152 (1964).

74. Soliday, C. L., Dean, B. B., Kolattukudy, P. E.: unpublished results.

75. Lin, T. S., Kolattukudy, P. E.: unpublished results.

76. Purdy, R. E., Kolattukudy, P. E.: Hydrolysis of plant cuticle by plant pathogens. Purification, amino acid composition, and molecular weight of two isozymes of cutinase and a nonspecific esterase from *Fusarium solani f. pisi*. Biochemistry 14:2824-2831 (1975).

77. Purdy, R. E., Kolattukudy, P. E.: Hydrolysis of plant cuticle by plant pathogens. Properties of cutinase I, cutinase II, and a nonspecific esterase isolated from *Fusarium solani pisi*. Biochemistry 14:2832-2840 (1975).

Synthesis and Secretion of Cell Wall Glycoprotein in Carrot Root Disks

David Sadava and Maarten J. Chrispeels

CELL WALL GLYCOPROTEINS

Analysis of the protein component of plant cell walls began in earnest in 1960 when two laboratories independently showed that hydroxyproline was a major nitrogenous component of cell walls isolated by differential centrifugation following cell disruption (13, 26). Since this otherwise rare amino acid is found in collagen and elastin, two extracellular structural glycoproteins of animal tissues, it was tempting to propose a similar structural glycoprotein for plant cell walls. This temptation yielded the term extensin (23).

The boldness of the extensin hypothesis caused some to doubt the very existence of a cell wall protein. Cell wall fractions free of cytoplasmic contamination are notoriously difficult to prepare, and it was possible that the hydroxyproline-rich component was cytoplasmic and adhered to the cell walls during isolation. In addition, an autoradiographic experiment in which cultured plant cells were incubated in 3,4-^{3}H-proline showed little ^{3}H label (presumably in proline and hydroxyproline) over the cell wall (17). The choice of isotopes here was unfortunate, since half of the label of this proline is lost when proline is hydroxylated (22); thus any cell wall labelling in this experiment would be reduced. Subsequent autoradiographic experiments using generally labelled ^{3}H-proline have confirmed the existence of cell wall protein. The strategy employed was to allow the cells to incorporate ^{3}H-proline and then to plasmolyze the cells prior to autoradiography. This would permit clear separation of the

Biochemistry of Wounded Plant Tissues

cytoplasmic and cell wall label. Such experiments clearly showed abundant cell wall labelling in disks of carrot root phloem parenchyma by light microscopy (39) and in cultured sycamore cells by electron microscopy (37).

The cell wall protein moiety can be divided into two classes: wall enzymes and glycoproteins that are probably structural. The wall enzymes include peroxidase, phosphatase, and invertase (23). Although isolated cell walls can adsorb cytoplasmic proteins, the existence of wall-specific enzymes is indicated by experiments in which cell walls were used to pack a chromatographic column. Elution of this column with a linear gradient of 0-1.0M NaCl reproducibly extracted a unique phosphatase (tobacco: ref.45) and peroxidase (carrot: Sadava, unpublished). These fractions were not observed when extracted cell walls were loaded with cytoplasm and re-extracted. Thus there appears to be phosphatase and peroxidase ionically bound to the cell wall.

Extensin, the structural glycoprotein of the cell wall, is characterized by two types of linkage between amino acids and sugars: hydroxyproline to arabinose and serine to galactose. We have found both of these linkages in cell walls of carrot root disks which were incubated in water after slicing (aged) for 24 hours (4). Lamport has demonstrated these linkages elsewhere in the plant kingdom (24, 25). Indeed, such glycopeptide bonds are even found in cell wall enzymes (28). This immediately raises the question of how extensin is built into the overall framework of the primary cell wall. Arabinose and galactose are prominent components of the hemicellulose fraction of the wall, and it has been proposed that it is in the wall matrix that extensin plays its supposed structural role (1). Indeed, it is possible to extract extensin from cell walls of sycamore (16) and a green alga (36) without breaking glycosidic bonds. This argues against any direct attachment of the glycoprotein to cellulose. However, in mung bean and lupin hypocotyls, a more direct extensin-cellulose interaction is indicated from chemical analyses (31). Thus there may be no universal structure for the cell wall and its protein component.

The role of extensin has been examined in few plant systems. In the pea epicotyl, it appears to act to stiffen the cell wall at the cessation of cell elongation. There is an abrupt increase in hydroxyproline synthesis and accumulation when elongation ceases (11, 35, 43). If this increase in hydroxyproline is blocked by the chelator dipyridyl (see below), elongation proceeds beyond the time at which it would normally stop (30, 42). This indicates a causal relationship between extensin deposition and the cessation of elongation of the cell. It is interesting that the degree of glycosylation of extensin increases as the rate of pea epicotyl elongation decreases (20); this may be a further reflection of extensin's role in wall stiffening.

The role of extensin in other tissues is much less clear than in the case of stem tissues. We have studied its synthesis and secretion in disks of carrot-root phloem parenchyma. When such disks (1 cm in diameter by 1 mm thick) are cut and incubated in an antibiotic solution to prevent bacterial growth, the disks undergo many metabolic and structural changes. Most of these are described elsewhere in this volume. For the purposes of our investigations, the most important change in the disks upon aging is an accumulation of cell wall hydroxyproline (Figure 1). Cell wall proteins are rather stable (23). Thus aging carrot disks have provided a convenient, homogeneous system for studies of extensin synthesis and secretion, as well as their induction during aging. But we have not approached the question of why these events occur. Since the disks have been produced by slicing, it is tempting to think that extensin is made as a response to wounding. Potato tuber lectin, possibly made as an agent to combat infections, is a glycoprotein very similar in composition to extensin (2). When melon plants are infected with a fungus, the plants respond by synthesizing and accumulating extensin in their cell walls (14, 29). We have examined carrot disks by light microscope histology, and they are about ten cells in thickness. It would be interesting to determine if extensin accumulates in the cell walls of cells near or at the cut surface, and not in the interior of the disk. Such a study may require the use of anti-extensin antibodies as a stain for the molecule. Generation of such antibodies will only be possible when extensin is fully purified.

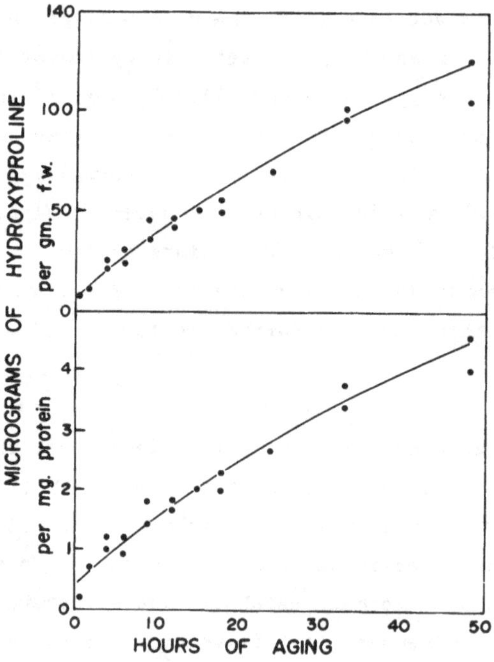

Figure 1. Hydroxyproline accumulation during aging of carrot disks. Phloem parenchyma tissue from carrot roots was cut into slices 1 cm in diameter by 1 mm thick and the slices were incubated in water at 30°C. Hydroxyproline was determined for whole slice homogenates. Virtually all of the hydroxyproline was in the cell wall fraction. From reference 10. Reprinted with permission of the Oxford University Press.

BIOSYNTHESIS OF EXTENSIN

In carrot disks, virtually all of the cellular hydroxyproline is in the cell wall fraction. The pioneering work of Steward (32) indicated that peptidyl proline is the source of peptidyl hydroxyproline. Taken together these two facts have allowed us to study the assembly of extensin in the cytoplasm and its secretion to the cell wall. In a typical experiment, we used carrot disks that had been aged for 24 hours, since such disks are

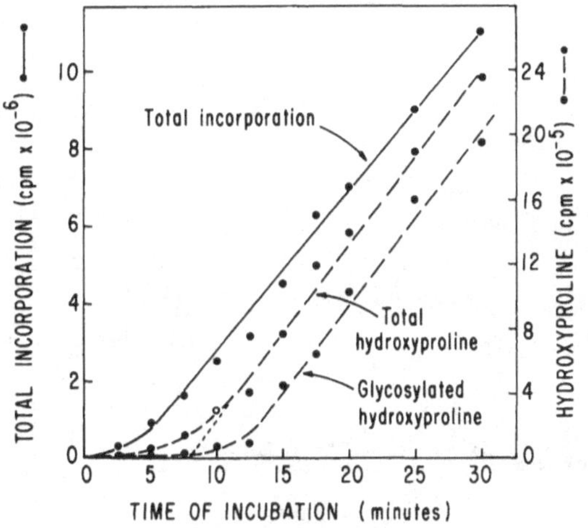

Figure 2. Sequence of events in the biosynthesis of extensin. Carrot disks aged for 24 hours were incubated in ^{14}C-proline. The tracer was rapidly incorporated into proteins (total incorporation). After several minutes, bound ^{14}C-hydroxyproline appeared, and after a further lag of several minutes, glycosylated hydroxyproline appeared. From reference 6. Reprinted with permission of Academic Press, Inc.

very active in extensin synthesis (Figure 1 and see below). These disks were incubated in ^{14}C-proline for a short time. The radioactive proteins made were analyzed for ^{14}C-hydroxyproline and glycosylated ^{14}C-hydroxyproline with the presumption that the latter two represent extensin in the process of being assembled. The kinetics of such an experiment are shown in Figure 2. Clearly, two proceeses occur sequentially: formation of peptidyl hydroxyproline and the glycosylation of that amino acid.

Formation of Hydroxyproline

We have reviewed the evidence that strongly indicates that hydroxyproline is formed from peptidyl proline, and that free proline is not a substrate for hydroxylation (9). At the level of a peptide, hydroxylation could

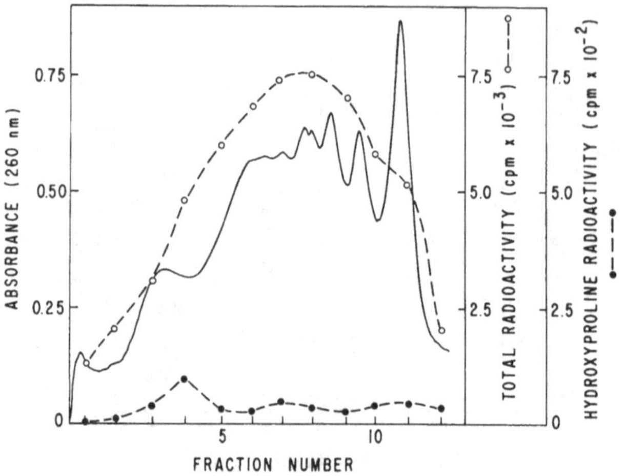

Figure 3. Sucrose density gradient analysis of polyribosomes from carrot disks aged for 24 hours. The disks were incubated for 10 minutes in ^{14}C-proline. Radioactivity bound to the polyribosomes as nascent protein was analyzed for hydroxyproline. From reference 40. Reprinted with permission of the American Chemical Society.

occur while the protein chain is still being made, attached to the ribosomal complex. Alternatively, hydroxylation could occur after the protein had been completed and released from the ribosome. During aging of carrot disks, polyribosomes, active in protein synthesis, are induced to form, possibly by the synthesis of new molecules of messenger RNA (27). To determine if nascent chains were hydroxylated, we incubated aged disks in ^{14}C-proline for a brief time period and then isolated nascent proteins attached to polyribosomes (Figure 3). Analyses of these chains showed no ^{14}C-hydroxyproline (40). This indicates that proline hydroxylation occurs after the protein chain has been released from the ribosome.

The cell can synthesize a protein of typical size (e.g. globin) in about 1-2 minutes (46). About 3 minutes after ^{14}C-proline is first detected in bound form, ^{14}C-hydroxyproline appears (Figure 2). Thus it is apparent that the proline-rich protein backbone of extensin is hydroxylated soon after its release from the ribosome. The period during which hydroxylation occurs is brief. After a time, the protein becomes unavailable to the hydroxylating system (7). Since the next step in extensin assembly, that of glycosylation, occurs within a membranous compartment of the cell, the timing is consistent with the idea that proline hydroxylation occurs in a cellular compartment separate from the site of glycosylation.

From in vivo experiments,several characteristics of an enzyme that catalyzes peptidyl proline hydroxylation are apparent. Molecular oxygen is the source of the oxygen of the hydroxyl group (21, 44) and the incoming hydroxyl group displaces the proton on the 4-trans position of proline (22). Ferrous ions appear to be involved, since the chelator dipyridyl blocks hydroxyproline formation and this blockage is reversed by ferrous ions (7). We have characterized an enzyme from cell-free extracts of carrot disks which has the above characteristics in its reaction (41). The enzyme, peptidyl proline hydroxylase, has been partially purified. Sub-cellular fractionation indicates that it is localized in the soluble cytoplasm, and this is where the reaction undoubtedly occurs.

Chemical analyses of the protein moiety of the immediate cytoplasmic precursor to cell wall extensin indicate that 80 per cent of the proline residues are hydroxylated (3). Why some proline residues are not substrates for the hydroxylase is not known. Some tryptic peptides of extensin have been sequenced, but the order of the amino acids shows no pattern consistent with recognition for the hydroxylase. The molecular weight of the extensin precursor in the cytoplasm indicates a protein chain several hundred amino acids in length.

Glycosylation of the Protein

Approximately five minutes after hydroxyproline formation begins, glycosylation of the extensin precursor commences (Figure 2). Both hydroxyproline-arabinose and serine-galactose bonds are formed (4). The concurrent

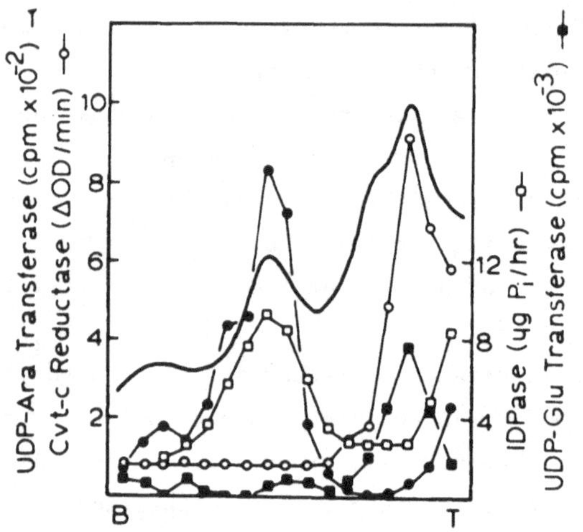

Figure 4. Rate zonal gradient analysis of membrane organelle marker enzymes from carrot disks aged for 24 hours. Inosine diphosphatase (IDPase) marks dictyosomes, cytochrome c reductase marks endoplasmic reticulum, and UDP-glucose glucosyl transferase marks plasma membrane. The enzyme for the glycosylation of hydroxyproline, UDP-arabinose arabinosyl transferase, occurs in the same fractions as the dictyosome marker. From reference 15. Reprinted with permission of the American Society of Plant Physiologists.

association of the macromolecule with membranous organelles (5) indicates that the glycosylation reaction may occur within such organelles. An enzyme that catalyzes the transfer of arabinose from UDP-arabinose to peptidyl hydroxyproline has been identified in extracts of sycamore cells in culture (19) and carrot disks (15). Careful cell fractionation studies have identified dictyosomes as the organelles with which the glycosylating enzyme is associated (Figures 4 and 5). The product of this glycosylation is a protein backbone with arabinosyl side-chains of varying lengths (usually 1-4 arabinose residues). The other glycosylation reaction, which puts single galactose residues onto certain serine residues, has not yet been characterized. Not all hydroxyprolines are glycosylated nor all serines. As is the case with proline hydroxylation, the reason for this

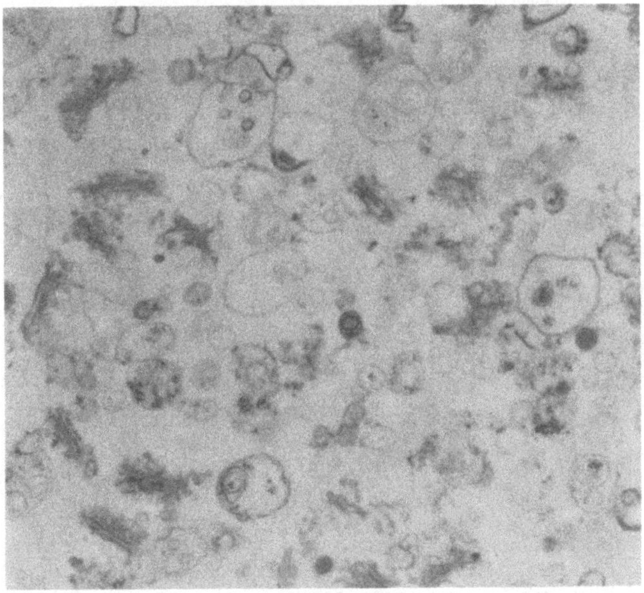

Figure 5. Electron micrograph of membranes isolated from the IDPase- and UDP-arabinose arabinosyl transferase-rich fraction shown in Figure 4. Most membranes are intact dictyosomes or appear to be derived from dictyosomes by vesiculation. From reference 15. Reprinted with permission of the American Society of Plant Physiologists.

enzyme specificity is unknown.

Glycosylation within the dictyosomes presumably completes the biosynthesis of extensin. Because the glycoprotein is soluble in cold, 5 per cent trichloroacetic acid (a condition which precipitates most proteins), it has been possible to isolate and characterize extensin on its way to the cell wall. The macromolecule is 60 per cent protein and 40 per cent carbohydrate by weight. The major amino acids are hydroxyproline (10 mole per cent), serine (19 per cent), glycine (12 per cent) and lysine (12 per cent). The major sugars are arabinose and galactose. Although sedimentation and gel chromatographic properties differ in their interpretation, the macromolecule appears to be larger than a molecular weight of 200,000.

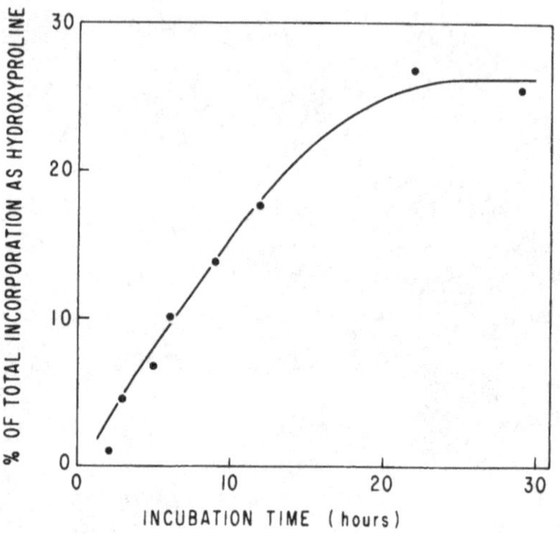

Figure 6. Increase in the synthesis of peptidyl hydroxyproline as a function of aging. Carrot disks were incubated in ^{14}C-proline for 1 hour and the extent of ^{14}C-hydroxyproline formation was determined. From reference 10. Reprinted with the permission of the Oxford University Press.

Induction of Extensin Synthesis in Aging Carrot Disks

Carrot disks incubated after slicing accumulate a great deal of extensin (Figure 1). That this represents an induction of the synthesis of extensin is shown by the data of Figure 6. As the aging increases, the cells hydroxylate an increasing proportion of their proline residues immediately after protein synthesis. This increase reflects an increased synthesis in the proportion of extensin made relative to other proteins: After 6 hours aging, the carrot cells synthesize the putative extensin precursor described above to the extent of 2 percent of overall protein synthesis; at 24 hours, this macromolecule represents 6 per cent of total protein synthesis (10).

To examine the control of extensin induction, we have studied the effect of aging on the two enzymes critical to its synthesis: peptidyl proline

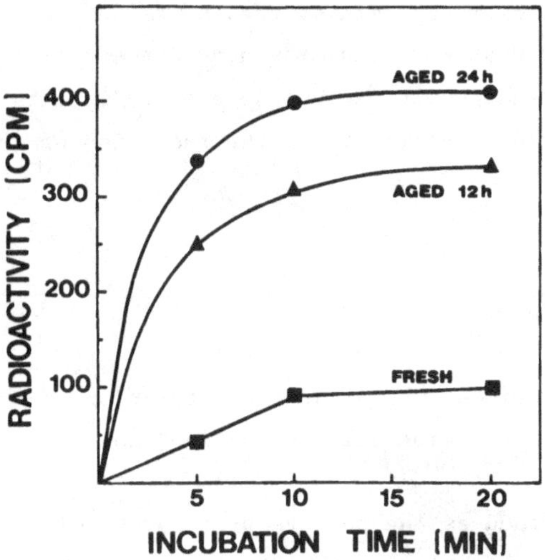

Figure 7. Effect of aging on UDP-arabinose arabinosyl transferase activity. Carrot disks were aged for various periods, homogenized, and equivalent amounts (about 5 mg protein) of membrane organelle fraction were used to test for the incorporation of ^{14}C-arabinose into macromolecules from ^{14}C-arabinose-UDP. From reference 10. Reprinted with permission of the Oxford University Press.

hydroxylase and UDP-arabinose arabinosyl transferase (10). The first enzyme does not change in specific activity during aging. The second enzyme does appear to increase, albeit from data of crude extracts (Figure 7). However, several points argue against this increase as being critical to extensin induction. First, the enzyme is associated with membranous organelles, and these are induced in general during aging (see elsewhere in this volume). Second, glycosylation is not a limiting step in the secretion of the protein to the wall, since it is secreted when little glycosylation occurs (see below). Third, the data of Figure 6 point to an increase in the protein chains available for hydroxylation. Thus control resides at the level of protein synthesis. A suggestive experiment along this line was one in which disks were incubated in actinomycin D during aging. This drug blocks messenger RNA formation, and in its presence the induction of extensin synthesis was not observed (10). It is possible to conclude that extensin messenger RNA synthesis is essential for induction of the macromolecule during aging. But this is by no means proved. A direct approach necessitates the isolation of extensin-specific messenger RNA.

SECRETION OF EXTENSIN

One of us recently reviewed the subject of secretion of macromolecules in plants (8). This discussion will be confined to extensin secretion.

Secretion can be defined as the passage of a large molecule out of the cell through the cell membrane. In the case of extensin, the secreted product ends up in the cell wall. We should note that in cultured cells from sycamore plants, a glycoprotein very similar to extensin is secreted into the culture medium. It is not known whether this is an artifact of the tissue culture situation (34).

We have studied the secretion of extensin in carrot disks by radioactive labelling experiments. Disks were incubated in ^{14}C-proline for a period sufficient fot the completion of the macromolecule in the cytoplasm. An excess of nonradioactive proline was then added to the medium around the disks. This ensured that subsequently made extensin would be non-radioactive. The labelled extensin was then observed to disappear from the cytoplasm and to appear in the cell wall fraction. Kinetic studies indicated that this secretory process from cytoplasm to cell wall took about 10 minutes (5). We have examined the metabolic requirements for, and organelle involvements in, this process.

Metabolic Requirements For Secretion

In carrot disks, secretion, as the final step in the extensin pathway, does not require continued protein synthesis to "push" it along. Once the macromolecular precursor of extensin has been synthesized in the cytoplasm, it will be secreted to the cell wall whether or nor there is any further extensin synthesis. This conclusion was reached from experiments in which carrot disks were incubated in ^{14}C-proline for a time and then secretion of the ^{14}C-extensin thus made was followed by the addition of an excess of nonradioactive proline. When this "chase" was done in the presence of the protein synthesis inhibitor cycloheximide, secretion occurred at the normal rate (12). Using a similar strategy, we also examined secretion in the presence of other metabolic inhibitors. When uncouplers of oxidative phosphorylation or inhibitors of electron transport were used, secretion was blocked; this indicates that extensin secretion requires a steady flow of metabolic energy and is not a mere diffusion process.

A surprising finding of our studies of extensin secretion is that the entire glycoprotein need not be synthesized in order to be secreted to the cell wall. If the formation of hydroxyproline is blocked by the use of the chelator dipyridyl, the protein backbone of extensin is synthesized at the normal rate (7, 43). Since there is no hydroxyproline in this backbone, the bulk of glycosylation, formation of hydroxyproline-arabinose bonds, does not occur. (It is not known whether serine-galactose bonds are formed in this situation). Nevertheless, this nonhydroxylated, underglycosylated macromolecule is secreted to the cell wall at the normal rate and to the normal extent. This situation is in contrast to that in animal cells synthesizing collagen. When such cells are incubated in dipyridyl, nonhydroxylated, underglycosylated protocollagen is formed and it remains in the cytoplasm, not being secreted to the extracellular medium. If ferrous ions are added to the medium to overcome the inhibition of hydroxylation, the cytoplasmic protocollagen is hydroxylated and subesquently secreted. Thus in animal cells, collagen must be completed in order to be secreted (38). In plant cells, the "protoextensin" secreted to the wall appears to be nonfunctional to the extent that a function can be assigned to extensin. In pea epicotyls, extensin made in the presence of dipyridyl does not act to stiffen the cell wall at the cessation of cell elongation (42). Unfortuantely, there are no data on the physical state of "protoextensin" in the cell wall (linkages to hemicellulose, etc.). One can speculate that the underglycosylated macromolecule may have little or no covalent attachments to the rest of the wall. That it is secreted may indicate that the bulk of glycosylation is not needed for secretion and the protein backbone is the key to this process.

Organelles and Secretion

Elegant studies by Palade and co-workers have established that the secretion of some glycoproteins in animal cells involves a number of cellular organelles. Proteins are glycosylated within the endoplasmic reticulum and/or dictyosomes. Vesicles containing the completed glycoprotein are then dispatched to the cell membrane where the macromolecule is secreted by a process which appears to be the reverse of pinocytosis (18).

In carrot disks, we found in our initial studies that most of the rapidly-secreted bound hydroxyproline was transiently associated with membranous organelles before arriving in the cell wall (5). Indeed, the macromolecular extensin precursor was purified from a heterogeneous membranous organelles fraction (3). This fraction has been further fractionated, and it now appears that dictyosomes are involved (see Figure 5). The dictyosomes contain the glycosylating enzyme, UDP-arabinose arabinosyl transferase. This situation in carrot disks has recently been confirmed for the case of cultured sycamore cells (33). Whether secretion then proceeeds via dictyosomes and vesicles remains to be determined.

ACKNOWLEDGEMENTS

We benefitted from fruitful collaboration with our colleagues, Drs. Y. Cho, M. Gardiner, M. Doerschug, and M. Brysk. Our research was supported by a contract from the US-ERDA E(04-3)-34/159.

REFERENCES

1. Albersheim, P., Bauer, W., Keegstra, K., Talmadge, K.: The structure of the wall of suspension-cultured sycamore cells. In: Biogenesis of Plant Cell Wall Polysaccharides. Ed., Loewus, F. New York, Academic Press. pp. 117-148 (1973).
2. Allen, A., Neuberger, A.: Purification and properties of the lectin from potato tubers, a hydroxyproline-containing glycoprotein. Biochem. J. 135: 307-314 (1973).
3. Brysk, M., Chrispeels, M.: Isolation and partial characterization of a hydroxyproline-rich cell wall glycoprotein and its cytoplasmic precursor. Biochim. Biophys. Acta. 257: 421-432 (1972).
4. Cho, Y.-P., Chrispeels, M.: Serine-O-galactosyl linkages in glycopeptides from carrot cell walls. Phytochem. 15: 165-169 (1976).
5. Chrispeels, M.: Synthesis and secretion of hydroxyproline-containing macromolecules in carrots. 1. Kinetic analysis. Plant Physiol. 44: 1187-1193 (1969).
6. Chrispeels, M.: Biosynthesis of cell wall protein: Sequential hydroxylation of proline, glycosylation of hydroxyproline, and secretion of the glycoprotein. Biochem. Biophys. Res. Commun. 39: 732-737 (1970)
7. Chrispeels, M.: Synthesis and secretion of hydroxyproline-containing macromolecules in carrots. 2. In vivo conversion of peptidyl proline to peptidyl hydroxyproline. Plant. Physiol. 45: 223-227 (1970).
8. Chrispeels, M.: Biosynthesis, intracellular transport, and secretion of extracellular macromolecules. Ann. Rev. Plant Physiol. 27: 19-38 (1976).
9. Chrispeels, M., Sadava, D.: Synthesis and secretion of proteins in plant cells; The hydroxyproline-rich glycoprotein of the cell wall. In: Macromolecules Regulating Growth and Development. Ed. Hay, E., King, T., Papaconstantiou, J. New York. Academic Press. pp. 131-152 (1974).
10. Chrispeels, M., Sadava, D., Cho, Y.-P.: Enhancement of extensin biosynthesis in ageing disks of carrot storage tissue. J. Exp. Bot. 25: 1157-1166 (1974).
11. Cleland, R., Karlsnes, A.: A possible role for hydroxyproline-contain-

ing proteins in the cessation of cell elongation. Plant Physiol. 42: 669-671 (1967).

12. Doerschug, M., Chrispeels, M.: Synthesis and secretion of hydroxyproline containing macromolecules in carrots. 3. Metabolic requirements for secretion. Plant. Physiol. 46: 363-366 (1970).
13. Dougall, D., Shimbayaishi, K.: Factors affecting growth of tobacco callus tissue and its incorporation of tyrosine. Plant Physiol. 33: 396-404 (1960).
14. Esquerre-Tugaye, M., Mazau, D.: Effect of a fungal disease on extensin the plant cell wall glycoportein. J. Exp. Bot. 25: 509-513 (1974).
15. Gardiner, M., Chrispeels, M.: Involvement of the golgi apparatus in the synthesis and secretion of the hydroxyproline-rich cell wall glycoprotein. Plant Physiol. 55: 536-541 (1975).
16. Heath, M., Northcote, D.: Glycoprotein of the wall of sycamore tissue culture cells. Biochem. J. 125: 953-961 (1971).
17. Israel, H., Salpeter, M., Steward, F.: The incorporation of radioactive proline into cultured cells. J. Cell Biol. 39: 698-715 (1968).
18. Jamieson, J., Palade, G.: Intracellular transport of secretory proteins in the pancreatic exocrine cell. J. Cell Biol. 34: 577-615 (1967)
19. Karr, A.: Isolation of an enzyme system which will catalyze the glycosylation of extensin. Plant Physiol. 50: 275-282 (1972).
20. Klis, F.: Glycosylated serine residues in wall protein of elongating pea stems. Plant Physiol. 57: 224-226 (1976).
21. Lamport, D.: Oxygen fixation into hydroxyproline of plant cell wall protein. J. Biol. Chem. 238: 1438-1440 (1963).
22. Lamport, D.: Hydroxyproline biosynthesis: Loss of hydrogen during the hydroxylation of proline. Nature. 202: 293-294 (1964).
23. Lamport, D.: The protein component of primary cell walls. Adv. Botan. Res. 2: 151-218 (1965).
24. Lamport, D., Katona, L., Roerig, S.: Galactoserine in extensin. Biochem. J. 133: 125-131 (1973).
25. Lamport, D., Miller, D.: Hydroxyproline-arabinosides in the plant kingdom. Plant Physiol. 48: 454-456 (1971).
26. Lamport, D., Northcote, D.: Hydroxyproline in primary cell walls of higher plants. Nature. 188: 665-666 (1960).

27. Leaver, C., Key, J.: Polyribosome formation and RNA synthesis during aging of carrot-root tissue. Proc. Nat. Acad. Sci. U.S. 57: 1338-1344 (1967)

28. Liu, E., Lamport, D.: An hydroxyproline-O-gyycosidic linkage in an isolated horseradish isozyme. Plant Physiol. 43: S-16 (1968).

29. Mazau, D., Esquerre-Tugaye, M.: Hydroxyproline and peroxidase in the cell wall and cytoplasm of melon plants infected with anthracnose fungus. Plant Sci. Lett. 7: 119-125 (1976).

30. Monro, J., Bailey, R., Penney, D.: Cell wall polysaccharide-hydroxyproline associations in Lupinus hypocotyls. Phytochem. 13:375-382 (1974).

31. Monro, J., Penney, D., Bailey, W.: The organization and growth of primary cell walls of lupin hypocotyls. Phytochem. 15: 1193-1198 (1976).

32. Pollard, J., Steward, F.: The use of ^{14}C-proline by growing cells; Its conversion to protein and to hydroxyproline. J. Exp. Bot. 10: 17-32 (1959).

33. Pope, D.: In preparation, 1977.

34. Pope, D., Lamport, D.: Hydroxyproline-rich material secreted by cultured Acer pseudoplanatus cells. Evidence for polysaccharide attached directly to hydroxyproline. Plant Physiol. 53: S-81 (1974).

35. Ridge, I., Osborne, D.: Hydroxyproline and peroxidases in cell walls of Pisum sativum: Regulation by ethylene. J. Exp. Bot. 21: 843-856 (1970).

36. Roberts, K., Gurney-Smith, M., Hills, G.: Structure, composition, and morphogenesis of the cell wall of Chlamydomonas reinhardi. J. Ultra. Res. 40: 599-613 (1972).

37. Roberts, K., Northcote, D.: Hydroxyproline: Observations on its chemical and autoradiographic localization in plant cell wall protein. Planta. 107: 43-51 (1972).

38. Rosenbloom, J., Prockop, D.: Biochemical aspects of collagen synthesis In: Repair and Regeneration. Eds. Dunphy, D., VanWinkle, N. New York. McGraw-Hill. pp. 117-135 (1969).

39. Sadava, D., Chrispeels, M.: Cell wall protein in plants: autoradiographic evidence. Science. 165: 299-300 (1969).

40. Sadava, D., Chrispeels, M.: Intracellular site of proline hydroxyla-

tion in plant cells. Biochemistry. 10: 4290-4294 (1971).
41. Sadava, D., Chrispeels, M.: Peptidyl proline hydroxylase from carrot disks. Biochim. Biophys. Acta. 227: 278-287 (1971).
42. Sadava, D. Chrispeels, M.: Hydroxyproline-rich cell wall protein (extensin): Role in the cessation of elongation in excised pea epicotyls. Develop. Biol. 30: 49-55 (1973).
43. Sadava, D., Walker, F., Chrispeels, M.: Hydroxyproline-rich cell wall protein (extensin): Biosynthesis and accumulation in growing pea epicotyls. Develop. Biol. 30: 42-48 (1973).
44. Stout, E., Fritz, G.: Role of oxygen fixation in hydroxyproline biosynthesis by etiolated seedlings. Plant Physiol. 41: 197-202 (1966).
45. Suzuki, T., Sato, S.: Cell wall column chromatographic identification of the native cell wall acid phosphatase of cultured tobacco cells. Plant and Cell. Physiol. 17: 847-849 (1976).
46. Winslow, R., Ingram, V.: Peptide chain synthesis of human hemoglobins A and A_2. J. Biol. Chem. 241: 1144-1149 (1966).

Membrane Systems and their Transformations in Aging Plant Storage Tissues

Pierre Benveniste

The excision of slices from storage tissues, such as the potato tuber, initiates many metabolic changes. When the slices are aged aerobically, these tissues are "activated" (1). Activation comprises the following events : the turning on of various genes, with subsequent synthesis of t, r, and mRNAs ; synthesis of many enzymes and structural proteins or glycoproteins ; synthesis of fatty acids, phospholipids, galactolipids, and sterols ; and synthesis of polysaccharides and precursors of suberin and lignin. This metabolic behaviour is commonly viewed as a cell differentiation process (2) comprising degradative as well as biosynthetic steps and accompanied by membrane transformations or rearrangements. Before discussing some of the recent observations related to the cytological and biochemical changes in cellular compartmentation and membrane structure occurring during this cell differentiation process, we will introduce the notion of endomembrane system.

THE ENDOMEMBRANE SYSTEM

According to D.J. MORRE (6-8), there is a membranous structural and functional continuity in cells between the nuclear envelope, the ER, the golgi apparatus, and various types of

Biochemistry of Wounded Plant Tissues

vesicles which are able to give rise to provacuoles (9) and/ or to export cell wall materials following fusion with plasmalemma (10) (fig 1). According to this theory, some components of the endomembrane system could give rise to glyoxysomes (11) and to the external membranes of plastids (8) and mitochondria (12). Moreover this membrane system is continuous from cell to cell through the presence of plasmodesmata (13), thus allowing a possibility of control between groups of cells (3).

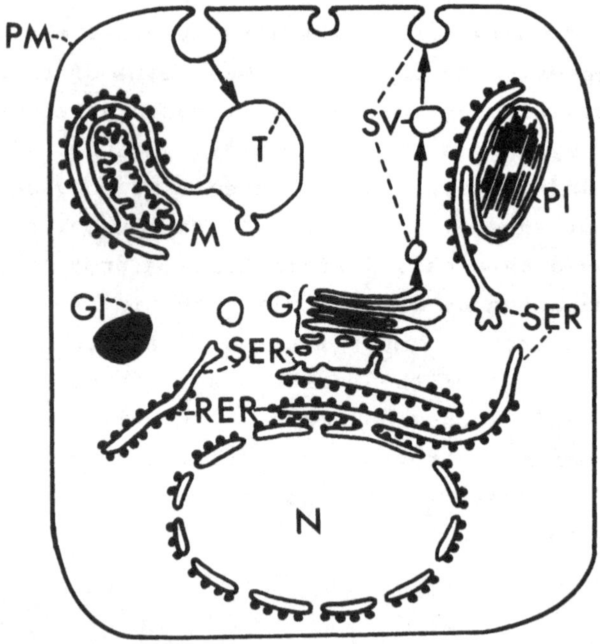

Fig.1. Representation of the endomembrane system (8).
G = golgi apparatus; RER = rough endoplasmic reticulum; SER = smooth endoplasmic reticulum ; PM = plasma membrane ; SV = secretory vesicle ; Gl = glyoxysomes ; N = nucleus ; M = mitochondria ; Pl = plastid ; T = tonoplast.

Profound changes appear in the endomembrane system of sliced storage tissues. Cytological studies on potato and Jerusalem artichoke tubers have revealed the following features : cells fixed directly from intact tubers exhibit only a rudimentary ER, and dictyosomes are practically absent; slices aged 24 hr show a strong increase of polyribosomes (2,5) rough ER (2,5), and golgi vesicles (14). In the case of beetroot tissue (4), aging leads first to some degradation of membranes from the ER a few hours after slicing, and then to some apparent development of the ER. Amongst other changes in potato tuber slices are the appearance, besides the preexisting typical amyloplasts, of some smaller organelles resembling proplastids (14), and the development of numerous peroxysomes often containing crystals (14). In contrast, the mitochondrial population appears to be stable (2,14), although some contradictory results have been reported (15,16). These ultrastructural changes reflect important biochemical transformations which have been studied concomitantly. The increase in ER has to be linked to the induction or activation of the enzymes involved in the biosynthesis of proteins, phospholipids, sterols, phenylpropanoïds, terpenoids, etc.. (1), since the ER is the site of biosynthesis of all these compounds (7) and since their synthesis increases considerably after the tissue is wounded (1). Increase of ER and especially of golgi cisternae could also be related to the production and exportation of cell wall polysaccharides (10), cutin and suberin (17), various phenylpropanoids (18), lignin (18), glycoproteins like extensin (19), etc.. As these secretory compounds are generally enclosed inside membrane vesicles, the whole secretory process would involve continuous synthesis of new membrane systems and/or rearrangement of preexisting ones.

The present paper gives a rapid survey of the main biochemical events occurring during membrane degradation, rearrangements, and biogenesis in held storage tissue slices.

EARLY MEMBRANE REARRANGEMENTS

It has been reported (4) that slicing of beetroot tissue leads to a drastic reduction of ER immediately after the slices are cut and rinsed. Such morphological observations have been confirmed biochemically, In particular it has been shown that the CO_2 from fresh slices is of lipid origin whereas that from tubers and aged slices is of carbohydrate origin (20), suggesting that early biochemical events following slicing involve extensive membrane degradation. Accordingly, a study was made of the breakdown of phospholipids and galactolipids in storage organs in response to cutting. Galactolipids of potato slices drop more than 50 percent in the first hour after cutting. The phospholipids PC, PE, and PI also drop in the first hour and are regenerated in the next 24 hr (20). Lipid and possibly membrane breakdown has been correlated with the cyanide sensitivity appearing in potato tissues after cutting (21). Lipid degradation in potato has been substantiated recently by the separation and characterization of lipid acylhydrolases (22,23). One of them would be an A-type phospholipase which could be involved in the deterioration of mitochondria during their preparation (24). Also, oxidative enzymes (lipoxygenase, peroxidases) are active in potato cells (25) and could perform the peroxidation of unsaturated fatty acids. There are reasons to believe that these membrane alterations on wounding have equally profound effects on salt transport characteristics of the tissue (3,21).

MEMBRANE BIOGENESIS AND DIFFERENTIATION.

The biochemical studies on membrane biogenesis and differentiation which have been reported so far concern mainly the biosynthesis of the structural and functional components of membranes, i.e., proteins and lipids.

Protein synthesis There is a strong synthesis of proteins in response to slicing and aerating storage organs (14). This proteosynthesis concerns membranes, since after homogenization and differential centrifugation an important increase in proteins is found in mitochondria (15,26,27) and in microsomes (28). Of the two types of membrane proteins i.e. structural membrane proteins and functional membrane proteins like enzymes, very little is known about structural membrane proteins, whereas considerable amount of work has been reported concerning membrane bound enzymes. It has been shown that microsomal and mitochondrial NADH-cytochrome c oxido-reductase (29), microsomal NADPH-cytochrome c oxido-reductase (30) and the enzymes involved in the biosynthesis of fatty acids (29), of sterols, and of phenylpropanoids (18) are much more active in held storage-tissue slices than in fresh slices suggesting that some *de novo* synthesis of these enzymes could occur in response to slicing and aerating. Spectacular development of ER is another indication of significant proteogenesis which occurs in wounded tissues, since it is now generally accepted that mRNAs are associated with membranes of the rough ER (32,33). As some membrane proteins are strongly hydrophobic, it would be reasonable for them to be synthesized within or near the membrane on a membrane-associated template (7). Also widely accepted is the thesis that attachment of ribosomes to membranes facilitates, or even may be obligatory for, the secretion of proteins destined for export (7).

Phospholipids ER contains most of the phospholipid (PC, PS, PE, PG, PI) synthesizing enzymes for example CDP-choline diglyceride transferase (34), CDP-diglyceride inositol transferase, PE-L-serine phosphatidyl transferase (35) and CDP-diglyceride glycerol phosphate transferase (36). Synthesis of membrane (microsomal, mitochondrial, etc..) phospholipids is strongly stimulated during aging of potato tuber slices (29,37,38); therefore, development of rough ER in held slices can be correlated with phospholipid biosynthesis (5). As phospholipids are present in all membrane systems of the cell, the results cited above raise the problem of the transport of phospholipids from sites of synthesis to the membranes where they have to be integrated. It has been shown (39) that a soluble supernatant could facilitate exchange of phospholipids between ER and mitochondria. This result points to the participation of cytoplasmic proteins in the exchange process. That this is so has been borne out by the isolation of a protein from the cytoplasmic supernatant fraction of potato cells that is involved in this process (40,41). Such a mechanism could be involved during biogenesis of other membrane systems (plasmalemma, peroxysomes, plastids, etc..) than ER and mitochondria since by this process, phospholipids could be distributed from the sites of biosynthesis to membranes lacking the enzymatic equipment needed for *de novo* biosynthesis. However recent results obtained with glyoxysomes suggest that this organelle could derive directly from the ER (11) during membrane flow (7); thus the phospholipid exchange mechanism may not be exclusive. As the fatty acid moiety of phospholipids from plants is mainly unsaturated (linoleic or linolenic), enzymes must exist which introduce double bonds into more saturated fatty acids (stearic or oleic). Such an enzyme, oleyl-CoA desaturase, has been studied in detail (31). It transforms oleyl-CoA into linoleic acid. This enzyme is probably associated with ER; absent in the freshly cut slices, it was induced during the aging of slices. It has been suggested (31) that the oleyl-CoA desaturase synthesized on ER membranes can desaturate oleyl-CoA into linoleyl-CoA which is then transferred to PC molecules, closely associated with the desaturation reaction.

Sterols Sterols are universal constituents of eukaryotic cells. Plant cell membranes contain a mixture of three or more sterols, commonly campesterol, sitosterol, and stigmasterol (42-45). Amongst tested membranes, plasmalemma is by far the richest in free sterols (46-48) whereas plastids are the poorest (44); separation of the envelope from the thylakoïds showed clearly that free sterols are restricted to the envelope but are practically absent from thylakoïd membranes (49). Mitochondria and ER contain intermediate amounts of sterols (46,50). ER has been proposed as the site of sterol biosynthesis, since four key enzymes involved in sterol biosynthesis : all trans farnesyl pyrophosphate-squalene synthetase (51), SAM-cycloartenol C-24 methyl transferase (52), 2,3-squalene-epoxide cyclase, and cycloeucalenol-obtusifoliol isomerase (53) were shown to be associated exclusively with ER. When fresh potato tuber slices were incubated with (1-^{14}C)-acetate, cycloartenol was heavily labelled but no radioactivity was recovered in 24-methylene cycloartanol or free sterols. If potato slices were aged for 0 to 24 hr before being fed with radioactive acetate, a rapid increase of the label in 24-methylene cycloartanol and the free sterols was observed (54). As the transformation of cycloartenol into 24-methylene cycloartanol has been shown to be catalyzed by a SAM C-24 methyl transferase (55), the induction of the synthesis of sterols in held slices could be related to the appearance of this enzyme (54). The induction of sterol biosynthesis in held potato tuber slices, as for phospholipid synthesis, correlates with the development of endomembranes observed cytologically (2,5). Whereas the biosynthetic capacities for phospholipids is the highest 10hr after slicing, 24 hr are necessary for maximal sterol synthesis (37,54). Therefore SAM-cycloartenol C-24 methyl transferase seems to be a likely marker of the synthesis of new membranes during the aging process. The problem of the transport of newly formed sterols from their site of synthesis to membranes where they have to be integrated (plasmalemma mainly) raises interesting questions, in particular, whether "sterol carrier proteins" are involved in the transport of sterols as in the case of phospholipids. Sterols exist in four main forms in higher plants : free sterols, steryl esters, steryl glycosides, and 6'-acyl steryl glycosides. In contrast to free sterols, which interact with phospholipids in biological membranes (56) and decrease the fluidity and passive permeability of membranes (56), the physiological role of the other forms is so far largely unknown. It has been recognized that potato tuber tissues are remarkably rich in steryl glycosides and in acyl steryl glycosides (57,58) but no studies of the evolution of these two forms in held slices have been reported. It may be expected that synthesis of sterols and their rearrangement into bound forms in held slices could have a key influence on properties of plasmalemma and more particularly on the development of absorption capacities for solutes (21). Finally it should be remembered that cholesterol is a precursor of steroidal alkaloids (solanine), which are synthesized actively in held potato slices (59).

SECRETION AND EXPORTATION OF COMPOUNDS

In many storage tissues, slicing and aging lead to regeneration of cell walls, and, depending on the species, to lignification and suberinization (1). These three processes would involve secretion and exportation of macromolecules and/or precursors of these macromolecules capable of polymerizing outside the cell. Enzymes also would have to be secreted in order to perform the assembly of the units. As wall constituents and enzymes are probably enclosed in membrane vesicles during the secretion process (7), this process is closely linked with membrane biogenesis, differentiation, and transformation. Although very little work has been done on this subject, it could be useful to point out some partial but promising aspects.

Cell wall Secretion of a hydroxy proline-rich cell wall glycoprotein (extensin) in held (24hr) disks of carrot tissues was studied (60); it could be shown that the golgi apparatus is involved in the synthesis and secretion of this protein. Biogenesis of golgi vesicles in such a system is suggested by Castelfranco's experiments showing that labelled choline (a precursor of PC) is incorporated most efficiently in a presumed golgi fraction from potato tuber slices held for 24hr. (61). Assembly of cell wall constituents (cellulose, hemicellulose, pectic acids, extensin, etc.) probably depends on wall enzymes. In this connection, it can be recalled that the activity of some wall enzymes, especially acid invertase (62-63), is strongly stimulated following slicing of storage tissues. Acid invertase was essentially cytoplasmic in fresh tissues of sweet potato, it appears in the cell wall a few hours after slicing (63); it is probably a glycoprotein, and it may be secreted via the endomembrane system in roots and hypocotyls of developing radish seedlings (64).

Lignin It is reasonable to assume that phenylpropanoid precursors of lignin are secreted via the endomembrane system and that lignin is assembled in the wall (18). A strong stimulation of phenylpropanoid synthesis in held slices of storage tissues has been demonstrated (1) by many authors and is probably related to lignification. Phenylpropanoid biosynthesis involves membrane-associated enzymes, such as cinnamic acid 4-hydroxylase (65,66) which is integrated into membranes of the ER (30) and which is strongly stimulated during the aging of disks of potato (67), sweet potato (68) and Jerusalem artichoke (30) tubers and of swede (69) roots. It can be suggested that association of cinnamic acid 4-hydroxylase with membranes of the ER may possibly play a role in secretion and export of phenylpropanoids and especially of lignin precursors. In that context, the observed increase of peroxidase activities in cell walls of held slices of sweet potato (70) may perhaps be related to lignin biosynthesis in this material, since these enzymes are probably involved in oxidative polymerization of phenylpropanoid units, leading to lignin.

Suberin Kolattukudy has shown that suberin is actively formed at the wound surface of potato slices and that it is identical to that in the natural skin of the tuber (17). During the first several days after excision of the potato slices, little deposition of suberin occured. Most of the suberin formed in these tissue slices was synthesized between day 5 and day 7. From structural studies made in Kolattukudy's laboratory, it appears that the major monomers of suberin are very long-chain fatty acids, ω-hydroxy acids, and dicarboxylic acids. Thus a ω-hydroxylation of free or more likely bound fatty acids is probably a key step during suberin biosynthesis (17). By analogy with the ω-hydroxylation of lauric acid in animal cells, it is reasonable to assume that the ω-hydroxylation reaction is located in the ER in held slices of storage tissues. It has been shown recently that fatty acid elongation took place at the ER in *Allium porrum* (71). As these suberin precursors are mainly hydrophobic compounds, membranes are probably involved during their synthesis (71), assembly into macromolecular compounds, and/or exportation outside the cell. The enzymology of the whole process is still largely unknown although actively studied in Kolattukudy's laboratory.

DISCUSSION

Excision of storage tissues first induces an extensive membrane degradation which is reflected in the activities of phospholipases and galactolipases (22,24). After these early events, the products of lipid hydrolysis are then catabolized and this seems to coincide with the onset of a regenerative metabolism which leads to membrane biogenesis and/or transformation involving an increased synthesis of membrane constituants i.e. proteins, phospholipids and sterols. This metabolic behaviour is commonly viewed as a dedifferentiation during which differentiated storage cells are rejuvenated and become undifferentiated. This state is rapidly followed by differentiation events during which lignin is synthesized and suberin or cutin may be deposited on the outside of the cell. At the same time, secondary metabolites are produced. The factors involved in the transition from degradative to regenerative metabolism are still unknown but it can be suggested that membrane degradation leads to profound changes in cellular compartmentation and that this process could lead to the release of some chemical mediators involved in gene activation.

We have pointed out above that lipid and protein hydrolysis and resynthesis which occur after the slicing of quiescent storage tissues are correlated to the membrane degradation and biogenesis. Changes in membrane composition must play a key role in the appearance of new properties in held slices. For example the profound effect of wounding on salt absorption characteristics may have some biochemical correlative in phospholipids or sterols of plasmalemma (21). The same would be true of mitochondria, since cutting leads to a loss of cyanide resistance, and metabolic recovery involves a reestablishment of cyanide-resistant respiration (21). Qualitative and quantitative changes in membrane composition as well as turnover of various constituents must be determined in morphologically and biochemically well defined membrane systems. A clear separation of nuclear membrane, ER, golgi vesicles, tonoplast, and plasmalemma is a prerequisite for determination of structural and functional changes occurring in these membrane systems in held slices. If technology for membrane purification can be improved, held slices from storage tissues will offer an ideal system for studying the dynamics of membrane biogenesis, the relationships between the endomembrane system and organelles and the mechanism of secretion and exportation of the cell wall constituents.

NOMENCLATURE

SAM = S-adenosyl methionine, ER = endoplasmic reticulum, PC = phosphatidylcholine, PE = phosphatidylethanolamine, PI = phosphatidylinositol, PG = phosphatidylglycerol, PS = phosphatidylserine, silosterol = 24 (R)-24-ethyl-cholest-5-en-3β-ol, campesterol = 24 (R)-24-methyl-cholest-5-en-3β-ol, stigmasterol = 24 (S)-24-ethyl-cholesta-5,E-22-dien-3β-ol.

REFERENCES

1 - KAHL G. Metabolism in plant storage tissue slices. Botanical Review 40, 263-314 (1974).

2 - VAN STEVENINCK, R.F.M. The "washing" or "aging" phenomenon in plant tissues. Ann. Rev. Plant Physiol. 26, 237-258 (1975).

3 - FOWKE L.C. and SETTERFIELD, G. Cytological responses in Jerusalem artichoke tuber slices during aging and subsequent auxin treatment, in Biochem. and Physiol. of Plant Growth Substances, Ed. WIGHTMAN F. and SETTERFIELD G. Ottawa (1968).

4 - JACKMAN, M.E., VAN STEVENINCK, R.F.M. Changes in the endoplasmic reticulum of beetroot slices during aging. Austr. J. Biol. Sci. 20, 1063-1068 (1967).

5 - BEN ABDELKADER A. AUDERSET G. Formation de polyribosomes dans des tranches de tubercules de Pomme de Terre maintenues en survie. C.R. Acad. Sci. Paris 274, 1311-1314 (1972).

6 - MORRE D.J., MOLLENHAUER, H.H. and BRACKER, C.E. in Results and problems in cell Differentiation II. Origin and continuity of Cell organelles. Ed. REINERT J. and URSPRUNG H., Berlin (1971).

7 - MORRE D.J. Membrane biogenesis. Ann. Rev. Plant Physiol. 26, 441-481 (1975).

8 - MORRE, D.J., YUNGHANS W.N., VIGIL, E.L. and KEENAN, T.W. Isolation of organelles and endomembrane components from rat liver : biochemical markers and quantitative morphometry in methodological developments, in Biochemistry vol.4 Ed. REID, E. Longman Group Ltd (1974).

9 - BUVAT R. Relations entre l'ergastoplasme et l'appareil vacuolaire C.R. Acad. Sci. Paris. 245, 350-352 (1957).

10 - NORTHCOTE, D.H. The golgi apparatus. Endeavour 30, 26-33 (1971).

11 - GONZALES, E. and BEEVERS, H. Role of the ER in glyoxysome formation in castor bean endosperm. Plant Physiol. 57, 406-409 (1976).

12 - MORRE, D.J. MERRITT, W.D. and LEMBI, C.A. Connections between mitochondria and ER in rat liver and onion stem. Protoplasma 73, 43-49 (1971).

13 - JUNIPER, B.E. and BARLOW, P.W. The distribution of plasmodesmata in the root tip of maize Planta 89, 352-360 (1969).

14 - KADER, J.C. Biogénèse des lipides membranaires pendant la "survie" de tranches de tubercules de Pomme de Terre Thèse Dr. D'Etat Paris (1972).

15 - VERLEUR, J.D. Observations on the induction of mitochondrial particles in potato tuber tissue after wounding. Z. Pflanzenphysiol. 61, 299-309 (1969).

16 - VAN STEVENINCK, R.F.M. and VAN STEVENINCK, M.E. Formation of new mitochondria in plant cells. Naturwissenschaften 56, 146 (1969).

17 - KOLATTUKUDY, P.E. The biochemistry of the lipid barriers. in Recent Advances in the Chemistry and Biochemistry of plant lipids. Ed. GALLIARD, I. and MERCER, E.I. Acad. Press (1975).

18 - STAFFORD, H.A. The metabolism of aromatic compounds. Ann. Rev. Plant Physiol. 25, 459-486 (1974).

19 - CHRISPEELS, M.I. Synthesis and secretion of hydroxyproline containing macromolecules in carrots. I. Kinetic analysis. Plant Physiol. 44, 1187-1193 (1969).

20 - LATIES, G.G. Membrane destruction and resynthesis in salt absorbing tissue,in Proceedings of the Workshop

on salt effects on plant structures and processes. Riverside Cal. April 1976. Agricultural Res. Service US OA December (1976).

21 - LATIES, G.G. Solute transport in relation to metabolism and membrane permeability in plant tissues, in Historial and Current Aspects of Plant Physiology. Published by New York State College of Agriculture and Life Sciences, Ithaca, N.Y. (1975).

22 - GALLIARD, T. The enzymic breakdown of lipid in potato tuber by phospholipid and galactolipid-acyl hydrolase activities and by lipoxygenase. Phytochemistry 9, 1725-1734 (1970).

23 - HASSON, E.P. and LATIES, G.G. Separation and characterization of potato lipid acyl hydrolases. Plant Physiol. 57, 142-147 (1976).

24 - HASSON, E.P. and LATIES, G.G. Purification and characterization of an A type phospholipase from potato and its effect on potato mitochondria. Plant Physiol. 57, 148-152 (1976).

25 - GALLIARD, T. WARDALE, D.A. and MATTHEW, J.A. The enzymic and nonenzymic degradation of colneleic acid, an unsaturated fatty acid ether intermediate in the lipoxygenase pathway of linoleic acid oxidation in potato. Biochem. J. 138, 23-31 (1974).

26 - NAKANO, M. and ASAHI, T. Biochemical studies on mitochondria formed during aging of sliced potato tissue. Plant Cell Physiol. 11, 499-502 (1970).

27 - ASAHI, T. HONDA, Y. and URITANI, I. Increase of mitochondrial fraction in sweet potato tissue after wounding or infection with C. fimbriata. Plant Physiol. 41, 1179-1184 (1966).

28 - HACKETT, D.P. HAAS, D.W., GRIFFITHS, S.K. and NIEDER-PRUEM D.J. Studies on the development of cyanide resistant respiration in potato tuber slices . Plant Physiol., 37, 8-19 (1960).

29 - BEN ABDELKADER A. Influence de la survie sur la biosynthèse des phospholipides dans les microsomes de tubercules de Pomme de Terre. C.R. Acad. Sci. Paris 268, 2406-2409 (1969).

30 - BENVENISTE, J. SALAÜN, J.P. and DURST F. Wounding induced CAH in jerusalem artichoke tuber. Phytochemistry 16, 69-73 (1977).

31 - BEN ABDELKADER, A. CHERIF, A. DEMANDRE, C. and MAZLIAK P. The oleyl-CoA desaturase of potato tubers. Enzymatic properties, intracellular localization and induction during aging of tuber slices. Eur. J. Biochem. 32, 155-165 (1973).

32 - LOENING, U.E. RNA structure and metabolism. Ann. Rev. Plant Physiol. 19, 37-70 (1968).

33 - SPARKUHL, J. GARE, R.L. and SETTERFIELD G. Metabolism of free and membrane-bound ribosomes during aging of jerusalem artichoke tuber slices. Planta 129, 97-104 (1976).

34 - LORD, J.M., KAGAWA, T. MOORE, T.S. and BEEVERS, M. ER as the site of lecithin formation in castor bean endosperm. J. Cell. Biol. 57, 659-667 (1973).

35 - MOORE, T.S. LORD, J.M., KAGAWA, T. and BEEVERS, H. Enzyme of phospholipid metabolism in the ER of castor bean endosperm. Plant Physiol. 52, 50-53 (1973).

36 - MOORE, T.S. PG synthesis in castor bean endosperm. Kinetics, requirements, and intracellular localization Plant Physiol. 54, 164-168 (1974).

37 - BEN ABDELKADER A. La lipogénèse dans le tubercule de Pomme de Terre . Analyse et biosynthèse des lipides dans le parenchyme central. Influence de la survie de rondelles de parenchyme sur cette biosynthèse. Physiol. Vég., 6, 417-422 (1968).

38 - WEN JING TANG and CASTELFRANCO, P.A. Phospholipid synthesis in aging potato tuber tissue. Plant Physiol. 43, 1232-1238 (1968).

39 - BEN ABDELKADER, A. and MAZLIAK, P. Echanges, *in vitro*, de phospholipides entre mitochondries, microsomes et surnageant cytoplasmique de cellules de Pomme de Terre. C.R. Acad. Sci., Paris. 269, 697-700 (1969).

40 - KADER, J.C. Isolement d'une fraction protéique stimulant les échanges de phospholipides entre mitochondries et microsomes de tubercules de Pomme de Terre. C.R. Acad. Sci. Paris. 277, 1455-1458 (1973).

41 - KADER, J.C. Exchange of phospholipids between membranes. in Cell Surface Reviews. Vol. 3, Dynamic Aspects of Cell Surface Organization. Ed. POSTE, G. and NICOLSON, G. San Diego (1976).

42 - KEMP, R.J. and MERCER, E.I. Studies on the sterols and sterol ester of intracellular organelles of maize shoots. Biochem. J., 110, 119-125 (1968).

43 - GRUNWALD, C. Sterol distribution in intracellular organelles isolated from tobacco leaves. Plant Physiol. 45, 663-666 (1970).

44 - BRANDT, R.D. and BENVENISTE P. Isolation and identification of sterols from subcellular fractions of bean leaves. Biochim. Biophys. Acta, 282, 85-92 (1972).

45 - DUPERON, R. BRILLARD M. and DUPERON, P. Localisation intracellulaire des différents composés stéroliques

dans le tubercule de Pomme de Terre avant sa germination. C.R. Acad. Sci. Paris 274, 2321-2323 (1972).

46 - HARTMANN, M.A., NORMAND, G. and BENVENISTE, P. Sterol compostiion of plasma membrane enriched fractions from maize coleoptiles. Plant Science Letters. 5, 287-292 (1975).

47 - HARDIN, J.W., CHERRY, J.H., MORRE D.J. and LEMBI, C.A. Enhancement of RNA polymerase activity by a factor realeased by auxin from plasma membrane. Proc. Natl. Acad. Sci. USA 69, 3146-3150 (1972).

48 - LEONARD, R.T. and MODGES, T.K. Characterization of plasma membrane-associated AT Pase activity of oat roots. Plant Physiol. 52, 6 (1973).

49 - JOYARD, J. HARTMANN, M.A., BOUVIER, P. and DOUCE, R. Isolation and identification of sterols from the envelope and thylakoïds from chloroplasts of spinach. Manuscript in preparation.

50 - MEANCE, J. DUPERON, P. and DUPERON, R. Repartition des substances stéroliques à l'intérieur des mitochondries de l'inflorescence de Chou-Fleur. Physiol. Vég. 14, 745-756 (1976).

51 - HARTMANN, M.A., FERNE, M., GIGOT, C. BRANDT, R. and BENVENISTE, P. Isolement, caractérisation et composition en stérols de fractions subcellulaires de feuilles étiolées de Haricot. Physiol. Vég. II, 209-230 (1973).

52 - HARTMANN, M.A. FONTENEAU, P. and BENVENISTE, P. Subcellular localization of sterol synthesizing enzymes in maize coleoptiles. Plant Science letters 8, 45-51 (1977).

53 - HARTMANN, M.A. and RAHIER, A. Unpublished results.

54 - HARTMANN, M.A. and BENVENISTE, P. Effect of ageing on sterol metabolism in potato tuber slices. Phytochemistry 13, 2667-2672 (1974).

55 - FONTENEAU, P. HARTMANN, M.A. and BENVENISTE, P. A 24-methylene lophenol C-28 methyl transferase from suspension cultures of bramble cells. Plant Science Letters (in Press).

56 - DEMEL, R.A. and DE KRUYFF, B. The function of sterols in membranes. Biochim. Biophys. Acta 457, 109-132 (1976).

57 - GALLIARD, I. Aspects of lipid metabolism in higher plants I. Identification and quantitative determination of the lipids in potato tubers. Phytochemistry 7, 1907-1914 (1968).

58 - DUPERON, P. and DUPERON, R. Evolution des différentes catégories de composés stéroliques au cours de la germination du tubercule de Pomme de Terre. Physiologie Végétale 11, 487-505 (1973).

59 - MCKEE, R.K. Factors affecting the toxicity of solanine and related alkaloïds to Fusarium caeruleum. J. Gen. Microbiol. 20, 686-696 (1959).

60 - GARDINER, M. and CHRISPEELS, M.J. Involvment of the golgi apparatus in the synthesis and secretion of hydroxyproline - rich cell wall glycoproteins. Plant Physiol. 55, 536-541 (1975) and ref. cited therein.

61 - CASTELFRANCO, P.A. TANG, W.J. and BOLAR, M.L. Membrane transformations in aging potato tuber slices Plant Physiol. 48, 795-800 (1971).

62 - VAUGHAN, D. and Mac DONALD, J. Development of soluble and insoluble invertase activity in washed storage tissue slices. Plant Physiol. 42, 456-458 (1967).

63 - MATSUSHITA, K. and URITANI, I. Change in invertase activity of sweet potato in response to wounding and purification and properties of its invertases. Plant Physiol. 54, 60-66 (1974).

64 - ZOUAGHI and ROLLIN, P. Unpublished results.

65 - RUSSELL, D.W. Properties of the CAH of pea seedlings and some aspects of its metabolic and developmental control. J. Biol. Chem. 246, 3870-3878 (1971).

66 - POTTS, J.R.M., WEKLYCH R. and CONN, E.E. The 4-hydroxylation of cinnamic acid by Sorghum microsomes and the requirement for cyt P-450. J. Biol. Chem. 249, 5019-5026 (1974).

67 - CAMM, E.L. and TOWERS, G.H.N. Effect of ageing on enzymes of phenyl propanoid metabolism in solanum tuberosum disks. Phytochemistry 12, 1575-1580 (1973).

68 - TANAKA, Y., KOJIMA, M. and URITANI, I. Properties, development and cellular localization of CAH in cut-injured sweet potato. Plant and Cell Physiol. 15, 843-854 (1974).

69 - HILL, A.C. and RHODES, M.J.C. The properties of CAH OF aged swede root disks. Phytochemistry 14, 2387-2391 (1975).

70 - SHANON, L.M., URITANI, I. and IMASEKI, H. *De novo* synthesis of peroxydase isozymes in sweet potato slices. Plant Physiol. 47, 493-498 (1971).

71 - CASSAGNE, Personal communication.

Lipid Metabolism in Aging Plant Storage Tissues

Paul Mazliak and Jean-Claude Kader

During the hours following the slicing of plant storage organs, many metabolic pathways are reactivated in the tissues artificially exposed to a medium much richer in oxygen than the internal atmosphere of the initial storage organs. The general increase in metabolic activity also affects lipid metabolism. E.g., biosynthesis of lipids has been observed in many laboratories in experimental systems consisting of slices of potato tubers maintained for several hours in an aerated aqueous medium ("aging system" of potato slices). These newly formed lipids are principally membrane lipids and this chapter is an attempt to summarize the facts concerning lipid synthesis in relation to the formation of new membranes or organelles in the cells of aging plant storage tissue slices. Many catabolic pathways involving lipid metabolism are also activated after slicing ; this aspect of lipid metabolism will not be dealt with in this chapter, since this is mainly the subject of T. Galliard's contribution to this volume. The particular degradation of fatty acids in slices of plant storage organs will be covered in G. Laties' contribution.

I BIOSYNTHESIS OF MEMBRANE LIPIDS DURING AGING.

The lipids of plant cell organelles have been analyzed recently in a great variety of plants or tissues (for a review, see [1]). Lipid analyses have very rarely been performed upon the different organelles present in the same plant tissue. This has been done in our laboratory with some organelles of potato tuber parenchyma and we shall first describe briefly the lipid composition of these organelles, either before or after aging of the

Biochemistry of Wounded Plant Tissues

slices from which they are extracted.

1) lipid composition of cellular membranes. Lipids are minor components of potato tubers, forming about 0.6% of the dry weight of mature tubers. Some data on the lipid composition of entire potato tubers, both immature and mature, are given in Table 1. It can be seen that the molecular classes of the lipids which are encountered in these organs are typical of all non-green plant organs : these are *phospholipids, galactolipids, glycerides, sterols* and *neutral lipids* [2-7]. The main phospholipids of the tissues are *phosphatidylcholine* and *phosphatidylethanolamine* ; the main fatty acids are *palmitic, stearic, oleic, linoleic* and *linolenic* acids. A remarkable stability in lipid composition must be noted throughout the maturation and storage of the tubers [8-11] ; no major differences between the different varieties of potato have been found. Analyzing 23 different varieties of potato, Galliard [6] has found that polyunsaturated fatty acids (*linoleic* and *linolenic*) form respectively 51-60% and 13-24% of the total fatty acid weight in all the varieties.

Ageing of potato slices provoked no qualitative changes in the lipid composition of the tissue [18]. There is, however, an increase in total lipid when "ageing" of potato slices takes place at room temperature (22°C).

Potato tubers are conveniently used as a starting material for preparing purified plant cell organelles. With this material, one can easily obtain kilogram amounts of homogenous plant tissue. The white parenchyma of peeled mature tubers is devoid of true chloroplasts and one removes most of the amyloplasts and proplasts of the tissue by preliminary low speed centrifugation of the homogenate. Thus, potato tuber homogenates have been used extensively to prepare isolated plasmalemma [12], microsomes [13], mitochondria [14] and peroxysomes [15]. These subcellular fractions have been obtained by a combination of differential centrifugation and gradient density centrifugation (purified plasmalemma have been obtained by flottation through a discontinuous sucrose density gradient [12]). The purity of all the fractions has been checked by electron microscopy, measurements of enzyme marker activities and assays of respiratory activities for isolated mitochondria. The inner and outer membranes of potato mitochondria have been separated by treatment of the intact organelle with digitonin [17]. Some difficulties had to be overcome. For instance, artificial degradation

Table 1. Lipid composition of entire potato tubers (central parenchyma)

Molecular classes	Majestic variety [4]	Bintje variety [5]
	% weight of total lipid	
Neutral lipid Triglyceride Cerebroside	25	20.6
Steryl lipid	12	9.1
Total galactolipid	28	22.9
MGDG	—	5.7
DGDG	—	14.2
SL	—	1.3
Total phospholipid	35	47.4
	% lipid phosphorus	
PC	50	55
PE	35	27.5
PI	7	11.8
PG	3	1.7
PS	2	1.2
PGP	2	1.1
PA	traces	0.7

Fatty acids	Bintje variety [9]		Arran Pilot variety [8]	
	fresh tubers	after six month storage at 10°C	75 days after planting	117 days after planting
	% total fatty acid weight			
<C16:0	4	1.2		—
C16:0	11.3	14.8	15.1	15.4
C16:1	traces	traces	—	—
C18:0	5	2	3.7	4.3
C18:1	3.1	0.9	4.9	6.5
C18:2	53.4	54.5	59.9	62.9
C18:3	23.2	26.6	16.4	10.9

of membrane lipids may occur during grinding and homogenization of the tissues because of the release of endogenous phospholipases or other acylhydrolases. Also, the physiological activities of isolated organelles may be impaired by the free fatty acids appearing in the medium as a result of lipid degradation. T. Galliard [16] has reviewed the procedures to minimise the effects of lipid degradation (in particular he has pointed out the advantages of using homogenization media enriched with bovine serum albumin). The composition of the membrane lipids of various organelles or membranes prepared from fresh potato tubers (Bintje Variety) are compared in Table 2. *Phospholipids* form about 40 to 60% of the total lipid in each subcellular fraction. *Glycolipids* (mainly galactolipids and sterylglucosides) have been found in plasmalemma, microsomes and peroxysomes. When comparing the lipid composition of different membranes, one observes an overall striking *uniformity* in the phospholipid composition of the membranes. The same phospholipids are present in all membranes, except for *diphosphatidylglycerol* which is clearly concentrated in the inner membranes of mitochondria. *Phosphatidylcholine* and *phosphatidylethanolamine* are the most abundant of the membrane phospholipids, together constituting more than 60% of the total in all membranes. There are, however, some variations from one membrane to another. The two mitochondrial membranes, for instance, do not have the same quantitative phospholipid composition, the outer membrane containing about twice as much phosphatidylcholine as phosphatidylethanolamine whilst the inner membrane has almost equal quantities of these lipids. Nevertheless, in spite of these minor variations, the overwhelming impression is one of great uniformity of lipid composition amongst the different cell membranes. This impression is reinforced by the comparison of the fatty acid composition of the various molecular classes in the different membranes (Table 2). Clearly, phosphatidylcholine has the same acyl residues (*linoleic acid* being the major one) in all membranes. The same is true for *phosphatidylethanolamine*. Again, in the case of *phosphatidylinositol* the fatty acids present are the same in all membranes and *palmitic acid* is clearly predominant. In short, detailed analyses of the membrane lipids of various subcellular fractions from potato parenchyma have shown that identical phospholipids, having the same fatty acid distribution, are present in all membranes. This result will be explained, in the second part of this chapter, by exchange processes

Table 2. Lipid composition of various potato organelles or membranes.

Molecular classes	Plasmalemma	Mitochondria: Whole organelles	Mitochondria: Inner Membrane	Mitochondria: Outer Membrane	Microsomes	Peroxysomes
	[12]	[19]	[17]	[17]	[19]	[15]
	% total lipid weight					
Neutral lipids	32	35.1	—	—	36.5	22
Glycolipids	30	—	—	—	16.5	10
Phospholipids	38	62.5	—	—	47	68
	% lipid phosphorus					
PC	32	43++	27	52.6	62	24
PE	46	30	29	25.0	23	60
PI	19	7	24.5	10.3	10	16
PGP	3	8	19.5	12.1	1	—
PS	—	3	—	—	1	—
PA	traces	5	—	—	2	—

Fatty acids	Plasmalemma			Mitochondria: Whole organelles			Mitochondria: Inner Membrane			Mitochondria: Outer Membrane			Microsomes			Peroxysomes		
	PC	PE	PI	PC	PE	PI	PC	PE	PGP	PC	PE	PI	PC	PE	PI	PC	PE	PI
	% total fatty acid weight																	
C16:0	23	13	31	20.3	21.5	55.2	18.5	15.5	12.6	26.4	14.7	56.5	32.2	29.2	58.6	17	9	49
C18:0	16	5	9	5.6	9.5	8.5	3.8	5	4.1	4.3	3.0	3.1	10.7	7.5	8.3	4	3	4
C18:1	6	17	11	8.7	6.2	4.2	9.6	11	17.5	1.5	2.1	2.6	5.5	2.4	3.3	11	3	8
C18:2	41	57	39	50.3	52.5	25.7	57.6	62	50.5	56.6	65	32.6	46.6	49.4	26.7	56	70	30
C18:3	10	8	9	6.5	3.4	1.5	10.5	6.5	15.3	11.2	15.2	5.2	5.0	9.1	2.5	11	14	8

++Compare with the results of another laboratory [21] : PC 43%, PE 39%, lyso PC 3%, unknown 15

between membranes and organelles within potato cells.

The membrane lipid composition of the various organelles does not change qualitatively during the 24 hours of aging of potato slices. Quantitatively, available data (table 3 and [20]) indicate that some increase in the total amount of membrane lipid per 100 g of fresh weight is regularly observed after 24 hours period of aging. However, the weight ratio of lipid to protein is unchanged in all cellular fractions in the case of the potato. In contrast, during "aging" of sweet potato slices, Asahi [20] has observed the appearance of new mitochondrial membranes, with a higher density than that of fresh mitochondrial membranes ; these new membranes are characterized by a low phospholipid to protein ratio. (The physiological properties of these newly-formed mitochondrial membranes are presented in T. Asahi's contribution to this book). Nevertheless, in potato as well as in sweet potato (and all other storage plant tissues examined so far), the aging period is accompanied by a net increase in membrane lipid in the cell, which means that active lipid biosynthesis (necessary for the biogenesis of new membranes) is occurring after slicing.

2) <u>lipid biosynthesis</u>. Labelled precursors employed to follow the lipid biosynthesis in plant storage tissues have been administered most frequently by incubating tissue slices, for different times, in an aqueous solution of the precursor. We call this kind of experiment, an "*in vivo*" experiment. In using this expression, we should point out that each subcellular organelle remains in its natural environment during the incubation period and that all the interactions which do occur naturally between the organelles within plant cells, can still occur in this kind of experiment. On the contrary, we call "*in vitro*" experiments, those experiments where isolated sub-cellular organelles are plunged directly into a solution of labelled precursor in conditions very different from those which they encounter naturally within plant cells.

a) <u>"*in vivo*" experiments</u>.

Fresh potato slices, incubated in aqueous solution of ^{14}C-acetate very actively incorporate, the labelled precursor into all molecular species of lipids [5, 22, 23]. The possible participation of bacterial contamination in the observed metabolic activities is ruled out by the systematic addition of chloramphenicol to the incubation medium.

Table 3. Membrane lipid content of various cellular fractions from fresh or 24 hour aged potato slices [19].

Cellular fractions	mg protein per 100 g fresh weight		mg phospholipid per 100 g fresh weight		$\frac{\mu g \text{ phospholipid}}{\text{mg protein}}$	
	fresh slices	aged slices	fresh slices	aged slices	fresh slices	aged slices
Purified mitochondria	3.5	6.4	0.51	0.97	145	150
Microsomes	38	52.8	11.5	14.2	303	271
1.500 g pellet	5.1	13.7	0.35	0.77	68	56
100.000 g supernatant	254	294	1.2	1.3	4	4

^{14}C-acetate incorporation increases regularly with time for the first ten hours : at this stage, about 25% of the administered radioactivity is incorporated into the lipids (only 5%, or less, of the ^{14}C is released as $^{14}CO_2$ and the remaining part (70%) is incorporated into water-soluble constituents.) After 10 hours, the radioactivity in the lipids reaches a plateau, all the labelled precursor being exhausted. The labelled acetate is incorporated mainly into *phospholipids* (46.4%) and *neutral lipids* (44%). Glycolipids are weakly labelled (11%), even after a 10 h incubation. The most heavily labelled phospholipids in fresh tissue slices are *phosphatidlycholine* and *phosphatidylglycerol*. Saturated (*palmitic, stearic*) and monounsaturated (*oleic*) acids are heavily labelled, even after short incubation times. Polyunsaturated fatty acids (*linoleic, linolenic* acids) are not labelled (or very little) even after long incubations (fig. 1 A). This incorporation of acetate into the tissue lipids may be explained by their physiological "turn-over". The "half-lives" of membrane lipids from potato slices have been calculated [24] by measuring over several days, the decrease in specific radioactivities of these lipids following an incubation of tissue slices in ^{14}C-acetate solutions and subsequent transfer of the slices into a medium without any radioactive precursor. The phospholipids show various turn over rates in the different cellular fractions. The measured "half-lives" vary from 3 to 32 days, very similar to the data obtained with animal cells [25]. This renewal of membrane lipids *in vivo* can be observed in each sub-cellular fraction. Extracting the organelles from tissue slices previously incubated for only one hour in a ^{14}C-acetate solution and analyzing the lipids of these organelles shows that all are labelled and roughly in the same proportions as in whole tissue (table 4). The labelling of membrane lipids in fresh storage plant tissues can be obtained not only from ^{14}C-acetate, but also from ^{32}P-phosphate, ^{14}C-glycerol and [^{14}C-Me]-choline (table 5). So, all parts of the complex lipid molecules are susceptible to renewal.

If labelling experiments are done with plant storage tissue slices previously aged for 24 hours, dramatic changes can be observed in the metabolism of membrane lipids. For instance, with aged potato slices a general increase in biosynthesis of all lipid molecular species can be observed (table 5). It must be pointed out that this general increase does

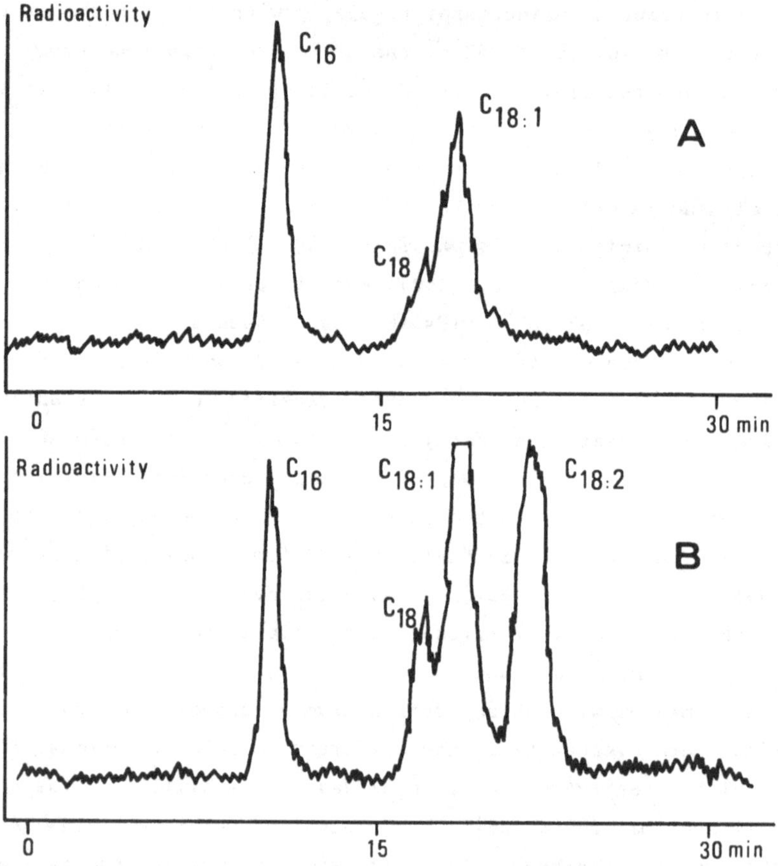

Fig. 1 - Fatty acid biosynthesis in fresh (A) or aged (B) potato slices. ^{14}C-acetate was administred, in aqueous solution, to potato slices for 120 mn. After incubation, the lipids were extracted and methyl esters of the fatty acids prepared. These esters were analyzed by radiogas chromatography [5].

Table 4. Labelling, *in vivo*, of the membrane lipids of organelles extracted from potato slices previously incubated for one hour in a ^{14}C-acetate solution recalculated from [19].

% total radioactivity

Lipids	Mitochondria		Microsomes	
	Fresh slices	24 hour aged slices	fresh slices	24 hour aged slices
Neutral lipid	64.7	37.7	54.9	37.8
Total glycolipid	3.2	7.3	2.6	4.2
Total phospholipid	32.0	55.0	42.5	58.0
PC	13.1	24.0	16.4	22.2
PG	8.6	7.3	14.4	15.3
PE	4.2	14.7	4.9	10.7
Fatty acids	% total radioactivity			
C16:0	32	29.7	26.1	18.7
C18:0	10.5	4.6	8.1	14.0
C18:1	57.0	35.6	63.0	33.8
C18:2	——	24.1	——	27.2
C20:0	——	6.0	4.6	——

Table 5. Incorporation of labelled precursors into the lipids of fresh or 24 hour aged potato slices, after a 10 hour incubation of the tissues in precursor solution [19].

Precursors	precursors utilized for lipid synthesis (as % of precursor up take)		
	Fresh slices	24 hour aged slices	utilized in aged slices / utilized in fresh slices
1-^{14}C-acetate	1.9	12	6.3
^{32}P-phosphate	2.1	6.3	3
1-^{14}C-glycerol	0.4	1.5	3.75
^{14}C-Me choline	2.0	3.2	1.6

not result simply from the increased permeability of the membranes of the tissues to labelled precursors. One can see, in table 5, that this increase is effectively observed when expressing the intensities of lipid synthesis as a percentage of the absorbed precursor molecules utilized for lipogenesis. From two to six times more precursor molecules are utilized for lipid synthesis after aging compared to the amount used by fresh tissue.

In potato slices this increase principally concerns the phospholipids of the tissue. Up to 70% of the newly formed lipids are represented by this class of lipid (compared to 46-51% before aging of the slices) [5, 19, 26]. On the other hand aged disks of apple tissue incorporate the greater part of the labelled precursors into neutral lipids and free fatty acids [27].

In potato slices, the more interesting changes observed after aging concern fatty acid metabolism. As early as 1966, it was shown that a several -fold increase in fatty acid synthetase activity occurs during aging in potato slices [22]. This is concommitant with a change in the pattern of the fatty acids synthesized. A marked synthesis of linoleic and C_{20} fatty acids from ^{14}C-acetate could be observed after aging although these acids were not formed in fresh tissue ([22] and fig. 1). Moreover, experiments with inhibitors of protein synthesis (actinomycin D, puromycin, cycloheximide) showed that the increase in and the change in fatty acid synthetase activity, observed during the aging of potato disks, were accompanied by and related to a temporary rise in the rate of protein and RNA synthesis. Those antibiotics which specifically inhibit protein synthesis also inhibited the rise in fatty acid synthetase activity [29] as well as the specific formation of linoleic acid [5, 18, 19].

Recently, the biosynthesis of saturated fatty acids has also been studied in aged potato slices [40]. Formation of very long chain fatty acids (esterified in waxes) was inhibited by the presence of fluoride or by high incubation temperatures. Arsenite caused an increase in the percentage incorporation of radioactivity from acetate-^{14}C into palmitic acid, apparently by inhibiting further elongation. These results indicate that aged potato tissue contains at least three enzymes responsible for saturated fatty acid synthesis : fatty acid synthetase forming palmityl-ACP, palmityl ACP elongase and a second acyl-elongase of unknown nature.

The changes observed in lipid metabolism during aging do not reflect simply the rise in general metabolic activity in the tissue. The qualitative changes, particularly, the high levels of phospholipid and polyunsaturated fatty acid biosynthesis, can be explained by the formation of new membranes in the 24-hour-aged cells. Studies by electron microscopy have shown the development during aging of a dense rough and smooth endoplasmic reticulum within the potato cell as well as numerous dictyosomes of the golgi apparatus [19]. Phospholipids and linoleic acid are the major components of these new membranes and in some unknown manner, the aging of potato slices triggers the biosynthesis of these lipids which are necessary for membrane biogenesis.

So, as expected from the preceding observations, all the lipid changes observed in aged potato slices are found also in the different cell organelles prepared from such slices. No significant changes in the phospholipid composition of microsomes or mitochondria can be detected, after aging of the slices even though the absolute amount of lipid (and protein) found in each sub-cellular fraction has increased (table 3). After aging the labelling of the membrane lipids of the organelles with ^{14}C-acetate results (table 4) in an increase in labelling of the phospholipids at the expense of neutral lipids and a remarkable accumulation of the newly synthesized linoleic acid in microsomal and mitochondrial membranes is observed (table 4). The same labelling pattern is observed in the lipids of plasma membranes [30] and peroxysomes [15] prepared from aged potato slices.

The biogenesis of the various membranes (or organelles) dœs not occur simultaneously. At any stage of the aging process, the amount of ^{14}C-acetate incorporated into the microsomal lipids is far greater than the amount incorporated into the lipids of any other membrane, mitochondrial, plasmalemmal or peroxisomal [15, 19, 30]. This would suggest that microsomal membranes are formed first. Castelfranco et al. [31] have reached similar conclusions from lipid labelling experiments "*in vivo*". They too incubated potato tuber slices with [^{14}C-Me] choline and found that labelled, membrane-bound phosphatidylcholine was formed. If potato slices were aged for 0 to 24 hours before exposure to radioactive choline, there was a marked increase in the labelling of the lipid with time of aging and the distribution of the radioactive phosphatidylcholine changed among the various sub-cellular fractions of the tissue. The authors observed a shift in the kind of sub-cellular fractions containing ^{14}C-PC ; fresh slices incorporated most of the radioactivity into the microsomes ; slices aged for 9 hours incorporated most of the label into a fraction consisting of single membrane-bound cisternae (presumably dictyosomal fragments) ; slices aged for 24 hours before incubation with radioactive choline incorporated the greater portion of the label into this same fraction but a significant portion of the label was found in the heavier, mitochondria-containing fraction. Such "*in vivo*" experimental results lead to the idea of lipid transfer from microsomes to other cellular membranes.

The pattern of the newly synthesized fatty acids, in aged tissues can be modified by altering the experimental conditions of the aging process. Aging potato slices under a N_2 atmosphere suppresses the biosynthesis of linoleic acid [18], whereas pure O_2 atmosphere stimulates the biosynthesis of hexadecenoic acid [18]. Finally, aging potato slices in a hypertonic medium (0.40 M mannitol) reduces the biosynthesis of linoleic acid, the proportion of which is greatly reduced in all membranes (table 6).

b) "*in vitro*" experiments.

In these experiments, sub-cellular organelles (mitochondria, microsomes, peroxisomes, plasmalemma) were isolated from plant storage tissues (potato tuber parenchyma, apple fruit parenchyma, root tissues, etc...) and these isolated organelles were given labelled precursors to determine what fatty acids and phospholipids they were capable of synthesizing. The following results were obtained :

1 - Isolated mitochondria, prepared either from fresh or aged tissues, are only capable of synthesizing saturated and mono-unsaturated fatty acids, despite the fact that poly-unsaturated fatty acids are the main fatty acids present in mitochondrial membranes ([32] and fig. 2).

2 - Isolated mitochondria, prepared either from fresh or aged tissues, are only capable of synthesizing acidic phospholipids such as *phosphatidic* acid, *phosphatidylglycerol*, *diphosphatidylglycerol* and *phosphatidylinositol*, despite the fact that the main phospholipids of mitochondrial membranes are phosphatidylcholine and phosphatidylethanolamine [33, 19].

3 - Isolated peroxisomes [15], and plasmalemma [30] from potato slices, either fresh or 24 hour aged, appear to be devoid of any lipid biosynthetic capacity.

Table 6. Incorporation of ^{14}C-acetate into the fatty acids of various subcellular fractions from 24 hour aged potato slices. Aging was restaured either in water or hypertonic (0.40M) mannitol solutions [30].

% total fatty acid radioactivity

Fatty acids	Microsomes		Mitochondria		Plasmalemma	
	OM	0.40 M	OM	0.40 M	OM	0.40 M
C16:0	18	22	21	23	26	24
C18:0	5	18	4	18	traces	14
C18:1	41	28	37	25	43	34
C18:2	36	15	38	22	31	14
C20:0	—	16	—	12	—	14
Total saturated	23	56	25	53	26	52

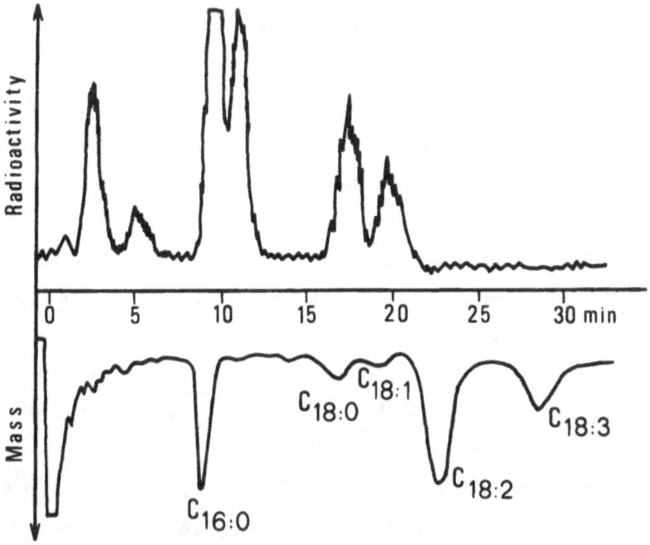

Fig. 2 - Radiogas chromatogram of potato mitochondrial fatty acid methyl esters labelled from ^{14}C-acetate *in vitro*. Isolated potato mitochondria, prepared either from fresh or aged potato slices, were incubated for one hour in a aqueous ^{14}C-acetate. The methyl esters of the mitochondrial fatty acids were then prepared. The mass trace (lower) shows that linoleic ($C_{18:2}$) and linolenic acids ($C_{18:3}$) are major components ; the radioactivity trace (upper) shows that only saturated and monounsaturated fatty acids are labelled.

4 - The fatty acid synthetase, producing mainly *palmitic acid* from acetyl ACP and malonyl ACP, is located in the cytoplasmic supernatant of potato tuber parenchyma [34]. This fatty acid synthetase is stimulated several-fold by aging but it has not been determined whether this stimulation corresponds to an increase in the enzymatic protein level or an increase in the activity of the synthetase.

5 - Finally, isolated microsomes contain many enzymes involved in lipid metabolism : CDP-choline and ethanolamine phosphotransferases responsible for the biosynthesis of phosphatidylcholine and phosphatidylethanolamine, and oleyl-CoA desaturase, producing linoleic acid [35]. Very important differences may be observed in microsomal synthetic activities, depending on whether the microsomes are prepared from fresh or aged potato slices. An active *oleyl-CoA desaturase* is associated with the microsomal fraction from aged slices of potato tuber, only [36]. When [1-^{14}C] oleyl-CoA is used as substrate in incubations with aged potato microsomes more than 40% of the labelled fatty acids is linoleic acid. (The presence of serum albumin in the reaction medium was essential for the desaturation reaction and for preventing the degradation of poly-unsaturated fatty acids). The pH optimum for oleyl-CoA desaturation was 7.0 - 7.2, the temperature optimum was 25°C. The apparent K_m was 25 µM. The newly formed linoleic acid was mainly incorporated into phosphatidylcholine. Oxygen and pyridine nucleotide were absolutely required for oleyl-CoA desaturation (NADH or NADPH were found to be nearly equally effective in the reaction). Although NADPH was more effective than NADP, NAD was found as effective as NADH for desaturation, suggesting that a NADH-regenerating system probably exists in the microsomal suspension.

As it is well known that plants grown at low temperatures are generally enriched in poly-unsaturated fatty acids, it was of interest to follow systematically the effects of varying oxygen concentrations and temperature on oleyl-CoA desaturation *in vitro*. The highest activity was observed at 30°C and at an oxygen concentration of about 20% [37]. So, the microsomal system of oleyl-CoA desaturase (implying electron transfer from NADH to molecular oxygen) dœs not present any special behaviour in response to oxygen or temperature changes. The classical hypothesis of Harris and James [38], attributing the rise in the unsaturation of fatty acids at low temperature to the increase in oxygen dissolved in cellular fluids, dœs not seem to be valid, at least within the potato tuber cell. The microsomal oleyl-CoA desaturase from aged potato slices is very susceptible to the presence of detergents [38] or heavy metals (Pb, Hg, Cd) [39] in the reaction medium. Both detergents and heavy metals are inhibitory at high concentrations, but very low concentrations of some detergents (e.g. cholate) or Cd provoked a stimulation of oleyl-CoA desaturase activity.

The oleyl-CoA desaturase activity, absent in the freshly cut slices, was *induced* during the aging of slices. A linear increase with the time of aging was found in many microsomal activities parallel to oleyl-CoA desaturase, particularly NADH-cytochrome c reductase and protein synthesis [36, 19]. The inhibition of all these activities by cycloheximide, or the partial inhibition by actinomycin D indicated that derepression of the genetic activity and synthesis of new proteins were involved in the induction of the oleyl-coenzyme A desaturase. With whole potato tubers, the same induction could be obtained, *in vivo*, without any aging treatment, by vapours of Rindite (a gaseous mixture of glycol monochlorhydrin, trichloroethane, and carbon tetrachloride [18]).

To summarize briefly the major effects of "aging" on lipid metabolism in plant storage tissues, one could say that an oleyl-CoA desaturase system is induced in the microsomes of the cell, following the appearance of new proteins. This induced enzymatic system produces large amounts of linoleic acid, which is immediately incorporated into microsomal phosphatidylcholine. These highly unsaturated phospholipids are rapidly found not only in microsomes, but also in all other sub-cellular membranes or organelles (plasmalemma, mitochondria, peroxisomes). Since these last mentioned organelles, when isolated even from aged tissues, have very limited lipid synthesizing capacities, (they are unable to synthesize phosphatidylcholine or linoleic acid) one is forced to the conclusion that in aged tissues, there must be some transfer of lipids, from microsomes to other intracellular organelles, in order to explain the presence of the same major phospholipids and poly-unsaturated fatty acids in all the membranes of the cell. The second part of this article will be devoted to the study of this lipid transfer phenomenon.

II EXCHANGE OF LIPIDS BETWEEN MEMBRANES.

The existence of lipid exchange between membranes in the potato cell was demonstrated by *in vivo* and *in vitro* experiments.

1) *In vivo* experiments.

Aged potato tuber slices were incubated for 30 to 120 minutes in a solution containing [1-^{14}C] acetate. After incubation, the different subcellular fractions (microsomes, mitochondria and cytoplasmic supernatant) were isolated and their lipids were extracted. The highest radioactivity was found associated with the microsomal fractions, from the beginning of the experiments.

Pulse-chase experiments were performed in aged tuber slices with [Me-^{14}C] choline or [1-^{14}C] acetate as precursors. In the first experiment, the slices were incubated for 30 min, with labelled choline, then transferred to a non-radioactive medium. As formerly, the microsomal phospholipids were found to be the most heavily labelled in the first minutes of the experiment. Then the specific radioactivity of the phospholipids of the microsomal fraction gradually decreased while an increase was noted in the mitochondrial fraction. The sum of the total radioactivity of the lipids of the two fractions remained constant during the chase.

A similar result was obtained when [1-^{14}C]-acetate was used as tracer, however, the decrease in the radioactivity of microsomal phospholipids was higher [19].

When the fatty acids from the slices incubated with ^{14}C-acetate, were analysed a net decrease was observed in the radioactivity of both total fatty acids and of linoleic acid of the microsomal fraction, whilst a concommitant increase was found in the mitochondrial fatty acids.

These pulse-chase experiments, showing a redistribution of the radioactivity of the lipids between microsomal and mitochondrial fractions, suggested that an exchange of lipids occurs between these two fractions. The transfer of labelled lipids, newly synthesized in the microsomes during the pulse, from the microsomal fraction to mitochondria during the chase, explains the increase in the amount of label recovered in mitochondrial lipids, while the reciprocal transfer of unlabelled lipids from mitochondria to the microsomal fraction causes a decrease in the radioactivity of the microsomal lipids.

Other arguments in favour of the existence of lipid exchange *in vivo* may be given by the study of the effect of cycloheximide on the induction of the lipid biogenesis during the aging process. As explained in the first part of this review, labelling experiments showed that lipids are newly synthesized and found to be associated with microsomal and mitochondrial fractions of slices aged for various times. The introduction into the medium of cycloheximide (actidione) (which specifically inhibits the protein synthesis in the endoplasmic reticulum but not in the mitochondria) suppressed the synthesis of new lipids not only in the microsomal fraction but also in the mitochondria [19]. Again, the hypothesis of lipid transfer explains this dependence of mitochondria on the microsomal fraction.

2) *In vitro* experiments.

a) Properties of the exchange process.

A direct demonstration of the existence of lipid exchange was given by the *in vitro* approach. In this method carefully isolated and purified sub-cellular fractions were incubated together *in vitro*. Organelles containing labelled lipids were incubated with unlabelled fraction and exchange medium (buffer or cytoplasmic supernatant). After incubation, the different fractions were separated by centrifugation and the radioactivities of the lipids in each fraction were determined. The appearance of radioactive lipids in the previously unlabelled fraction indicated the transfer of labelled lipids to this fraction.

In a typical experiment microsomal fractions containing lipids labelled from [1-^{14}C] acetate were mixed with unlabelled mitochondria in the presence of unlabelled cytoplasmic supernatant. After incubation for 30 to 120 min, the organelles were separated by density - gradient centrifugation and the specific radioactivities of the phospholipids were determined. The decrease in the specific radioactivity of the microsomal phospholipids and the appearance of labelled phospholipids in mitochondria strongly suggested an exchange of phospholipids, since no significant change in the concentration of the membrane lipids occurred during the incubation (fig.3).

The possibility of cross-contamination of one fraction by another can be minimized in the following ways. Firstly, it is necessary to use density-gradient centrifugation or to wash many times the re-isolated fractions. Secondly, it is possible to determine the maximum amount of cross-contami-

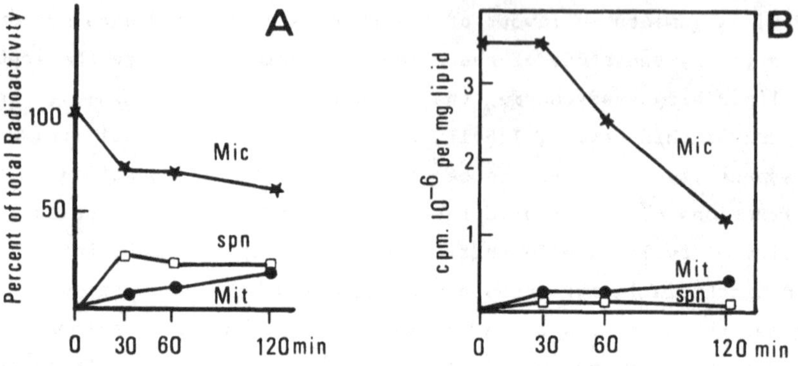

Fig. 3 - Lipid exchange between unlabelled mitochondrial (Mit) and microsomal (Mic) lipids labelled from [$1-{}^{14}C$] acetate. Cytoplasmic supernatant (spn) was present in the exchange mixture. After incubation, the radioactivity of the lipids of the re-isolated sub-cellular fractions was determined. The radioactivity was expressed as % of the total radioactivity recovered (A) as the total radioactivity per mg of phospholipid (B) (Abdelkader and Mazliak, [41]).

nation by starting from [^{14}C]-leucine-labelled microsomes and by incubating these microsomes with unlabelled mitochondria under the same conditions. Counting of the ^{14}C label recovered in the washed mitochondrial pellets indicates the % contamination, assuming that no leucine-labelled protein is exchanged between microsomes and mitochondria [42]. The % of the radioactivity of the initial microsomal phospholipids recovered in the mitochondria at the end of the incubation was corrected for cross-contamination and for loss of mitochondria during the incubation. Finally, a corrected figure for the amount of exchange was obtained [43].

Similar experiments were performed using various "donor" (initially labelled) and "acceptor" (initially unlabelled) organelles with various "exchange media" (buffer or cytoplasmic supernatant) [41].

From these experiments, the following properties of the lipid exchange were determined.

a_1 - bidirectional characteristics of the exchange.

The transfer of labelled lipids was shown equally starting from labelled microsomes mixed with unlabelled mitochondria or from labelled mitochondria mixed with non-radioactive microsomes. In fig. 3 the decrease in the specific radioactivity of the microsomal lipids is explained by the transfer of unlabelled lipids from mitochondria to microsomes.

a_2 - effect of time and temperature on the exchange.

The transfer of phospholipids from [^{32}P] labelled microsomes to mitochondria increases with the time of incubation and the temperature. However, appreciable transfer was noted at + 4°C [43].

a_3 - importance of the process.

The exchange involves a major part of the lipid pool ; up to 40% of the microsomal lipids are transferred to mitochondria (fig. 3). The amount of lipid exchanged between two membranes is equal to the a/d ratio, where a (in dpm) is the total radioactivity of the lipids of the "acceptor", after incubation with a "donor" fraction and d (in dpm/µg phospholipid) is the specific radioactivity of the lipids of this latter fraction before incubation. a is corrected for loss of lipid during the experiment and for cross-contamination. It is of interest ot note that total depletion of the labelled phospholipid in the initially radioactive fraction was never found. This may reflect the reverse transfer of non-radioactive lipids originating from the acceptor fraction, although it may be suggested that only a part of the lipid pool of the membrane is avalaible for exchange.

a_4 - nature of the lipid exchanged.

The exchange process involves intact phospholipid molecules since this process was studied with various lipid precursors [^{32}P]phosphate, [1-^{14}C]acetate, [1-^{14}C]glycerol. In addition, the [^{14}C]acetate labelling indicates that fatty acids are exchanged more intensively than entire phospholipids [42]. The phospholipids exchanged most actively are those containing nitrogen, i.e., phosphatidylcholine and phosphatidylethanolamine, which are also the most abundant phospholipids in the potato cell, as described earlier. In contrast, diphosphatidylglycerol is not exchanged between mitochondria and microsomes. The exchanged lipids are not metabolized in the membranes, the distribution of the radioactivity among the fatty acids exchanged between mitochondria and microsomes remained constant during the experiment [44].

a_5 - nature of the membranes.

Although the exchange process was essentially demonstrated in the microsome-mitochondria assay, lipid exchange can be easily studied by replacing one of the membranes by liposomes (a sonicated solution of emulsified lipids). Lipid exchange was thus observed between [^{14}C] or [^{32}P]-labelled liposomes and the following unlabelled organelles : mitochondria [47], plasmalemma (Douady and Demandre, unpublished), or peroxisomes (Tchang and Mazliak, unpublished).

a_6 - influence of the cytoplasmic supernatant.

The finding that the cytoplasmic supernatant contained labelled lipids after incubation with labelled microsomes and mitochondria (fig. 3) suggested that the supernatant may play a role in the exchange process. As confirmation, labelled supernatant was incubated with non radioactive mitochondria or microsomes (table 7).

Labelled lipids were recovered from the organelles after incubation. This indicates that the lipids associated with the supernatant are exchangeable with sub-cellular membranes. A more direct demonstration was given by the increase in the intensity of the lipid exchange caused by the

Table 7. Exchange of lipids between cytoplasmic supernatant and unlabelled organelles.
10 ml of $[^{32}P]$ phosphate-labelled cytoplasmic supernatant (400 μg of phospholipids) were incubated with 0.5 ml of microsomes (350 μg of phospholipids) or 0.5 ml of mitochondria (400 μg of phospholipids) for 120 min at 25°C.

Acceptor organelle	Time of incubation (min)	Specific radioactivity (dpm/μg phospholipid)		
		supernatant	mitochondria	microsomes
——	0	1300	—	—
mitochondria	120	1200	65	
microsomes	120	911		41

replacement of buffer by cytoplasmic supernatant in an exchange assay. These observations suggested that lipid-protein interactions may be involved in the exchange process and that soluble proteins, present in the cytoplasmic supernatant, may participate to this phenomenon.

b) <u>Isolation and properties of the phospholipid-exchange protein from potato tuber.</u>

Adjustement of the pH of the post-mitochondrial supernatant to pH 5.1, followed by centrifugation, gives a membrane-free supernatant which is almost completely depleted of its lipids, but still contains proteins (technique of Wirtz and Zilversmit [45]). These proteins, precipitated with ammonium sulphate and then dialyzed, were separated by column chromatography using Sephadex (fig. 4).

Two main groups of fractions were obtained, differing by their molecular weight. The protein fractions of group II, when introduced into an exchange assay greatly increased (more than 80%) the amount of $[^{32}P]$ labelled microsomes. The stimulation of the exchange was suppressed by boiling the fractions ; thus these fractions seemed to be protein in nature. These active fractions were named "phospholipid-exchange protein", (PLEP) in accordance with similar proteins discovered in mammalian cells [46]. When increasing amounts of this phospholipid-exchange protein were introduced into the exchange assay, the % of exchange is increased conco-

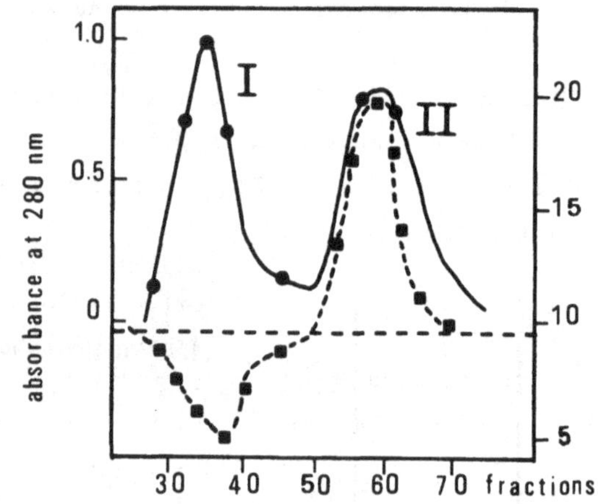

Fig. 4 - Elution of the total protein of the potato cytoplasmic supernatant from a Sephadex G-75 column. 750 mg of protein were applied to a column (4.5 x 65 cm) ; the eluting solution was 0.01 M mercaptoethanol in 0.05 M sodium phosphate buffer (pH 7.5) (flow rate : 48 ml/h). 10 ml fractions were collected ●——● absorbance at 280 nm ; ■——■ % of phospholipid exchange between microsomal fraction and mitochondria when 2 ml of the protein fractions collected from the column were added to the organelle mixture. The horizontal dashed line indicates the % of exchange for the control assay performed with 0.01 M mercaptoethanol in 0.05 M sodium phosphate buffer (pH 7.5) (Kader, [43]).

mitantly .

Assuming that this is globular protein, a molecular weight of 22,000 ± 1000 may be attributed to it. This exchange protein is able to stimulate the exchange of phosphatidlycholine, phosphatidylethanolamine and phosphatidylinositol between microsomes and mitochondria, but without marked specificity for one particular lipid.

The activity of the potato PLEP is difficult to detect in the crude protein solution before the gel filtration step, since lipid acyl hydrolases, very active in the potato cell, are present in the crude protein extract (see Galliard's contribution). The fact that potato lipolytic-acyl hydrolases have a molecular weight of about 100,000 facilitated their separation from the potato phospholipid-exchange protein.

Another difficulty in the study of the potato PLEP is the instability of its activity. Keeping the active fraction at - 80°C or in 50% glycerol at - 20°C only preserves a maximum of 40% of the initial activity after 7 days storage. In spite of this instability further purification was made on Sephadex-purified fractions. Using ion-exchange chromatography, (DEAE cellulose), partially purified active proteins were obtained. However, the gain in purification was not higher than 4.0. Recent experiments indicate that these fractions present a slight specificity to phosphatidylcholine (Douady and Kader, unpublished).

The potato PLEP is also able to stimulate the exchange of phospholipids between mitochondria and liposomes made from [^{32}P] labelled phospholipids (Kader and Douady, unpublished). It is thus possible to control the composition of one of the two membrane fractions involved in the exchange assay. The use of liposomes facilitates the study of the properties of PLEP.

It is of interest to note that progress was made in the study of animal PLEP by the use of liposome-membrane or liposome-liposome exchange assay (see reviews of Dawson [48], Wirtz [46], Kader [49]). The animal phosphatidylcholine exchange protein of beef liver (PC - PLEP) is specific for phosphatidylcholine [50] while the beef brain phosphatidylinositol exchange protein, is specific for phosphatidylinositol and phosphatidylcholine as demonstrated by the experiments of Helmkamp et al. [51]. These exchange proteins are now well characterized [46] [49]. The mechanism of their action begins to be understood, particularly for PC - PLEP [49] [50]; one molecule of phosphatidylcholine is bound to each molecule of PC - PLEP. This molecule can be exchanged with a molecule of membrane-bound phosphatidylcholine. Interaction of membrane with PC - PLEP or PI - PLEP seems to play an essential role in the exchange or transfer process. The mechanism of the action of potato PLEP - which may be similar to that of PC - PLEP - is under investigation.

It remained to determine whether the physiological stage of the potato cell influenced the intensity of the exchange mediated by potato PLEP. The aging process is an interesting way of studying this correlation.

c) Effect of the "aging" treatment on lipid exchange.

The intracellular membranes - mitochondria and endoplasmic reticulum in particular - are transformed during the aging treatment. It was of interest to determine if the lipids of mitochondrial or microsomal fractions from aged slices were more easily exchangeable than those of organelles from fresh tissue.

Microsomal fractions, isolated from aged slices, containing [1-^{14}C] acetate labelled lipids, were incubated respectively with unlabelled mitochondria isolated from freshly-cut slices or aged slices [19]. The radioactivity transferred to aged mitochondria is slightly greater than that transferred to fresh mitochondria. This data suggests that the incorporation of lipids into a mitochondrial membrane is partly controlled by the nature of the membrane by itself. As indicated above, interaction between membrane and lipid-carrying protein, seems essential for the exchange process [49]. In these experiments the microsomal fractions contained labelled linoleic acid, which is not synthesized by isolated mitochondria, (as described earlier). When the fatty acids of mitochondria, incubated with these labelled microsomes, were analyzed by radio-gas chromatography, labelled linoleic acid was found in the mitochondrial membrane. The hypothesis of a transfer of this fatty acid - synthesized by the microsomal oleyl-CoA desaturase - from microsomal to mitochondrial membranes is thus clearly demonstrated.

The aging treatment also influences the activity of the phospholipid exchange protein. This treatment modifies the protein pattern of the cytoplasmic supernatant treated at pH 5.1. The supernatant from aged slices is richer than that of fresh tissue in high-molecular weight proteins (Kader, unpublished). The physiological significance of the increase in high-molecular weight proteins is not known, but a similar phenomenon was found by Stegemann [52] for soluble protein, during the storage of potato tubers. It is of interest to note that the specific activity of the potato PLEP, when expressed as units of PLEP (one unit catalyses the transfer of 1 μg of phospholipid per 15 min at 30°C.) increased during the aging period (Kader, unpublished).

3) Conclusions.

The exchange process, by distributing the lipids synthesized in limited sites to the different membranes of the cell, may result in the homogeneity of lipid composition of the intracellular membranes. This homogeneous composition may be found, even in the membranes unable to achieve a complete lipid biosynthesis. This process involves pure physical interactions between lipids and proteins. This was reinforced by the discovery of PLEP, which is able to carry phospholipids from one membrane to another. The precise nature of these interactions has yet to be elucidated in order to understand how the intensity of the exchange is controlled. The exchange phenomenon probably plays an important role in membrane biogenesis and evidence for a net transfer of phospholipids from one membrane to another, mediated by animal PLEP, reinforces this idea. The formation of new organelles during the aging treatment implies the appearance of newly synthesized lipids in these organelles. The aging potato cell is certainly a suitable model for the study of the regulation of the exchange process.

III GENERAL CONCLUSIONS AND PERSPECTIVES.

The aging treatment is one of the best ways to study the biogenesis of many membrane constituents, and especially membrane lipids. It may be suggested that during the aging process, a derepression of the genetic activity provokes an induction of the enzymes involved in the biosynthesis of lipids, for instance oleyl-CoA-desaturase. The newly formed lipids participate in the biogenesis of new membranes. A flow of lipids from the sites of biosynthesis (i.e. endoplasmic reticulum for linoleic acid) allows the biogenesis of various membrane structures (mitochondria, plasmalemma, etc). The aging treatment, by modifying the intracellular membranes, seems to change the intensity of the lipid exchange mediated by phospholipid-exchange protein. The biogenesis of membrane lipids can be conveniently studied in the biological model provided by the aging potato slices. It may be predicted that many investigations will be carried out in the future in attempts to elucidate the various events related to the regulation of lipid biogenesis in the potato cell after induction by aging.

REFERENCES.

1 MAZLIAK, P. - Glyco- and phospholipids of Biomembranes in higher plants in *Lipids in higher plants* M. TEVINI, editor, Springer-Verlag, Heidelberg (in the press) (1977).

2 LEPAGE, M. - The separation and identification of plant phospholipids and glycolipids by two dimensional thin-layer chromatography. *J. Chromatog.*, *13*, 99-103 (1964).

3 LEPAGE, M. - The lipid components of white potato tubers (*Solanum tuberosum*, L.). *Lipids*, *3*, 477-481 (1968).

4 GALLIARD, T. - Aspects of lipid metabolism in higher plants. I. Identification and quantitative determination of the lipids in potato tubers. *Phytochem.*, *7*, 1907-1914 (1968).

5 BEN ABDELKADER, A. - La lipogenèse dans le tubercule de Pomme de terre (*Solanum tuberosum*, L.) I. Analyse et biosynthèse des lipides dans le parenchyme central. Influence de la "survie" (ageing) de rondelles de parenchyme sur cette biosynthèse. *Physiol. Vég.*, *6*, 417-442 (1968).

6 GALLIARD, T. - Lipids of potato tubers. I. Lipid and fatty acid composition of tubers from different varieties of potato. *J. Sc. Food Agric.*, *24*, 617-622 (1973).

6bis SPENGLER, H. and SCHORMÜLLER, J. - Beitrage zur analyse der polaren lipoids in der Kartoffel. *Z. lebensmittel Untersuchung Forschung*, *149*, 145-153 (1972).

7 FRICKER, A., DUDEN, R. and KOLLER, W.D. - Über die Zusammensetzung einiger Lipid fraktionen von Kartoffeln. *Potato Res.*, *15*, 365-373 (1972).

8 GALLIARD, T. - Fatty acid composition of immature potato tubers. *Phytochem.*, *11*, 1899-1903 (1972).

9 CHERIF, A. et BEN ABDELKADER, A. - Analyse quantitative des acides gras présents dans différentes régions du tubercule de Pomme de terre. Variation au cours de la conservation à 10°C. *Potato Res.*, *13*, 284-295 (1970).

10 CHERIF, A. - Métabolisme des lipides dans le tubercule de Pomme de terre (*Solanum tuberosum*, L.) 1 - Evolution des lipides au cours de la conservation des tubercules. *Potato Res.*, *16*, 126-147 (1973).

11 BOLLING, H. and ELBAYA, A.W. - Changes in lipid composition during the development of the potato plant with regard to the biosynthesis of fatty acids. *Z. Pflanzen physiol., 69*, 402-408 (1973).

12 DEMANDRE, C. - Activités enzymatiques et composition lipidique de fractions enrichies en membranes cytoplasmiques isolées de tubercules de Pomme de terre. *Physiol. Vég., 13*, 103-114 (1975).

13 BEN ABDELKADER, A. - Influence de la survie (ageing) sur la biosynthèse des phospholipides dans les microsomes de tubercules de Pomme de terre. *C. R. Ac. Sc., 268*, 2406-2409 (1969).

14 BEN ABDELKADER, A., MAZLIAK, P. et CATESSON, A.M. - Biogenèse des lipides mitochondriaux au cours de la survie (ageing) de disques de parenchyme de tubercule de Pomme de terre. *Phytochem., 8*, 1121-1133 (1969).

15 TCHANG, F. - Ph. D. Thesis, <u>in preparation</u> (1977).

16 GALLIARD, T. - Techniques for overcoming problems of lipolytic enzymes and lipoxygenases in the preparation of plant organelles, *Methods in enzymology* - Vol. 31 A *Biomembranes*, S. FLEISCHER and L. PACKER editors, p. 520-528.

17 MEUNIER, D. and MAZLIAK, P. - Différence de composition lipidique entre les deux membranes des mitochondries de Pomme de terre. *C. R. Ac. Sc., 275*, 213-216 (1972).

18 CHERIF, A. - Ph. D. Thesis, University of Paris 6. 94 pp. (1972).

19 BEN ABDELKADER, A. - Doctorate thesis, University of Paris 6. 184 pp. (1972).

20 ASAHI, T. - Role of phospholipid in the biogenesis of plant mitochondria in relation ot the development of cyanide-insensitive respiration - *Seminar on Metabolism and Functions of lipids in higher plants*, M. YAMADA and P.K. STUMPF editors, Tokyo, p. 114-119 (1976).

21 YAMAKI, S. and URITANI, A. - The mechanism of chilling injury in sweet potato VI changes of lipid components in the mitochondrial membrane during chilling storage. *Plant and Cell Physiology, 13*, 67-81 (1972).

22 WILLEMOT, C. and STUMPF, P.K. - Fat metabolism in higher plants. XXXIII Development of fatty acid synthetase during the aging of storage tissue slices. *Can J. Bot.*, *45*, 579-584 (1967).

23 CHERIF, A. - Métabolisme des lipides dans le tubercule de Pomme de terre (*Solanum tuberosum*, L.). 2. Evolution des lipides au cours de la "survie" de rondelles de parenchyme de tubercules de Pomme de terre. *Potato Res.*, *16*, 148-158 (1973).

24 BEN ABDELKADER, A. and MAZLIAK, P. - Renouvellement des lipides dans diverses fractions cellulaires de parenchyme de Pomme de terre ou d'inflorescence de Chou-fleur. *Physiologie Végétale*, *9*, 227-240 (1971).

25 MAZLIAK, P., BEN ABDELKADER, A., TREMOLIERES, A. - Le renouvellement des lipides dans les membranes animales et végétales. *Cahiers Nutrition Diététique*, *7*, 51-65 (1972).

26 TANG, W.J. and CASTELFRANCO, P. - Phospholipid synthesis in aging potato tuber tissues. *Plant Physiology*, *43*, 1232-1238 (1968).

27 GALLIARD, T., RHODES, M.J.C., WOOLTORTON, L.S.C. and HULME, A.C. - Metabolic changes in excised fruit tissues. II. The development of a lipid synthesizing system during the ageing of peel disks from preclimacteric apples. *Phytochemistry*, *7*, 1453-1464 (1968).

28 Reference deleted

29 WILLEMOT, C. and STUMPF, P.K. - Fat metabolism in higher plants. XXXIV. Development of fatty acid synthetase as a function of protein synthesis in aging potato tuber slices. *Plant Physiol.*, *42*, 391-397 (1967).

30 DEMANDRE, C. - Ph. D. Thesis, University of Paris 6. 75 pp. (1976).

31 CASTELFRANCO, P.A., TANG, W.J., BOLAR, M.L. - Membrane transformations in aging potato tuber slices. *Plant Physiol.*, *48*, 795-800 (1971).

32 MAZLIAK, P., OURSEL, A., BEN ABDELKADER, A. and GROSBOIS, M. - Biosynthèse des acides gras dans les mitochondries végétales isolées. *Eur. J. Biochem.*, *28*, 399-411 (1972).

33 MAZLIAK, P., STOLL, U. and BEN ABDELKADER, A. - Coopération entre les mitochondries et le reste de la cellule pour le renouvellement des lipides mitochondriaux du parenchyme de Pomme. *Biochim. Biophys. Acta, 152,* 414-417 (1968).

34 HUANG, K.P. and STUMPF, P.K. - Fat metabolism in higher plants. XLIV. Fatty acid synthesis by a soluble fatty acid synthetase from *Solanum tuberosum. Arch. Biochem. Biophys., 193,* 412-427 (1971).

35 MAZLIAK, P. - Lipid metabolism in plants. *Ann. Rev. Plant. Physiol., 24,* 287-310 (1973).

36 BEN ABDELKADER, A., CHERIF, A., DEMANDRE, C. and MAZLIAK, P. - The oleyl-coenzyme A desaturase of potato tubers. *Eur. J. Biochem., 32,* 155-165 (1973).

37 CHERIF, A. and KADER, J.C. - Effect of temperature and oxygen on the activity of microsomal oleyl-CoA desaturase of potato tubers. *Potato Research, 19,* 21-26 (1976).

38 MAZLIAK, P. - Effets de quatre détergents sur l'oleyl-CoA desaturase des microsomes de Pomme de terre. *C. R. Ac. Sc., 281,* 1471-1474 (1975).

39 MAZLIAK, P., DECOTTE, A.M. and KADER, J.C. - Effects of heavy metals (Lead, Cadmium, Mercury) upon oleyl-CoA desaturase. *Chemico-Biological Interactions* (in the press) (1977).

40 BOLTON, P. and HARWOOD, J.L. - Fatty acid synthesis in aged potato slices. *Phytochemistry, 15,* 1501-1506 (1976).

41 BEN ABDELKADER, A. and MAZLIAK, P. - Echanges de lipides entre mitochondries, microsomes et surnageant cytoplasmique de cellules de Pomme de terre ou de Chou-fleur. *Eur. J. Biochem., 15,* 250-262 (1970).

42 WIRTZ, K.W.A. and ZILWERSMIT, D.B. - Exchange of phospholipids between liver mitochondria and microsomes *in vitro. J. Biol. Chem., 243,* 3596-3602 (1968).

43 KADER, J.C. - Protein and the intracellular exchange of lipid. I. The stimulation of phospholipid exchange between mitochondria and microsomal fractions by proteins isolated from potato tuber. *Biochim. Biophys. Acta, 380,* 31-44 (1975).

44 MAZLIAK, P. et BEN ABDELKADER, A. - Echanges d'acides gras *in vitro* entre mitochondries, microsomes et surnageants cytoplasmiques de cellules de Pommes de terre et de Chou-fleur. *Phytochemistry, 10,* 2879-2890 (1971).

45 WIRTZ, K.W.A. and ZILVERSMIT, D.B. - Participation of soluble liver proteins in the exchange of membrane phospholipids. *Biochim. Biophys. Acta, 193,* 105-116 (1969).

46 WIRTZ, K.W.A. - Transfer of phospholipids between membranes. *Biochim. Biophys. Acta, 344,* 95-117 (1974).

47 DOUADY, D. and MAZLIAK, P. - Etude des transferts d'acides gras entre liposomes et mitochondries de Chou-fleur. *Physiol. Vég., 13,* 383-392 (1975).

48 DAWSON, R.M.C. - The exchange of phospholipids between cell membranes. *Sub. Cell Biochem., 2,* 69-89 (1973).

49 KADER, J.C. - Exchange of phospholipids between membranes in *Cell Surface Reviews,* Vol. 3 (G. POSTE and G.L. NICOLSON) (1977).

50 KAMP, H.H., WIRTZ, K.W.A. and VAN DEENEN, L.L.M. - Some properties of phosphatidylcholine exchange proteins purified from beef liver. *Biochem. Biophys. Acta, 318,* 313-325 (1973).

51 HELMKAMP, Jr.G.M., HARVEY, M.S., WIRTZ, K.W.A. and VAN DEENEN, L.L.M.- Phospholipid exchange between membranes. Purification of bovine brain proteins that preferentially catalyze the transfer of phosphatidylcholine. *J. Biol. Chem., 249,* 6382-6389 (1974).

52 STEGEMANN, H. - Properties and physiological changes in storage proteins. in *The chemistry and biochemistry of plant proteins,* J.B. HARBORNE and G.F. VAN SUMERE eds, Academic Press, London p. 71-88 (1975).

ACKNOWLEDGEMENTS.

The authors are much indebted to Dr. T. Galliard (Food Research Institute, Norwich, U.K.) for critical reading of the manuscript. The authors thank Mrs M. Poittevin for typing and illustrating the text.

Lipolytik and Lipoxygenase Enzymes in Plants and their Action in Wounded Tissues

Terence Galliard

INTRODUCTION

The membranes of plant cells and of subcellular organelles are lipoprotein structures in which the lipid components are predominantly phospholipids and galactolipids. It has recently been recognized that the membranes and their lipid components are susceptible to enzymic degradation when the cells of intact tissues are subjected to stress factors, e.g. mechanical damage and microbial infection.

At the Karlsruhe symposium, where this volume was conceived, Laties (1) stated "The slicing of a variety of fleshy roots and storage organs (e.g. potato tubers) leads to an immediate and extensive degradation of membrane phospholipids and galactolipids." Cytological evidence for this extensive damage to membranes on slicing of storage tissue had been provided by the work of Jackman and van Steveninck (2) with beetroot tissue.

Unpublished work in Laties' laboratory (quoted in ref. 3) has shown that slicing of potato tubers results in 20% loss of endogenous membrane lipid within seconds and a 40% loss within minutes of cutting. If it is assumed that maximum loss occurs at, and adjacent to, the cut tissue and that a gradient of effects due to damage exists across a slice, then the extent of lipid degradation close to the area of damage might be expected to be of even greater magnitude. Certainly, the loss of membrane lipid caused by total disruption (homogenization) of potato, even at 0°, is very extensive and virtually instantaneous (4).

Deacylation of membrane phospholipids and galactolipids liberates fatty acids. Free fatty acids are not found in significant amounts in healthy intact plant cells and, in fact, are toxic to many cellular processes. In damaged tissues, the free fatty acids that are released by hydrolytic enzymes may be attacked by oxidative enzymes. Amongst the products of such oxidations may be found fatty acid hydroperoxides, carbonyl compounds and other oxygenated derivatives that are potentially destructive towards membrane structures and enzyme processes. Thus disruption of intact plant cells can lead to sequential degradative processes affecting both structure and function of membranes and enzymes.

The purposes of this contribution to the volume are: (a) to provide information on lipid degrading enzymes so far identified in plants, with particular reference to storage tissues used for studies on ageing and wound healing phenomena, and (b) to attempt to discuss the very limited information available to date on the possible involvement of these enzymes in damage, ageing and wound healing processes.

HYDROLYTIC ENZYMES

The enzymes of importance to this discussion are those that attack glycerolipids with the general formula given in Fig. 1. The two major groups of enzymes to be discussed are those that (a) remove fatty acid residues from glycerolipids, i. e. acyl hydrolases, or (b) attack the phosphodiester bonds of phospholipids, especially phospholipase D.

Acyl Hydrolase Enzymes

In addition to possible functions in metabolic turnover of lipids, acyl hydrolase activity is manifested under conditions in which net loss of acyl lipid occurs. This may be accompanied by a corresponding increase in free fatty acids, although in many instances the further oxidation of the fatty acids may limit their accumulation per se. Some clear examples of the net loss of lipids, indicating acyl hydrolase activity are: the metabolism of seed oils during seed germination (5); loss of membrane (especially chloroplast) lipids during fruit ripening (6) and senescence of leaves (7); the decrease of lipid during storage of edible plant materials (8); loss of endogenous lipid on infection or damage to plants (9).

Evidence for the presence of acyl hydrolase activity in bulky storage organs was provided in 1963 by Dalgarno and Birt (10) who showed that cell-free extracts of carrot roots resulted in the formation of free fatty acids which inhibited the activity of mitochondria in their extracts. Addition of bovine serum albumin reduced this inhibition. (A similar process in homogenates of leaves results in fatty acid inhibition of chloroplasts (11) and mitochondria (12)). An unpublished survey in the author's laboratory of many plant tissues has shown that most contain some enzyme activity which leads to autolysis of endogenous lipids when the tissues are homogenized. Potato tubers were by far the most active in this respect (4) and much of our earlier work on lipid degradation therefore concentrated on this tissue.

Fig. 1. General formulae of the major glycerolipids in plants

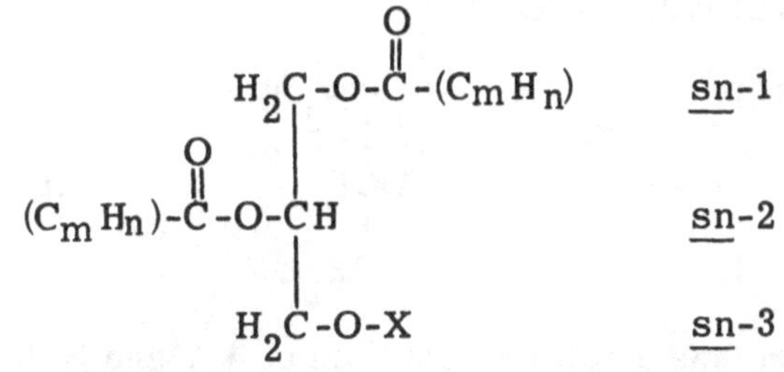

where $C_m H_n$ represents the hydrocarbon chains of fatty acids - usually 16 or 18 carbons with 0-3 cis-olefinic bonds.

In lyso-glycerolipids, the sn-1 or sn-2 position of glycerol is unesterified.

Nature of X	Glycerolipid Class
-C(=O)-(CmHn)	triacylglycerol (triglyceride)
-P(=O)(-O⁻)-O-Y	phosphatidyl-Y (phospholipid)
-galactose(s)	galactosyldiacylglycerol (galactolipid)
-sulphoquinovose	sulphoquinovosyldiacylglycerol (plant sulpholipid)

Although a range of different lipid acyl hydrolase enzymes are known in animal and microbial organisms, the number of these so far recognised in higher plants is limited. The considerable confusion in the literature on the nomenclature of such enzymes has recently been discussed (9). Pertinent details of lipid acyl hydrolases in plants are given below.

True lipases (triacylglycerol acyl hydrolase, EC 3. 3. 3. 3) will attack triacylglycerols containing long chain fatty acyl esters and they are much less active towards water-soluble esters or polar lipids. True lipases may therefore be distinguished from esterases and phospholipid- or galactolipid-acyl hydrolases. Lipases have been characterized in several plants (9) but, naturally, attention has primarily been focussed on germinating seeds where lipases are involved in mobilization of seed oils.

The most well characterized plant lipase, from castor bean, catalyses the sequential, but complete, hydrolysis of triacyl glycerols to free fatty acids and glycerol (13) e. g. :-

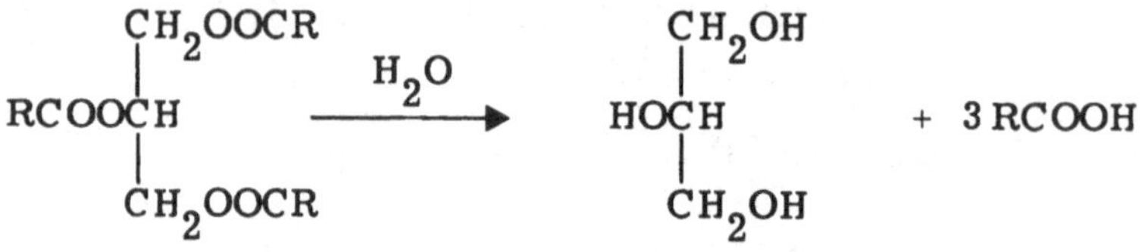

The castor bean lipase has a low pH optimum at 4. 1 and is located in the membranes of the oil storage bodies (spherosomes) (14).

Less attention has been paid to lipases in bulky storage organs which have relatively low levels of storage lipids. For example, the triacylglycerol content of potato tubers is only 15% of the total lipid of this tissue or 0. 2% of the dry weight (15). This contrasts with, e. g., castor beans in which the triacylglycerol accounts for some 60% of the dry weight (5). Our studies with potato have indicated that true lipase activity is extremely low in comparison with the potato lipid acyl hydrolase activity of this tissue (see later). A low but measurable activity towards triolein was shown in potato tubers by Gatfield and Woelk (16). Hasson and Laties (3) have recently isolated a true lipase from potato tubers although the specific activity of the lipase towards triolein was 2-3 orders of magnitude lower than that of a phospholipid acyl hydrolase enzyme which they also isolated from the same tissue. Thus there is evidence for low levels of true lipase activity in bulky storage tissues but the more active acyl hydrolases, discussed below, are probably of more significance in the wounding process.

Polar lipid acyl hydrolases attack the phospholipids and galactolipids which are found predominantly as membrane components. In animals and microorganisms acyl hydrolases acting on phospholipids (phospholipases A) specifically attack one of the fatty acyl ester groups at either the *sn*-1 or *sn*-2 position. However, plants contain lipid acyl hydrolase activities that hydrolyse both acyl groups to liberate free fatty acids and water-soluble glycerol derivatives.

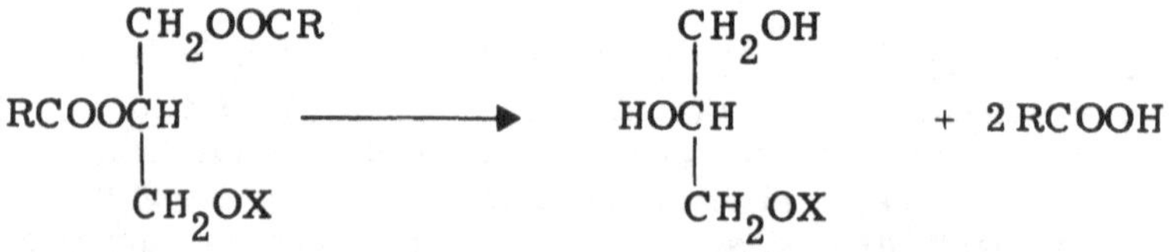

A lipid acyl hydrolase was identified in leaves of *Phaseolus* sp. by Sastry and Kates (17) and subsequently purified (18). This enzyme was called a "galactolipase" (EC 3. 1. 1.26)because, of the substrates tested, mono- and di-galactosyl diglycerides were hydrolysed most readily. Subsequent work (19, 20) has shown that this enzyme in Phaseolus sp. leaves is more active towards lysophospholipids and monoacylglycerols than to galactolipids and appears to be virtually identical to the lipolytic acyl hydrolase of potato tuber.

The demonstration of rapid and extensive hydrolysis of phospholipids and galactolipids resulting from disruption of potato tuber tissue (4) led to the isolation and characterization of a lipolytic acyl hydrolase enzyme (LAH) from this tissue (21). Subsequent studies in the author's laboratory and elsewhere (3, 22-26) have confirmed the high LAH activity in potato tubers. Close examination of the LAH activity of potato in four laboratories has produced some doubt as to whether the activity towards polar lipids is due to one (19, 21, 24) or two (3, 25, 26) enzymes in addition to a neutral lipid acyl hydrolase. However, all laboratories agree on the potency of the polar lipid acyl hydrolase activity. Details of the properties of the enzyme(s) are given in a recent review (9) and in subsequent publications (3, 24-26). For the purpose of the present discussion, the number of enzymes or isoenzymes is less important than the effects produced by such activity.

The action of the LAH of potato towards some natural glycerolipid substrates is shown in Table 1 which presents results with the potato enzyme from two laboratories and (for comparison) with a purified enzyme from Phaseolus multiflorus leaves (i. e. the "galactolipase"). Despite quite different assay conditions, the relative substrate specificities are similar. In particular, all the preparations were active on phospholipids, galactolipids and monoacylglycerols and inactive with triacylglycerols. Also the order of activities generally compares well, i. e. monoacylglycerol lysophosphatidylcholine $>$ monogalactosyldiacylglycerol $>$ digalactosyldiacylglycerol $>$ phospholipids $>$ sulpholipids $\gg$ triacylglycerol.

Thus the enzyme will hydrolyse all the major glycerolipids in plant tissues with the exception of triacylglycerols. Furthermore, the high activity in potato tuber (e. g. 5 - 50 µmol galactolipids hydrolysed/min/g fresh weight of tuber in over 20 commerical cultivars investigated (22)) is sufficient to hydrolyse all the endogenous lipid of potato (potato contains 0. 3 µmoles of galactolipids/g fresh weight of tuber (14)). It should be noted that one European commercial variety of potato (Desirée) contained much lower levels of LAH activity (0. 06 - 0. 2 µmoles/min/g fresh weight (22)) which was genetically determined (27) as shown by breeding experiments (S. Svensson, personal communication). However, even with this low LAH variety, the enzyme activity is still sufficient to cause rapid autolysis of endogenous lipid in homogenized tissue.

Table 1. Substrate specificities of lipid acyl hydrolase activity in potato

Substrate	specific activities (relative to phosphatidylcholine)		
	potato tuber (ref. 14)	potato tuber (ref. 24)	runner bean leaf (ref. 20)
phosphatidylcholine	100	100	100
phosphatidylethanolamine	82	33	-
phosphatidylglycerol	94	151	-
lysophosphatidylcholine	550	304	600
sulphoquinovosyldiacylglycerol	-	52	-
monogalactosyldiacylglycerol	235	105	420
digalactosyldiacylglycerol	134	71	255
monoleyolglycerol	770		680
dioleoylglycerol	160		-
trioleyolglycerol	1	-	-
tristearoylglycerol	-	-	-

The phospholipids and galactolipids are mainly membrane bound. Thus enzymic hydrolysis must occur in a lipoprotein phase. The potato LAH deacylated endogenous phospholipids in homogenates (4) but, when purified enzyme was incubated with aqueous dispersions of phosphatidylcholine or with a mitochondrial preparation from cauliflower, little activity was detected (28). This suggested that crude homogenates contained an activating agent which was subsequently identified as free fatty acid (28). When the purified enzyme was incubated with phospholipid liposomes or mitochondria in the presence of linoleic acid, phospholipid acyl hydrolase activity was restored and electron microscope studies showed that the free fatty acid was causing a disruption of bilayer structures (28), presumably allowing access of enzyme to the susceptible acyl ester bonds. The stimulatory effects of fatty acids and their reversal by bovine serum albumin are demonstrated in Table 2. Hasson and Laties (26) recently showed that their purified phospholipid acyl hydrolase enzyme from potato catalysed extensive hydrolysis of potato mitochondrial lipids.

Table 2. Stimulation of LAH* activity on membrane-bound lipid by linoleic acid

	Hydrolysis of lipid (relative values)	
	cauliflower mitochondria	phosphatidylcholine liposomes
LAH + linoleic acid (1 mM)	100	100
LAH	30	4
LAH + BSA[+] (1%)	12	-
LAH + BSA + linoleic acid	9	3

*LAH = purified lipolytic acyl hydrolase from potato

[+]BSA = bovine serum albumin; (Data from ref. 28)

Because free fatty acids - the products of LAH activity - also stimulate the action of the enzyme on membrane-bound lipids, the process is autocatalytic. However, in vitro studies (21) have demonstrated that the enzyme is inhibited by excess substrate (>1-2 mM). Although the acyl hydrolase activity is the important property for this discussion, it should be pointed out that the affinity of the potato LAH for methanol as acyl acceptor is a factor of 10 higher than for water (19), thus the enzyme should be considered as an acyl transferase. In addition the potato LAH will catalyse reverse hydrolysis in forming wax esters from long chain acids and alcohols (29).

The extremely rapid and extensive hydrolysis of potato lipids resulting from tissue disruption indicates that the LAH enzyme and the membrane bound substrates readily come into contact. In all our studies on subcellular localisation of LAH in potato tubers, the enzyme occurs almost entirely in the particle-free supernatant fraction (4). It seemed unlikely that such a potentially destructive enzyme could be located in the cytosol in intact cells and localisation within a discrete but fragile organelle appeared to be more likely. The high activity of the enzyme in potato tubers and its autolytic properties forced us to look at sub-cellular localisation of the enzyme in tissues with relatively low levels of activity. From studies with brassica florets, potato shoots and pea roots, the LAH activity was always found coincident with other acid hydrolase enzymes in a particulate fraction with the properties of a "lysosome" or vacuole (30, 31). We therefore conclude that the enzyme is located within vacuoles and is released when the cells are disrupted. Using different organs of a potato cultivar with a genetically-determined low LAH activity, we were able to demonstrate that the proportion of LAH activity obtained in a particulate fraction was inversely proportional to the total activity of the enzyme in the homogenate. This is illustrated in Fig. 2 and suggests that the enzyme acts autolytically in degrading organelle membranes during isolation.

Fig. 2. Relationship between total lipolytic acyl hydrolase (LAH) activity and the proportion of that activity in a particulate form in extracts of potato tubers. From Wardale and Galliard (30).

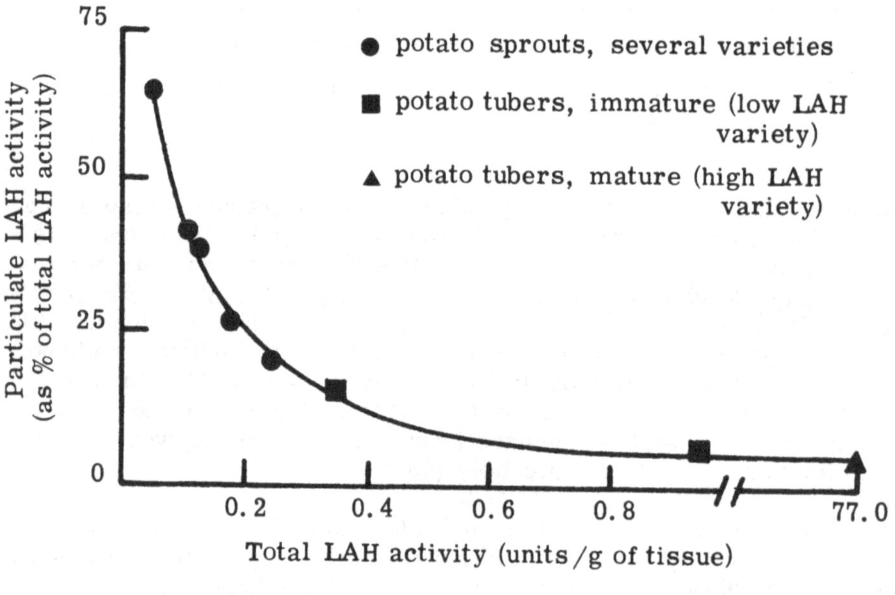

Other acyl hydrolase enzymes. There is little information available on other enzymes attacking acyl ester bonds in plants. There is circumstantial evidence for the presence of phospholipase A (EC 3.1.1.4) which removes one fatty acyl group from phospholipids to produce lysophospholipids. Accumulation of lysophospholipids in preparations of cauliflower mitochondria (32) and the relatively high concentrations of lysophospholipids in some tissues, e.g. starch grains of cereal seeds (33), indicate the presence of phospholipase A activity. Although phospholipase A activity is not detected in crude extracts of potato tuber, a 15,000 g particulate fraction showed some lysophosphatidylcholine formation from phosphatidylcholine in the author's laboratory (unpublished) but Hasson and Laties found no phospholipase A activity in potato mitochondria (26). An enzyme that deacylates sulpholipid has been isolated from *Phaseolus multiflorus* leaves (34) but this activity is probably due to the LAH enzyme described above (20).

Hydrolytic Enzymes Attacking Polar Groups of Glycerolipids

Phospholipase D (EC 3.1.4.4) occurs widely in the plant kingdom and, like the LAH enzyme described above, attacks membrane-bound phospholipids when plant tissues are disrupted. In 1947 it was demonstrated that virtually all the nitrogenous phospholipids of carrot root tissue were lost during extraction of fresh tissue (35) unless the enzyme was first denatured. Since that time, the enzyme has received much attention and is discussed in detail in several recent reviews (9, 36-38). The general reaction catalysed by phospholipase D may be written as a transacylation reaction:-

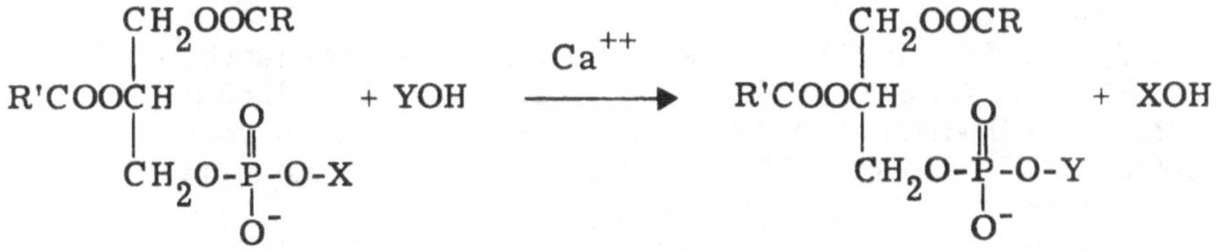

in which YOH may be water or one of a number of primary alcohols. The usual product of phospholipase D action in aqueous systems is *phosphatidic acid* (Y = H in above formula) but, when primary alcohols are present, e.g. methanol or ethanol used in extraction of plant materials, then phosphatidyl alcohol artefacts may be produced (39).

Phospholipase D was thought to be unique to higher plants, but recently it has been detected in microorganisms (see ref. 9). It requires Ca^{++} ions for maximal activity and is activated by organic solvents. The physical state of its substrates is critical. The enzyme has been shown to hydrolyse, under appropriate conditions, most of the common phospholipids, with the exception of phosphatidylinositol.

The physiological role of the enzyme is uncertain and in fact a recent paper (40) has questioned whether phospholipase D has any physiological function at all, or whether its activity is due to an "accidental" property of a membrane protein in plants. However, there is no doubt that phospholipase D activity and the resulting breakdown of membrane phospholipids is clearly manifested in disrupted tissues from a wide range of plants.

Fig. 3. Thin layer separation of polar lipids from intact and homogenized tissues of some storage organs

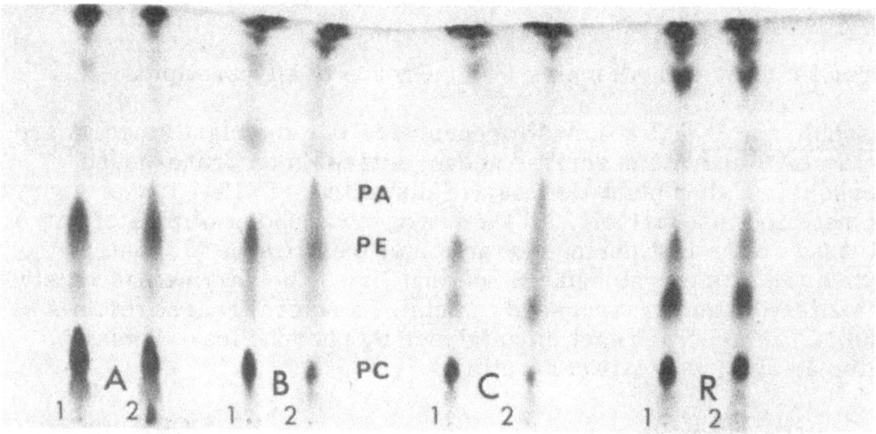

Lipid extracts from intact tissue (1) or homogenates incubated at 25° for 10 min. at pH 5.5 (2) were separated by t.l.c. in $CHCl_3$-MeOH-HAc-H_2O (170:30:20:5, by vol.) Lipids were detected in I_2 vapour. The plants used were Jerusalem artichoke (A), sugar beet (B), carrot root (C), and rose hip (R). The two major phospholipids were phosphatidylcholine (PC) and phosphatidylethanolamine (PE); phosphatidic acid (PA) was the product of phospholipase D activity in homogenates.

Fig. 3 shows thin-layer separations of polar lipids from four storage tissues and compares the profiles from intact and homogenized materials (T. Galliard and J. A. Matthew, unpublished observations). The loss of the major phospholipids, phosphatidylcholine and phosphatidyl ethanolamine, in homogenates is accompanied by the appearance of phosphatidic acid, causing streaking of the chromatograms; this was particularly marked with roots from beet and carrot and occurred without addition of Ca^{++} to the media.

Quarles and Dawson (41) have studied the distribution of the enzyme. On the basis of their survey (selected results are given in Table 3) they noted that highest activities were found in swollen storage tissues of certain plants: stems or swollen leaf stalks of brassica sp. and celery, carrot root, some seeds and tubers of Jerusalem artichoke. However, the enzyme was not found in all storage tissues; potato tubers, many fruits and seeds had very low levels of phospholipase D (Table 3). Studies on changes of enzyme levels during germination and development led to the conclusion (41) that "phospholipase D in plant storage tissues and seeds may be related to the rapid growth involved in their formation rather than being necessary for the utilization of food reserve substances."

Table 3. Distribution of phospholipase D activity in extracts of some plant storage organs

Tissue	activity*	Tissue	activity*
apple fruit	0	cereal seeds	0.6 - 1
Helianthus tuberosus tuber	18	oat roots	10
		onion bulb	0 - 0.3
Brassica sp.		stem	4
cultivars			
leaf	2-32	pea seed	20
swollen stem	27-64	leaf, stem, root	< 1
carrot root	14		
stem	0.1	potato tuber	0 - 0.3
		rose hips	0
celery stem	66-80		
leaf	0	Phaseolus sp.	
		seed	8
marrow		pod	0.3
mature fruit	30		
immature fruit	9	tomato fruit	0 - 0.1

*μg choline released from phosphatidylcholine min^{-1} mg dry wt.$^{-1}$
From Quarles and Dawson (41)

Failure to detect phospholipase D activity in crude extracts of certain plants is not absolute evidence for the absence of potential activity. Crude extracts from some plants appear to contain high molecular weight inhibitors of phospholipase D (42). Although phospholipase D activity in crude extracts of potato was not detected (41, 43), significant conversion of sonicated preparations of phosphatidylcholine to phosphatidic acid was catalysed by a 100,000 g supernatant fraction from potato tubers when LAH activity was reduced in the absence of surface-active agents (43). In crude extracts, the presence of free fatty acids which stimulate LAH activity on phosphatidylcholine (29) and possibly other inhibitory factors, resulted in the absence of apparent phospholipase D activity.

The subcellular localization of phospholipase D has been the subject of conflicting views (see 36) but recent studies have indicated that plastids and mitochondria, prepared under conditions which eliminated contamination with other cell fractions, had no phospholipase D activity, whereas plastids prepared by conventional techniques had activity (44). The enzyme is probably localized in 'lysosomes' or vacuoles (45) in common with many other hydrolases; disruption of cells would result in release of the enzyme into a non-particulate fraction and perhaps some non-specific association with particulate fractions in homogenates.

Other hydrolytic enzymes attacking polar groups of glycerolipids. Evidence for plant enzymes, other than phospholipase D, acting on the polar groups of phospholipids, or on the glycosyl linkage of glycolipids, is scarce. Kates (46) obtained evidence for phospholipase C (EC 3.1.4.3)- which splits phospholipids to diacylglycerol and phosphate esters - in carrot root and spinach leaves. No further information is available on this enzyme in plants although it is widespread in micro-organisms. Galactolipids are completely degraded to galactose, glycerol and fatty acids by extracts of *Phaseolus* sp. leaves but the initial attack is by a lipid acyl hydrolase, releasing mono- or digalactosyl glycerol which is subsequently hydrolysed by a galactosidase (17).

OXIDATIVE ENZYMES

Enzymes which catalyse the oxidation of acyl lipids generally act, not on the intact lipids, but on free fatty acids or fatty acyl thioester derivatives. Thus, enzymic oxidative degradation of lipids is preceded by acyl hydrolase or acyl transferase activities. This is in contrast to non-enzymic oxidations of lipids, e.g. metal-, light-or free radical-induced oxidation of triacylglycerols or membrane peroxidation. The preceding discussion on hydrolytic enzymes demonstrated the means by which net loss of lipid can occur. Acyl hydrolase activity results first in liberation of free fatty acids. However, in experimental systems, the loss of acyl lipid is frequently much greater than the corresponding accumulation of free fatty acids, indicating that further metabolism of the fatty acids takes place. Plant enzymes responsible for oxidative degradation or oxygenation of fatty acids have been reviewed recently (9, 47-49). The most common oxidative enzyme systems in plants are summarized in Table 4.

α-oxidation and peroxidation reactions will be discussed in detail below. β-oxidation is a universal process among higher organisms associated with the mitochondrial energy production system. In oil seeds, an additional glyoxysomal β-oxidation process is an essential component in the conversion of oil reserves to carbohydrate in germinating seeds. β-oxidation requires fatty acyl thioesters as substrates. There is no evidence for direct involvement of the process in disrupted plant tissues and it will not be included in the following discussion (for details see refs 9 and 48).

ω-oxidation and in-chain hydroxylation processes are involved in the formation of the lipid polymers of cutin and suberin and are therefore of major importance in wound healing in plant tissues. This is the subject of a separate contribution by P.E. Kolattukudy in this volume and no further discussion is necessary here.

Table 4. Summary of major processes for enzymic oxidation of fatty acids in plants

	Substrate		Cofactors etc.	Initial products
	free acid	thio-ester		
α-oxidation	+ +		a) O_2 b) $O_2 + NAD^+$	2-D-hydroxy acid C_{n-1}-fatty acid + CO_2
β-oxidation		+	NAD^+, flavo-protein	acetyl CoA
ω-oxidation	+		a) O_2, NADPH b) " + $NADP^+$	ω-hydroxy acid dicarboxylic acid
in-chain hydroxylation		+	O_2, NADPH	in-chain hydroxy acids
in-chain epoxidation		+		epoxy acids
peroxidation*	+		O_2 (Fe-enzyme)	fatty acid hydroperoxides

*specific for fatty acids with cis, cis, 1,4-diene groups

Alpha Oxidation

The universal nature of β-oxidation and its early demonstration in plants - in oil seeds (50) - has perhaps overshadowed the importance of α-oxidation in plant tissues (51). It is now known that β-oxidation in oil seeds is a specialized system for fat mobilization (52) and, although few comparative studies have been made, fatty acids are more rapidly utilized by α-oxidation than by β-oxidation in young leaves (47). Laties et al. (53) have demonstrated the importance of α-oxidation in the respiration of potato slices (see later). A role for α-oxidation in the synthesis of the 2-hydroxy acid compoents of plant cerebrosides and of odd-chain length fatty acids has been assumed and a role in the formation of long chain volatile flavour carbonyl compounds has been demonstrated (54, 55). However, further study of the possible role for α-oxidation in respiration and general metabolism in plants could be illuminating.

Because α-oxidation proceeds with free fatty acids (cf. thiolesters in β-oxidation), it can play an important part in the further degradation of free fatty acids released from glycerolipids by acyl hydrolase action in disrupted plant tissues. Until recently, it was thought that at least two α-oxidation mechanisms existed (see 9) but recent work from Stumpf's laboratory (56, 57) has attempted to reconcile these in the following proposed scheme:-

Fig. 4. Proposed reactions of α-oxidation (from Shine and Stumpf (57))

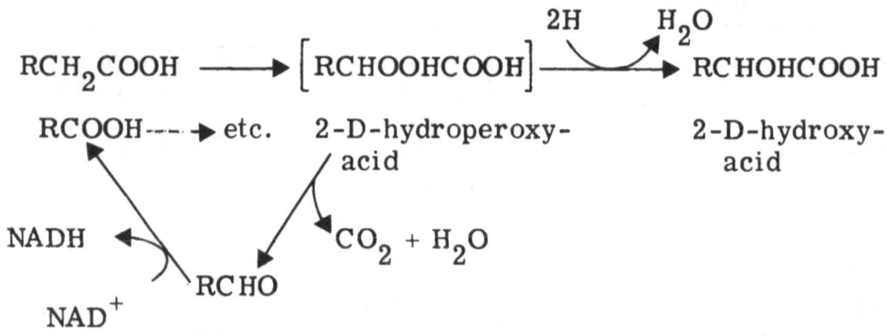

The above scheme involves a 2-D-hydroperoxy acid intermediate which may be either reduced to the 2-D-hydroxy acid (as found in cerebrosides) or decarboxylated to the C_{n-1} aldehyde, which may subsequently be oxidized to the C_{n-1} acid to continue the α-oxidation cycle. Shine and Stumpf suggested that a flavoprotein enzyme was involved in the peroxidation reaction in leaves and seeds (57) but recent studies in the author's laboratory (55) produced no evidence for flavoproteins in an α-oxidation system from cucumber fruit,but rather indicated a possible metallo-enzyme involvement.

The subcellular localization of α-oxidation enzymes has not been fully established. Recent cell fractionation work (57) has indicated that in leaves the enzyme is mainly in a non-particulate fraction whereas in peanut cotyledons 80% of the activity was associated with a microsomal fraction.

Fatty Acid Peroxidation : Lipoxygenase

The fatty acid composition of the membrane glycerolipids in plant tissues generally shows a high proportion of polyunsaturated fatty acids, principally linoleic (18:2) and linolenic (18:3) acids. For example, in potato tuber lipids, 18:2 + 18:3 represent about 75% of the total fatty acids (58). Some classes of membrane lipid are particularly rich in polyunsaturated acids; in potato 18:2 and 18:3 represent 58% and 40% respectively of the fatty acids of monogalactosyldiacylglyerol (15).

Polyunsaturated fatty acids are notoriously susceptible to oxidative breakdown catalysed by a range of agents including light (singlet oxygen) metals or metalloproteins (e.g. heme) and free radicals in addition to enzymes. However, these fatty acids *in situ*, as components of membrane lipids, are more stable in *healthy* tissues but damage or a breakdown in cellular control may render the lipids more susceptible to peroxidation. Accumulation of lipid peroxidation products in mammalian cells during physiological ageing is well established and, in plants, a similar production of lipid peroxidation products during senescence of fruit has recently been demonstrated (59).

In both enzymic and non-enzymic lipid peroxidation systems, the first product that can be isolated is a hydroperoxide. Non-enzyme systems will not be discussed further here; they are the subject of intensive research, although few comprehensive and recent reviews are available. Gardner (60) has discussed important aspects of both enzymic and non-enzymic formation and breakdown of lipid hydroperoxides. The *enzymic* formation of hydroperoxides from polyunsaturated fatty acids is catalysed by lipoxygenase enzymes.

Lipoxygenase (EC 1.13.11.12, previously named lipoxidase) is a generic term for a group of enzymes that catalyse the peroxidation of polyunsaturated acids with a *cis, cis*, 1,4-pentadiene structure to form conjugated hydroperoxides:-

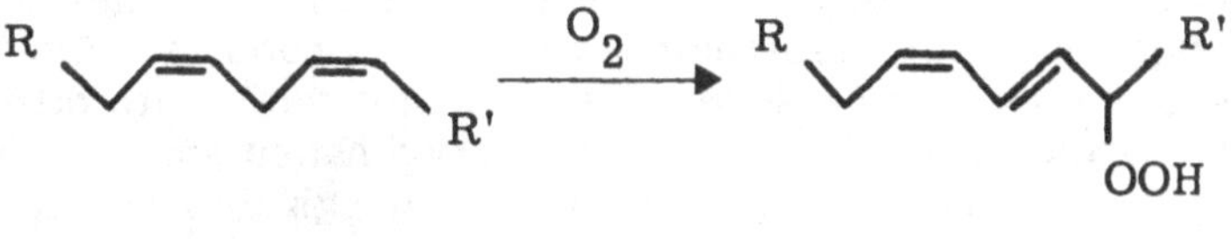

Although previously thought to be found only in leguminous seeds and certain cereals, lipoxygenases are now known to be present in very many plants and in various plant organs. Table 5 presents a selection of results from a survey by Pinsky *et al.* (61) to illustrate the wide range of lipoxygenase activities between *different* edible plants. It can be seen that, although the leguminous seeds contain high levels of enzyme, potato tubers and artichoke and egg plant fruits are also very active and significant amounts of activity are found in many other tissues.

Generalisations are difficult because some storage tissues are high in activity (leguminous seeds, potato tubers, egg plant fruits), whereas no activity was observed in roots from beet and carrot or fruits of pear and date. Lipoxygenase has received considerable attention recently, particularly in laboratories concerned with edible plants, because of its involvement inthe conversion of unsaturated fatty acids to volatile compounds with either desirable or undesirable flavour properties (see later).

Table 5. Oxidation of linoleic acid by aqueous extracts of edible plants Selected data from Pinsky et al. (61)

Tissue	lipoxygenase activity*	Tissue	lipoxygenase activity*
legume seeds		date fruit	0
lettuce leaves	120	carrot root	0
spinach "	80	beet "	0
strawberry fruit	40	turnip "	0
pear "	0	potato tuber	4560
pumpkin "	120	celery stalk	120
avocado "	720	artichoke heart	3360
egg plant "	4320	cauliflower florets	1440
tomato "			

*ul O_2 consumed/min/10g fresh weight with linoleic acid substrate

Lipoxygenases from various sources differ in several respects including substrate specificity, pH optima, effects of inhibitors and, most significantly, in the isomeric form of the products. Even in a single tissue, there may be several isoenzymes with quite different properties. Some general reviews on lipoxygenases and their properties have been published recently (9, 62-64). Although lipoxygenase was one of the first enzymes to be purified, only very recently has any progress been made in understanding its essential structure and mechanism of action.

Until 1972, the enzyme was an enigma because, unlike all other oxygenases, it appeared to lack any prosthetic groups or co-factors. Chan (65, 66) then reasoned that, from the principles of conservation of electronic spin angular momentum, a transition metal should be involved in the interaction between an organic compound and ground state oxygen in an enzymic process. He was able to demonstrate that soyabean lipoxygenase did in fact contain one atom of Fe per mol (10^5 daltons) of enzyme. This was subsequently confirmed in other laboratories and the mechanism is now under intensive investigation.

Additional problems to be explained by any reasonable proposal for lipoxygenase action were:- (a) the formation of alternative isomeric products, e. g. linoleic acid may be converted to 13-L- or 9-D-hydroperoxy derivatives, depending on the lipoxygenase isoenzyme; (b) both activation and inhibition of lipoxygenase by its product (fatty acid hydroperoxide) and (c) its activity under anaerobic conditions. Current thinking on the mechanism of the lipoxygenase stereospecific action may be summarized as in Fig. 5.

Fig. 5. Stereospecific formation of alternative isomeric hydroperoxides from linoleic acid by lipoxygenases. Based on Hamberg and Samuelson (67) and Egmond et al. (68).

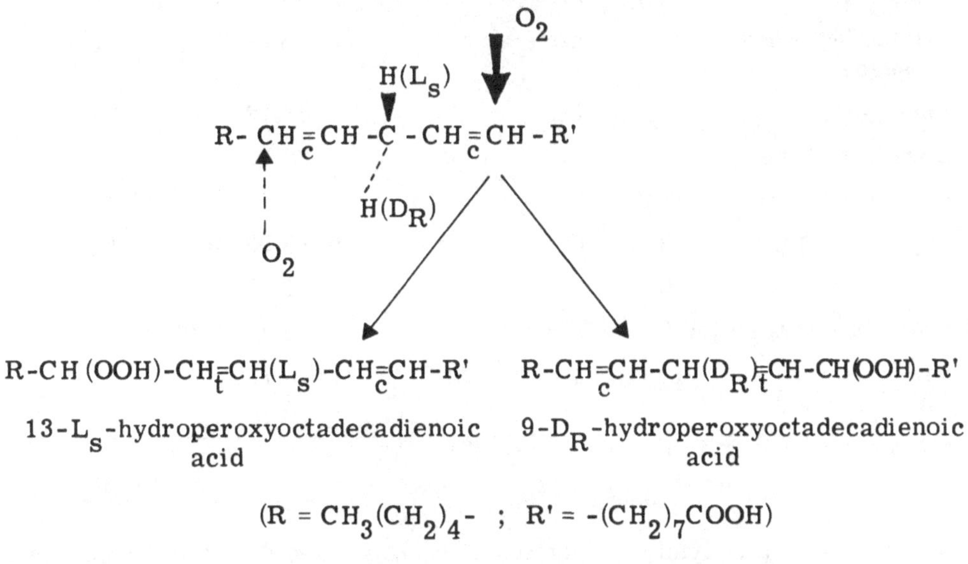

The initial event appears to be stereospecific abstraction of a H atom at the doubly allylic methylene group from one or other side of the hydrocarbon chain to produce a radical; electronic rearrangment and stereospecific attack by molecular oxygen results in a conjugated hydroperoxydienoic acid. These processes take place on the enzyme surface and the stereo specific orientation of the substrate determines the isomeric structure of the product. Thus the classical enzyme from soyabean (Type 1 isoenzyme) produces exclusively 13-L-hydroperoxides from linoleic acid (67) whereas that from corn germ forms the 9-D-isomer (68, 69). In tissues where the positional specificity of oxygenation has been studied, most show predominant specificity for one isomer. The observed isomeric ratios, however, may be affected by experimental conditions favouring one isoenzyme (9 and Table 6), or by isomerization which can readily occur (70). Table 6 presents information currently available on the isomeric distribution of products by lipoxygenases from a range of plants.

Table 6. Isomeric product specificity of lipoxygenase preparations from different plants acting on linoleic acid

Source of lipoxygenase	Conditions			Ratio of 9:13 hydroperoxide	Reference*
	pH	atmos.	temp.		
potato tuber	5.5	air	25°	95:5	*
tomato fruit	5.5	air	25°	95:5	79
oat, barley, wheat seeds	6-7	air	ambient	80-96:4-20	*, 78
corn germ	6.5	O_2	25°	88:12	*
soyabean (Type II)	7	air	25°	70:30	*
pea seed	6.8	O_2	0°	55:45	*
soyabean (Type I)	9	O_2	0°	0-10:90:100	*
corn germ	9	O_2	0°	15:85	*
water melon seedling	6	air	ambient	ca. 0: ca. 100	77
Dimorphotheca sinuata seed	6.9	air	25°	0:100	*

*References given in recent review (9)

Recent work on enzymic properties of lipoxygenase (9, 71 and refs. loc. cit; 72-76) has indicated that the enzyme (the soyabean Type I isoenzyme) has at least two binding sites, one for activation and one for substrate oxygenation; a third site is probably involved in oxygen binding. Hydroperoxide products activate the enzyme by converting it from an inactive to an active conformation; hydroperoxide products can also compete with substrate at the active site to cause inhibition. Activation of the enzyme also involves a change int he electronic state of the Fe-enzyme interaction to convert inactive native enzyme (Fe^{++})

to active high-spin-Fe^{+++} form. According to de Groot et al. (71), the essential aerobic process, which is similar to an earlier proposal (9), can be summarized as in Fig. 6.

Fig. 6. Proposed mechanism for the aerobic reaction of lipoxygenase From de Groot et al. (71)

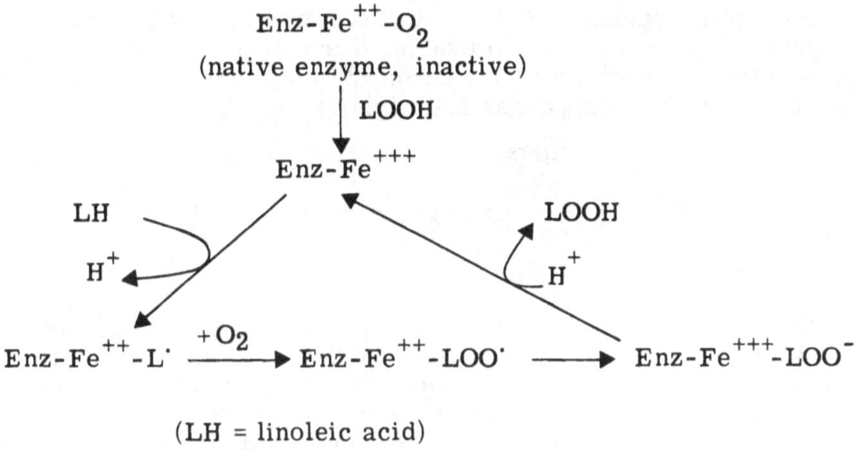

Under anaerobic conditions, lipoxygenase (once activated by hydroperoxide) can catalyse conversion of substrates to a range of polymeric or cleavage products. Relevant information is available in a recent paper by Verhagen et al. (75) and will not be discussed further here.

Despite early recognition of lipoxygenase and its widespread distribution, little is yet known about the physiological significance of the enzyme. Studies on changes of enzyme activity during growth and development of tissues have not yielded strong evidence for any defined role. Attempts to determine the subcellular localization have been equally frustrating. Lipoxygenase activity has been ascribed to different organelles in studies with various plants (see 9). In the author's laboratory, recent studies using differential and density-gradient centrifugation techniques with a range of plant tissues have also shown that the enzyme occurs in different fractions depending upon the plant material (30, 31). For example, in potato tuber extracts lipoxygenase activity was non-particulate (possibly due to release from a fragile organelle as discussed earlier for the lipid acyl hydrolase activity in potato); in brassica florets (cauliflower and calabrese) lipoxygenase was present in a dense fraction (similar to, but not identical to plastids); in pea roots the enzyme appeared to be 'lysosomal' and was probably in vacuoles. In no cases was lipoxygenase activity localized in mitochondria, microbodies or plastids (30, 31; cf. earlier studies reviewed in ref. 9).

Lipoxygenase-catalysed co-oxidation processes. The action of lipoxygenases in the co-oxidation of pigments has been recognized for many years (an early name for lipoxygenase was "carotene oxidase) and is employed commercially to bleach wheat flour carotenoids in bread making. However, the mechanism is not understood. Co-oxidation requires lipoxygenase, fatty acid substrate (e. g. 18:2 or 18:3) and oxygen together with the oxidizable co-substrate. Fatty acid hydroperoxides do not cause co-oxidation and the reaction appears to be coincident with lipoxygenase activity, i. e.

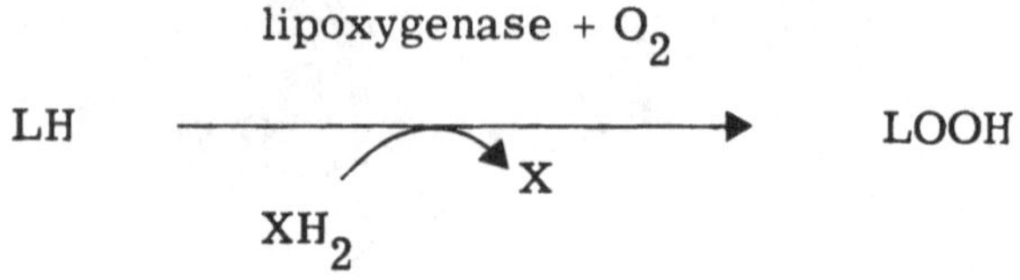

In this process, XH_2 may represent pigment (e. g. carotenoid, chlorophyll, thiols (including protein-SH) or lipid. Thus, concurrent with lipoxygenase activity may be oxidative attack on cellular components resulting, for example, in membrane damage, enzyme inactivation, etc. Information on lipoxygenase-catalysed co-oxidation reactions may be obtained from recent reviews (9, 60, 62). Weber and Grosch (80) have proposed a scheme which can explain the known facts. The basis of their proposal is the formation during lipoxygenase action of a peroxy radical intermediate which can be reduced to hydroperoxide stereospecifically in the normal lipoxygenase reaction or, alternatively, can interact with oxidizable co-substrate. A modified representation of their proposals can be written as in Fig. 7 (cf. Fig. 6)

Fig. 7. Proposed mechanism for lipoxygenase-catalysed co-oxidation reactions. Adapted from Weber and Grosch (80) to the mechanism given in Fig. 6.

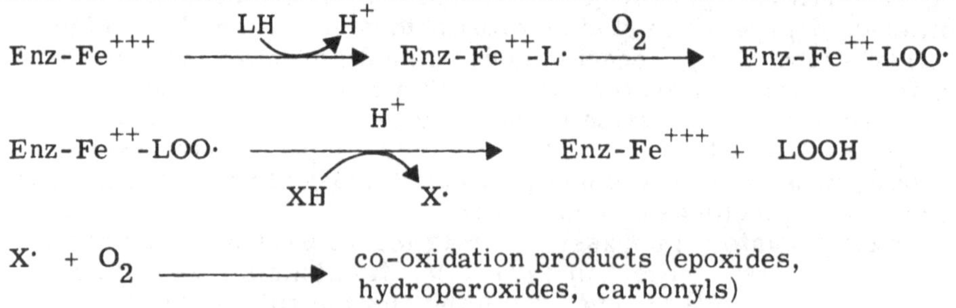

When XH = fatty acid (Fig. 7), the product could be an isomeric mixture of hydroperoxides (80) thus perhaps partly explaining the frequent absence of absolute stereo-specificity for products observed with lipoxygenase preparations (see e.g. Table 6).

Further Metabolism of Fatty Acid Hydroperoxides

Hydroperoxides are potentially unstable structures and are cytotoxic, particularly affecting protein and membrane structures. It is therefore not surprising that fatty acid hydroperoxides, the initial products of lipoxygenase action, are not found in healthy tissues. Even in damaged tissues or cell homogenates, one is more likely to find hydroperoxide reaction products than hydroperoxides *per se*. Fatty acid hydroperoxides are susceptible to non-enzymic reactions, e.g. isomerization, reduction, homolytic and heterolytic cleavage and free radical polymerization reactions. In addition it is now apparent that in plants there are a variety of enzymic processes which catalyse further metabolism of the hydroperoxides. A summary of enzymic and non-enzymic reactions is given in Table 7. Further discussion of the non-enzymic reactions is beyond the scope of this chapter but Gardner (60) has recently reviewed the area and further details may be obtained from refs 9, 62, 81-85. The enzymic conversions of hydroperoxides have also been reviewed recently (9, 60, 62) and the following discussion relates mainly to developments over the last few years.

Reduction of fatty acid hydroperoxides to hydroxydienoic acid occurs in *crude extracts* from plant tissues. A variety of non-enzymic agents will reduce hydroperoxides but, in addition, a 'lipoperoxidase' activity has been identified in seeds from cereals (86, 87), peas (88) and soya-beans (85). This differs from the glutathione peroxidase activity of mammalian systems and there is evidence that lipoxygenase itself, in the presence of electron donors (guaiacol, *p*-phenylenediamine), is responsible for the peroxidase activity. In this case, the hydroxydiene product appears to be formed by reduction, not of hydroperoxide, but of a radical intermediate of the lipoxygenase reaction (88).

Isomerization. Conversion of fatty acid hydroperoxides to ketols by hydroperoxide isomerase activity has been identified in several plants. The products formed by isomerases from different sources can be represented as in Fig. 8.

Table 7. Summary of known enzymic transformation of C_{18} fatty acid hydroperoxides in plant extracts

Reactions	Enzyme	Major Products
reduction	'lipoperoxidase' (? lipoxygenase)	C_{18}-hydroxy acids
isomerization	hydroperoxide isomerase	C_{18}-α and γ-ketol acids
isomerization/ dehydration	enzyme in potato tuber	C_{18}-divinyl ether acids
chain cleavage	cleavage enzyme (hydroperoxide lyase)	C_6- or C_9-aldehydes C_{12}- or C_9-aldehydo acids
isomerization-hydration	?	C_{18}-epoxyhydroxy acids C_{18}-trihydroxy acids
anaerobic reactions	lipoxygenase	C_{18}-keto acids C_{13}-keto acids + pentane

Table 8. Products formed by isomerases from different sources

Source of isomerase (seeds)	Product (see Fig.8)	Reference
flax	II	*
wheat	II	*
barley	II + III	*, 89
corn germ	II + III	*, 90
alfalfa	IV	*

For references marked * see recent reviews (9, 60)

Fig. 8. Products formed by hydroperoxide isomerases

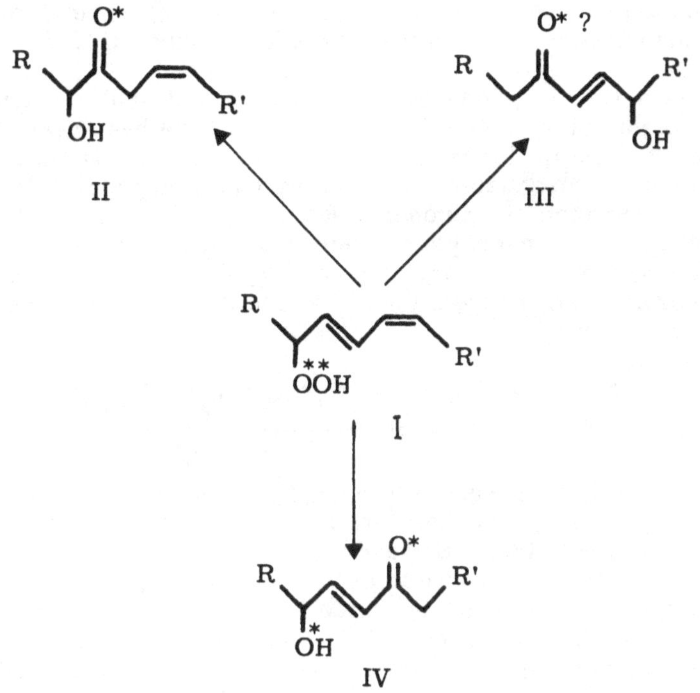

Some sources of isomerase activity and the products obtained are shown in Table 8. Thus most isomerases found so far produce α-ketols (II in Fig. 8) or both α- and γ-ketols (III) and most are active on both 9- and 13-hydroperoxy-18:2 fatty acid isomers, although the flax enzyme may be specific for the 13-isomer (91). In addition to sources listed in Table 8, evidence for isomerase activity has been obtained in soyabean, mung beans and peanuts (see 9, 60). Claims that isomerase activity was present in seedlings of water melon and sunflower and in cauliflower buds (92) were based on a non-specific assay. Subsequent work (77) has shown that water melon seedlings convert hydroperoxides by a cleavage process, not by isomerase action.

The mechanism of isomerase action is not fully understood. Studies with ^{18}O-labelled hydroperoxide substrates have shown that, in formation of the α-ketols (II) only the oxygen atom in the keto group is derived from the hydroperoxide oxygens (asterisk in Fig. 8) and the hydroxyl group is assumed to derive from water (91). This was supported by findings of Christianson and Gardner (93) who showed that various reagents (HX) could substitute the X group in place of the hydroxl group of the ketol. On the other hand, a different system must operate in alfalfa which produces γ-ketols of type IV (Fig. 8) in which both hydroperoxide oxygens are retained in the ketol function (94).

Epoxyhydroxy acids are produced from linoleic and linolenic acids in a variety of plant extracts and these processes have been reviewed in detail by Gardner (60). Lipoxygenase is involved in the initial step of this system but the mechanism of epoxyhydroxy acid production is uncertain. Conversion of hydroperoxides to epoxyhydroxy derivatives is catalysed by several non-enzymic processes and by cereal flour-water extracts (see 60) but purified lipoxygenase (type I) from soyabean (85, 95) and an enzyme extract from potato tubers (96) have also been shown to catalyse the reaction:

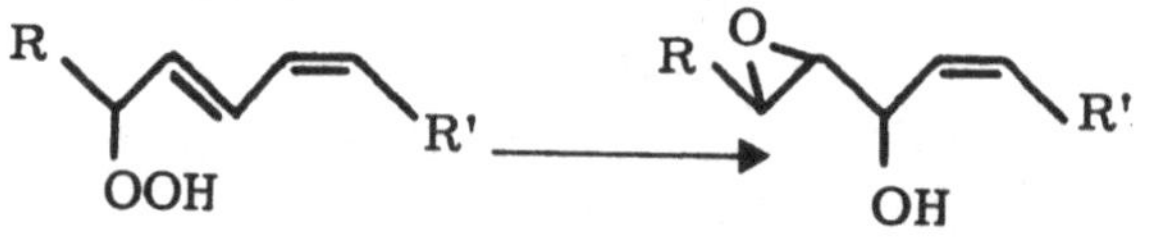

The conversion of hydroperoxide to epoxyhydroxy derivative with soyabean lipoxygenase proceeded with high (70%) retention of the hydroperoxide oxygens (95). However, in the conversion of linoleic acid to the same product with a purified pea lipoxygenase preparation (88) it has been claimed that the hydroperoxide is not an intermediate. All the systems described here convert both 9- and 13-hydroperoxide isomers to corresponding 9,10-epoxy-11-hydroxy or 12,13-epoxy-11-hydroxy derivatives.

Under conditions where epoxyhydroxy acids are produced, trihydroxy monoenoic acids are also found (see 9, 60 for reviews and 88, 95, 96). It has been proposed that hydrolytic ring opening of unstable epoxy-hydroxy acids is the source of trihydroxy acids but this is not yet certain.

Vinyl ether derivatives are formed from fatty acid hydroperoxides by an enzyme in potato tubers (97-99). So far, this reaction has not been found in other plants. The conversion of hydroperoxide may be considered as a dehydration/isomerization reaction:-

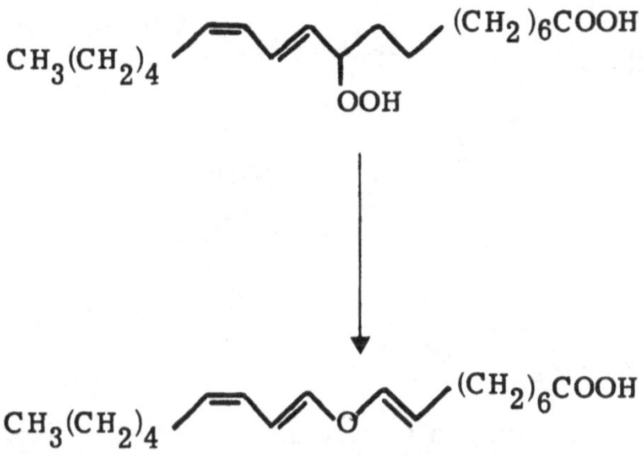

The product from linoleic acid hydroperoxide is a butadienyl vinyl ether derivative (named colneleic acid). An analogous product (colnelenic acid) is formed from linolenic acid hydroperoxide. A significant point in this reaction is that an atom of oxygen is inserted into the hydrocarbon chain although ^{18}O-labelling studies (99) have shown that the oxygen atoms of the hydroperoxide group are not incorporated into the ether oxygen.

These unusual ether derivatives were found to represent major lipids in homogenates of potato tubers when the tissue was homogenised at pH > 6.5 (97). They are not present in intact tissue and are not observed in homogenates prepared at pH < 5.5. They are formed in a sequential enzyme process from endogenous lipid via free 18:2 and 18:3 and their hydroperoxides:-

Endogenous lipid	acyl hydrolase →	free 18:2 + 18:3
18:2 + 18:3	lipoxygenase →	9-hydroperoxy-18:2/18:3
9-hydroperoxy 18:2 9-hydroperoxy 18:3	enzyme →	colneleic acid colnelenic acid

That the formation of the ether derivatves was enzymic was established by the usual criteria plus the fact that the reaction was specific for the 9-hydroperoxide isomers of 18:2 or 18:3. No conversion was obtained when the isomeric 13-hydroperoxides were tested as substrates (99).

Subsequent work in our laboratory (100) has shown that colneleic and colnelenic acids are cleaved in an aerobic reaction to form carbonyl fragments, e.g. for colneleic acid:-

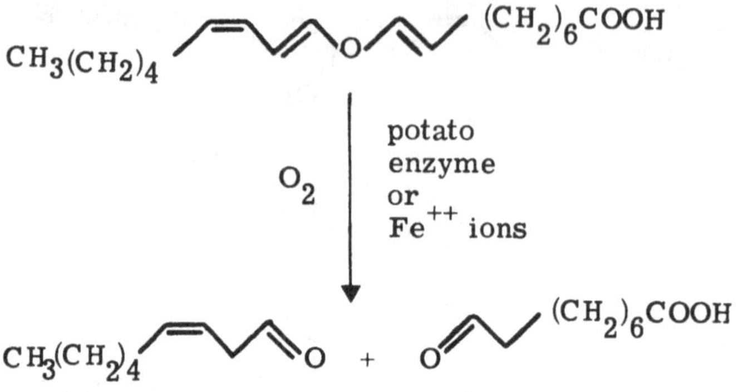

An analogous reaction with colnelenic acid produces a volatile C_9-dienal. The C_9 volatile fragments give a characteristic 'cucumber' odour. The cleavage reaction has the properties of a free radical reaction and partial purification of the enzyme from potato has failed to separate this from lipoxygenase activity in potato extracts (J. C. Kader and T. Galliard, unpublished results.) Also, a partially purified preparation of soyabean lipoxygenase and ferredoxin would also catalyse the cleavage reaction (100). It is possible that a lipoxygenase-ether formation-ether cleavage complex is present in potato tubers but this has not been fully investigated.

Hydroperoxide cleavage. It has long been recognized that unsaturated fatty acids are the source of volatile compounds giving desirable flavour or undesirable off-flavour in foods and in the so-called 'leaf aldehydes' and'alcohols'. Whilst in many products, e. g. fats and oils, the production of volatile cleavage products is clearly chemically- or light-catalysed, there was also good evidence for enzymic production of volatile aldehydes and alcohols in plant tissues (for recent reviews see 101-103). The possible role of lipoxygenases and fatty acid hydroperoxides in these reactions had been postulated in some instances and discounted in others. However, only very recently has any direct evidence been obtained for hydroperoxides and their enzymic cleavage to volatile compounds.

Work from the author's laboratory during the past two years has demonstrated the existence of enzymic hydroperoxide cleavage reactions in fruits of cucumber (104-106) and tomato (107, 108) and a similar process was recently discovered in water melon seedlings (77). A particulate enzyme from cucumber fruits catalyses the anaerobic cleavage of both 9- and 13-hydroperoxide isomers from linoleic and linolenic acids. The optimal activity is at pH 6. 4 and the enzyme is very heat-sensitive. High yields (60-70%) of cleavage products are produced. The reaction is shown in Fig. 9.

Fig. 9. Enzymic cleavage of fatty acid hydroperoxides to carbonyl fragments. From Galliard et al. (104, 105)

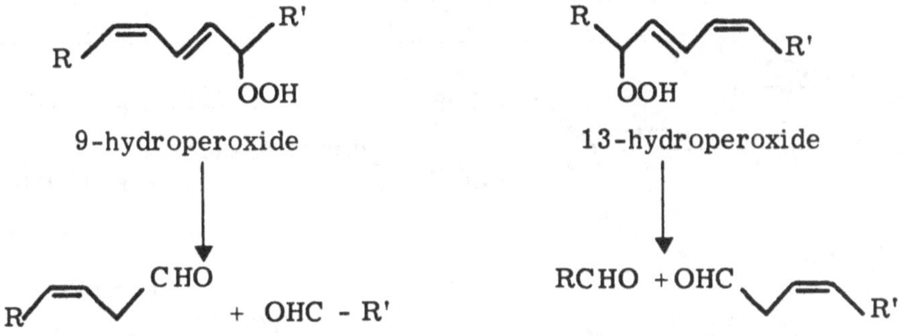

$R = CH_3(CH_2)_4-$ (18:2 derivative) ; $R' = (CH_2)_7COOH$
$R = CH_3CH{=}CH\text{-}CH_2\text{-}CH-$ (18:3 derivative)

Thus the volatile products from the 9-hydroperoxides of 18:2 and 18:3 fatty acids are cis-3-nonenal and cis-3, cis-6-nonadienal respectively. (these are the two main components of the characteristic cucumber odour); the 13-hydroperoxides give hexanal and cis-3-hexenal. The non-volatile fragments are $C_{9:0}$- or $C_{12:1}$-aldehydo acids. The cis-3-enals are converted by enal isomerase activity to the isomeric trans-2-enals. Subcellular localization studies (D. A. Wardale, E. Lambert and T. Galliard, in preparation) have shown that the cleavage activity is particulate and associated with low density membrane fractions. The hydroperoxide cleavage (or lyase) activity in water melon seedlings (77) appears to be similar to that of cucumber with respect to 13-hydroperoxide cleavage; its action on 9-hydroperoxides has not been reported. In water melon seedlings the cleavage activity was located principally at the hypocotyl-root junction and achieved maximal activity after 6 days germination (77).

In tomato fruits, lipoxygenase and hydroperoxide cleavage activity catalyse the degradation of linoleic and linolenic acids when the tissue is disrupted (79, 107, 108). However, the cleavage activity in tomato appears to differ in one important respect from that in cucumber, i. e. in substrate specificity. Although the major products of lipoxygenase action in tomato are the 9-hydroperoxide isomers of 18:2 and 18:3, these are not attacked by the cleavage enzyme. Rather, it is the minor, 13-hydroperoxides that are cleaved to carbonyl fragments. The process in tomato extracts acting on linoleic acid can be summarized and compared with that in cucumber as in Fig. 10.

Fig. 10. Comparison of the enzymic formation of volatile fragments by hydroperoxide cleavage in cucumber and tomato fruits. From Galliard *et al*. (104, 107)

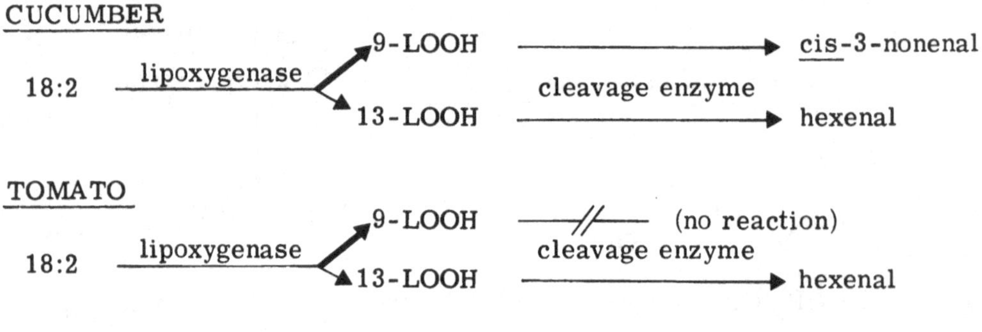

LOOH = linoleic acid hydroperoxide

Analogous reactions to those above can be written for linolenic acid.

It should be noted that the initial non-volatile carbonyl fragment from the 13-hydroperoxide of 18:2 or 18:3 is a *cis*-9-monoenoic C_{12}aldehydo acid although the *trans*-10-isomer is produced in cucumber (102) and water melon (77) extracts. Oxidation of the aldehyde -group would produce $C_{12:1}$ dibasic acids identical to traumatic acid (*trans*-10-$C_{12:1}$ the so-called "wound hormone" - see later) or the *cis*-9-$C_{12:1}$ dibasic acid formed from linoleic acid in young pea leaves (109). The probable enzymic pathways are shown in Fig. 11.

Many plants contain active alcohol dehydrogenase activities which catalyse interconversion of aldehydes and alcohols (101) e. g. the "leaf aldehydes" - hexanal and *cis*-3-hexenal - may be reduced to the corresponding "leaf alcohols" - hexanol and *cis*-3-hexenol. Studies in the author's laboratory on the peroxide cleavage systems have so far concentrated on tissues in which the volatile products are important in flavour production in cucumber and tomato. However, it should be noted that both volatile and non-volatile products obtained in these studies (and closely related compounds) may also be important from other aspects. For example, compounds of this type are known to have fungicide, insect attractant/repellant activities and may be involved in wound healing (see later).

Fig. 11. Proposed pathways for the enzymic formation of traumatic acid by hydroperoxide cleavage

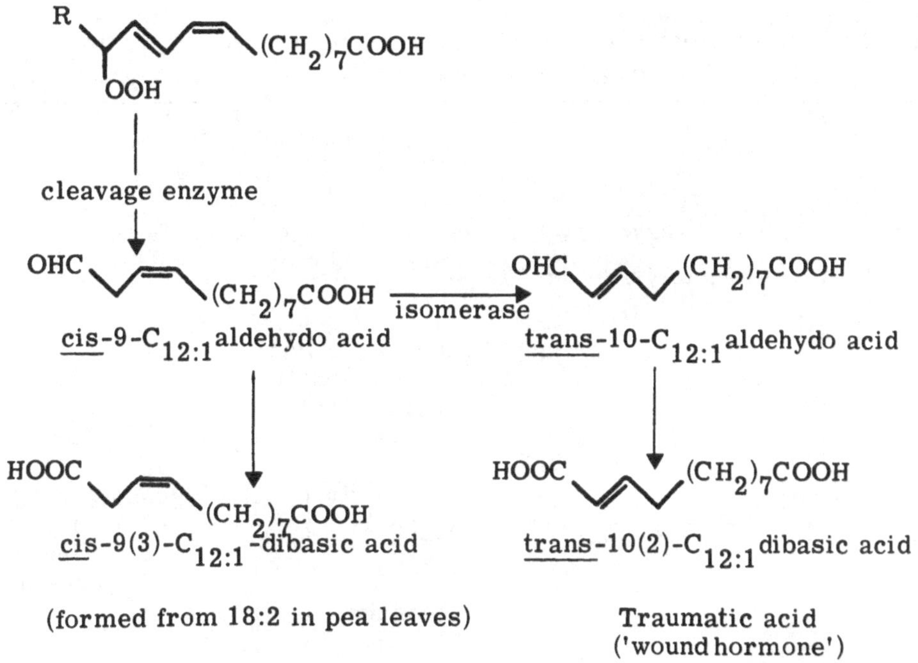

ACTION OF LIPID DEGRADING ENZYMES IN WOUNDED TISSUES

Although the nature and activities of lipid-degrading enzymes are becoming increasingly recognized, most of these enzymes suffer from a common, embarrassing lack of a defined physiological role. Of the enzymes discussed in detail in the first part of this chapter, no adequate explanation has yet been given for the in vivo function in healthy tissues of polar lipid acyl hydrolases, phospholipase D or lipoxygenase; even the full significance of α-oxidation in plants (in comparison with β-oxidation) has not been properly determined.

However, there is no doubt about the manifestation of these enzyme activities in disrupted tissues. Unfortunately, there is little information available on the levels or involvement of these enzymes in subsequent stages of wound response processes following the initial trauma. The following discussion will therefore be somewhat superficial and disjointed. An attempt is made to fit the limited information into the general framework of plant wound responses by dealing first with the immediate action on wounding and then short-term and longer term effects during wound healing.

Fig. 12. Interactions between cellular damage and lipid degrading enzymes in potato tuber

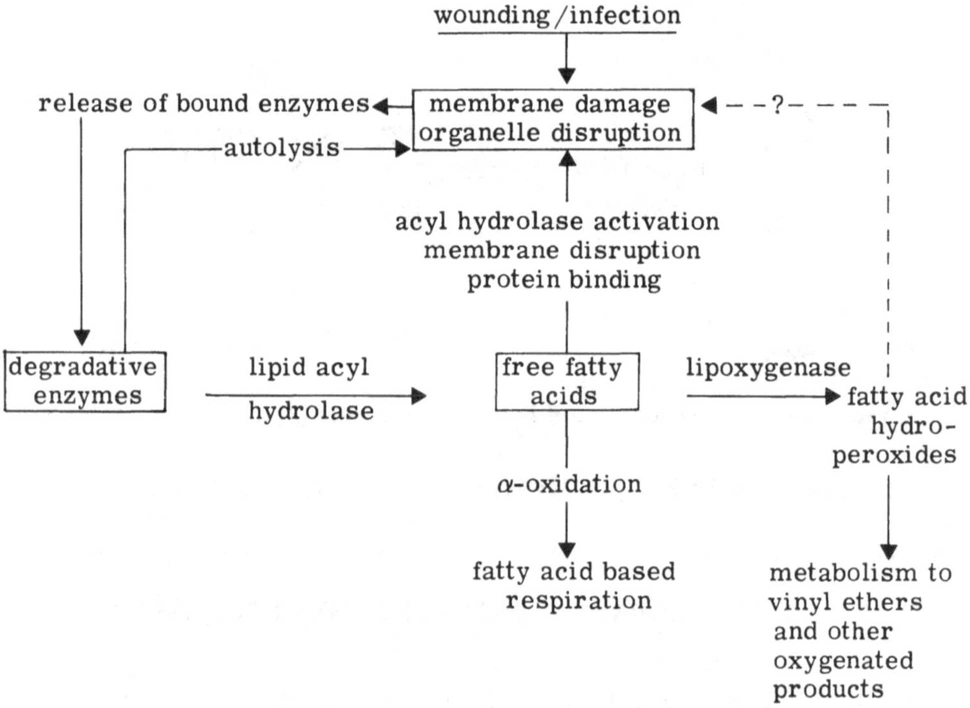

Immediate Effects of Wounding

The Introduction and previous sections have illustrated the rapid and extensive breakdown of lipid which occurs on wounding certain plant tissues. In storage tissues, most of the relevant information comes from studies with potato tubers. A summary of the sequential processes involved in disrupted potato tissues is given in Fig. 12.

Two important points should be emphasised. Firstly, damage to membrane lipids of disrupted cells on wounding is virtually instantaneous, even at 0° and, secondly, the degradation is autocatalytic. Thus, as indicated in Fig. 12, mechanical (or microbial) disruption of tissue leads to rupture of cells and their organelles; degradative enzymes released from, e.g., vacuoles can digest membrane lipids and, at least in the case of the acyl hydrolase of potato, this digestion is stimulated by free fatty acids (i.e. the products of hydrolase action). Moreover, the fatty acids themselves can have a detergent effect on membranes and will also bind to proteins. Fatty acid hydroperoxides, formed by lipoxygenase action, could also have a toxic effect on enzymes and membranes. On the other hand, further metabolism of fatty acids and hydroperoxides could direct these away from any direct action although we know little about possible effects of the final reaction products.

Most of the studies on potato enzymes in the author's laboratory have been with cell-free systems and there is little information on the action of lipid-degrading enzymes in cells adjacent to cut surfaces. However, Laties (see ref. 3) has observed that 60% of the total membrane lipid of potato tissue is lost in fresh slices. Assuming a slice thickness of at least 10 cells, then cells other than those at the cut surface must be affected. Thus subsequent metabolism of the wounded tissue must be influenced by the disruptive nature of the degradative enzymes and by the products of their action.

It should be mentioned at this point that the effect of these enzymes in cell-free extracts can explain the commonly observed inhibition of membrane-bound enzymic processes in crude extracts (e.g. the loss of mitochondrial and chloroplast activities). In potato tubers, the microsomal system which desaturates oleic acid, and which is induced during ageing of potato disks, is inhibited by endogenous lipid acyl hydrolase activity (110). The presence of bovine serum albumin during cell fractionation usually improves recovery of activity. The general problems associated with lipid-degrading enzymes in the preparation of organelles and techniques for overcoming these have been reviewed recently (111).

Fungal infection is analogous to physical wounding in the disruption of tissue. Recent studies with potato tubers have indicated that degradative enzymes of both host and parasite are involved in the interaction. Phospholipases of Erwinia carotovora will attack, e.g., protoplasts from potato tuber but apparently do not penetrate the cell wall in the absence of cell wall degrading enzymes (112). On the other hand, a phospholipase from Botrytis cinerea (113) will penetrate cell walls of potato tuber or beetroot and attacks the tonoplast membrane but apparently does not cause disruption of isolated protoplasts. Shephard and Pitt (113) have concluded that "Provided a pathogen possesses the enzymic ability to initiate host autolysis, there is logistically no need for it to possess enzymes capable of degrading all structural components of the host since many of these can be hydrolysed by the lysosomal enzymes of the host."

Fatty Acid Oxidation in Wounded Tissues

A series of papers from Laties' laboratory have provided several lines of evidence to establish the importance of α-oxidation of fatty acids in the respiration of freshly cut slices of potato tuber. Respiration can be accounted for by the process of α-oxidation (for details, see earlier in this chapter):-

$$RCH_2COOH + O_2 + NAD^+ \longrightarrow \longrightarrow \quad RCOOH + CO_2 + NADH + H^+$$

Thus oxygen uptake and CO_2 evolution is accompanied by NADH formation. From studies using ^{13}C:^{12}C ratio measurements (114, 115) α-oxidation inhibitors (53) and comparison of fatty acids with other respiratory substrates (53, 116), it has been concluded that the major component of respiratory CO_2 in fresh slices is due to lipid oxidation and, of this, α-oxidation of long-chain fatty acids accounts for up to 30% of the CO_2 produced. On the other hand, α-oxidation makes little contribution to respiration of slices aged for 24 hours. [Further discussion of this topic is given in Laties' chapter in this volume.] α-Oxidation requires free fatty acid substrates; lipid acyl hydrolase enzymes liberated on slicing and acting on endogenous membrane lipids would provide these substrates. Unpublished studies in the author's laboratory confirm the finding of Hasson and Laties (26) that the lipid acyl hydrolase activity in potato slices remains constant during ageing.

Lipoxygenases also attack free polyunsaturated (18:2 + 18:3) fatty acids - i.e. the major (75%) fatty acids in potato tuber. Lipoxygenase might then be expected to compete with α-oxidation for unsaturated (but not saturated) fatty acids. Hydroperoxide formation and subsequent conversions would involve O_2 uptake but there is no evidence to date for metabolism of the peroxidation products through to CO_2 in potato tuber. In a single preliminary experiment with 1mm disks of potato tuber we observed that the lipoxygenase activity of disks aged for 18 hours was only 25% that of the freshly-cut disks (unpublished).

Lipid Degradation and Wound Responses

In addition to the immediate action of lipid-degrading enzymes at the time of wounding, as described above, the enzymes and the products of their activity may function in the longer-term processes of wound responses and wound healing. The levels of many enzymes increase in response to physical wounding or infection of storage tissues (see e.g. chapters by Rhodes, Uritani and Kahl in this volume). Several oxidative enzymes, e.g. peroxidase and phenol oxidases, are known to increase after wounding but there is little information on lipoxygenase or other lipid-degrading enzymes. [There is more information on lipid-

Fig. 13. Changes in enzyme levels in (A) wounded and (B) Phoma-infected potato tubers.

Plugs of tissue 5 mm and 6 mm diam. removed. Cavity inoculated with Phoma or sterile agar and covered. Tubers were then stored at 5° and 90% RH. Enzyme assays on uninfected tissue surrounding the wound.

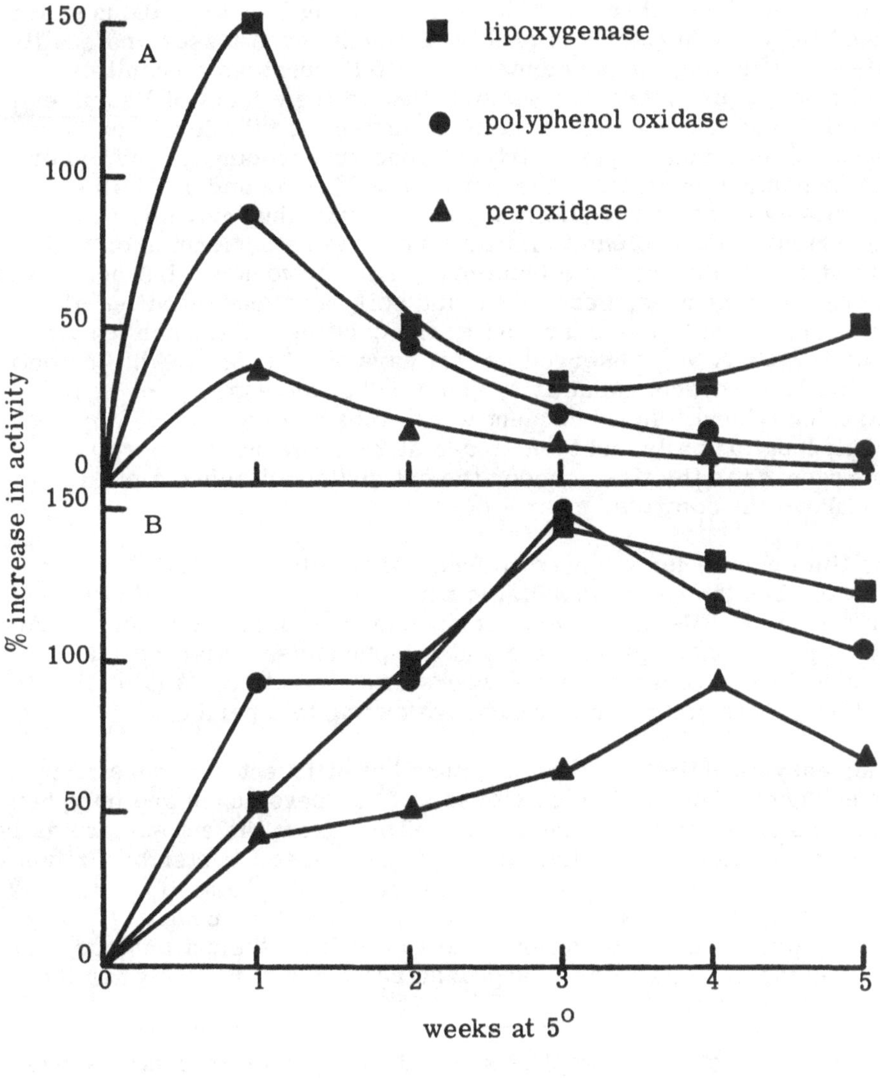

synthesising enzyme systems and this is reviewed by Kader and Mazliak and by Kolattukudy in this volume. It has been suggested (117, 118) that lipoxygenase and other fatty acid oxidation systems are involved in cutin synthesis of injured leaf tissue. However, this should be examined in relation to the recent studies on cutin and suberin formation where a specific group of fatty acid oxygenation enzymes have been characterised (see Kolattukudy's chapter).

In the potato (and other vegetables) processing industry, damage and fungal infection are serious problems which cause losses and quality defects, including the development of off-flavours as a result of lipid oxidation. Some years ago we investigated the effects of Phoma exigua infection and of wounding on the lipid content and the levels of lipid-degrading enzymes in potato tuber tissue surrounding inoculated or sterile control cavities. This work (M. J. Oliver and T. Galliard, unpublished) was of a preliminary nature only, but may be of interest in the context of this volume. Initial studies were performed on material kept at 5° and 90% relative humidity (RH). In wounded (but uninfected) tubers, lipoxygenase, peroxidase and polyphenolase activities all increased in the tissue surrounding the cavities. For each enzyme, a peak of activity was observed after 7 days at 5° (Fig. 13A). In contrast, the lipid acyl hydrolase activity remained at the same level as in untreated tubers. In subsequent work, tubers were wounded by cutting in half longitudinally and then stored at 25°. At this temperature, lipoxygenase in the tissue under the cut surface reached a maximum, 60% above the controls, after 4 days.

The time curves for wounded/infected tissue (i. e. a plug of tissue removed and the cavity inoculated with Phoma) showed quite different profiles (Fig. 13B) from those of wounded but uninfected tubers. At 5°C, lipoxygenase, peroxidase and polyphenolase activities all continued to increase up to 3-4 weeks after infection. Again, the lipid acyl hydrolase activity did not change during this period.

When enzyme activities were measured at different distances from the cut surfaces of potato halves stored at 5°C, peroxidase and polyphenol oxidase showed maximal increases 0-2 mm from the cut surface as is usual in wounded or infected storage tissues (see Chapter by Uritani in this volume). Lipoxygenase also showed a peak of activity in the 0-2mm section but, unlike the other two enzymes, a second peak of lipoxygenase activity was observed between 9 and 13mm from the cut surface. These two regions of increased lipoxygenase activity were found when the wounded tubers were stored at 5° or 25°.

Although it must be re-emphasised that the results presented above should be investigated further for confirmation, they do indicate that lipoxygenase in potato tubers responds to wounding and infection in a similar manner to the well known activities of peroxidase and polyphenolase, whereas the lipid acyl hydrolase activity is unaffected by these stress factors. The physiological basis for increased lipoxygenase activity is not known and no direct involvement of the enzyme in wound healing response has yet been established. The increase in lipoxygenase at a large (1cm) distance from the cut surface (if confirmed) is even more puzzling since this would indicate a transmitted wound response to a part of the tissue not obviously involved in the process.

Traumatic Acid and Wound Hormones

It has long been known that damage to plant tissues can result in the formation of large amounts of a substance which induces cell proliferation and in tumescence formation or wound periderm formation in bioassays with, e.g. potato tubers, bean pods or tomato fruit. The "wound hormone" was subsequently identified as traumatic acid (*trans*-2-$C_{12:1}$-dicarboxylic acid). The literature on traumatic acid is covered by a recent review by Lipetz (119). Although much interest in traumatic acid and wound hormones was generated up to 1950, the outcome was not generally satisfactory and further studies were eclipsed by the more promising auxins and gibberellins. Nevertheless, a few isolated studies in more recent years have indicated that a re-investigation of "wound hormones" may be necessary (see 119). For example, traumatic acid appears to be the active principle in foliar abscission of cotton plants following injury by *Lygus hesperus* (120); in this case traumatic acid could not be substituted by other hormones.

Traumatic acid is formed by oxidative degradation of linoleic or linolenic acids (121). An early proposal that lipoxygenase is involved in its formation (122) is supported by the recent work in the author's laboratory and elsewhere (described earlier in this chapter) showing that enzymic cleavage of C_{18} hydroperoxides produces $C_{12:1}$ aldehydoacids, from which traumatic acid may be derived by aldehyde dehydrogenase activity. Current studies in the author's laboratory are involved in elucidation of this problem in *Phaseolus* sp.

Ethane and Ethylene

The evolution of ethylene and ethane by damaged plant tissue is well known although the biogenesis of these is still not fully understood. The formation of ethylene and its involvement in wounded tissues is reviewed elsewhere in this volume by Yang and Pratt. Ethane and ethylene are products of fatty acid hydroperoxide decomposition in model systems and in plant extracts (123-126). The involvement of linolenic acid and lipoxygenase in ethylene production in plants is controversial (see chapter by Yang and Pratt) and, although methionine is generally recognised as the precursor of ethylene, this is still not fully resolved (G. Laties, personal communication) and the possibility remains that ethane (and in some cases ethylene also) may be derived from lipoxygenase-mediated reactions in disrupted tissues. In a recent and interesting paper Elstner and Konze (127) have demonstrated that, in damaged leaf tissue, ethane production is proportional to the number of damaged cells, whereas ethylene production is proportional to the number of undamaged cells adjacent to disrupted tissue. This suggested that ethane production was a function of disrupted cells whereas ethylene evolution was a typical wound response.

REFERENCES

1. Laties, G.G. : Wound respiration in retrospect and prospect. Abstracts of Int. Symp. on Biochemistry and Cytology of wounded Plant Tissues, Karlsruhe (1975).

2. Jackman, M.E., van Steveninck, R.F.M. : Changes in the endoplasmic reticulum of beetroot slices during aging. Aust. J. biol. Sci. 20, 1063-1068 (1967).

3. Hasson, E.P., Laties, G.G. : Separation and characterization of potato lipid acyl hydrolases. Plant Physiol. 57, 142-147 (1976).

4. Galliard, T. : The enzymic breakdown of lipids in potato tuber by phospholipid- and galactolipid acyl hydrolase activities and by lipoxygenase. Phytochemistry 9, 1725-1734 (1970).

5. Beevers, H. : Metabolic production of sucrose from fat. Nature, Lond. 191, 433-436 (1961).

6. Galliard, T. : Aspects of lipid metabolism in higher plants. II. The identification and quantitative analysis of lipids from the pulp of pre- and post-climacteric apples. Phytochemistry 7, 1915-1922 (1968).

7. Draper, S.R. : Lipid changes in senescing cucumber cotyledons. Phytochemistry 8, 1641-1647 (1969).

8. Bengtsson, B., Bosund, I. : Lipid hydrolysis in unblanched frozen peas. J. Food Sci. 31, 474-481 (1966).

9. Galliard, T. : Degradation of plant lipids by hydrolytic and oxidative enzymes. In Recent Advances in the Chemistry and

Biochemistry of Plant Lipids (T. Galliard & E. I. Mercer, eds.) Academic Press, London, pp. 319-357 (1975).

10. Dalgarno, L., Birt, L. M. : Free fatty acids in carrot-tissue preparations and their effect on isolated carrot mitochondria. Biochem. J. 87, 586-596 (1963).

11. McCarty, R. E., Jagendorf, A. T. : Chloroplast damage due to enzymatic hydrolysis of endogenous lipids. Plant Physiol. 40, 725-735 (1965).

12. Baddeley, M. S., Simon, E. W. : An inhibitor of mitochondrial oxidations in extracts prepared from senescing leaves. J. Exp. Bot. 20, 94-99 (1969).

13. Noma, A., Bergström, B. : The acid lipase of castor beans. Positional specificity and reaction mechanism. Biochim. Biophys. Acta 227, 106-115 (1971).

14. Ory, R. L. : Acid lipase of the castor bean. Lipids 4, 177-185 (1969).

15. Galliard, T. : Identification and quantitative determination of the lipids in potato tubers. Phytochemistry 7, 1907-1914 (1968).

16. Catfield, I. L., Woelk, H. U. : Lipoxidase and lipase in the common potato, Solanum tuberosum. Lebensm. Wiss. u. Technol. 4, 129-130 (1971).

17. Sastry, P. S., Kates, M. : Hydrolysis of monogalactosyl and digalactosyl diglycerides by specific enzymes in runner bean leaves. Biochemistry 3, 1280-1287 (1964).

18. Helmsing, P. J. : Purification and properties of galactolipase. Biochim. Biophys. Acta 178, 519-533 (1969).

19. Galliard, T., Dennis, S. : Phospholipase, galactolipase and acyl transferase activities of a lipolytic enzyme from potato. Phytochemistry 13, 1731-1735 (1974).

20. Burns, D. D., Galliard, T., Harwood, J. L. : in preparation.

21. Galliard, T. : The enzymic deacylation of phospholipids and galactolipids in plants. Biochem. J. 121, 379-390 (1971).

22. Galliard, T., Matthew, J.A. : Lipid degrading enzymes in different varieties of potato tuber. J. Sci. Food Agric. 24, 623-627 (1973).

23. Berkeley, H.D., Galliard, T. : Effect of growth and storage on lipid degrading enzymes of the potato tuber. J. Sci. Fd Agric. 25, 869-873 (1974).

24. Hirayama, O., Matsuda, H., Takeda, H., Maenaka, K., Takatsuka, H. : Purification and properties of a lipid acyl hydrolase from potato tubers. Biochim. Biophys. Acta 384, 127-137 (1975).

25. Shepard, D.V., Pitt, D. : Purification and physiological properties of two lipolytic enzymes of Solanum tuberosum. Phytochemistry 15, 1471-1474 (1976).

26. Hasson, E.P., Laties, G.G. : Purification and characterization of an A type phospholipase from potato and its effect on potato mitochondria. Plant Physiol. 57, 148-152 (1976).

27. Galliard, T., Dennis, S. : Isoenzymes of lipolytic acyl hydrolase and esterase in potato tuber. Phytochemistry 13, 2463-2468 (1974).

28. Galliard, T. The effects of free fatty acids on the hydrolysis of phospholipids by the lipolytic acyl hydrolase of potato tubers. Eur. J. Biochem. 21, 90-98 (1971).

29. Dennis, S., Galliard, T. : Wax ester formation catalysed by isoenzymes of lipolytic acyl hydrolase. Phytochemistry 13, 2469-2473 (1974).

30. Wardale, D.A., Galliard, T. : Subcellular localization of lipoxygenase and lipolytic acyl hydrolase enzymes in plants. Phytochemistry 14, 2323-2329 (1975).

31. Wardale, D.A., Galliard, T. : Further studies on the subcellular localization of lipid-degrading enzymes. Phytochemistry 16, 333-338 (1977).

32. Douce, R., Lance, C. : Alteration des activitées oxydatives et phosphorylantes des mitochondries de chou-fleur sous l'action de phospholipases et du vieillissement. Physiol. Veg. 10, 181-198 (1972).

33. Acker, L., Schmitz, H.J. : Uber die Lipide der Weizenstärke nachweis und Isolierung von Lysolecithin. Die Stärke 19, 233-239 (1967).

34. Burns, D. D., Galliard, T., Harwood, J. L. : Catabolism of sulphoquinovosyl diacylglycerol by an enzyme preparation from Phaseolus multiflorus. Phytochemistry 16 in press (1977).

35. Hanahan, D. J., Chaikoff, I. L. : The phosphorus-containing lipids of carrot. J. Biol. Chem. 168, 233-240 (1947).

36. Galliard, T. : Phospholipid metabolism in photosynthetic plants. In Form and Function of Phospholipids (G. B. Ansell, R. M. C. Dawson & J. N. Hawthorne, eds), Elsevier 1973, pp. 252-288.

37. Kates, M. : Plant phospholipids and glycolipids. Adv. Lipid Res. 8, 225-265 (1970).

38. Nolte, D., Acker, L. : Phospholipase-D - a review. Z. Lebensm. Unters. Forsch. 159, 225-233 (1975).

39. Douce, R., Faure, M., Marechal, J. : Le phosphatidylmethanol, un phospholipide isolé d'une souche de crown-gall de vigne. Comptes Rend. Acad. Sci. Paris D, 262, 1549-1552 (1966).

40. Roughan, P. G., Slack, C. R. : Is phospholipase D really an enzyme? A comparison of in situ and in vitro studies. Biochim. Biophys. Acta 431, 86-95 (1976).

41. Quarles, R. H., Dawson, R. M. C. : The distribution of phospholipase D in developing and mature plants. Biochem. J. 112, 787-794 (1969).

42. Tookey, H. L., Balls, A. K. : Plant phospholipase D. I. Studies on cottonseed and cabbage phospholipase D. J. Biol. Chem. 218, 213-224 (1956).

43. Galliard, T. : unpublished observations.

44. Clermont, H., Douce, R. : Localization de l'activité phospholipasique dans les tissues végétaux. FEBS Lett. 9, 284-286 (1970).

45. Clermont, H. : Sur une methode rapide de dosage de l'activité phospholipasique D chez les végétaux superieurs. Physiol. Vég. 108, 153-158 (1972).

46. Kates, M. : Hydrolysis of lecithin by plant plastid enzymes. Can. J. Biochem. Physiol. 33, 575-589 (1955).

47. Hitchcock, C., Nichols, B.: Plant Lipid Biochemistry. Academic Press, London (1971).

48. Mazliak, P.: Lipid metabolism in plants. Ann. Rev. Plant. Physiol. 27, 287-310 (1973).

49. Galliard, T.: Unusual fatty acids in plants. In Metabolism and Regulation of Secondary Plant Products (V.C. Runeckles, E.E. Conn, eds.) Academic Press, New York, pp. 209-241 (1974).

50. Stumpf, P.K., Barber, G.A.: β-oxidation of fatty acids by peanut mitochondria. Plant Physiol. 31, 304-308 (1956).

51. Martin, R.O., Stumpf, P.K.: α-Oxidation of long-chain fatty acids. J. Biol. Chem. 234, 2548-2554 (1959).

52. Beevers, H.: Organelles from castor bean seedlings: biochemical roles in glucogenesis and phospholipid biosynthesis. In Recent Advances in the Chemistry and Biochemistry of Plant Lipids. (T. Galliard, E.I. Mercer, eds.) Academic Press, London. pp. 287-299 (1975).

53. Laties, G.G., Hoelle, C., Jacobson, B.S.: α-Oxidation of endogenous fatty acids in fresh potato slices. Phytochemistry 11, 3403-3411 (1972).

54. Kemp, T.R.: Characterization of some new C_{16} and C_{17} unsaturated fatty aldehydes. J. Am. Oil Chem. Soc. 52, 300-302 (1975).

55. Galliard, T., Matthew, J.A.: The enzymic formation of long chain aldehydes and alcohols by α-oxidation of fatty acids in extracts of cucumber fruit. Biochim. Biophys. Acta 424, 26-35 (1976).

56. Markovetz, A.J., Stumpf, P.K.: Characterization of the product of the peanut α-oxidation system. Lipids 7, 159-164 (1972).

57. Shine, W.E., Stumpf, P.K.: Recent studies on plant α-oxidation systems. Arch. Biochem. Biophys. 162, 147-157 (1974).

58. Galliard, T.: Lipid and fatty acid composition of tubers from different varieties of potato. J. Sci. Food Agric. 24, 617-622 (1973).

59. Maguire, Y.P., Haard, N.F.: Fluorescent product accumulation in ripening fruit. Nature, Lond. 258, 599-600 (1975).

60. Gardner, H. W. : Decomposition of linoleic acid hydroperoxides. Enzymic reactions compared with non-enzymic. J. Agric. Food Chem. 23, 129-136 (1975).

61. Pinsky, A., Grossman, S., Trop, M. : Lipoxygenase content and antioxidant activity of some fruits and vegetables. J. Food Sci. 36, 571-572 (1971).

62. Grosch, W. : Das Enzym Lipoxygenase. Eigenschaften und Wirkung in Lebensmittelm. Fette Seifen Anstrichtm. 74, 375-381 (1972).

63. Axelrod, B. W. : Lipoxygenases. In Food Related Enzymes (J. R. Whitaker, ed.) Amer. Chem. Soc. pp. 324-348 (1974).

64. Hamberg, M., Samuelsson, B., Björkhem, I., Danielsson, M. : Oxygenases in fatty acids and sterol metabolism. In Molecular Mechanisms of Oxygen Activation (O. Hayaishi, ed.). Academic Press, New York, pp. 29-85 (1974).

65. Chan, H. W-S. : Dioxygenation reactions related to lipoxygenase action. Int. Soc. Fat Res. 11th World Congress, Göteborg, Sweden. Abst. p. 211 (1972).

66. Chan, H. W-S. : Soyabean lipoxygenase - an iron-containing dioxygenase. Biochim. Biophys. Acta 327, 32-35 (1973).

67. Hamberg, M., Samuelsson, B. : On the specificity of the oxygenation of unsaturated fatty acids catalysed by soybean-lipoxidase. J. Biol. Chem. 242, 5329-5335 (1967).

68. Egmond, M. R., Vliegenthart, J. F. G., Boldingh, J. : Stereospecificity of the hydrogen abstraction at carbon atom n-8 in the oxygenation of linoleic acid by lipoxygenase from corn germs and soya beans. Biochem. Biophys. Res. Commun. 48, 1055-1060 (1972).

69. Gardner, H. W., Weisleder, D. : Lipoxygenase from Zea mays : 9-D-hydroperoxy-trans-10, cis-12-octadecadienoic acid from linoleic acid. Lipids 5, 678-683 (1970).

70. Chan, H. W-S., Costaras, C. T., Prescott, F.A.A., Swoboda, P. A. T. : Specificity of lipoxygenases. Thermal isomerizations of linoleate hydroperoxides, a phenomenon affecting the determination of isomeric ratios. Biochim. Biophys. Acta 398, 347-350 (1975).

71. de Groot, J. J. M. C., Veldink, G. A., Vliegenthart, J. F. G., Boldingh, J., Wever, R., van Gelder, B. F. : Demonstration by EPR spectroscopy of the functional role of iron in soybean lipoxygenase. Biochim. Biophys. Acta 377, 71-79 (1975).

72. Gibian, M. J., Galway, R. A. : Steady-state kinetics of lipoxygenase oxygenation of unsaturated fatty acids. Biochemistry 15, 4209-4214 (1976).

73. Lagocki, J. W., Emken, E. A., Law, J. H., Kezdy, Kinetic analysis of the action of soybean lipoxygenase on linoleic acid. J. Biol. Chem. 251, 6001-6006 (1976).

74. Egmond, M. R., Brunori, M., Fasella, P. M. : The steady-state kinetics of the oxygenation of linoleic acid catalysed by soybean lipoxygenase. Eur. J. Biochem. 61, 93-100 (1976).

75. Verhagen, J., Bouman, A. A., Vliegenthart, J. F. G., Boldingh, J.: Conversion of 9-D- and 13-L-hydroperoxy linoleic acids by soybean lipoxygenase-1 under anaerobic conditions. Biochim. Biophys. Acta 486, 114-120 (1977).

76. Pistorius, E. K., Axelrod, B., Palmer, G. : Evidence for participation of iron in lipoxygenase reaction from optical and electron spin resonance studies. J. Biol. Chem. 251, 7144-7148 (1976).

77. Vick, B. A., Zimmerman, D. C. : Lipoxygenase and hydroperoxide lyase in germinating watermelon seedlings. Plant Physiol. 57, 780-788 (1976).

78. Yabuuchi, S. : Occurrence of a new lipoxygenase isoenzyme in germinating barley embryos. Agr. Biol. Chem. 40, 1987-1992 (1976).

79. Matthew, J. A., Chan, H. W-S., Galliard, T. : A simple method for the preparation of pure 9-D-hydroperoxides of linoleic acid and methyl linoleate based on the specificity of lipoxygenase in tomato fruit. Lipids, in press (1977).

80. Weber, F., Grosch, W. : Co-oxidation of a carotenoid by the enzyme lipoxygenase : Influence on the formation of hydroperoxides. Z. Lebensm. Unters. -Forsch. 161, 223-230 (1976).

81. Gardner, H. W., Kleiman, R., Weisleder, D. : Homolytic decomposition of linoleic acid hydroperoxides. Identification of fatty acid products. Lipids 9, 696-706 (1974).

82. Hamberg, M. : Decomposition of unsaturated fatty acids by hemoglobin. Structures of major products of 13-L-hydroperoxy-octadeca-9,11-dienoic acid. Lipids 10, 87-92 (1975).

83. Grosch, W. : Breakdown of linoleic acid hydroperoxides. Formation of volatile carbonyl compounds. Lebensm. Unters.-Forsch. 160, 371-375 (1976).

84. Grosch, W. : Abbau von linol- und linolensäurehydroperoxyden in Gegenwart von Ascorbinsäure. Analyse der flüchtigen Aldehyde. Z. Lebensm. Unters. -Forsch. 163, 4-7 (1977).

85. Streckert, G., Stan, H.J. : Conversion of linoleic acid hydroperoxide by soyabean lipoxygenase in the presence of guaiacol : identification of reaction products. Lipids 10, 847-854 (1975).

86. Heimann, W., Dresen, P., Klaiber, U. : Über die Bildung und den Abbau von Linolsaurehydroperoxiden in Cerealien. Quantitative Bestimmung der Reactions-produkte. Z. Lebensm.Unters. -Forsch. 153, 1-3 (1973).

87. Heimann, W., Dresen, P., Schreier, P. : Über das Lipoxygenase-Lipoperoxidase System in Cerealien. Abtrennung von zwei Proteinkomplexen mit lipoxygenase und Linolsäurehydroperoxid Abbau. Aktivität aus Hafer und Sojabohnen. Z. Lebensm. Unters. Forsch. 153, 147-151 (1973).

88. Arens, D., Grosch, W. : Non-volatile reaction products from linoleic acid. Comparison of a ground pea suspension with a purified pea lipoxygenase. Z. Lebensm. Unters. -Forsch. 156, 292-299 (1974).

89. Yabuuchi, S., Amaha, M. : Partial purification and characterization of the linoleate hydroperoxide isomerase from grains of Hordeum distichum. Phytochemistry 15, 387-390 (1976).

90. Gardner, H.W., Kleiman, R., Christianson, D.D., Weisleder, D.: Positional specificity of γ-ketol formation from linoleic acid hydroperoxides by a corn germ enzyme. Lipids 10, 602-608 (1975).

91. Veldink, G.A., Vliegenthart, J.F.G., Boldingh, J. : The enzymic conversion of linoleic acid hydroperoxide by flax-seed hydroperoxide isomerase. Biochem.J. 120, 55-60 (1970).

92. Zimmerman, D.C., Vick, B.A. : Intracellular distribution of hydroperoxide isomerase in plants. Plant Physiol. 53, 1-4 (1974).

93. Christianson, D.D., Gardner, H.W. : Substitution reactions of linoleic acid hydroperoxide isomerase. Lipids 10, 448-453 (1975).

94. Esselman, W.J., Clagett, C.O. : Products of a linoleic hydroperoxide-decomposing enzyme of alfalfa seed. J. Lipid Res. 15, 173-178 (1974).

95. Garssen, G.J., Veldink, G.A., Vliegenthart, J.F.G., Boldingh, J. : The formation of threo-11-hydroxy-trans-12:13-epoxy-9-cis-octadecenoic acid by enzymic isomerization of 13-L-hydroperoxy-9-cis, 11-trans-octadecadienoic acid by soybean lipoxygenase-1. Eur. J. Biochem. 62, 33-36 (1976).

96. Galliard, T., Phillips, D.R., Matthew, J.A. : Enzymic reactions of fatty acid hydroperoxides in extracts of potato tuber. II. Conversion of 9- and 13-hydroperoxy-octadecadienoic acids to monohydroxydienoic acid derivatives. Biochim. Biophys. Acta 409, 157-171 (1975).

97. Galliard, T., Phillips, D.R. : The enzymic conversion of linoleic acid into 9-(nona-1'3'-dienoxy)-non-8-enoic acid, a novel unsaturated ether derivative isolated from homogenates of Solanum tuberosum tubers. Biochem. J. 129, 743-753 (1972).

98. Galliard, T., Phillips, D.R., Frost, D.J. : Novel divinyl ether fatty acids in extracts of Solanum tuberosum. Chem. Phys. Lipids 11, 173-180 (1973).

99. Galliard, T., Matthew, J.A. : Enzymic reactions of fatty acid hydroperoxides in extracts of potato tuber. I. Comparison of 9-D- and 13-L-hydroperoxydienoic acids as substrates for the formation of butadienyl vinyl ether derivatives. Biochim. Biophys. Acta 398, 1-9 (1975).

100. Galliard, T., Wardale, D.A., Matthew, J.A. : The enzymic and non-enzymic degradation of colneleic acid, an unsaturated fatty acid ether intermediate in the lipoxygenase pathway of linoleic acid oxidation in potato (Solanum tuberosum) tubers. Biochem. J. 138, 23-31 (1974).

101. Eriksson, C. : Aroma compounds derived from oxidized lipids. Some biochemical and analytical aspects. J. Ag. Food Chem. 23, 126-128 (1975).

102. Drawert, F. : Biochemical formation of aroma compounds. Proc. Int. Symp. Aroma Res. Zeist. Pudoc, Wageningen, 13-39 (1975).

103. Tressl, R., Holzer, M., Apetz, M. : Biogenesis of volatiles in fruits and vegetables. Proc. Int. Symp. Aroma Res. Zeist. Pudoc, Wageningen, 41-42 (1975).

104. Galliard, T., Phillips, D.R. : The enzymic cleavage of linoleic acid to C_9 carbonyl fragments in extracts of cucumber (Cucumis sativus) fruits and the possible role of lipoxygenase. Biochim. Biophys. Acta 431, 278-287 (1976).

105. Galliard, T., Phillips, D.R., Reynolds, J. : The formation of cis-3-nonenal, trans-2-nonenal and hexanal from linoleic acid hydroperoxide isomers by a cleavage enzyme system from cucumber (Cucumis sativus) fruits. Biochem. Biophys. Acta 441, 181-192 (1976).

106. Galliard, T., Matthew, J.A., Fishwick, M.J., Wright, A.J. : The enzymic degradation of lipids resulting from physical disruption of cucumber (Cucumis sativus) fruit. Phytochemistry 15, 1731-4 (1976).

107. Galliard, T., Matthew, J.A. : Lipoxygenase-mediated cleavage of fatty acids to carbonyl fragments in tomato fruits. Phytochemistry 16, 339-343 (1977).

108. Galliard, T., Matthew, J.A., Wright, A.J., Fishwick, M.J. : The enzymic breakdown of lipids to volatile and non-volatile carbonyl fragments in disrupted tomato fruit. Submitted to J. Sci. Food Agric.

109. Tremolières, A., Dubacq, J.P. : Formation of a Δ^9-dodecenoic dibasic acid from linoleic acid by young pea leaves. Phytochemistry 15, 1123-1124 (1976).

110. Mazliak, P., Kader, J.C. : Lipid metabolism in ageing plant storage tissues. This volume.

111. Galliard, T. : Techniques for overcoming problems of lipolytic enzymes and lipoxygenases in the preparation of plant organelles. In Methods in Enzymology Vol. 31 (Biomembranes; S. Fleischer, L. Packer, eds.) Academic Press, pp. 520-528 (1974).

112. Tseng, T.C., Mount, M.S. : Toxicity of endopolygalacturonate trans eliminase, phosphatidase and protease to potato and cucumber tissue. Phytopathology 64, 229-236 (1974).

113. Shephard, D.V., Pitt, D. : Purification of a phospholipase from Botrytis and its effects on plant tissues. Phytochemistry 15, 1465-1470 (1976).

114. Jacobson, B. S., Laties, G. G., Smith, B. N., Epstein, S., Laties, B. : Cyanide-induced transition from endogenous carbohydrate to lipid oxidation as indicated by the carbon-13 content of respiratory CO_2. Biochim. Biophys. Acta 216, 295-304 (1970).

115. Jacobson, B. S., Smith, B. N., Epstein, S., Laties, G. G. : The prevalence of carbon-13 in respiratory carbon dioxide as an indicator of the type of endogenous substrate. The change from lipid to carbohydrate during the respiratory rise in potato slices. J. Gen. Physiol. 55, 1-17 (1970).

116. Laties, G. G., Hoelle, C. : The α-oxidation of long chain fatty acids as a possible component of the basal respiration of potato slices. Phytochemistry 6, 49-57 (1967).

117. Heinen, W., Brand, I. : Enzymatische Aspekte zur Biosynthese des Blatt-Cutins bei Gasteria verricuosa - Blättern nach Verletzung. Zeit. Naturforsch. 18B, 67-79 (1963).

118. Bredemeijer, G., Heinen, W. : Cutin synthesis in plants. I. Free fatty acid movement during cutin synthesis in injured leaves. Acta Bot. Neerl. 17, 15-26 (1968).

119. Lipetz, J. : Wound healing in higher plants. Int. Rev. Cytol. 27, 1-28 (1972).

120. Strong, F. E., Kruitwagen, E. : Traumatic acid : An accelerator of abscission in cotton explants. Nature 215, 1380-1381 (1967).

121. Hall, S. W., Morris, L. J. : Unpublished results quoted in Hitchcock, C. and Nichols, B. W. Plant Lipid Biochemistry Academic Press, London, pp. 276-277 (1971).

122. Haagen-Smit, A. J., Viglierchio, D. R. : Investigations of plant wound hormones. Rec. Trav. Chem. 74, 1197 (1955).

123. Meigh, D. F. : Problems of ethylene metabolism. Nature 196, 345-347 (1962).

124. Lieberman, M., Mapson, L. W. : Genesis and biogenesis of ethylene. Nature 204, 343-345 (1964).

125. Galliard, T., Hulme, A. C., Rhodes, M. J. C., Wooltorton, L. S. C. Enzymic conversion of linolenic acid to ethylene by extracts of apple fruits. FEBS Lett. 1, 283-286 (1968).

126. Curtis, R. W. : Oxygen requirement for ethane production in vitro by Phaseolus vulgaris. Plant Physiol. 44, 1368-1370 (1969).

127. Elstner, E. F., Konze, J. R. : Effect of point freezing on ethylene and ethane production by sugar beet leaf disks. Nature 263, 351-352 (1976).

Terpenoids and their Role in Wounded and Infected Plant Storage Tissue

Joe Kuć and Norberto Lisker

Terpenoids are compounds that have as their basic structural components branched five carbon (5C) building blocks. These blocks are derived from the common 5C precursor isopentenyl pyrophosphate (IPP). The terpenoids are often referred to as isoprenoids due to their structural but not biosynthetic relationship to the 5C compound isoprene. They are ubiquitous in nature and are found and synthesized in plants and animals. They range in complexicity from single 5C units as part of non-terpenoid molecules, mixed terpenoids, to complex ring structures, the steroids, and polymeric open chain structures, the polyterpenes. The terpenoids

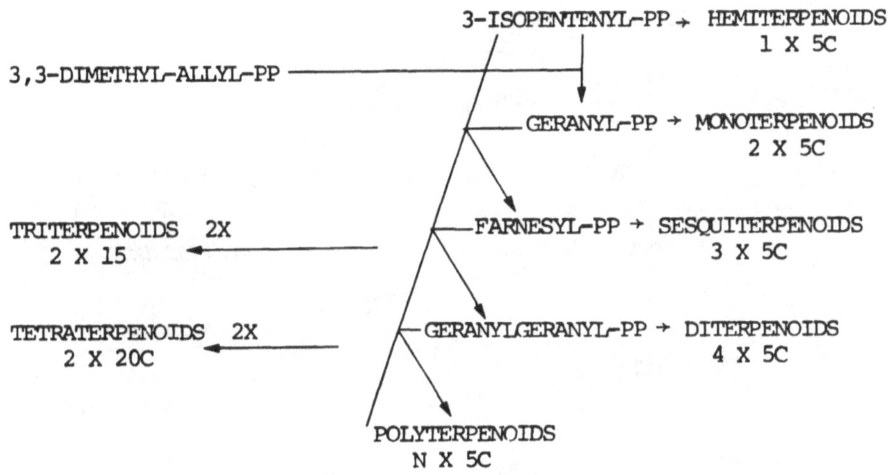

Figure 1. A scheme for the derivation of terpenoids from isopentenyl pyrophosphate.

Biochemistry of Wounded Plant Tissues

are arranged in subgroups according to the number of 5C units from which they are derived. The hemiterpenoids are derived from one 5C unit, the monoterpenoids two 5C units, the sesquiterpenoids three 5C units, etc. (Figure 1). Occasionally carbon atoms are lost during metabolism, e.g., the norsesquiterpenoid rishitin has fourteen rather than fifteen carbon atoms. The terpenoids are normal constituents of all cells, some being unique to plants others to animals. Their biosynthesis is not limited to wounded or infected tissues and some terpenoids have profound regulatory effects on animal metabolism, e.g., steroid hormones, and plant metabolism, e.g., the gibberellins and abscisic acid. The position of terpenoids in metabolism is schematically illustrated in figure 2.

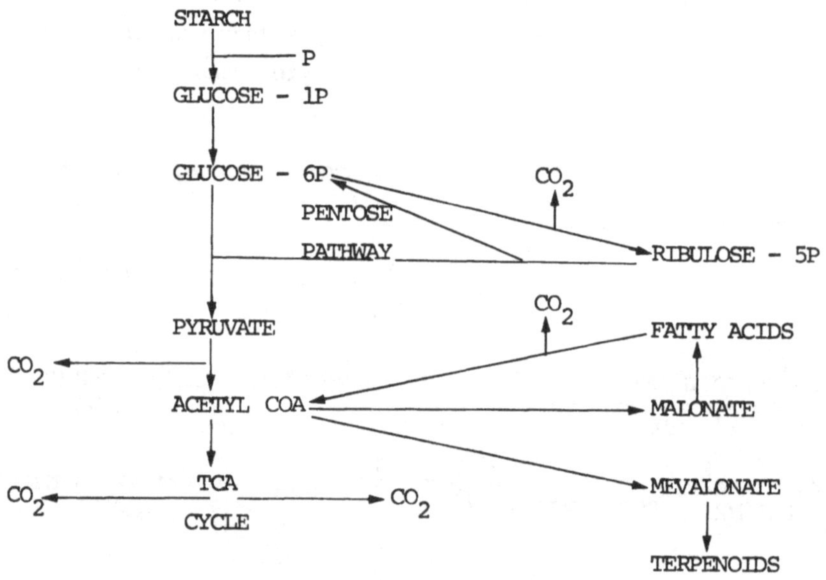

Figure 2. Position of terpenoids in metabolism.

Wounding or infection often disrupts the normal patterns of terpenoid synthesis and degradation in plant cells giving rise to profound qualitative and quantitative changes in terpenoids. These changes may be related to wound repair or disease resistance mechanisms in plants. Some of the terpenoids accumulating after wounding or infection inhibit the development of bacteria and fungi and are included in a class of compounds called phytoalexins, though the term "stress metabolites" more accurately describes their origin and function (33, 96, 99, 128). Numerous reviews on phytoalexins and their role in disease resistance are available (34, 71, 93, 96, 97, 99-103, 173). An excellent review of sesquiterpenoid stress compounds of the Solanaceae has recently appeared (176).

METABOLISM

Injury and infection in plants markedly affect metabolism. Since metabolic processes are interdependent and regulated, a process that was unaffected by injury or infection would be the exception rather than the rule. Infection and/or injury of plants lead to the accumulation of numerous norsesquiterpenoid, sesquiterpenoid and triterpenoid derivatives in potato, eggplant and Datura stramonium (Fig. 3, 4); sweet pepper, and Nicotiana spp. (Fig. 5); cotton (Fig. 6); sweet potato (Fig. 7) and the Pinacea (Fig. 8). The role of individual terpenoids in wound repair and disease resistance will be considered later in this chapter. Some general considerations, however, appear pertinent concerning their metabolism and its regulation.

A high concentration of the steroid glycoalkaloids α-solanine and α-chaconine (Fig. 3) is localized in the peel below the primary periderm of potato tubers. Analyses of four cultivars indicated an average of 0.55 and 0.03 mg/g fresh weight in the peel and peeled tubers, respectively (161). Soon after removal of the peel, secondary periderm formation occurs and the steroid glycoalkaloids accumulate to a depth of one or two mm (72). The concentration which accumulates in the top mm of peeled tissue may reach or surpass that in the peel (162, 164). Slices inoculated with incompatible races of Phytophthora infestans or some nonpathogens of potato accumulate low levels of the steroid glycoalkaloids and high levels of numerous sesquiterpenoids and a norsesquiterpenoid (Fig. 4). Sweet potato roots inoculated with Ceratocystis fimbriata accumulate extremely high levels of furanoterpenoids and the level of the major furanoterpenoid, ipomeamarone, may reach 1-2% of the fresh weight of the upper mm of an infected slice (186). A considerable number of carbon skeletons and a large amount of energy are required to accomplish the biosyntheses described, and the initiation, direction and termination of synthesis appear under metabolic control. It appears unlikely that fatty acids are the principal source of acetyl CoA and energy required for terpenoid synthesis in potato tubers. The tubers are low in storage lipids (0.1% total lipid) and, though acyl hydrolases and lipoxygenases are liberated and α and β oxidation are active in cut tissue (48, 49, 50, 60, 79, 106, 202), overall fatty acid metabolism becomes rapidly directed toward the synthesis of suberin and various cytoplasmic and mitochondrial membranes (79).

The synthesis of suberin and new cells necessary for periderm formation require an increased availability of fatty acids and the incorporation of acetate -1-^{14}C into lipids is enhanced several fold during the first four hours after cutting. This increase follows the respiratory rise (205, 206). The incorporation of acetate -1-^{14}C into total lipids, polar lipids and neutral fats reaches a maximum within 10-12 hours and is subsequently appreciably slowed down. Thus, though injury releases acyl hydrolases and lipoxygenases which may be destructive to membranes, the rapid synthesis of fatty acids and their incorporation into new membranes may be a key factor of the repair mechanism to return the tissue to "normalcy". The contribution of lipids to respiration of potato slices appears significant only during the first 24 hours after cutting, and subsequently the source of carbon is predominantly carbohydrate, probably starch (73, 106).

Little free glucose is available in intact tubers and it appears that the bulk of the carbon and energy required for terpenoid synthesis under stress conditions are derived from glucose stored as starch. Starch degradation is evident 9-12 hours after slicing potato tubers and after 48 hours most of the starch in the phellogen and phellem cells is degraded (142, 170). Starch degradation is largely due to enhanced phosphorylase activity (79), therefore, glucose-1-P (G-1-P) is likely to be the major carbon precursor entering the metabolic pool from which terpenoids are synthesized (Fig. 2). The rate of starch hydrolysis, glycolysis and pentose pathway activity would all influence the production of acetyl CoA, and the consumption of acetyl CoA in the tricarboxylic acid cycle and fatty acid synthesis would limit its availability for terpenoid synthesis (Fig. 2).

What is the signal, and perhaps there is more than one, which initiates the profound metabolic changes associated with the "wound response" and which have as one consequence the increased accumulation of terpenoids? Slicing a potato tuber has been reported to immediately release CO_2 trapped in the tuber (79). The periderms of the potato tuber and sweet potato root may function as a barrier to diffusion of CO_2 and levels above 5% have been reported in potato tuber and sweet potato root (24, 26, 52, 79, 204). The release of CO_2 is temporary and within 1-2 hours is ended. At this time respiratory activity of cut tuber discs is 5-10 times that of tissue within the intact tuber. The signal for terpenoid biosynthesis and accumulation may be the sudden release of CO_2 which removes a brake to metabolic activity and activates phosphorylase. Within 10-12 hours fatty acid synthesis markedly declines (1), and if the activity of glucose metabolism via the pentose pathway and glycolysis has not declined, acetyl CoA may accumulate and be shunted off to the synthesis of terpenoids. Indeed, the accumulation of steroid glycoalkaloids beneath a slice surface in potato tubers requires a lag period of 12-24 hours and doesn't reach a maximum until 96-144 hours after slicing (164). The impaired oxidation of acetyl CoA due to the inhibition of TCA oxidation by short chain fatty acids (106), which arise from lipid degradation, would provide appreciable carbon skeletons for terpenoid synthesis only during the first 12-24 hours after tissue is sliced. The steroid glycoalkaloids may represent a trap for acetyl CoA until periderm formation is sufficient to once again brake metabolism. The steroid glycoalkaloids may, however, be more than a "metabolic safety valve". Their antibiotic activity (2) and localization at the wounded surface may make them an important part of the wound repair process.

Pathogens generally cause physical and metabolic alterations in plants which resemble wounding, i.e., destruction of permeability barriers, positive traumatotaxis of nuclei, enhanced cytoplasmic streaming, increase in the number of mitochondria and ribosomes, synthesis of enzymes, accumulation of phenolics and terpenoids, and activation of the pentose pathway and tricarboxylic acid cycle (79). Wounding and infection have, however, some very significant differences. Plant metabolism is undoubtedly modified by extracellular toxins and enzymes produced by pathogens and by the metabolic activity of pathogens. The developing pathogen in a host has a continuous effect on host metabolism which is quite different from the transient trauma elicited by slicing or mechanical injury.

The acetate-mevalonate pathway is activated following the slicing of sweet potato roots (132, 137, 138, 139, 181) and potato tubers (162, 179). The pathway for synthesis of α-solanine and α-chaconine in potato tubers probably follows the general pathway HMGCoA → mevalonate → isopentenyl PP → farnesyl PP → squalene → cycloartenol → cholesterol → α-solanine and α-chaconine (55, 74).

Though the pathways for the biosynthesis of the norsesquiterpenoids and sesquiterpenoids of potato and furanoterpenoids of sweet potato are not completely elucidated, several have been suggested. Dehydroipomeamarone may be an immediate precursor for ipomeamarone on the pathway from farnesylpyrophosphate (135). Two schemes suggested by Stoessl, Stothers and Ward (174, 176) accomodate spirovetiva-1 (10), 11-diene-2-one (solavetivone) spirovetiva-1(10), 3,11-triene-2-one (anhydro-β-rotunol), lubimin, rishitin, capsidiol, and phytuberin. In a preliminary report, Kalan and Osman (80) observed that potato slices treated with solavetivone yielded lubimin and rishitin within 24 hr. A vetispirane, isolubimin, which had not been previously detected in potatoes infected with fungi, was also isolated and it may be an intermediate in the biosynthesis of rishitin and lubimin. The concentration of solavetivone decreased concommitantly with increased concentration of isolubimin, and isolubimin was detected prior to lubimin and rishitin and decreased with increasing rishitin and lubimin concentration. The suppression of the hypersensitive reaction and accumulation of rishitin, phytuberin and lubimin by compatible races of P. infestans and the concommitant accumulation of nonfungitoxic terpenoids (40, 180, 190, 191) suggests a block which may occur in the conversion of solavetivone to lubimin.

The qualitative and quantitative response of plants within the Solanacea to stress and infection varies among plant species. Some terpenoids are common to more than one species, e.g., rishitin in potato and tomato and lubimin in potato, eggplant and Datura stramonium. Within a plant species, different fungi, bacteria and viruses differ qualitatively and quantitatively in their influence on terpenoid accumulation. Conditions of incubation also have a profound influence on terpenoid accumulation, and a recent report indicated that phytuberin accumulation in potato was markedly enhanced by Ethrel, a source of ethylene (107). It is not surprising that different microorganisms and conditions of incubation influence terpenoid accumulation since a small alteration in environment can have a profound effect on metabolic processes, and caution is advised in interpreting these differences to the presence of "specific elicitors" in the cell wall of microorganisms.

The concentration of steroid glycoalkaloids remains constant at the surface of a cut potato slice after ca 144 hr which suggests there is little metabolism of the compounds. The norsesquiterpenoids and sesquiterpenoids, however, appear to be rapidly metabolized and the level of rishitin markedly drops in potato tubers infected with P. infestans or treated with cell-free sonicates of the fungus 120-144 hr after inoculation or treatment (72, 98). Accumulation of the sesqui- and norsesquiterpenoids may be dependent on the relative rates of biosynthesis and degradation.

The formation of terpenoids in plants as a response to wounding or infection indicates a pattern in which families of plants form similar terpenoids (172, 174, 176). Individual terpenoids will, therefore, be considered as they occur in various plant families.

SOLANACEAE

Potato, Tomato, Eggplant, and Jimsonweed

Of the four plants in this group, the Irish potato has been studied most thoroughly. Rishitin (185), lubimin (123), phytuberin (192), phytuberol (35), hydroxylubimin (84), rishitinol (83), anhydro β-rotunol (32), and solavetivone (32) have been isolated from potatoes following infection with fungi or bacteria (Fig. 4). Isolubimin, a possible precursor of lubimin, was obtained by applying spirovetiva-1 (10), 11 diene 2-one to potato tuber slices (80).

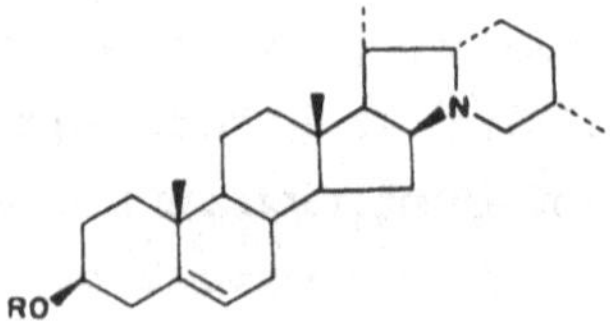

R = H = SOLANIDINE

R = β-SOLATRIOSYL = α-SOLANINE(P)

R = β-CHACOTRIOSYL = α-CHACONINE(P)

R = H = TOMATIDENOL

R = β-SOLATRIOSYL = α-SOLAMARINE(P)

R = β-CHACOTRIOSYL = β-SOLAMARINE(P)

TOMATIDINE = 5,6 DIHYDROTOMATIDENOL

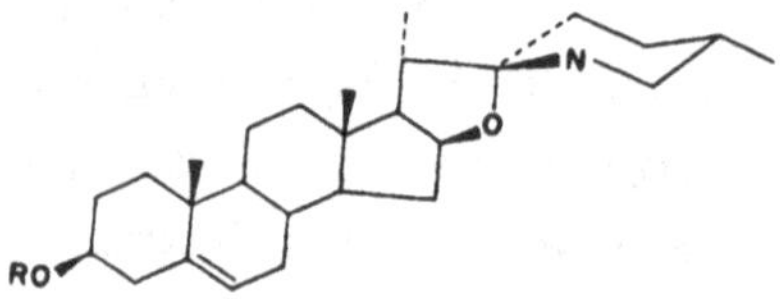

TOMATINE(T) = TOMATIDINE(R = β-LYCOTETRAOSYL)

Figure 3. Major steroid glycoalkaloids of potato (P) and tomato (T).

Phytuberin and phytuberol accumulation was enhanced by Ethrel applied prior to infection with Phytophthora infestans (107). Rishitin is the only stress terpenoid reported in tomato (157, 176, 183), whereas lubimin, a bicyclic sesquiterpenoid, and several other acyclic sesquiterpenoids accumulate in infected eggplants (174, 176, 197). The accumulation of lubimin, hydroxylubimin, germacrenediol and a small amount of capsidiol has been reported in infected jimsonweed, Datura stramonium (175, 198). Dihydroxgermacrene (Fig. 4) is a possible precursor of hydroxylubimin and lubimin (198).

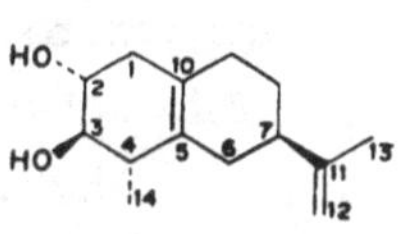

RISHITIN (P,T)

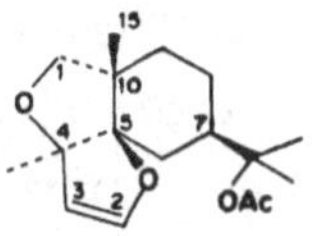

PHYTUBERIN (P)

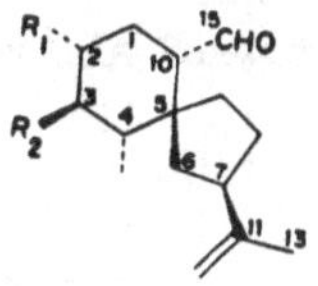

R_1 = OH,R_2 = H: LUBIMIN (P,E,D)

R_1, R_2 = OH: HYDROXYLUBIMIN (P,D)

R_1= = O,R_2=H: ISOLUBIMIN (P)

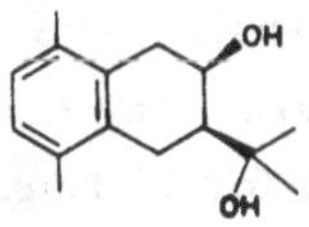

RISHITINOL(P)

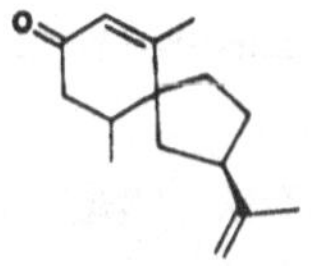

SOLAVETIVONE(P)

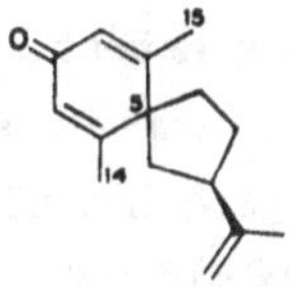

ANHYDRO-β-ROTUNOL(P)

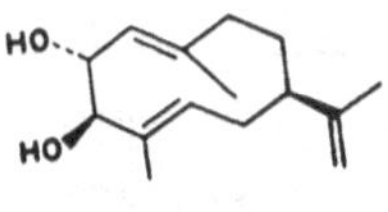

2,3-DIHYDROXYGERMACRENE(D)

Figure 4. Major sesqui and norsesquiterpenoids of potato (P), tomato (T), eggplant (E) and Datura stramonium (D).

Rishitin was the first terpenoid isolated from potato and tomato infected with an incompatible race of P. infestans (157, 185). This substance was characterized as a bicyclic norsesquiterpene alcohol (85). Phytuberin, an aliphatic unsaturated sesquiterpene acetate (66, 188) and lubimin, a bicyclic sesquiterpene aldehyde (123) were characterized shortly thereafter. These substances will be considered here in depth because of their toxicity toward microorganisms. The ED_{50} for mycelial growth of P. infestans for rishitin, phytuberin and lubimin is 50, 40, and 60 μg/ml, respectively. Rishitin is also toxic to some bacteria (112). In a study with 14 fungi, phytuberin was toxic only to P. infestans, whereas anhydro β-rotunol and solavetivone showed general toxicity toward all the fungi (58).

Rishitin, lubimin, and phytuberin accumulated in potato tuber slices inoculated with incompatible races of P. infestans or non-pathogens of potato that have glucan-cellulose or glucan-chitin (12) as their major cell wall component (108, 122, 141, 189, 192). The fungi belonging to the first group, even after they were sonicated and autoclaved for 15 minutes, did not lose elicitor activity (108). Compatible races of P. infestans elicited the accumulation of appreciable amounts of these terpenoids only after death of the fungus (89, 108, 189). Fungi with glucan-mannan, chitin-mannan or chitin-chitosan as major cell wall components (12), either alive or autoclaved, failed to elicit the accumulation of appreciable quantities of terpenoids (108). Some bacteria (13, 111, 113) elicited the accumulation of rishitin, phytuberin and to a lesser extent lubimin. Heat-killed bacteria (65 C for 20 min) did not elicit terpenoid accumulation (108). Lisker and Kuć (108) reported that 15 polysaccharides (including 2 commercial laminarins) as well as 9 lipids did not elicit terpenoid accumulation. Currier (35) suggested that saponins were part of the elicitor molecule of P. infestans; however, two commercial saponins tested did not elicit the accumulation of terpenoids (108). Varns et al (189) tested several physical methods and chemical substances that cause injury to the potato tuber slices, but all of them were ineffective in eliciting terpenoid accumulation. Some authors reported that mercuric chloride, a nematocide chloramphenicol or sodium fluoride elicited accumulation of terpenoids (92, 124, 184), but others report that chloramphenicol, streptomycin or sodium fluoride did not elicit rishitin accumulation (44, 89, 213, and our unpublished results). The amounts of rishitin accumulated after treating with mercuric chloride (189) or the nematocide (92) were considerably less than that in an incompatible reaction with fungus. Numerous substances extracted from the fungus have been reported as elicitors: proteins (124), lipids (43) and saponin-polysaccharide complexes (35). Currier (35) working with a high molecular component which was separated from sonicated mycelium of P. infestans, demonstrated that elicitor activity was not lost by autoclaving or treatment with trypsin, amylase, cellulase, pronase, chymotrypsin, RNAase and DNAase. Treatment with lipase or laminaninase also did not destroy elicitor activity (unpublished results). DNA from a resistant potato cultivar was reported to enhance resistance when applied to a susceptible cultivar (211).

Rishitin was detected in potato tuber 7-12 hours after inoculation with incompatible races of P. infestans and rapidly accumulated to levels which inhibited growth of the fungus in vitro (153, 154, 156). The substance is synthesized in the healthy tissue around dead cells and diffuses to the dead cells where it accumulates (65). A similar pattern for movement of rishitin and lubimin has been reported in potato leaves (122). Pure rishitin added to healthy tuber tissue is degraded to 5 other compounds in a few hours (65, 130). The amounts of terpenoids accumulated in potato tissue is a function of the potato cultivar, pathogen, inoculation time, and temperature of incubation (36, 107, 141, 156, 192). Though potato cultivars containing major "R" genes for resistance rapidly accumulate numerous sesquiterpenoids when infected with incompatible but not compatible races of P. infestans, all cultivars rapidly accumulated terpenoids when treated with cell-free sonicates from any race of the fungus. Rishitin and phytuberin accumulation were markedly suppressed when inoculation with a compatible race of P. infestans preceded inoculation with a incompatible race or treatment with cell-free sonicates of the fungus (190). Non-fungitoxic terpenoids accumulated in compatible interactions or when rishitin and phytuberin accumulation were suppressed (180).

The involvement of rishitin in the defense mechanism of potato has been questioned (44, 89, 189). Varns et al (189) reported that the infection of etiolated potato sprouts with a compatible race of P. infestans lead to the rapid formation of large brown lesions and accumulation of rishitin. In the incompatible interaction, the lesions were restricted and little or no rishitin was detected. A reinvestigation of this phenomenon by the authors showed that brown lesions were as rapidly formed (36-48 hrs) in the compatible as in the incompatible interaction and rishitin rapidly accumulated in both interactions. Sprouts of susceptible potato cultivars reacted as sprouts of resistant cultivars. The high amounts of steroid glycoalkaloids present in uninfected sprouts of all cultivars of potato may play an important role in the hypersensitive response of the sprouts to compatible and incompatible races of P. infestans (161). Recently Dorozhkin et al (41) reported that more rishitin and lubimin accumulated in infected resistant than infected susceptible sprouts.

The suggestion has been made that the hypersensitive reaction and rishitin accumulation may be a consequence and not a cause of resistance (44, 89). When P. infestans was killed in a compatible interaction, the affected potato tissue rapidly browned and terpenoids accumulated as in the incompatible interaction. The investigators concluded that the death of the fungus released cell materials which in turn caused the death of the host cells, and a chain of events was initiated which culminated with the production of terpenoid phytoalexins. This argument was made less tenable by electron microscopic studies which indicated that the death of potato cells preceded that of the fungus (165) and by studies with infected pepper (75, 76) and lettuce (116). The contention that rishitin accumulation is a secondary event in resistance is also supported by the observation that the greatest accumulation is detected after host cell collapse (153). Histochemical studies are necessary to verify this observation. It is possible that rishitin and/or other terpenoids are formed before they can be detected chemically. The higher and more rapid accumulation of terpenoids in resistant interactions of tubers and leaf petioles, however, has been reported to be a direct function of the degree of resistance in numerous investigations (153, 154, 155).

The involvement of terpenoid phytoalexins in diseases of tomatoes and eggplants deserves more research. Although terpenoid accumulation appears higher in resistant than susceptible tissues of tomato plants infected with Verticillium or Fusarium wilt, the total amount detected was very low (119, 183).

The steroid glycoalkaloids are another group of fungitoxic terpenoids reported to have a role in disease resistance and the wound repair mechanism of plants. α-Solanine and α-chaconine are the major steroid glycoalkaloids in potato (Fig. 3). They are found in healthy tubers as well as in the foliage (57, 140) and their concentration increases under stress conditions in tubers (109). They are largely restricted to the peel in healthy tubers, but their concentration increases markedly around sites of injury. They are the major fungitoxic compounds accumulating around sites of mechanical injury in potato (2, 120). The accumulation of the glycoalkaloids is suppressed by inoculation with various pathogens or nonpathogens (162, 164). In resistant interactions, the suppression is concomittant with an increase in accumulation of simpler fungitoxic terpenoids, e.g. rishitin and phytuberin, which appear synthesized de novo via the acetate-mevalonate pathway (162, 164). The concentration of steroid glycoalkaloids is higher in young than old leaves of potato (140, 210). With all the fungi tested, α-chaconine showed greater fungitoxity than α-solanine (2, 121, 168). The role of these substances in the defense mechanism of the plant is still in question. Allen and Kuć (2) and Locci and Kuć (109) suggested the glycoalkaloids are part of a general defense mechanism of potato tubers. Steroid glycoalkaloid content was reported as unrelated to the resistance of potato to P. infestans and Alternaria solani (37, 46), however, another paper suggests a direct relationship between steroid glycoalkaloid content and resistance to A. solani (168). Recently, two additional steroid glycoalkaloids, α and β-solamarine (Fig. 3) were isolated from leaves and aged potato tuber slices of the cultivar Kennebec. These compounds were not found in 20 other cultivars investigated (163). Kennebec is derived from a cross between a wild Mexican species, Solanum demissum, and Solanum tuberosum. The glycosylation pattern of α and β-solamarine is derived from the latter and the steroid alkaloid from the former parent. Tomatidenol (Fig. 3), is the major steroid alkaloid aglycone found in the former parent. Tomatine (Fig.3), the major steroid glycoalkaloid in tomato, may be a factor in the resistance of tomato to disease caused by Septoria lycopersici (7, 8) and Pseudomonas solanacearum (127). A correlation was not found between tomatine content and the resistance of tomato plants to either Verticillium or Fusarium wilts (105, 183).

Pepper and Nicotiana spp.

Monoterpenes are found in peppers as normal metabolites (27) but the fungitoxic sesquiterpene, capsidiol (Fig. 5), is formed only following inoculation with fungi and to a lesser extent with at least one bacterium (56, 78, 177, 178, 199).

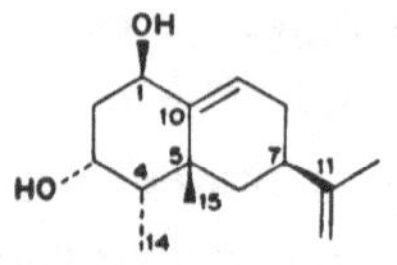

CAPSIDIOL (P,T)

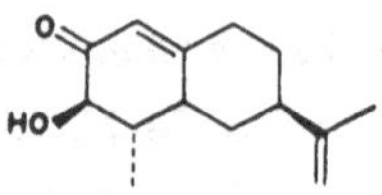

GLUTINOSONE (T)

Figure 5. Major sesquiterpenoids of infected or stressed sweet pepper (P) and tobacco (T).

Capsidiol is synthesized via the acetate-mevalonate pathway, (11) and, as rishitin, readily accumulates in infection droplets on foliage or fruit (77). Capsidiol is more toxic than rishitin, and at a concentration of 5×10^{-4} to 10^{-5} M it inhibited a wide range of pathogenic and nonpathogenic fungi (200). Twenty four hours after inoculation with a nonpathogen, capsidiol accumulated in pepper fruit to levels that prevented the growth of fungi in vitro (75-78). It also accumulated around sites of infection in leaves as part of the hypersensitive response (195). Capsidiol is fungistatic, and other factors are probably responsible for fungal death in resistant interactions (77). Capsidiol is degraded to the less fungitoxic capsenone by fungi (11, 196). This, and the capacity to tolerate high concentrations of capsidiol, may allow slow-growing pathogenic fungi to develop in the fruit. Tomato plants sprayed with solutions of 2.5 X 10^{-3} to 5 X 10^{-1} M capsidiol were protected against disease caused by P. infestans (201).

Though mono-, sesqui- and diterpenoids have been isolated from N. tabacum, it is not apparent whether they are stress compounds (176). In this specie and in N. clevelandii, capsidiol was formed following a virus infection (9), and another fungitoxic terpenoid, glutinosone, (Fig. 5), accumulated in N. glutinosa after virus infection (25). These compounds were formed at the time lesions appeared. Capsidiol is eight times more effective in inhibiting fungal germination than glutinosone. Glutinosone was only detected within lesions where it accumulated to a concentration of 8,000 μg/g fresh weight. The terpenoids may have a role in plant defense against fungi and bacteria, but their role in restricting virus disease is not apparent (10).

COTTON

Gossypol and six related compounds, desoxy-6-methoxyhemigossypol [formerly incorrectly characterized as vergosin (215)], 6-methoxyhemigossypol, 6-desoxyhemigossypol, hemigossypol, 6-methoxygossypol and 6,6'-dimethoxygossypol, are the major terpenoids (Fig. 6) accumulating in cotton (14, 19, 20, 171, 172). Desoxy-6-methoxyhemigossypol and 6-desoxyhemigossypol are non-aldehyde naphthofurans that autoxidize to hemigossypol and 6-methoxyhemigossypol, respectively (171). Hemigossypol is further enzymatically transformed to gossypol (19, 193).

Gossypol is a triterpenoid aldehyde that occurs in seeds, foliar parts and roots of healthy Gossypium (28, 45, 145, 169) and related genera

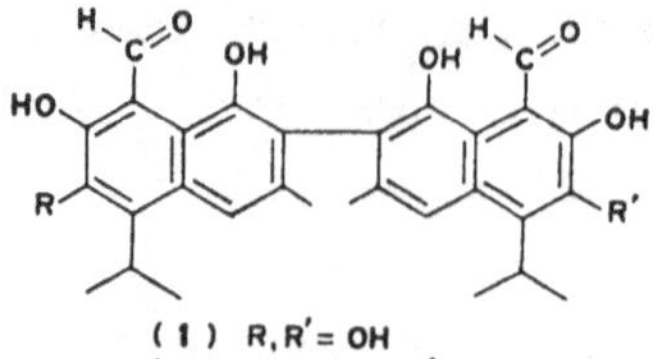

(1) R, R' = OH
(2) R = OMe, R' = OH
(3) R, R' = OMe

1) GOSSYPOL

2) 6-METHOXYGOSSYPOL

3) 6-6'-DIMETHOXYGOSSYPOL

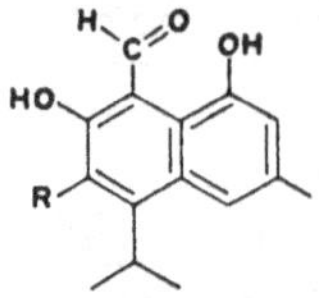

(4) R = OH
(5) R = OMe

4) HEMIGOSSYPOL

5) 6-METHOXYHEMIGOSSYPOL

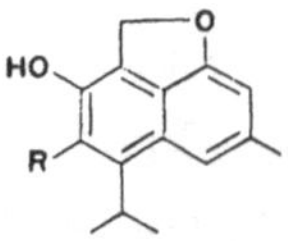

(6) R = OH
(7) R = OMe

6) DESOXYHEMIGOSSYPOL

7) DESOXY-6-METHOXYHEMIGOSSYPOL

Figure 6. Major terpenoids of infected or stressed cotton.

(47, 110, 172). In the seeds and in the aerial parts it is localized in the lysigenous glands (21). Quantities in the roots are higher than in the seeds (145) where it may exist in two optical isomers (38). It is found together with other terpenoids in the roots and is localized in the epidermal and cortical parenchyma cells but not within the first 3 cm of the healthy root tips (115). The methoxylated terpenoid aldehydes form a large percentage of the total aldehydes in the roots (19, 172). Terpenoid aldehydes have not been detected in healthy stelar stem tissue (15, 152, 214).

The accumulation of gossypol and related terpenoids in stems was elicited by inoculating the plants with either living or heat-killed spores of Verticillium, a purified protein-lipolyssacharide from the same fungus (14, 18, 86, 214), chilling, wounding, or treatment with toxic chemicals (14). Other fungi (14, 67) and one bacterium (16) have also been reported to elicit terpenoid accumulation in cotton stems and the sesquiterpenoid aldehydes were in higher concentration than the triterpenoids (19, 172). In the stem these terpenoids are initially formed in scattered, solitary paratracheal parenchyma cells appressed to infected xylem vessel walls, and these cells are unable to form tyloses (114).

The gossypol terpenoids are fungitoxic and believed to function as phytoalexins (15, 19, 152, 214). The ED_{50} of gossypol and desoxy-6-methoxyhemigossypol depend on the species tested, and toward Verticillium is 100 µg/ml and 60 µg/ml, respectively (214). The involvement of gossypol in the protection of cotton against insects was also reported (118). Gossypol is toxic to nonruminants and is removed from cottonseed processed for consumption as foods or feeds (21, 42).

The involvement of these terpenoids in disease resistance was strengthened by the reports of enhanced accumulation of gossypol and its derivatives in resistant but not susceptible cultivars treated with heat-killed conidia, a solution of cupric chloride, a capsular slime preparation from Xanthomonas malvacearum or a protein-lypolyssacharide of Verticillium (16, 17, 18, 86). Increasing temperatures (up to 30 C) increased terpenoid accumulation which paralleled an increase in resistance (17). The terpenoids accumulated faster in cultivars after infection with mildly virulent as compared to highly virulent strains of Verticillium (15, 214). The pathogens Verticillium albo-atrum and Meloidogyne spp. (51, 126) penetrate through the root cap zone which does not have these terpenoids.

SWEET POTATOES

Ipomeamarone (Fig. 7) was first detected in sweet potato roots infected with Ceratocystis fimbriata (69) and its chemical structure was determined by Kubota and Matsuura, ten years later (94).

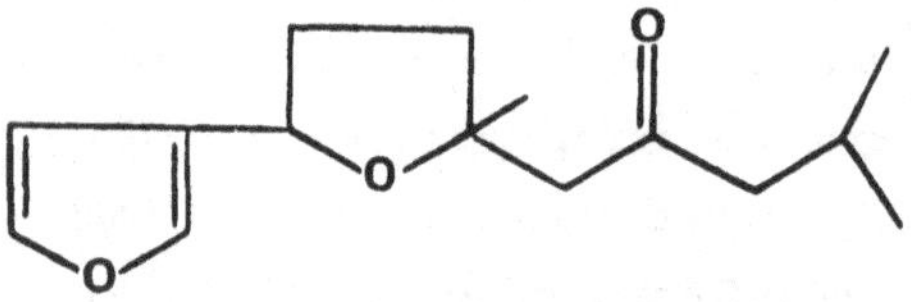

IPOMEAMARONE

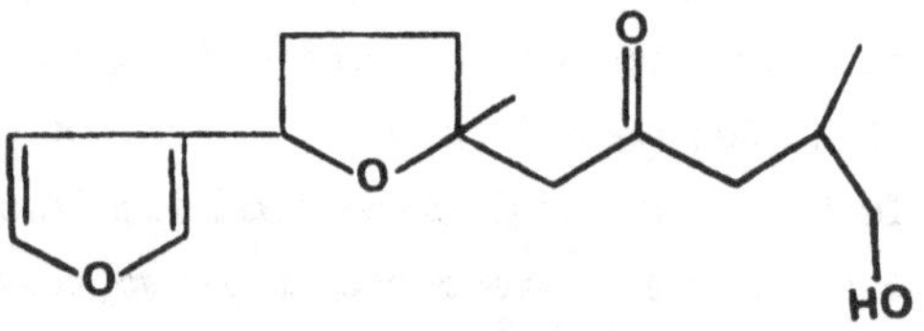

IPOMEAMARONOL

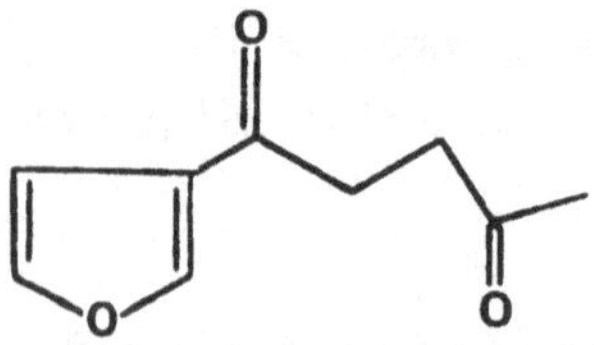

IPOMEANINE

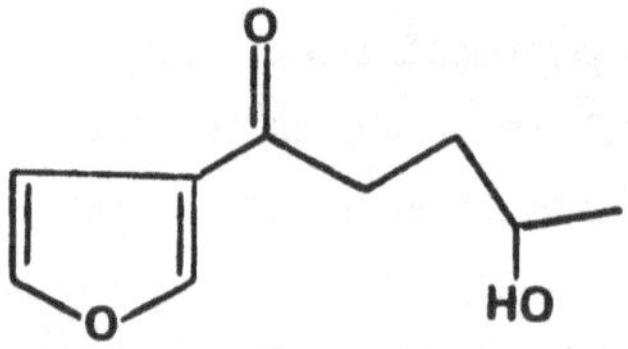

4-IPOMEANOL

Figure 7. Furanoterpenoids of infected or stressed sweet potato.

Several fungi (69, 117, 187, 207, 208), fungal extracts (88), chemical substances (87, 133, 187), and an insect (weevil) (5) elicit accumulation of ipomeamarone in sweet potato roots. A nematode and Streptomyces ipomoea did not elicit the accumulation of ipomeamarone and the internal cork virus elicited low amounts (117). The total concentration of the terpenoids accumulating in the infected tissue depends on the sweet potato cultivar, the pathogen, and the duration of infection (91, 117). Up to 20 mg/g of furanoterpenoids accumulate/g. fresh weight in infected tissue (6).

In addition to ipomeamarone, sweet potato roots accumulate lesser quantities of 4-ipomeanol (209), ipomeamaronol (81, 82, 212), dehydroipomeamarone (134) ipomeanine (95)(Fig. 7) and at least 5-7 additional furanoterpenoids (3, 91). Dehydroipomeamarone is an intermediate in the biosynthesis of ipomeamarone (135). Besides their fungitoxic activity, ipomeamarone and its hydroxyderivative, ipomeamaronol, are toxic to animals. Ipomeanine and ipomeanol are potent lung edemagenic agents (207, 208). Ipomeamarone is formed via the acetate-mevalonate pathway (4, 70, 137). Its formation is preceeded by an increase in activity of the enzymes between mevalonate and isopentynyl pyrophosphate, especially pyrophosphomevalonate decarboxylase (132).

The furanoterpenoids are synthesized in cells invaded by hyphae as well as surrounding noninvaded cells, (91, 182). There is little or no accumulation of ipomeamarone in healthy tissue distant from infected cells or tissue adjacent to such cells (117).

There is evidence that the furanoterpenoids may play an important role in the defense mechanism of the sweet potato root. Developing fungal hyphae are contained in the area in which the furanoterpenoids accumulate (91), the amount accumulating _in vivo_ is much higher than that required to inhibit germination and growth of fungi _in vitro_ (97), ipomeamarone is more toxic to nonpathogenic fungi than to pathogenic ones (91, 131), and it accumulates faster in resistant than susceptible cultivars (6, 186). When ipomeamarone accumulation was inhibited by applying cycloheximide to roots, an avirulent strain of _C. fimbriata_ developed in the tissue (91). Since inhibition by cyclohexamide is not specific, many other processes related to mechanisms for resistance may also have been inhibited. Some reports do not support the involvement of the terpenoids in the defence mechanism. Nonpathogenic strains of _C. fimbriata_ were reported to elicit less accumulation of furanoterpenoids than pathogenic strains (68). However, when the furanoterpenoid content was calculated on the basis of affected cells accumulating the terpenoids, there were no significant differences between the isolates tested (91). The report that ipomeamarone was not formed when resistance to _C. fimbriata_ was induced in sweet potatoes by a previous inoculation with a pathogenic strain of the fungus (203), suggests that factors other than terpenoid accumulation are also involved in resistance (90, 99).

PINACEAE

Many different terpenoids are produced by trees in response to wounding or infection. This chapter will consider only the oleoresin produced by the Pinaceae (Fig. 8).

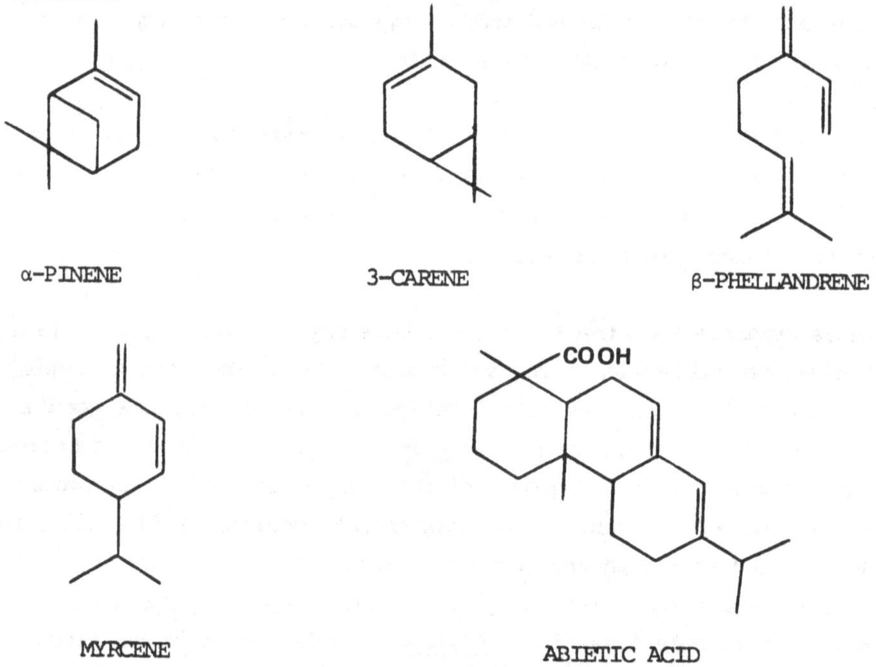

Figure 8. Terpenoids of infected or stressed Pinaceae.

Oleoresin is a viscous hydrophobic liquid secreted by the resin ducts and consists of a solution of resin and fatty acids in a volatile oil. The resin is composed by diterpenoids and consists mainly of two types of resin acids: abietic and pimaric. Mono and sesquiterpenoids are the major compounds found in the volatile oil. The monoterpene α-pinene is the most common (129). Nonetheless, there are large differences in the composition of the oil between genera of the same family, species of the same genera and different parts of the same tree (104).

The exact role of the oleoresins in the defence mechanism is still unclear. Infection and growth of Fomes annosus in roots was not prevented by the oleoresin that accumulated at the infection site (125). Infected varieties susceptible to F. annosus exuded more resins than did resistant ones (64). Resin flow as a consequence of mechanical wounding ceased after 24 hours, but after fungal infection it lasted much longer probably because of the constant stimulation resulting from the slow growth of the pathogen (30, 61). This may explain the higher total amounts of resin in the susceptible interaction. Similarly, it was reported that because of continuous synthesis promoted by the continuous growth of Verticillium in a susceptible cotton variety, no differences in the total amounts of terpenoids was found between the susceptible and resistant interaction 14 days after infection (15, 114). On the other hand, several authors related the increased resistance of conifers to a more intense exudation of resin (22, 23, 53, 136, 143, 147, 148, 159, 194) as that which also occurs on acidic soil (53, 147, 194). Insect injury also causes the exudation of oleoresin. A primary flow was noted as soon as the beetle, Dendroctonus ponderosa, penetrated the tree and if the beetle was not immediately killed, a secondary flow began (146, 158). Resistance to some diseases was related to resistance to insect vectors which were rapidly engulfed in the resinosis tissue (62, 146, 166, 167). Successful infections were associated with minimal secondary resinosis. Pathogenic fungi were rarely isolated from resinous wood behind the infection site (53, 194). The only reported successful isolation was done by Shain (159) in old infections but a decrease in resin content was also observed. Resistance to the fusiform rust was correlated to the higher amounts of the monoterpene β-phellandrene in the resistant trees (149). High amounts of another monoterpene, 3-carene, was correlated to the resistance to artificial inoculation with F. annosus (29). Decay caused by this fungus (159) or by Coriolus versicolor and by Poria monticola (59) was much slower in resin-soaked woods than in non-soaked treatments. The resistance of Abies grandis to fungi-vector insects was explained by the formation of the toxic and repellent monoterpenes, myrcene and 3-α carene at the expenses of the highly attractive (150) and less toxic terpenes already present in the preformed resin system (151). Suggestions have been made concerning the mode of action of oleoresin in resistance. The volatile terpenes inhibit growth of many fungi but their effect is fungistatic rather than fungicidal (31, 39, 54, 63). There is a great variation in the toxicity of oleoresin to different fungi. Low inhibitory activity of oleoresins to some fungi does not support a role for the compounds in the resistance of pine (144). Much less attention has been given to the nonvolatile compounds of the oleoresin. Abietic and dehydroabietic acids are the major components of the resin soaked resistant wood reported earlier (59). The toxicity of nonvolatile resin to _F. annosus_ (160) was reviewed by Prior (144), and he concluded that these compounds were not inhibitory. Pure resin acids are often very toxic to fungi, but they lose their activity after mixing with the whole oleoresin fraction as occurs in nature. Since the pure resin acids are not available to the invading fungus, Prior (143, 144) suggested that the action of the oleoresin is as a physical barrier that prevents the fungus from spreading inside the tracheids.

SUMMARY

Profound changes occur in terpenoid accumulation following injury or infection of some plant storage tissues. The initiation of these changes, metabolic pathways involved, and the nature of the metabolic control remain to be elucidated. The response to infection may be basically a response to injury or stress as modified by the interaction of plant and infectious agent. Modifications of stress or wound metabolism may be key factors in determining disease resistance or susceptibility in plants.

Journal paper 77-10-100 of the Kentucky Agricultural Experiment Station, Lexington, Kentucky 40506. The authors' work described in this chapter was supported in part by a grant from the Herman Frasch Foundation and grant 316-15-51 of the Cooperative States Research Service of the United States Department of Agriculture.

REFERENCES

1. Abdelkader, A. B., Mazliak, P., Catesson, A. M.: Biogenese des lipides mitochondriaux aux cours de la "survie" (ageing) de disques de parenchyme de tubercule de pomme de terre. Phytochem. 8, 1121-1133 (1969).

2. Allen, E., Kuć, J.: α-Solanine and α-chaconine as fungitoxic compounds in extracts of Irish potato tubers. Phytopathol. 58, 776-781 (1968).

3. Akazawa, T.: Chromatographic isolation of pure ipomeamarone and reinvestigation of its chemical properties. Arch. Biochem. Biophys. 90, 82-89 (1960).

4. Akazawa, T., Uritani, I., Akazawa, Y.: Biosynthesis of ipomeamarone. I. The incorporation of acetate-2-C^{14} and mevalonate-2-C^{14} into ipomeamarone. Arch. Biochem. Biophys. 99, 52-59 (1962).

5. Akazawa, T., Uritani, I., Kubota, H.: Isolation of ipomeamarone and two coumarin derivatives from sweet potato roots injured by the weevil, Cycles formicarius elegantulus. Arch. Biochem. Biophys. 88, 150-156 (1960).

6. Akazawa, T., Wada, K.: Analytical study of ipomeamarone and chlorogenic acid alterations in sweet potato roots infected by Ceratocystis fimbriata. Plant Physiol. 36, 139-144 (1961).

7. Arneson, P., Durbin, R.: Hydrolysis of tomatine by Septoria lycopersici: a detoxification mechanism. Phytopathology 57, 1358-1360 (1967).

8. Arneson, P., Durbin, R.: The sensitivity of fungi to α-tomatine. Phytopathology 58, 536-537 (1968).

9. Bailey, J., Burden, R., Vicent G.: Capsidiol: an antifungal compound produced in Nicotiana tabacum and Nicotiana deselandii following infection with tobacco necrosis virus. Phytochemistry 14, 597 (1975).

10. Bailey, J. A., Vincent, G. G., Burden, R. S.: The antifungal activity of glutinosone and capsidiol and their accumulation in virus infected tobacco species. Physiol. Plant Pathol. 8, 35-41 (1976).

11. Baker, F., Brooks, C. J. W.: Biosynthesis of the sesquiterpenoid, capsidiol, in sweet pepper fruits inoculated with fungal spores.

Phytochemistry 15, 689-694 (1976).

12. Bartnicki-Garcia, S.: Cell wall chemistry, morphogenesis, and taxonomy of fungi. Annu. Rev. Microbiol. 22, 87-108 (1968).
13. Beczner, J., Lund, B.: The production of lubimin by potato tubers inoculated with Erwinia carotovora var. atroseptica. Acta Phytopathol. Acad. Scien. Hungaric. 10, 269-274 (1975).
14. Bell, A.: Formation of gossypol in infected or chemically irritated tissues of Gossypium species. Phytopathology 57, 759-764 (1967).
15. Bell, A.: Phytoalexin production and Verticillium wilt resistance in cotton. Phytopathology 59, 1119-1127 (1969).
16. Bell, A. A.: Biochemical bases of resistance of plants to pathogens. In Biological Control of Plant insects and Diseases (Maxwell, F. G. and F. A. Harris eds) pp. 403-462. Univ. Press Mississippi - Jackson (1974).
17. Bell, A., Presley, J.: Temperature effects upon resistance and phytoalexin synthesis in cotton inoculated with Verticillium albo-atrum. Phytopathology 59, 1141-1146 (1969).
18. Bell, A., Presley, J.: Heat-inhibited or heat-killed conidia of Verticillium albo-atrum induce disease resistance and phytoalexin synthesis in cotton. Phytopathology 59, 1147-1151 (1969).
19. Bell, A. A., Stipanovic, R. D., Howell, C. R., Fryxell, P. A.: Antimicrobial terpenoids of Gossypium: hemigossypol, 6-methoxyhemigossypol and 6-deoxyhemigossypol. Phytochemistry 14, 225-231 (1975).
20. Bell, A. A., Stipanovic, R. D., Howell, C. R., Mace, M. E.: Terpenoid aldehydes of Gossypium: isolation, quantitation and occurence. Proc. Beltwide Cotton Prod. Res. Conf. pp. 40-41 (1974).
21. Berardi, L. C., Goldblatt, L. A.: Gossypol. In Toxic Constituents of Plant Foodstuffs. (I. E. Liener, ed.) pp. 211-266. Academic Press, New York (1969).
22. Berryman, A. A.: Response of Abies grandis to attack by Scolytus ventralis (Coleoptera: Scolytidae). Can. Entomol. 101, 1033-1041 (1969).
23. Berryman, A. A.: Resistance of conifers to invasion by bark beetle - fungus associations. BioScience 22, 598-602 (1972).
24. Boswell, J. G., Whiting, G. C.: Observations on the anaerobic respiration of potato tubers. Ann. Bot. 4, 257-268 (1940).

25. Burden, R., Bailey, J., Vincent, G.: Glutinosone, a new antifungal sesquiterpene from Nicotiana glutinosa infected with tobacco mosaic virus. Phytochemistry 14, 221-223 (1975).

26. Burton, W. G.: Studies on the dormancy and sprouting of potatoes. II. The carbon dioxide content of the potato tuber. New Phytol. 50, 287-296 (1951).

27. Buttery, R., Seifert, R., Guadagni, D., Ling, L.: Characterization of some volatile constituents of bell peppers. J. Agric. Food. Chem. 17, 1322-1327 (1969).

28. Carter, F. L., Castillo, A. E., Frampton, V. L., Kerr, T.: Chemical composition studies of seeds on the genus Gossypium. Phytochemistry 5, 1103-1112 (1966).

29. Chudnyi, A. V., Krangauz, R. A., Gundalva, E. I.: Effect of oleoresin (composition) on the resistance of Scots Pine to Fomes annosus. Lesnoe Khozyaistvo 7, 60-61 (1972) (In Russian) through Forest Abstr. 34, 4654 (1973).

30. Clapper, R. B.: Stimulation of pine oleoresin flow by fungus inoculation. Econ. Bot. 8, 269-284 (1954).

31. Cobb, F. W., Krstic, M., Zauarin, E., Barber, H. W. Jr.: Inhibitory effects of volatile oleoresin components of Fomes annosus and four Ceratocystis species. Phytopathology 58, 1327-1335 (1968).

32. Coxon, D., Price, K., Howard, B., Osman, S., Kalan, E., Zaccharium, M.: Two new vetispirane derivatives: stress metabolites from potato (Solanum tuberosum) tubers. Tetrahedron Letters No. 34, 2921-2924 (1974).

33. Cruickshank, I.: Phytoalexins. Annu. Rev. Phytopathol. 1, 351-374 (1963).

34. Cruickshank, I., Biggs, D., Perrin, D.: Phytoalexins as determinants of disease reaction in plants. J. Indian Botan. Soc. 50A, 1-11 (1971).

35. Currier, W.: Characterization of the induction and suppression of terpenoid accumulation in the potato - Phytophthora infestans interaction. Ph.D. Thesis, Purdue University. 114 p. (1974).

36. Currier, W., Kuć, J.: Effect of temperature on rishitin and steroid glycoalkaloid accumulation in potato tuber. Phytopathology 65, 1194-1197 (1975).

37. Deahl, K., Young, R., Sinclen, S.: A study of the relationship of

late blight resistance to glycoalkaloid content in fifteen potato clones. Am. Potato J. 50, 248-253 (1973).

38. Dechary, J. M., Pradel, P.: Occurrence of (+)-gossypol in Gossypium species. J. Am. Oil Chem. Soc. 48, 563-564 (1971).

39. De Groot, R. C.: Growth of wood-inhabiting fungi in saturated atmospheres of monoterpenoids. Mycologia 64, 863-870 (1972).

40. Doke, N.: Prevention of the hypersensitive reaction of potato cells to infection with an incompatible race of Phytophthora infestans by constituents of the zoospores. Physiol. Plant Pathol. 7, 1-7 (1975).

41. Dorozhkin, N. A., Remneva, Z. I., Ivanyuk, V. G., Psareva, V. V., Sergeev, V. V.: Formation of phytoalexins in sprouts resistant and susceptible to diseases of varieties of potatoes (in Russian). Dokl. Akad. Nauk B. SSR, 18, 931-933 (1974).

42. Eagle, E.: A review of some physiological effects of gossypol and cottonseed pigment glands. J. Am. Oil Chem. Soc. 37, 40-43 (1960).

43. Ersek, T.: Lipoid component from Phytophthora infestans inducing resistance and phytoalexin accumulation in potato tuber. Abst. Symp. Current Topics Plant Pathol. Budapest p. 15 (1975).

44. Ersek, T., Barna, B., Kiraly, Z.: Hypersensitivity and the resistance of potato tuber tissues to Phytophthora infestans. Acta Phytopath. Acad. Scient. Hungar. 8, 3-12 (1973).

45. Frampton, V. L., Pons, W. A. Jr., Kerr, T.: A comparison of chemical properties of seeds of Gossypium species. Economic Bot. 14, 197-199 (1960).

46. Frank, J., Wilson, J., Webb, R.: The relationship between glycoalkaloids and disease resistance in potatoes. Phytopathology 65, 1045-1049 (1975).

47. Fryxell, P. A.: A redefinition of the tribe Gossypieae. Botan. Gaz. 129, 296-308 (1968).

48. Galliard, T.: The enzymic breakdown of lipids in potato tuber by phospholipid and galactolipid-acyl hydrolase activities and lipoxygenase. Phytochem. 9, 1725-1734 (1970).

49. Galliard, T.: Lipids of potato tubers. I. Lipid and fatty acid composition of tubers from different varieties of potato. J. Sci. Fd. Agric. 24, 617-622 (1973).

50. Galliard, T.: Degradation of plant lipids by hydrolytic and oxidative

enzymes. In-Recent Adv. in the Chem and Biochem. of Plant Lipids. (T. Galliard and E. I. Mercer, eds.) Academic Press, London. pp. 319-357 (1975).

51. Garber, R. H., Houston, B. R.: Penetration and development of Verticillium albo-atrum in the cotton plant. Phytopathology 56, 1121-1126 (1966).

52. Gerhardt, F.: Simultaneous measurement of the carbon dioxide and organic volatiles in the internal atmosphere of fruits and vegetables. J. Agric. Res. 64, 207-219 (1942).

53. Gibbs, J.: Resin and the resistance of conifers to Fomes annosus. Ann. Bot. 32, 949-665 (1968).

54. Gibbs, J.: Tolerance of Fomes annosus isolates to pine oleoresin and pinosylvins. Eur. J. For. Path. 2, 147-151 (1972).

55. Goodwin, T. W.: Recent developments in the biosynthesis of plant triterpenes. Recent Adv. Phytochem. 6, 97-115 (1973).

56. Gordon, M., Stoessl, A., Stothers, J.: Post infectional inhibitors from plants. IV. The structure of capsidiol, an antifungal sesquiterpene from sweet peppers. Canad. J. Chem. 51, 748-752 (1973).

57. Gull, D., Isenberg, F.: Chlorophyll and solanine content and distribution in four varieties of potato tubers. Proc. Am. Soc. Hortic. Sci. 75, 545-556 (1960).

58. Harris, J. E., Dennis, C.: Antifungal activity of post-infectional metabolites from potato tubers. Physiol. Plant Pathol. 9, 155-165 (1976).

59. Hart, J., Wardell, J., Hemingway, R.: Formation of oleoresin and lignans in sapwood of white spruce in response to wounding. Phytopathology 65, 412-417 (1975).

60. Hasson, E. P., Laties, G. G.: Separation and characterization of potato lipid acylhydrolases. Plant Physiol. 57, 142-147 (1976).

61. Hepting, G. H.: Stimulation of oleoresin flow in pines by a fungus. Science, N.Y. 105, 209 (1947).

62. Hillis, W. E., Inque, T.: The formation of polyphenols in trees. IV. The polyphenols formed in Pinus radiata after Sirex attack. Phytochemistry 7, 13-22 (1968).

63. Hintikka, V.: Selective effects of terpenes on wood-decomposing fungi. Karstenia 11, 28-32 (1970).

64. Hodges, C. S.: Relative susceptibility of loblolly, longleaf and slash pines to infection by Fomes annosus. Phytopathology 59, 1031 (Abstr.) (1969).

65. Horikawa, T., Tomiyama, K., Doke, N.: Accumulation and transformation of rishitin and lubimin in potato tuber tissue infected by an incompatible race of Phytophthora infestans. Phytopathology 66, 1186-1191 (1976).

66. Hughes, D., Coxon, D.: Phytuberin. Devised structure from the x-ray crystal analysis of dihydrophytuberin. Chem. Comm. 20, 822-823 (1974).

67. Hunter, R. E., Halloin, J. M., Veech, J. A., Carter, W. W.: Terpenoid aldehydes produced by hypocotyls of cotton seedlings in response to infection by Rhizoctonia solani. Beltwide Cotton Prod. Res. Conf. Proc. p. 23 (1975).

68. Hyodo, H., Uritani, I., Akai, S.: Production of furanoterpenoids and other compounds in sweet potato root tissue in response to infection by various isclates of Ceratocystis fimbriata. Phytopathol. Z. 65, 332-340 (1969).

69. Hiura, M.: Studies on storage and rot of sweet potato. Rep. Gifu Agric. Coll. 50, 1-5 (1943).

70. Imaseki, H., Uritani, I.: Ipomeamarone accumulation and lipid metabolism in sweet potato infected by the black rot fungus. II. Accumulation mechanism of ipomeamarone in the infected region with special regard to contribution of the non-infected tissue. Plant and Cell Physiol. (Tokyo) 5, 133-143 (1964).

71. Ingham, J.: Phytoalexins and other plant natural products as factors in plant disease resistance. Botan. Rev. 38, 343-424 (1972).

72. Ishizaka, N., Tomiyama, K.: Effect of wounding or infection by Phytophthora infestans on the contents of terpenoids in potato tubers. Plant and Cell Physiol. 13, 1053-1063 (1972).

73. Jacobson, B. S., Smith, B. N., Epstein, S., Laties, G. G.: The prevalence of carbon-13 in respiratory carbon dioxide as an indicator of the type of endogenous substrate. J. Gen. Physiol. 55, 1-17 (1970).

74. Jadhav, S. J., Salankhe, D. K.: Formation and control of chlorophyll and glycoalkaloid in tubers of Solanum tuberosum L. and evaluation of glycoalkaloid toxicity. Adv. Food Res. 21, 307-354 (1975).

75. Jones, D., Graham, W., Ward E.: Ultrastructural changes in pepper cells in an incompatible interaction with Phytophthora infestans. Phytopathology 65, 1274-1285 (1975).

76. Jones, D., Graham, W., Ward E.: Ultrastructural changes in pepper cells in interactions with Phytophthora capsici (isolate 18) and Monilinia fructicola. Phytopathology 65, 1409-1416 (1975).

77. Jones, D., Unwin, C., Ward, E.: The significance of capsidiol induction in pepper fruit during an incompatible interaction with Phytophthora infestans. Phytopathology 65, 1286-1287 (1975).

78. Jones, D., Unwin, C., Ward, E.: Capsidiol induction in pepper fruit during interactions with Phytophthora capsici and Monilinia fructicola. Phytopathology 65, 1417-1418 (1975).

79. Kahl, G.: Metabolism in plant storage tissue slices. Bot. Rev. 40, 263-314 (1974).

80. Kalan, E. B., Osman, S. F.: Isolubimin: A possible precursor of lubimin in infected potato slices. Phytochem. 15, 775-776 (1976).

81. Kato, N., Imaseki, H., Nakashima, N., Uritani, I.: Structure of a new sesquiterpenoid, ipomeamaronol, in diseased sweet potato root tissue. Tetrahedron Lett. 13, 843-846 (1971).

82. Kato, N., Imaseki, H., Nakashima, N., Akazawa, T., Uritani, I.: Isolation of a new phytoalexin-like compound, ipomeamaronol, from black-rot fungus infected sweet potato root tissue, and its structural elucidation. Plant Cell Physiol. 14, 597-606 (1973).

83. Katsui, N., Matsunaga, A., Imaizumi, K., Masamune, T., Tomiyama, K.: The structure and synthesis of rishitinol, a new sesquiterpene alcohol from diseased potato tubers. Tetrahedron Letters No. 2, 83-86 (1971).

84. Katsui, N., Matsunaga, A., and Masamune, T.: Studies on the phytoalexins. XI. Structure of lubimin and oxylubimin, antifungal metabolites from diseased potato tubers. Tetrahedron Letters 4483-4486 (1974).

85. Katsui, N., Murai, A., Takasugi, M., Imaizumi, K., Masamune, T.: The structure of rishitin, a new antifungal compound from diseased potato tubers. Chem. Comm. 43-44 (1968).

86. Keen, N., Long, M., Erwin, D.: Possible involvement of a pathogen-produced protein - lipopolysaccharide complex in Verticillium wilt

of cotton. Physiol. Plant Pathol. 2, 317-331 (1972).

87. Kim, W. K., Oguni, I., Uritani, I.: Phytoalexin induction in sweet potato roots by amino acids. Agr. Biol. Chem. 38, 2567-2568 (1974).

88. Kim, W. K., Uritani, I.: Fungal extracts that induce phytoalexins in sweet potato roots. Plant Cell Physiol. 15, 1093-1098 (1974).

89. Kiraly, Z., Barna, B., Ersek, T.: Hypersensitivity as a consequence not the cause of plant resistance to infection. Nature 239, 456-458 (1972).

90. Kojima, M., Uritani, I.: The possible involvement of a spore agglutinating factor(s) in various plants in establishing host specificity by various strains of black rot fungus, Ceratocystis fimbriata. Plant and Cell Physiol. 15, 733-737 (1974).

91. Kojima, M., Uritani, I.: Possible involvement of furanoterpenoid phytoalexins in establishing host-parasite specificity between sweet potato and various strains of Ceratocystis fimbriata. Physiol. Plant Pathol. 8, 97-111 (1976).

92. Komai, K., Sato, S.: Induction of rishitin formation and its accumulation in potato-tuber tissues by DBCP treatment. J. Agric. Chem. Soc. Japan 48, 599-604 (1974).

93. Kosuge, T.: The role of phenolics in host response to infection. Annu. Rev. Phytopathol. 7, 195-222 (1969).

94. Kubota, T., Matsuura, T.: Chemical studies on the black rot disease of sweet potato. VI. On the chemical structure of ipomeamarone. J. Chem. Soc. Jap. Ind. Chem. Sect. 74, 248-251 (1953).

95. Kubota, T., Ichikawa, N.: Chemical studies on the black-rot disease of sweet potato. IX. The chemical constitution of ipomeanine. J. Chem. Soc. Japan 7, 450-456 (1954).

96. Kuć, J.: Phytoalexins. Annu. Rev. Phytopathol. 10, 207-232 (1972).

97. Kuć, J.: Compounds accumulating in plants after infection. In - Microbial Toxins Vol. VIII (S. Ajl, G. Weinbaum and S. Kadis, eds.) Academic Press, N. Y., pp. 211-247 (1972).

98. Kuć, J.: Teratogenic constituents of potatoes. In - Recent Advances in Phytochemistry Vol. 9 (V. C. Runeckles, ed.) Plenum Press, New York and London pp. 139-150 (1975).

99. Kuć, J.: Phytoalexins. In - Encyclopedia of Plant Physiology Vol. 4 Physiological Plant Pathology (R. Heitefuss and P. H. Williams, eds.)

Springer-Verlag, Berlin-Heidelberg-New York pp. 632-652 (1976).

100. Kuć, J.: Phytoalexins in the specificity of plant-parasite interaction. In - Specificity in Plant Diseases (R.K.S. Wood and A. Graniti, eds.) Plenum Press, N. Y. pp. 253-268 (1976).

101. Kuć, J., Currier, W.: Phytoalexins, plants and human health. Adv. in Chem. 149, 356-368 (1976).

102. Kuć, J., Currier, W., Elliston, J., McIntyre, J.: Determinants of plant disease resistance and susceptibility: A perspective based on three plant-parasite interactions. In - Biochemistry and Cytology of Plant-Parasite Interaction. (K. Tomiyama, J. Daly, I. Uritani, H. Oku, S. Ouchi, eds.) Kodahansha Ltd. Tokyo and Elsevier, N. Y. pp. 168-180 (1976).

103. Kuć, J., Currier, W., Shih, M. J.: Terpenoid phytoalexins. In - Biochemical Aspects of Plant-Parasite Relationships (J. Friend and D. R. Threlfall, eds.) Academic Press, N. Y. pp. 225-237 (1976).

104. Kurth, E. F.: Volatile oils in Wood Chemistry (Wise, L. E., Jahn, E. L., eds.) Vol. I, pp. 548-589. Reinhold Publ. Corp., New York (1952).

105. Langcake, P., Drysdale, R., Smith, H.: Post-infectional production of an inhibitor of Fusarium oxysporum f. lycopersici by tomato plants. Physiol. Plant Pathol. 2, 17-25 (1972).

106. Laties, G. G., Hoelle, C., Jacobson, B. S.: α-Oxidation of endogenous fatty acids in fresh potato slices. Phytochem. 11, 3403-3411 (1972).

107. Lisker, N., Henfling, J., Kuć, J.: The effect of ethrel on the accumulation of terpenoid phytoalexins in potato tubers. Proc. Amer. Phytopathol. Soc. (Abstr) 3, 356 (1976).

108. Lisker, N., Kuć, J.: Specificity in the elicitation of terpenoid accumulation in potato tubers. Proc. Am. Phytopathol. Soc. 3, 357 (1976).

109. Locci, R., Kuć, J.: Steroid alkaloids as compounds produced by potato tubers under stress. Phytopathology 57, 1272-1273 (1967).

110. Lukefahr, M. J., Fryxell, P. A.: Content of Gossypol in plants belonging to genera related to cotton. Economic Bot. 21, 128-131 (1967).

111. Lyon, G.: Occurrence of rishitin and phytuberin in potato tubers

inoculated with Erwinia carotovora var. atroseptica. Physiol. Plant Pathol. 2, 411-416 (1972).

112. Lyon, G., Baylis, C.: The effect of rishitin on Erwinia carotovora var. atroseptica and other bacteria. Physiol. Plant Pathol. 6, 177-186 (1975).

113. Lyon, G., Lund, B., Bayliss, C., Wyatt, G.: Resistance of potato tubers to Erwinia carotovora and formation of rishitin and phytuberin in infected tissue. Physiol. Plant Pathol. 6, 43-50 (1975).

114. Mace, M. E., Bell, A. A., Beckman, C. H.: Histochemistry and identification of disease-induced terpenoid aldehydes in Verticillium-wilt resistant and - susceptible cottons. Can. J. Bot. 54, 2095-2099 (1976).

115. Mace, M. E., Bell, A. A., Stipanovic, R. D.: Histochemistry and isolation of gossypol and related terpenoids in roots of cotton seedlings. Phytopathology 64, 1297-1302 (1974).

116. MacLean, D. J., Sargent, J. A., Tommerup, I. C., Ingram, D. S.: Hypersensitivity of plant cells as the primary event in resistance to a fungal parasite. Nature (London) 249, 186-187 (1974).

117. Martin, W. J., Hasling, V. C., Cataleno, E. A.: Ipomeamarone content in diseased and nondiseased tissues of sweet potatoes infected with different pathogens. Phytopathology 66, 678-679 (1976).

118. Maxwell, F. G., Jenkins, J. N., Parrott, W. L.: Resistance of plants to insects. Adv. Agron. 24, 187-265 (1972).

119. McCance, D., Drysdale, R.: Production of tomatine and rishitin in tomato plants inoculated with Fusarium oxysporum f. sp. lycopersici. Physiol. Plant Pathol. 7, 221-230 (1975).

120. McKee, R.: Host-parasite relationship in the dry-rot disease of potatoes. Ann. Appl. Biol. 43, 147-148 (1955).

121. McKee, R.: Factors affecting the toxicity of solanine and related alkaloids to Fusarium caeruleum. J. Gen. Microbiol. 20, 686-696 (1959).

122. Metlitskii, L., Ozeretskovskaya, O., Vasyukora, N., Davydova, M., Dorozhkin, N., Remneva, Z., Ivanyuk, V.: Potato resistance to Phytophthora infestans as related to leaf phytoalexin activity. Acad. Nauk SSSR Priklad. Biok. i Microbiol. 6, 568-573 (1970).

123. Metlitskii, L., Ozeretskovskaya, O., Vulfson, N., Chalova, L.: The

role of lubimin in potato resistance to Phytophthora infestans and its chemical identification. Mikologiya i Fitopatol. 5, 439-443. (1971).

124. Metlitskii, L., Yu, T., Dijakov, O. Ozeretskovkaya, L. Yurganova, L. Chalova, L., and Vasyukova, N.: Induction of Phytoalexins (in Russian). Osvestiya Akad. Nauk SSR. S. Biolog. No.3, 399-407 (1971).

125. Miller, T.: Host factors influencing infection and growth of Fomes annosus in roots of Pinus taeda. Phytopathology 54, 901 (Abstr.) (1964).

126. Minton, N. A.: Factors influencing resistance of cotton to root-knot nematodes (Meloidogyne spp.) Phytopathology 52, 272-279 (1962).

127. Mohanakumaran, N., Gilbert, J., Buddenhagen, I.: Relationship between tomatine and bacterial wilt resistance in tomato. Phytopathology 59, 14 (1969).

128. Muller, K. O., Borger, H.: Experimentelle untersuchungen uber die Phytophthora - resistenz der kartotoffel. Arb. Biol. Reichsanstalt. Land-u Forstiv. Berlin 23, 189-231 (1940).

129. Mutton, D. B.: Wood Resins In Wood Extractives (Hillis, W. E. ed.) Academic Press - New York and London. pp. 331-363 (1962).

130. Nakajima, T., Tomiyama, K., Kimkawa, M.: Distribution of rishitin and lubimin in potato tuber tissue infected by an incompatible race of Phytophthora infestans and the site where rishitin is synthesized. Ann.Phytopathol. Soc. Japan 41, 49-55 (1975).

131. Nonaka, F., Yasui, K.: On selective toxicity of ipomeamarone towards the phytopathogens. Agric. Bull. Saga Univ. 22, 39-49 (1966).

132. Oba, K., Tatematsu, H., Yamashita, K., Uritani, I.: Induction of furano-terpene production and formation of the enzyme system from mevalonate to isopentenyl pyrophosphate in sweet potato tissue injured by Ceratocystis fimbriata and by toxic chemicals. Plant Physiol. 58, 51-56 (1976).

133. Oguni, I., Suzuki, K., Uritani, I.: Terpenoid induction in sweet potato roots by cyclic-3'-5'-Adenosine monophosphate. Agr. Biol. Chem. 40, 1251-1252 (1976).

134. Oguni, I., Uritani, I.: Dehydroipomeamarone from infected Ipomoea batatas root tissue. Phytochemistry 13, 521-522 (1974).

135. Oguni, I., Uritani, I.: Dehydroipomeamarone as an intermediate in

the biosynthesis of ipomeamarone, a phytoalexin from sweet potato root infected with Ceratocystis fimbriata. Plant Physiol. 53, 649-652 (1974).

136. Olson, A. J.: A root disease of Jeffrey and Ponderosa pine reproduction. Phytopathology 31, 1063-1067 (1941).

137. Oshima, K., Uritani, I.: Participation of mevalonate in the biosynthetic pathway of ipomeamarone. Agric. Biol. Chem. 32, 1146-1152 (1968).

138. Oshima, K., Uritani, I.: Enzymatic synthesis of a β-hydroxy-β-methylglutaric acid derivative by a cell-free system from sweet potato with black rot. J. Biochem. (Japan) 63, 617-625 (1968).

139. Oshima-Oba, K., Uritani, I.: Enzymatic synthesis of isopentenyl pyrophosphate in sweet potato root tissue in response to infection by black rot fungus. Plant and Cell Physiol. 10, 827-843 (1969).

140. Paseshnichenko, V. A.: Content of solanine and chaconine in the potato during the vegetation period. Biochemistry (USSR) 22, 929-931 (1957).

141. Price, K. R., Howard, B., Coxon, D. T.: Stress metabolite production in potato tubers infected by Phytophthora infestans, Fusarium avenaceum and Phoma exigua Physiol. Plant Pathol. 9, 189-197 (1976).

142. Priestly, J. H., Woffenden, L. M.: The healing of wounds in potato tubers and their propagation by cut sets. Ann. Appl. Biol. 10, 96-115 (1923).

143. Prior, C.: Resin production and susceptibility to Heterobasidium annosum in Corsican pine. Ann. Bot. 39, 1103-1109 (1975).

144. Prior, C.: Resistance by Corsican pine to attack by Heterobasidium annosum. Ann. Bot. 40, 261-279 (1976).

145. Rao, K. V., Sarma, R. V.: Occurrence of gossypol. Curr. Sci. (Bangalore) 14, 270-271 (1945).

146. Reid, R. W., Whitney, G. S., Watson, J. A.: Reactions of lodgepole pine to attack by Dendroctonus ponderosae Hopkins, and blue-stain fungi. Can. J. Bot. 45, 1115-1126 (1967).

147. Rishbeth, J.: Observations on the biology of Fomes annosus with particular reference to East Anglian pine plantations. III. Natural and experimental infection of pines, and some factors affecting the

severity of the disease. Ann. Bot. 15, 221-246 (1951).

148. Rishbeth, J.: Resistance to fungal pathogens of tree roots. Proc. R . Soc. B 181, 333-351 (1972).

149. Rockwood, D.: Cortical monoterpene and fusiform rust resistance relationships in slash pine. Phytopathology 64, 976-979 (1974).

150. Rudinsky, J. A.: Scolytid beetles associated with Douglas fir: response to terpenes. Science (Wash.), 152, 218-219 (1966).

151. Russell, C. E., Berryman, A. A.: Host resistance to the fir engraver beetle. 1. Monoterpene composition of Abies grandis pitch blisters and fungus-infected wounds. Can. J. Bot. 54, 14-18 (1976).

152. Sadykov, A. S., Metlitskii, L. V., Karimdzhanov, A. K., Ismailov, A. I., Mukhamedova, R. A., Avazkhodzhaev, M. K., Kamalo, F. G.: Isohemigossypol - the phytoalexin of the cotton plant. Dokl. Akad. Nauk SSSR. 218, 1472-1475 (1974).

153. Sato, N., Kitazawa, K., Tomiyama, K.: The role of rishitin in localizing the invading hyphae of Phytophthora infestans in infection sites at the cut surface of potato tubers. Physiol. Plant Pathol. 1, 289-295 (1971).

154. Sato, N., Tomiyama, K.: Localized accumulation of rishitin in the potato tuber tissue infected by an incompatible race of Phytophthora infestans. Ann. Phytopathol. Soc. Japan 35, 202-207 (1969).

155. Sato, N., Tomiyama, K.: Relation between rishitin accumulation and degree of resistance of potato tuber tissue to infection by an incompatible race of Phytophthora infestans. Am. Phytopath. Soc. Japan 42, 431-435 (1976).

156. Sato, N., Tomiyama, K., Katsui, N., Masamune, T.: Isolation of rishitin from tubers of interspecific potato varieties containing different late blight resistance genes. Ann. Phytopathol. Soc. Japan 34, 140-142 (1968).

157. Sato, N., Tomiyama, K., Katsui, N., Masamune, T.: Isolation of rishitin from tomato plants. Am. Phytopathol. Soc. Japan 34, 344-345 (1968).

158. Schubert, K.: Das problem der harzbildung in holz und rinde von Pinus nigra Arn. III Die harzbildung im wundnahem gewebe. Angew. Bot. 41, 205-236 (1967).

159. Shain, L.: Sapwood resistance to Fomes annosus in loblolly pine.

Phytopathology 57, 1034-1045 (1967).

160. Shain, L.: The response of sapwood of Norway spruce to infection by Fomes annosus. Phytopathology 61, 301-307 (1971).

161. Shih, M. J.: The Accumulation of Isoprenoids and Phenols and its Control as Related to the Interaction of Potato (Solanum tuberosum L.) with Phytophthora infestans. Ph.D. Thesis, Purdue Univ., Lafayette, Indiana (U.S.A.) 172 p. (1972).

162. Shih, M., Kuć, J.: Incorporation of ^{14}C from acetate and mevalonate into rishitin and steroid glycoalkaloids by potato slices inoculated with Phytophthora infestans. Phytopathol. 63, 826-829 (1973).

163. Shih, M., Kuć, J.: α- and β-solamarine in Kennebec Solanum tuberosum leaves and aged tuber slices. Phytochemistry 13, 997-1000 (1974).

164. Shih, M., Kuć, J., Williams, E. B.: Suppression of steroid glycoalkaloid accumulation as related to rishitin accumulation in potato tubers. Phytopathol. 63, 821-826 (1973).

165. Shimony, C., Friend, J.: Ultrastructure of the interaction between Phytophthora infestans and leaves of two cultivars of potato (Solanum tuberosum L.) Orion and Majestic. New Phytol. 74, 59-65 (1975).

166. Shrimpton, D. M.: Extractives associated with the wound response of lodgepole pine attacked by the mountain pine beetle and associated micro-organisms. Can. J. Bot. 51, 527-534 (1973).

167. Shrimpton, D. M., Whitney, H. S.: Inhibition of growth of blue stain fungi by wood extractives. Can. J. Bot. 46, 757-761 (1968).

168. Sinden, S., Goth, R., O'Brien, M.: Effect of potato alkaloids on the growth of Alternaria solani and their possible role as resistance factors in potatoes. Phytopathology 63, 303-307 (1973).

169. Smith, F. H.: Biosynthesis of gossypol by excised cotton roots. Nature 192, 888-889 (1961).

170. Steward, F. C., Wright, R., Berry, W. E.: The respiration of cut discs of potato tuber in air and immersed in water with observations upon surface: volume effects and salt accumulation. Protoplasma 16, 576-611 (1932).

171. Stipanovic, R. D., Bell, A. A., Howell, C. R.: Naphthofuran precursors of sesquiterpenoid aldehydes in diseased Gossypium. Phyto-

chemistry 14, 1809-1811 (1975).

172. Stipanovic, R. D., Bell, A. A., Mace, H. E., Howell, C. R.: Antimicrobial terpenoids of Gossypium: 6-methoxygossypol and 6,6'-dimethoxygossypol. Phytochemistry 14, 1077-1081 (1975).

173. Stoessl, A.: Antifungal compounds produced by higher plants. Rec. Advan. Phytochem. 3, 143-180 (1970).

174. Stoessl, A., Stothers, J. B., Ward, E. W. B.: Lubimin. A phytoalexin of several Solanaceae. Structure revision and biogenetic relationships. J. Chem. Soc. Chem. Commun. 709-710 (1974).

175. Stoessl, A., Stothers, J., Ward, E.: A 2.3-dihydroxygermacrene and other stress metabolites of Datura stramonium. J. Chem. Soc. Commun. 431-432 (1975).

176. Stoessl, A., Stothers, J. B., Ward, W. B.: Sesquiterpenoid stress compounds of the Solanaceae. Phytochem. 15, 855-872 (1976).

177. Stoessl, A., Unwin, C., Ward, W.: Post-infectional inhibitors from plants. I. Capsidiol, an antifungal compound from Capsicum frutescens. Phytopathol. Z. 74, 141-152 (1972).

178. Stoessl, A., Unwin, C., Ward, W.: Post infectional fungus inhibitors from plants: Fungal oxidation of capsidiol in pepper fruit. Phytopathology 63, 1225-1230 (1973).

179. Stoessl, A., Ward, W. B., Stothers, J. B.: Incorporation of doubly labelled sodium acetate-$^{13}C_2$ into phytuberin and other sesquiterpenes in potatoes: Experimental conformation of postulated C-C cleavages. Tetrahedron Letters No. 37, 3271-3274 (1976).

180. Subramanian, S., Varns, J., Kuć, J.: Biochemical response in the compatible potato tuber - Phytophthora infestans interaction. Proc. Indiana Acad. Sci. 80, 367 (1970).

181. Suzuki, H., Oba, K., Uritani, I.: Occurrence of 3-hydroxy-3-methyl glutaryl CoA reductase in sweet potato. Agric. Biol. Chem. 38, 2053-2055 (1974).

182. Takiguchi, Y., Uritani, I., Akai, S.: The relation between invasion of Ceratocystis fimbriata and accumulation site of ipomeamarone in sweet potato. Ann. Phytopathol. Soc. Japan 36, 316 (1970).

183. Tjamos, E. C., Smith, I. M.: The role of phytoalexins in resistance of tomato to Verticillium wilt. Physiol. Plant Pathol. 4, 249-259 (1974).

184. Tomiyama, K., Fukaya, M.: Accumulation of rishitin in dead potato tuber tissue following treatment with $HgCl_2$. Ann. Phytopath. Soc. Japan 41, 418-420 (1975).

185. Tomiyama, K., Sakuma, T., Ishizaka, N., Sato, N., Katsui, N., Takasugi, M., Masamune, T.: A new antifungal substance isolated from potato tuber tissue infected by pathogens. Phytopathology 58, 115-116 (1968).

186. Uritani, I.: The biochemical basis of disease resistance induced by infection. Connecticut Agr. Expt. Sta. Bull. 663, 4-19 (1963).

187. Uritani, I., Uritani, M., Yamada, H.: Similar metabolic alterations induced in sweet potato by poisonous chemicals and by Ceratocystis fimbriata. Phytopathology 50, 30-34 (1960).

188. Varns, J.: Biochemical response and its control in the Irish potato tuber (Solanum tuberosum L.) - Phytophthora infestans interactions. Ph.D. Thesis, Purdue Univ., Lafayette, Indiana (1970).

189. Varns, J., Currier, W., Kuć, J.: Specificity of rishitin and phytuberin accumulation by potato. Phytopathology 61, 968-971 (1971).

190. Varns, J. L., Kuć, J.: Suppression of rishitin and phytuberin accumulation and hypersensitive response in potato by compatible races of Phytophthora infestans. Phytopathology 61, 178-181 (1971).

191. Varns, J. L., Kuć, J.: Suppression of the resistance response as an active mechanism for susceptibility in the potato - Phytophthora infestans interaction. In - Phytotoxins in Plant Diseases. (R.K.S. Wood, A. Graniti, eds.) Academic Press, N. Y. 465-468 (1972).

192. Varns, J., Kuć, J., Williams, E.: Terpenoid accumulation as a biochemical response of the potato tuber to Phytophthora infestans. Phytopathology 61, 174-177 (1971).

193. Veech, J. A., Bell, A. A.: Hemigossypol dimerase from cotton. Beltwide Cotton Prod. Res. Conf. Proc. p. 23 (1975).

194. Wallis, G.: Infection of Scots pine roots by Fomes annosus. Can. J. Bot. 39, 109-121 (1961).

195. Ward, E. W. B.: Capsidiol production in pepper leaves in incompatible interactions with fungi. Phytopathology 66, 175-176 (1976).

196. Ward, E., Stoessl, A.: Post infectional inhibitors from plants. III. Detoxification of capsidiol, an antifungal compound from peppers. Phytopathology 62, 1186-1187 (1972).

197. Ward, E., Unwin, C., Hill, J., Stoessl, A.: Sesquiterpenoid phytoalexins from fruits of eggplants. Phytopathology 65, 859-863 (1975).

198. Ward, E., Unwin, C., Rock, G., Stoessl, A.: Postinfectional inhibitors from plants. XXIII. Sesquiterpenoid phytoalexins from fruit capsules of Datura stramonium. Can. J. Bot. 54, 25-29 (1976).

199. Ward, E., Unwin, C., Stoessl, A.: Postinfectional inhibitors from plants. VI. Capsidiol production in pepper fruit infected with bacteria. Phytopathology 63, 1537-1538 (1973).

200. Ward, E., Unwin, C., Stoessl, A.: Postinfectional inhibitors from plants. XIII. Fungitoxicity of the phytoalexin, capsidiol, and related sesquiterpenes. Can. J. Bot. 52, 2481-2488 (1974).

201. Ward, E., Unwin, C., Stoessl, A.: Experimental control of late blight of tomatoes with capsidiol, the phytoalexin from peppers. Phytopathology 65, 168-169 (1975).

202. Wardale, D. A., Galliard, T.: Subcellular localization of lipoxygenase and lipolytic acyl hydrolase enzymes in plants. Phytochem. 14, 2323-2329 (1975).

203. Weber, D., Stahmann, M.: Induced immunity to Ceratocystis fimbriata infection in sweet potato root tissue. Phytopathology 56, 1066-1070 (1966).

204. Whiteman, T. M., Schomer, H. A.: Respiration and internal gas content of injured sweet potato roots. Plant. Physiol. 20, 171-182 (1945).

205. Willemot, C., Stumpf, P. K.: Fat metabolism in higher plants. XXXIV. Development of fatty acid synthetase as a function of protein synthesis in aging potato slices. Plant Physiol. 42, 391-397 (1967).

206. Willemot, C., Stumpf, P. K.: Fat metabolism in higher plants. XXXIII. Development of fatty acid synthetase during the "ageing" of storage tissue slices. Canad. J. Bot. 45, 579-584 (1967).

207. Wilson, B. J.: Toxicity of mold-damaged sweet potatoes. Nutr. Rev. 31, 73-78 (1973).

208. Wilson, B. J., Boyd, M. R.: Toxins produced by sweet potato roots infected with Ceratocystis fimbriata and Fusarium solani. in Mycotoxins (I.F.H. Purchase Ed.) Elsevier Scientific Publishing Co.

Amsterdam-Oxford-New York (1974).

209. Wilson, B. J., Boyd, M. R., Harris, T. M., Yang, D. T. C.: A lung vedema factor from mouldy sweet potatoes (Ipomoea batatas). Nature 231, 52-53 (1971).

210. Wolf, M., Duggar, B.: Estimation and physiological role of solanine in the potato. J. Agr. Res. 73, 1-32 (1946).

211. Yamamoto, M.: Potato late blight, with special reference to the resistance of potatoes to the invasion of Phytophthora infestans. Rev. Plant Protec. Res. 7, 45-56 (1974).

212. Yang, D. T. C., Wilson, B. J., Harris, T. H.: The structure of ipomeamaronol: A new toxic furanosesquiterpene from moldy sweet potatoes. Phytochemistry 10, 1653-1654 (1971).

213. Zacharius, R., Kalan, E., Osman, S., Herb, S.: Solanidine in potato (Solanum tuberosum) tuber tissue disrupted by Erwinia atroseptica and by Phytophthora infestans. Physiol. Plant Pathol. 6, 301-305 (1975).

214. Zaki, A., Keen, N., Erwin, D.: Implication of vergosin and hemigossypol in the resistance of cotton to Verticillium albo-atrum. Phytopathology 62, 1402-1406 (1972).

215. Zaki, A., Keen, N., Erwin, D., Sims, J.: Vergosin and hemigossypol, antifungal compounds produced in cotton plants inoculated with Verticillium albo-atrum. Phytopathology 62, 1398-1401 (1972).

The Biosynthesis of Phenolic Compounds in Wounded Plant Storage Tissues

J. Michael Rhodes and L. S. C. Wooltorton

Exposure of plant tissues to factors causing stress or injury such as mechanical damage, chemical treatments with heavy metals or ethylene or infection by fungi, bacteria or viruses can stimulate the metabolism of phenolic compounds (59). The injury or stress caused by these factors can lead to the loss of contents from broken cells following mechanical damage, to metabolic responses in cells close to the site of injury which do not themselves show signs of major damage, and to the premature death of some cells in the vicinity of the wound or infection (the hypersensitive response). It is in stressed but unbroken cells that repair processes are set in motion following injury and it is here that many of the responses to wounding in phenylpropanoid metabolism occur.

In general, there are three types of responses in phenolic metabolism to stress or injury. The first involves the oxidation of pre-formed phenolic compounds to yield quinones and hence polymeric materials. The other two types of response involve either the synthesis of monomeric or of polymeric phenolic compounds. The synthesis of monomeric phenols may lead to the accumulation of greater quantities of compounds already present in undamaged tissue or to the appearance of novel compounds which may play a special role in the defence mechanisms of the tissue in preventing infection. Compounds which play a role in the defence mechanisms of the tissue are termed phytoalexins (74) and it has been shown that many phytoalexins are phenolic compounds (62). In many cases the role played by the stimulation of the phenolic metabolism in the events which follow wounding is not at all clear. However, it is generally agreed that some role is played by these compounds in the defence mechanisms either by forming physical barriers to invasion, as in the case of polymeric products such as lignin, or as inhibitors of microbial growth in the case of the quinone products of phenolic oxidation and the phytoalexins.

Biochemistry of Wounded Plant Tissues

I. RESPONSES INVOLVING THE OXIDATION OF PRE-EXISTING PHENOLIC COMPOUNDS

The copper containing enzyme complex, phenolase (71), is thought to be the main enzyme involved in the oxidation of endogenous phenols that occurs when many plant tissues are injured. The requirements for this response are the presence of the enzyme, the availability of a suitable substrate and of oxygen. Mason (71) showed that the phenolase complex was composed of two separate activities, one responsible for the o-hydroxylation of monophenols and the other for the oxidation of o-diphenols to the corresponding o-quinones. The enzyme will thus oxidise diphenols such as chlorogenic acid, caffeic acid and catechol directly in the diphenolase reaction (o-diphenol : O_2 oxidoreductase) or will oxidise monophenols such as p-coumaric acid and tyrosine by first hydroxylating them and subsequently, in the presence of oxygen, oxidising the resultant o-diphenols, caffeic acid and 3, 4 dihydroxyphenylalanine (l-dopa) to give the quinone products in the monophenolase reaction. The o-quinone products of the oxidation reaction are very powerful oxidising agents (81) and can react with the α-amino, amine and thiol groups of proteins and are inhibitors of microbial growth (59). They can also undergo polymerisation reactions to produce the brown or black products which are characteristic of enzymic browning reactions following injury in plants. The polymers have properties similar to the endogenous tannins of some plant cells which can precipitate proteins by forming protein : tannin complexes and such reactions can slow or stop the rate of infection of a wounded tissue. It is generally thought that the reactions beyond the hydroxylation and oxidative reactions catalysed by phenolase proceed non-enzymically. A proposed pathway for the oxidation of tyrosine to the black product, melanin, is shown in Fig. 1. This figure is based on the work of Raper (88) and Mason (70) and is still accepted as representing the major pathway of the formation of melanin in animals (46). This shows that the monophenolase enzyme catalyses the oxidation of tyrosine via l-dopa to dopaquinone which, by a very rapid series of chemical interconversions, is converted to indole 5, 6 quinone which then polymerises to melanin. A similar pathway is thought to operate in plant tissues even though plant melanins are less well characterised than the corresponding materials in animals (81). Similar steps occur in the oxidation of chlorogenic acid with the phenolase catalysing the first step, the formation of chlorogenoquinone and the subsequent polymerisation steps occurring non-enzymically.

These enzymic browning reactions do not occur to a measurable extent in undamaged tissue but are activated on damage. Generally de novo synthesis is not involved but it is assumed that the enzyme and substrate are held in separate cell compartments in vivo and that, on injury, the compartmentalisation is destroyed. However, there is no really clear

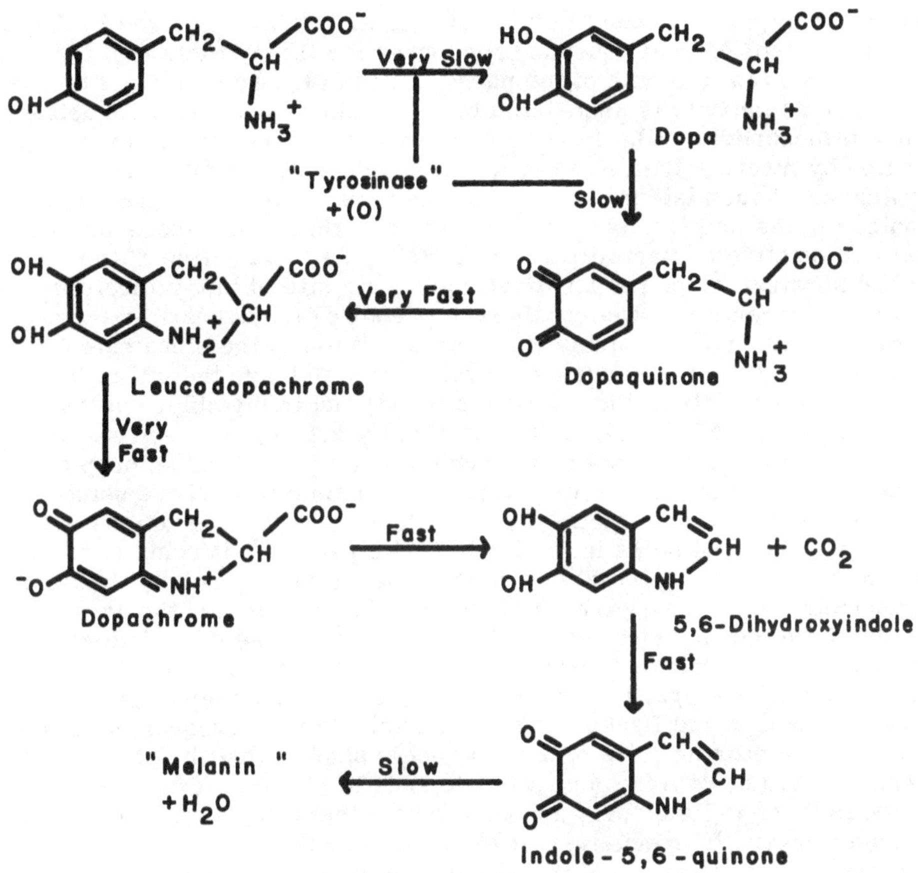

Figure 1. Proposed pathway for the oxidation of tyrosine to melanin by phenolase

evidence on the localisation of either phenolase or its substrates in the cell. The use of normal aqueous procedures for the extraction of the enzyme shows that the bulk of the phenolase is in the soluble fraction although some activity is associated with organelles such as chloroplasts (7) and mitochondria (40). In vitro phenolase shows latency and is activated by treatment with acids, alkali and detergents (59). The significance of such latency in vivo is uncertain but it may suggest that phenolase in the uninjured cell is held either within an organelle or is held in an inactive or partially active state. The localisation of the phenolic substrates and the localisation of their site of biosynthesis in vivo is also uncertain. The enzymes of phenolic biosynthesis are generally classified as 'soluble' with the exception of the cinnamate 4-hydroxylase and the p-coumarate hydroxylase which are thought to be localised respectively in the microsomes (98) and in the chloroplasts (118). Czichi and Kindl (24) have shown that a number of reactions of phenylpropanoid biosynthesis can occur on thylakoids of the chloroplasts of Dunaliella marina. However, although a number of workers using other membrane bound systems (3) have shown that these fractions will catalyse a number of steps in the biosynthetic pathway, it remains a fact that the bulk of important enzymes in the pathway, such as PAL and p-coumarate-CoA ligase, are found in the soluble form and the precise location of the biosynthetic pathway is unresolved. The site of storage of the potential phenolase substrates is often thought to be the cell vacuole although generally there is little direct evidence for this. However, Mueller and Beckman (73), in an electron microscopical study of the phenolic storing cells of banana roots, showed that the phenolic material is laid down in the cell vacuole and the presence of large plastids in these cells closely associated with the endoplasmic reticulum led these workers to speculate that the chloroplasts are the principal site of phenolic synthesis and that the products are transported via the endoplasmic reticulum to the vacuole for storage.

Whatever the mechanism by which phenolase is activated following injury, the course of development may well be affected by the specificity of the enzyme. Enzymes from different sources show marked differences in their relative monophenolase and diphenolase activities and even in their relative activities with substrates such as chlorogenic, caffeic, p-coumaric acids or tyrosine (20). Phenolases are very heterogeneous in molecular weight (41) and are prone to protein:protein association and dissociation reactions (52). Figure 2 shows a separation of potato phenolase by gel filtration on a column of Ultragel AcA 34. Three peaks of o-diphenol oxidase activity with apparent molecular weights of 288,000, 180,000 and 81,000 were found and these were partially separated from an enzyme catalysing the hydroxylation of a monophenol, p-coumaric acid, which had a molecular weight of 102,000.

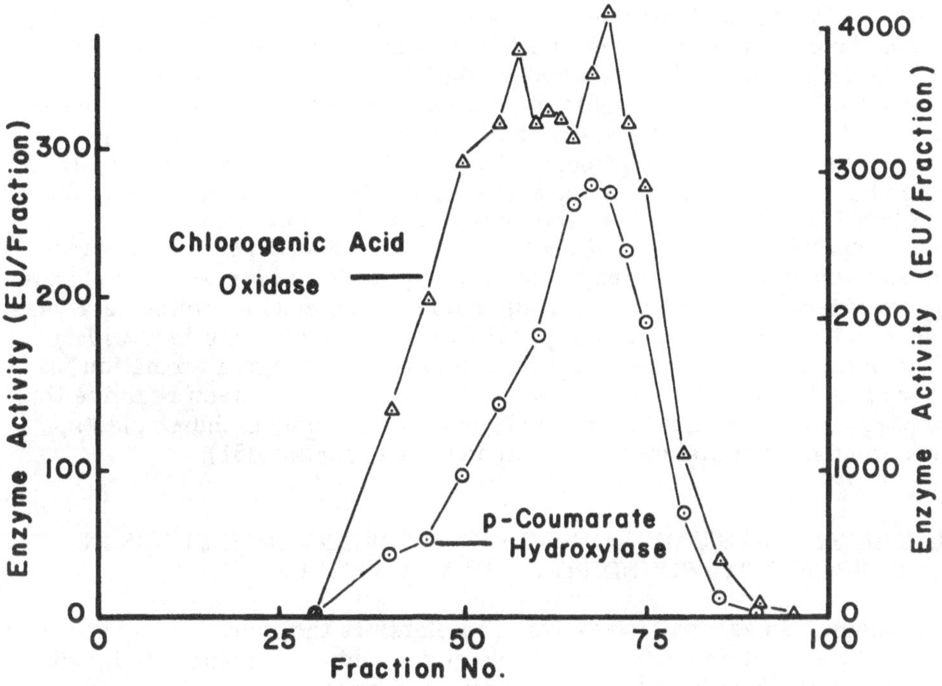

Fig. 2. Separation of potato phenolase on a column of Ultra gel AcA 34 (40 x 2. 5 cm)

The activity of diphenol oxidase using chlorogenic acid as substrate (△—△ right hand scale) and the activity in hydroxylating a monophenol, p-coumaric acid (○——○, left hand scale) were measured. The column was calibrated for molecular weight by determining the elution volumes of proteins of known molecular weight.

Frequently, phenolase substrates are present in vivo as glycosides and often, on wounding, browning reactions are initiated by the release of glycosidases which act on phenolic glycosides to produce the aglycones which are the substrates for phenolase activity. An example of this is the β-glucoside found in apple leaves, phloridzin, which, on wounding, is hydrolysed by a β-glucosidase to the aglycone, phloretin, which is hydroxylated and oxidised by phenolase to yield 3-hydroxyphloretin and hence the corresponding quinone which has antimicrobial activity (59).

Peroxidase is another enzyme capable of oxidising phenolic compounds. In the presence of H_2O_2, peroxidase will catalyse the oxidation of phenylpropanoid compounds to yield a product with lignin-like properties (31). Peroxidase, in the presence of dihydroxyfumaric acid, can act as a hydroxylase and catalyse the hydroxylation of aromatic compounds (14). Peroxidase is present in most plant tissues in a variety of isoenzymic forms which differ in their substrate specificities (103). In response to injury, the total activity of peroxidase has been shown to increase in a number of cases (54) and, in addition, the isoenzymic pattern changes and there is evidence for the synthesis of new isoenzymes. The physiological significance of such changes is not understood. Another enzyme capable of oxidizing phenolic compounds is laccase, p-diphenol O_2 oxidoreductase. This is a copper containing enzyme which catalyses the oxidation of p-diphenols to p-quinones. Laccases are less widely distributed than phenolases and may play a role in lignin formation (31) as well as in oxidation reactions. p-Quinones are far less reactive than the o-quinones but are fairly widely distributed among higher plants, even though their metabolic significance is uncertain (81).

II. THE SYNTHESIS OF MONOMERIC PHENOLIC COMPOUNDS IN RESPONSE TO WOUNDING OF PLANT TISSUES

A widespread response to wounding in plants is the increased synthesis of cinnamic acid derivatives. These compounds are present in intact tissue and their accumulation is stimulated by wounding. Among the major cinnamic acid derivatives which accumulate under these circumstances are chlorogenic acid [+](caffeyl-5-quinate) and caffeic acid. The role, if any, played by these compounds in the wound repair processes is unclear. Chlorogenic acid has been reported to have some antimicrobial properties but these are less important than those of its oxidation products, the quinones (62). Other workers (106) have suggested that chlorogenic acid acts as a mobilizible reserve for the synthesis of other phenylpropanoid derivatives. Steck (110) showed that chlorogenic acid was not an end product of metabolism and was degraded by tobacco tissue to form caffeic acid and this was eventually incorporated into scopolin and lignin. More recently, Rhodes and Wooltorton (95) have shown that tomato fruits have an enzyme which will cleave chlorogenic acid in the presence of CoA to yield caffeyl CoA and quinic acid and which could account for the first step in such inter-conversions.

The pathway of the synthesis of cinnamic acids and their derivatives has been known in outline for some time (78) but it is only fairly recently that enzymes capable of catalysing the postulated steps have been

[+]The IUPAC system for numbering the quinic acid ring is used in this paper.

Figure 3. Biosynthetic relationships between the major groups of phenolic compounds

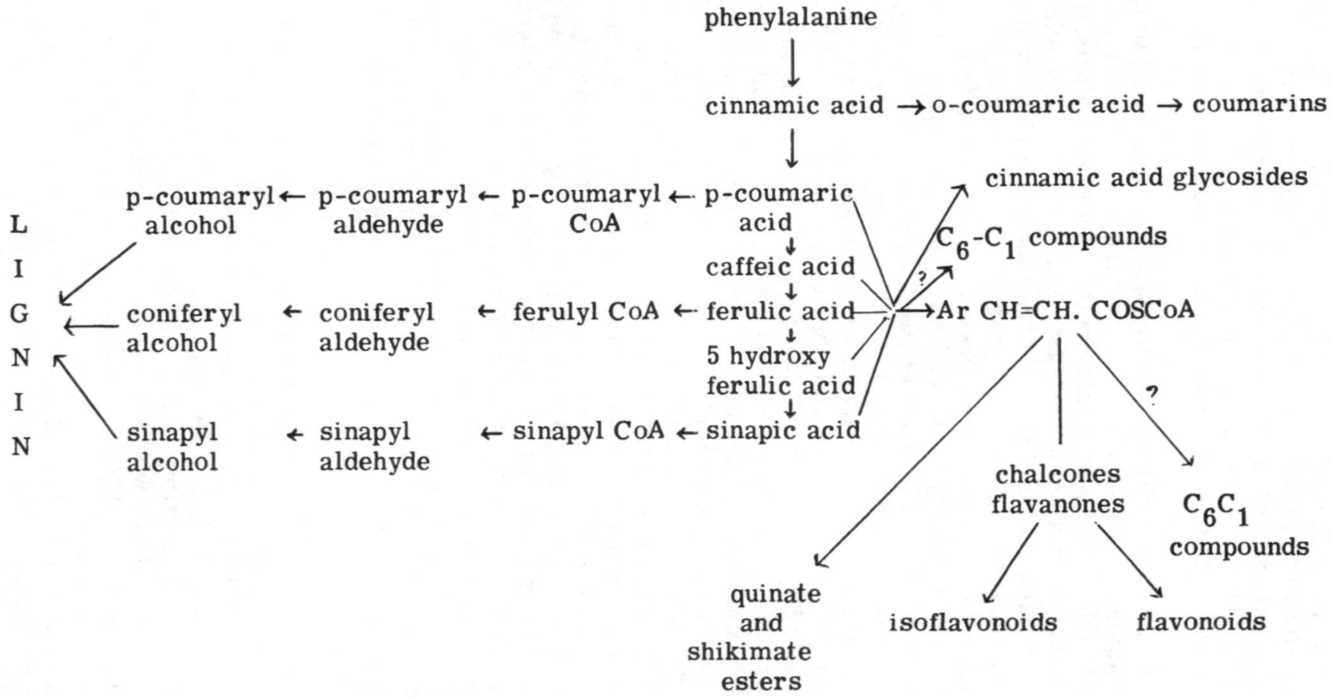

isolated from plant tissues and their properties described (107). More recent research has established the CoA thioesters of cinnamic acids as important intermediates in the synthesis of phenylpropanoid compounds (113). The interrelationships of the synthesis of the main groups of phenolic compounds are illustrated in Figure 3 (see also Figure 9).

The synthesis of cinnamic acids originates in the shikimate pathway of aromatic amino acid synthesis from phosphoenol pyruvate and erythrose 4 phosphate (78). The immediate precursors of phenylpropanoid formation are phenylalanine and tyrosine. Of these, phenylalanine is by far the most important and it is only in monocotyledonous species that tyrosine is incorporated to any degree into phenylpropanoids. The enzyme acting on phenylalanine, phenylalanine ammonia lyase (PAL), was discovered in barley by Koukol and Conn (60) and has since been widely studied (16). This enzyme, which has no cofactor requirements, catalyses the deamination of phenylalanine with the stereospecific elimination of the pro-S hydrogen yielding trans cinnamic acid (38). The enzymes from maize and potato (molecular weights 306, 000 - 320, 000) are thought to be composed of 4 identical polypeptide chains, each with a molecular weight of 83, 000 (43) while the enzyme from wheat has a molecular weight of 320, 000 and has two identical subunits each composed of two polypeptide chains of molecular weights 75, 000 and 85, 000 (77). The active site of PAL involves a dehydroalanine imino system (44). Enzymes isolated from monocotyledons have activity with both phenylalanine and tyrosine. Tyrosine is deaminated in a reaction analogous to that of PAL to yield trans-p-coumaric acid (tyrosine ammonia lyase, TAL). Detailed studies with highly purified PAL from monocotyledons suggest that a common catalytic site is responsible for both PAL and TAL activity (45). PAL from a number of sources shows important deviations from Michaelis-Menton kinetics giving increases in the apparent K_m and V_m values as the L-phenylalanine concentration is raised from 0.01 - 6.7 mM (42). This is interpreted by Nari et al. (77) in terms of conformational changes between the subunits of the PAL molecule.

The step in the pathway following PAL is the first hydroxylation in the sequence leading to the formation of trans p-coumarate, the first phenolic product of the pathway. The enzyme involved, cinnamate 4-hydroxylase (CA 4H) is membrane bound and sediments in the microsomal fraction (98). It is a mixed function oxidase involving $NADPH_2$ and O_2 and it is thought to be associated with the microsomal electron transport system and cytochrome P450. The involvement of such a system was suggested by the early work of Russell (98) who found that the hydroxylation was inhibited by CO and this inhibition was reversed by light. In later studies it was shown that the action spectrum for the light reversibility of CO inhibition showed a peak at 450 nm which suggested the involvement of cytochrome P450 (82). This cytochrome has been found in the microsomes of a number of higher plants and levels of 0.3 nmole/mg protein have been found in microsomes from Arum tissue (122) and 0.2 nmoles/mg protein in microsomes from aged swede root disks (49). In the hydroxylation reaction, electrons are passed from $NADPH_2$ to cinnamic acid which is hydroxylated by a reaction involving the incorporation of atmospheric oxygen. This hydroxylation occurs specifically in the para-position and during hydroxylation the hydrogen originally the the para-position migrates to the meta-position, the so-called 'NIH shift' which also occurs in hydroxylation reactions in animal tissues (99). The CA 4H activity proceeds in a lipid environment and there is evidence that lipid factors are required for its activity (13).

The next step in the pathway of cinnamic acid metabolism is another hydroxylation reaction in which the second hydroxyl group is inserted in the ortho position relative to the first hydroxyl group. This enzyme, p-coumarate hydroxylase, converts p-coumaric acid to caffeic acid and is thought to be catalysed by a phenolase type of enzyme. Molecular oxygen is involved in the reaction which is carried out in the presence of excess of a reducing agent such as ascorbic acid to prevent further oxidation of the diphenol product. This type of reaction is thought to be involved not only in the hydroxylation of p-coumarate to caffeic acid (118) but also in the hydroxylation of p-coumaryl esters such as p-coumaryl quinic acid to chlorogenic acid (67). The work of Stafford (109) suggests that p-coumarate hydroxylase is associated with membranes in greening leaves of Sorghum bicolor and that it can be dissociated into lower molecular weight forms by treatment with DTE (108) and in this work the p-coumarate hydroxylase was separated from the bulk of the chlorogenic acid oxidase activity.

The further steps in cinnamic acid formation involve a combination of a further hydroxylation and a series of methylation steps (see Figures 3 and 9). The hydroxylation step from ferulic acid to 5 hydroxy ferulic acid has not yet been enzymically characterised. The methoxylation steps between caffeic and ferulic acids and between 5 hydroxy ferulic and sinapic acids are catalysed by a single non-specific enzyme (104), an O-methyltransferase (OMT) which acts with S-adenosylmethionine as the methyl donor. The enzyme is, however, specific for methylating hydroxyl groups in the meta-position. For instance, in the methylation of caffeic acid, ferulic acid is formed while methylation does not occur at the para-position yielding isoferulic acid (29). A number of O-methyltransferases have been purified (47, 84, 104) and, although separate isoenzymes acting on flavonoid and cinnamic acid substrates have been separated and studied (83, 84), the isoenzymes acting on cinnamic acids are not specific and catalyse both methoxylation steps in sinapic acid formation.

The enzymological evidence suggests that the free cinnamic acids are metabolic intermediates in the biosynthetic pathway of phenylpropanoid formation. However, cinnamic acids are normally stored as glycosides or esters rather than the free acids. The occurrence of the glycosides of cinnamic acids has been described by Harborne (39) and enzymes which catalyse the formation of such glycosides in the presence of free cinnamic acids and UDP-sugars have been described (85). Other important storage products of cinnamic acids are the quinate esters of which chlorogenic acid, which is the 5'quinate ester of caffeic acid, is the most abundant, although the 4' and 3' esters also occur in nature (120).

The biosynthesis of chlorogenic acid has been a matter for conjecture for some time and three routes of biosynthesis from cinnamic acid derivatives have been suggested. These are illustrated in Figure 4. Levy and Zucker (67) proposed that in the potato esterification occurs at the cinnamic acid level followed by successive hydroxylations to form chlorogenic acid. Kojima et al. (57, 58) produced evidence which suggested that the glucose ester of cinnamic acid was an intermediate between cinnamic and cinnamylquinic acid in the pathway (Figure 4, route 1). Steck (110) proposed two additional pathways, one in which esterification occurs at the level of p-coumaric acid (route 2) and the other in which it occurs at the level of caffeic acid (route 3) and his data suggested that the route via p-coumarylquinic acid was the quantitatively more important route in these tissues. The route via p-coumarylquinic acid could be operative since crude extracts of potato tissue catalysed the hydroxylation of p-coumarylquinic acid to chlorogenic acid and evidence suggests that a phenolase type of enzyme is involved in this step (67). Nagels and Parmentier (76) in a kinetic study of the labelling of various postulated intermediates of chlorogenic acid biosynthesis, using cinnamate 3-^{14}C as the supplied substrate, showed that the labelling of cinnamylquinic and caffeic acid was slower than that of chlorogenic acid while p-coumaric and p-coumarylquinic acids possessed the required kinetic qualities of precursors of chlorogenic acid. Stockigt and Zenk (111) showed that a crude enzyme preparation from tobacco tissue catalysed the synthesis of chlorogenic acid from caffeyl CoA and quinic acid. In our work (95) a hydroxycinnamyl CoA : quinate hydroxycinnamyltransferase (HQT) was isolated from tomato fruits and its properties studied in some detail. This enzyme not only catalyses the synthesis of chlorogenic acid from caffeyl CoA, it also catalyses the synthesis of p-coumarylquinic acid from p-coumaryl CoA and moreover the affinity of the enzyme for p-coumaryl CoA was greater than that for caffeyl CoA. This suggests that there is the enzymic capacity for chlorogenic acid synthesis via either caffeyl CoA or p-coumaryl CoA but that the affinities of the enzyme suggest that the route via p-coumaryl CoA, and hence p-coumarylquinic acid, is the more important. In addition to the

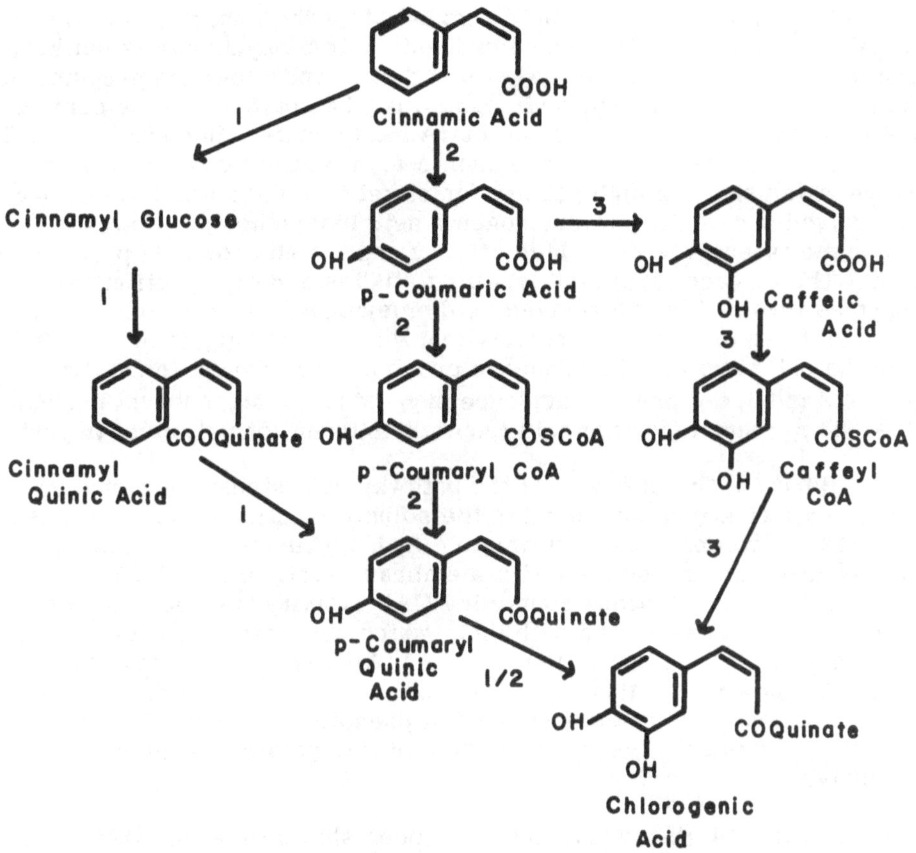

Figure 4. Proposed routes for the biosynthesis of chlorogenic acid from cinnamic acid.

---

transferase, the other enzyme involved in the formation of the quinate esters of caffeic and p-coumaric acids is the hydroxycinnamate CoA ligase which, in the presence of ATP, CoA and Mg^{2+}, catalyses the formation of the CoA thioesters from the corresponding cinnamic acids. This enzyme has been purified from a number of sources, including swede roots (92), Forsythia (34) and soybean cell suspension cultures (56). The ligase enzyme exists in a number of isoenzymic forms (56, 87) which may be related to the variety of pathways in which the CoA thioesters of cinnamic acids play a role. In our work on tomatoes, the ligase enzyme shows high activity and affinity towards both caffeic and p-coumaric acids but no activity towards cinnamic acid (95). This inactivity towards cinnamic acid suggests that the formation of cinnamylquinate as an intermediate in chlorogenic acid biosynthesis is unlikely. The tomato ligase, however, shows higher affinity towards p-coumaric than caffeic acid which suggests again that the route via p-coumaryl CoA and p-coumarylquinic acid is the more likely route. Ranjeva et al. (87) have, however, isolated an isoenzymic form of the ligase from Petunia tissue which shows a high affinity for caffeic acid and which they suggest is involved specifically in chlorogenic acid biosynthesis. However, this same isoenzyme has a high affinity and activity towards p-coumaric acid. The evidence from the labelling studies and the specificities of the enzymes involved in chlorogenic acid biosynthesis appear to rule out route 1 shown in Figure 4, namely that via cinnamylquinic acid. While both the pathways and the requisite enzymes are present to catalyse both routes 2 and 3, the present evidence suggests that the route via p-coumaryl CoA and p-coumarylquinic acid (route 2) is quantitatively more important.

The majority of the enzymes of the pathway of synthesis of cinnamic acids and their derivatives are found in the soluble fraction of aqueous tissue extracts. Two enzymes, cinnamate 4 hydroxylase and p-coumarate hydroxylase, are associated with membrane systems (7, 25, 98). Membrane bound fractions containing CAH activity have been shown to have some PAL associated with them which accounted for about 5% of the total cellular content of the enzyme. Alibert et al. (1) and Czichi and Kindl (25) have shown that membrane bound fractions can catalyse a number of reactions in the sequence of phenolic acid biosynthesis. However, the precise cellular location of this pathway is as yet unresolved.

The formation of chlorogenic acid has been shown to occur when potato (124), sweet potato (114) and carrot (19) tissues are injured. The form of injury may be mechanical, chemical or microbial. The production of chlorogenic acid in response to wounding has been studied in some detail in the potato where, following wounding, there is a rise in the activity of PAL to a maximum after 24 hours followed by a fall to 50% of the maximal value by 40 hours, and again followed by a steady phase (124). The synthesis of chlorogenic acid in aged disks of potato tuber is stimulated by light as is the increase in PAL activity that accompanies the synthesis (124). In subsequent inhibitor studies, Zucker (125) suggested that there was a sequential induction of PAL followed by the induction of a PAL inactivating system. More recent work has suggested that a number of enzymes in the pathway of chlorogenic acid biosynthesis are stimulated in response to wounding but not all the enzymes respond. Camm and Towers (17) showed that there was a large increase in CAH activity during ageing of potato disks but that this increase in activity was not stimulated by light. Similarly, Rhodes and Wooltorton (93) showed there was a similar large increase in the activity of the hydroxycinnamate CoA ligase. In a number of other situations in which phenolic biosynthesis is stimulated there is a large increase in these three enzymes activities, PAL, CAH and hydroxycinnamate CoA ligase (28, 37, 90) and the view has been put forward that there is a co-ordinated synthesis of these enzymes and that the control of the carbon flow through the pathway was mainly effected at the steps controlled by these enzymes (126). Zucker (126) proposed the operation of a 'PAL operon' consisting of the structural genes for these three enzymes of the pathway. Lamb and Rubery (64, 65, 66) showed differences in the induction of PAL and CAH by light and in their sensitiveness to inhibitors of protein synthesis such as cycloheximide. They argued that the concept of the co-ordinated 'PAL operon' as proposed by Zucker (126) is too simple a hypothesis to explain their data and they stress the role of cinnamic acid intermediates in the modulation of the PAL level. There seems general agreement that it is at the PAL level that the major control on the flow of carbon into chlorogenic acid is controlled and that the light effect is expressed at this level. The evidence for the mechanism of the increased activity of PAL and the other enzymes which are activated, CAH and hydroxycinnamate CoA ligase, is that de novo protein synthesis is involved but the evidence for this is by no means conclusive as it is based largely on the use of inhibitors of RNA and protein synthesis whose specificity is at least suspect (68).

We have studied the role of the transferase (HQT) enzyme in chlorogenic acid biosynthesis in potato disks aged in the light. Figure 5 shows a separation of a desalted $(NH_4)_2SO_4$ fraction on DEAE-cellulose eluted with a linear gradient of KCl in 5mM Tris pH 7.45 between 0 - 0.4M. The Figure shows the separation of the transferase enzyme from both PAL and the p-coumarate CoA ligase. Figure 6 shows the separation of transferase from tomatoes on DEAE-cellulose showing that the peak has four catalytic activities, namely the capacities to synthesise and break down both p-coumarylquinic acid and chlorogenic acid. We have shown that the transferase enzyme in potato has essentially the same properties as the enzyme described in the tomato (95). Further work in this laboratory (Lourenco, unpublished data) has shown that the molecular weight of the enzyme (38,000) as estimated from its elution volume from a column of Ultragel AcA54 is identical in tomato and potato. As was shown with the tomato enzyme, the transferase is specific for the formation and cleavage of the 5' isomers. Table 1 shows the levels of PAL, p-coumarate CoA ligase and transferase in potato disks aged for 24 hours in the light and in the dark. It can be seen that there is a large increase on ageing in both PAL and ligase both in the light and in the dark but very little change in transferase activity. In other experiments small rises (up to 4 fold) in transferase activity on ageing were observed but always these changes were small compared with those in PAL and ligase. Both PAL and ligase are present, if at all, in only very low levels in unaged tissue compared with the transferase

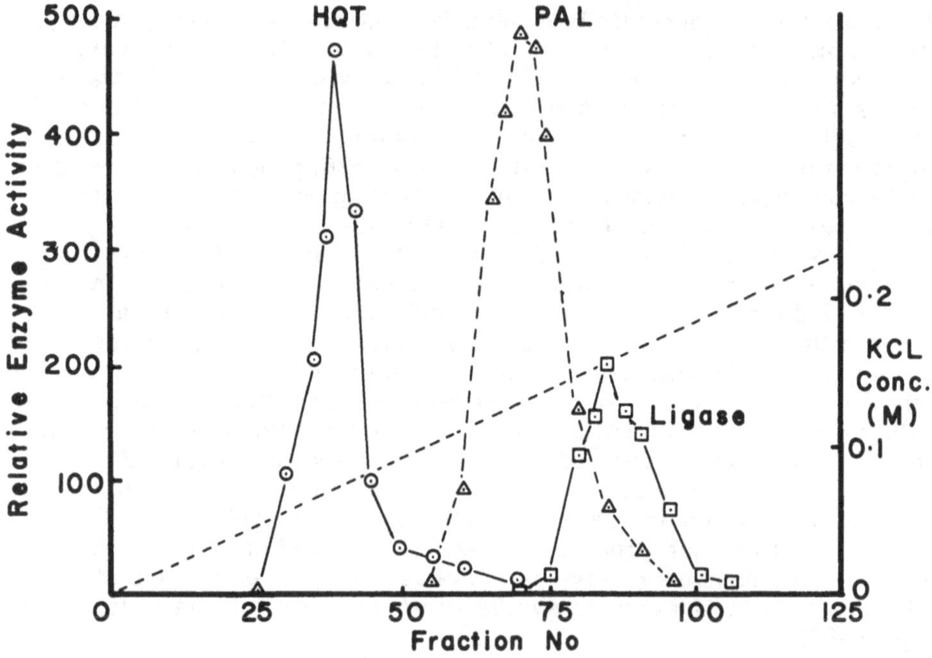

Figure 5. Separation of HQT, PAL and p-coumarate CoA ligase from extracts of aged potato tuber disks.

which is initially present in substantial amounts. Light has little or no effect on the transferase or ligase but causes a doubling of the activity of PAL. This supports the view that PAL is the major factor controlling the pathway and that, although there may be a secondary level of control on the ligase, very little is exerted at the level of transferase. It must be pointed out that the physiological role of the transferase is in some doubt as it is capable of catalysing the breakdown of chlorogenic acid as well as its synthesis. However, the high affinity of this enzyme for hydroxycinnamyl CoA substrates suggests that the synthetic reactions are the more important although this does not rule out the possibility that in some physiological situations the degradation reaction may be important.

The significance of the rise in chlorogenic acid which follows wounding may be related to the ultimate formation of a wound periderm and suberisation. Kolattukudy and his co-workers (97) have shown that suberin has covalently linked p-coumaric and ferulic acids as part of its structure and chlorogenic acid could provide a reservoir for the synthesis of phenylpropanoid units for suberin formation.

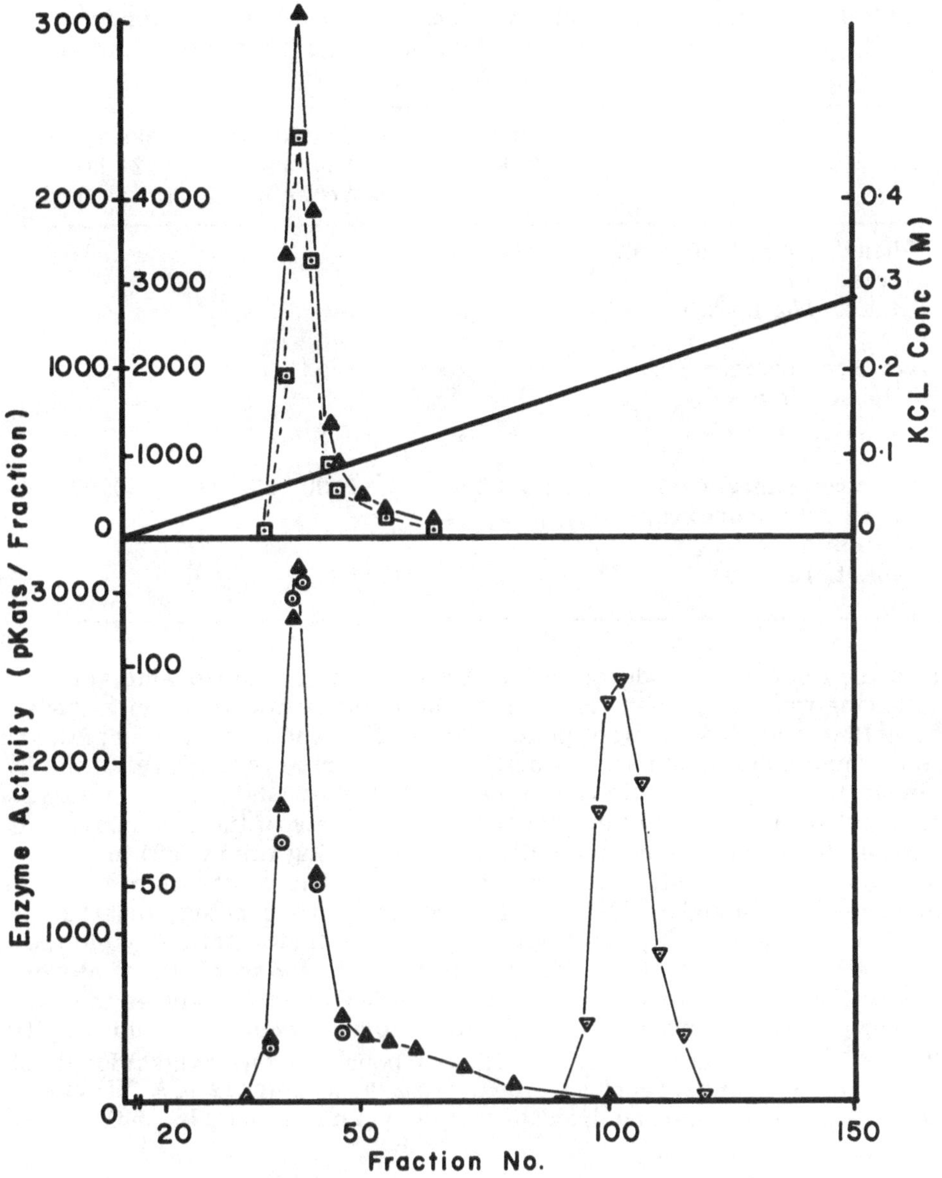

Figure 6. Separation of HQT assayed in the synthesis (upper curves) and breakdown (lower curves) of chlorogenic acid (△) and p-coumarylquinic acid (□, ○). The separation of HQT from the p-coumarate CoA ligase (▽—▽) is also shown.

Table 1. Changes in the activity of enzymes involved in chlorogenic acid biosynthesis in potato disks aged in the light and in the dark

	Initial disks	Disks aged for 24 hours in the light	Disks aged for 24 hours in the dark
Phenylalanine ammonia lyase (PAL) (nKat/10g f. wt.)	∼0	2.1	1.02
Hydroxycinnamate CoA ligase ("ligase") (nKat/10g f. wt.)	∼0	1.43	1.43
Hydroxycinnamyl CoA 5-quinate hydroxycinnamyl transferase (HQT) (nKat/10g f. wt.)	0.83	1.00	0.93

In many tissues, in addition to the accumulation of chlorogenic acid following wounding, there is an increase in other cinnamic acids such as caffeic acid, as in sweet potato tissue (59), and also of coumarin and isocoumarin derivatives. In potato there is a rise in the level of the coumarin, scopolin, following infection with Phytophthora infestans (50). In sweet potato, there is an increase in the levels of the coumarins, scopoletin, esculetin and umbelliferone, following injury (59) in addition to chlorogenic acid and caffeic acid. The synthesis of these compounds is thought to proceed from their corresponding cinnamic acids via ortho hydroxyl derivatives which, on lactonisation, yield the corresponding coumarins. An enzyme from the chloroplasts of clover has been described by Gestetner and Conn (32) which catalyses the 2 hydroxylation of trans cinnamic acid to form o-coumaric acid (CA 2H). It was found that 50% of the CA 2H was bound to the thylakoids of the chloroplasts. Subsequently, cinnamate 2 hydroxylases (CA 2H) have been found in membrane fractions from potato tissue (25, 63).

In carrot tissues, in addition to the stimulation of chlorogenic acid biosynthesis, a range of other phenolic compounds are formed in response to injury resulting from either mechanical damage or fungal infection. These compounds are isocoumarins and the related chromones. The major isocoumarin is 8-hydroxy 6-methoxy 3-methyl 3,4-dihydroisocoumarin (6 methoxymellein) while 6 hydroxymellein is a minor constituent. In addition to the isocoumarins, two chromones were formed, 5-hydroxy-7-methoxy-2-methyl chromone (eugenin) and 5, 7-dihydroxy-2-methyl chromone following either infection with Ceratocystis fimbriata (21) or treatment with ethylene (100). Evidence from labelling experiments showed that, while phenylalanine is not a good precursor of either the isocoumarins or the chromones, acetate is a very good precursor (102). Thus it would appear that these compounds are synthesised by the acetate pathway via malonyl CoA (10). However, in carrots in response to wounding, treatments with ethylene or to fungal infection, PAL activity is stimulated (18, 101). This increase in PAL is possibly related to the accumulation of chlorogenic acid that occurs but could also relate to the stimulation of scopoletin formation noted by Coxon et al. (21). We have studied changes in the activity of enzymes of phenylpropanoid metabolism in disks of the cortex and pith tissue of carrot roots and some of our results are set out in Table 2. These results show that there is a large increase in PAL and a smaller increase in the hydroxycinnamate CoA ligase on ageing in disks of both the pith and cortex tissue. Similar increases in the activity of enzymes of both the aromatic amino acid and phenylpropanoid biosynthetic pathways have been shown in disks of sweet potato roots following mechanical damage (51).

Table 2. Changes in the activity of two enzymes of phenylpropanoid biosynthesis in disks of the cortex and pith of carrot roots aged for 24 hours

		'PAL' (nKat/10g f. wt.)	'Ligase' (nKat/10g f. wt.)
Pith tissue	Initial Disks	0.0062	0.26
	Aged Disks	12.0	1.50
Cortex tissue	Initial Disks	0.0055	0.23
	Aged Disks	11.0	1.43

In apple fruit, infection by Nectria galligena stimulates the accumulation of benzoic acid (12). This accumulation, which was associated with a rise in PAL activity, could also be produced by treatment of apple tissue with a protease extracted from apple tissue under attack by Nectria. Experiments using ^{14}C labelled glucose and L-phenylalanine showed that the benzoic acid was synthesised de novo in response to the fungal attack (112). The biosynthesis of benzoic acid derivatives is via the normal cinnamic acid pathway and the benzoates are formed by a chain

Figure 7. Structure of isoflavonoid and pterocarpan compounds produced by leguminous plants on injury

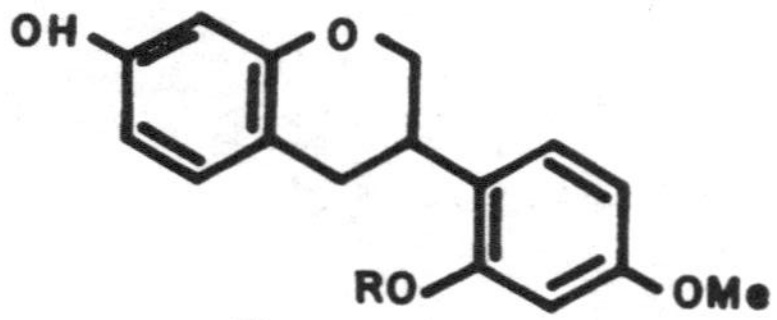

Sativan (I) R = Me

Vestitol (II) R = H

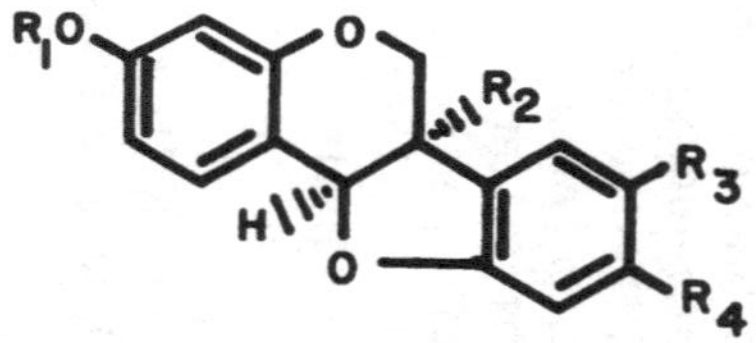

Maackiain (V) R_1 = H, R_2 = H, $R_3 \cdot R_4$ = OCH_2O

Medicarpin (VI) R_1 = H, R_2 = H, R_3 = H, R_4 = OMe

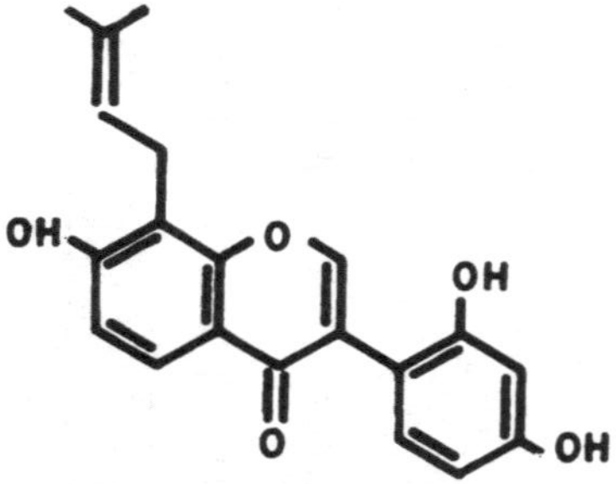

Kievetone (III)

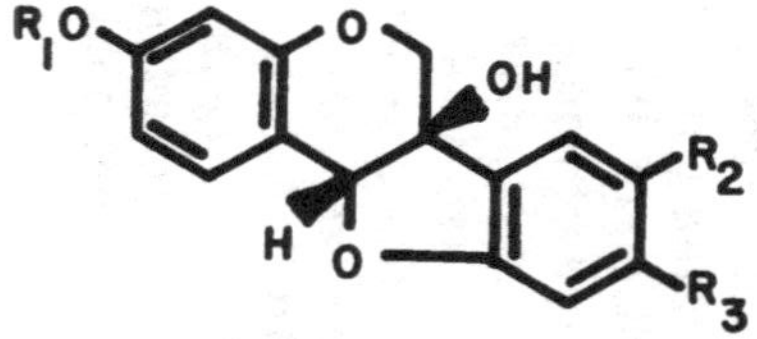

Pisatin (VII) R_1 = Me, $R_2 \cdot R_3$ = OCH_2O

Homopisatin (VIII) R_1 = Me, R_2 = H, R_3 = OMe

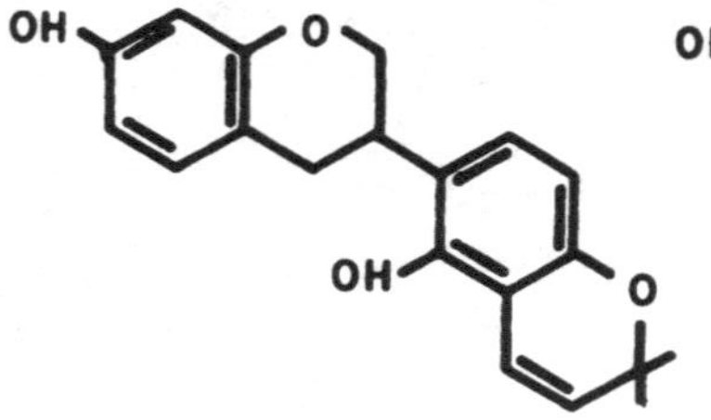

Phaseolin isoflavan (IV)

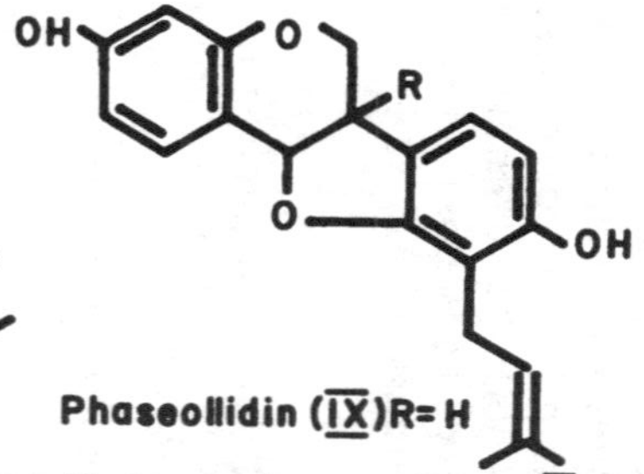

Phaseollidin (IX) R = H

6α Hydroxyphaseollidin (X) R = OH

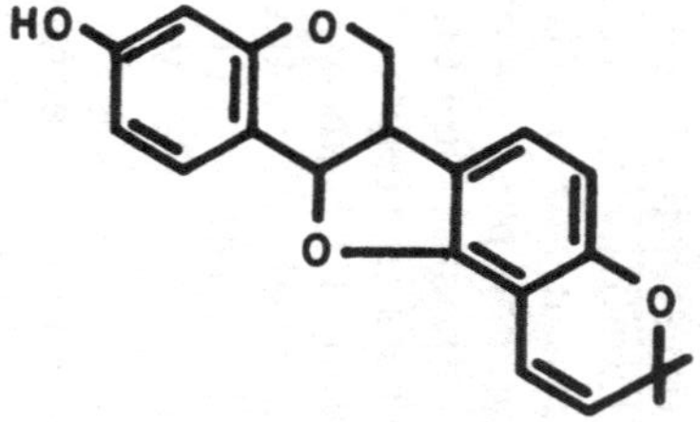

Phaseollin (XI)

shortening reaction (see Figure 3). French et al.(30) have shown the presence of a system capable of converting p-coumaric to p-hydroxy benzoic acid in extracts of potato tuber. The reaction is independent of CoA, ATP and NAD. Alibert and Ranjeva (1) have shown the formation of benzoic acid from cinnamic acid using a cell free extract of Quercus pedunculata. However, in this system CoA and ATP stimulate the formation of benzoic acid even though there is a considerable amount of activity in their absence. This stimulation by ATP and CoA could suggest the involvement of CoA thioesters in a chain shortening reaction analogous to the β-oxidation of fatty acids as suggested by Zenk (123) although pathways not involving the CoA thioesters cannot be ruled out by these studies. Further work by Alibert et al. (2) suggested that the pathway of benzoate biosynthesis is separated compartmentally from the pathway of cinnamic acid formation and that two isoenzymes of PAL are involved; a cinnamate sensitive isoenzyme present in the microsomes involved in cinnamate formation and a benzoate sensitive form in a 10,000 g fraction involved in benzoic acid formation.

In plants of the family Leguminosae, a major response to wounding is the production of isoflavonoids and related compounds called pterocarpans. The synthesis of these compounds is not normally induced by the injury caused by mechanical damage but is induced by chemical treatments with heavy metals or with ethylene or by infection of the tissues by micro-organisms. In tissues of the bean (Phaseolus vulgaris) there is a large production of isoflavonoids and pterocarpans under these conditions and the compounds formed (see Fig. 7) include phaseollin (XI), phaseollidin (IX), phaseollinisoflavan (IV) and kievitone (III). The first and most important of these compounds, phaseollin, was isolated and characterised by Cruickshank and Perrin (23) and Perrin (80) using detached bean pods infected with Monolinia fructicola. These same four compounds are produced in response to chemical treatment (62) or infection with fungi (105) or viruses (6). Phaseollin is produced in response to some, but not all, bacterial infections(62). Rathmell (89) showed that, in healthy hypocotyls of Phaseolus, flavonoid glycosides (such as kaempferol and quercetin), leucoanthocyanins and hydroxycinnamic acids were present but that their levels were not significantly higher when the hypocotyls were infected with Colletotrichum lindemuthianum. However, on infection, the levels of phaseollin and the other isoflavonoids rose together with that of a related pterocarpan derivative, coumestrol (XIV) (see Figure 8).

In soybean hypocotyls, Keen et al. (55) showed that, on infection with a pathogenic fungus, Phytophthora megasperma, there was an increase in the concentration of four isoflavonoids, hydroxyphaseollin (XVI), daidzein (XIII), coumestrol (XIV) and sojagol (XV). Of these, daidzein, coumestrol and sojagol are present in healthy soybean hypocotyls while hydroxyphaseollin is absent and only appears on infection. The structure of hydroxyphaseollin was re-investigated by Burden and Bailey

(15) and shown to be 6 α hydroxyphaseollin. In peas, the major product formed as a result of infection of the green pods was shown to be pisatin (VII) (22) although in more recent work additional pterocarpans have been shown to be produced when the pea is infected with Fusarium solani. These are 4-hydroxy-2, 3-9-trimethoxy pterocarpan, 3-hydroxy-2, 9-dimethoxy pterocarpan and 2, 3, 9-trimethoxy pterocarpan (86). The formation of pisatin in the pea does not occur as a result of either mechanical damage or bacterial infection. However, infection with fungi or treatment with ethylene or a broad spectrum of metabolic inhibitors did elicit the response in the synthesis of pterocarpans. In other leguminous species there is a similar response in the synthesis of isoflavonoids and pterocarpans following certain types of injury and infection. These include the production of kievitone in cow pea (75), maackiain (V), medicarpin (VI), pisatin and homopisatin (VIII) in red clover (9) and sativan (I) and vestitol (II) in Lotus corniculatus (11). The formation of these pterocarpans is normally restricted to the necrotic areas which result from the hypersensitive reaction in which cells close to an area of damage or infection die prematurely. However, they are not always so restricted (79). Many of these pterocarpans have been shown to have significant antifungal activity (62, 116).

The isoflavonoids are synthesised via the normal pathway of flavonoid formation involving an aryl shift mechanism (36). The A ring arises as in normal flavonoid biosynthesis from acetate and the B ring from shikimate and condensation is catalysed enzymically from malonyl CoA and a hydroxycinnamyl CoA by the enzymic mechanism similar to that described by Kreuzaler and Hahlbrock (61) to produce a flavanone product. Labelling experiments have shown that phenylalanine and cinnamic acids are good precursors of isoflavonoids such as daidzein (55) and pterocarpans such as pisatin (35) and there is evidence that the aryl migration occurs at or after the chalcone/flavanone stage (36). Keen et al. (55) in a study of the biosynthesis of hydroxypheaseollin and related isoflavonoids in fungi-inoculated soybean hypocotyls showed that phenylalanine-U-^{14}C and the chalcone, isoliquiritigenin 9 ^{14}C (XII) were readily incorporated into the hydroxyphaseollin. Radioactivity was incorporated initially into daidzein and coumestrol in short periods of incubation but tended to decrease during longer periods when the bulk of the radioactivity appeared in hydroxyphaseollin. Accompanying this increase in synthesis of isoflavonoids there is an increased activity of PAL and chalcone-flavanone isomerase suggesting that there is activation of the pathway of isoflavonoid biosynthesis following infection. On the basis of their work and the earlier work of Dewick et al. (27) Keen et al. (55) proposed the pathway for the synthesis of hydroxyphaseollin and the related isoflavonoids shown in Figure 8. Earlier work had suggested two possible intermediates in this pathway, a 4-hydroxy-3-arylcoumarin or a pterocarp-6-en (8). More recent work confirmed that the pathway via pterocarp-6-en is the important one

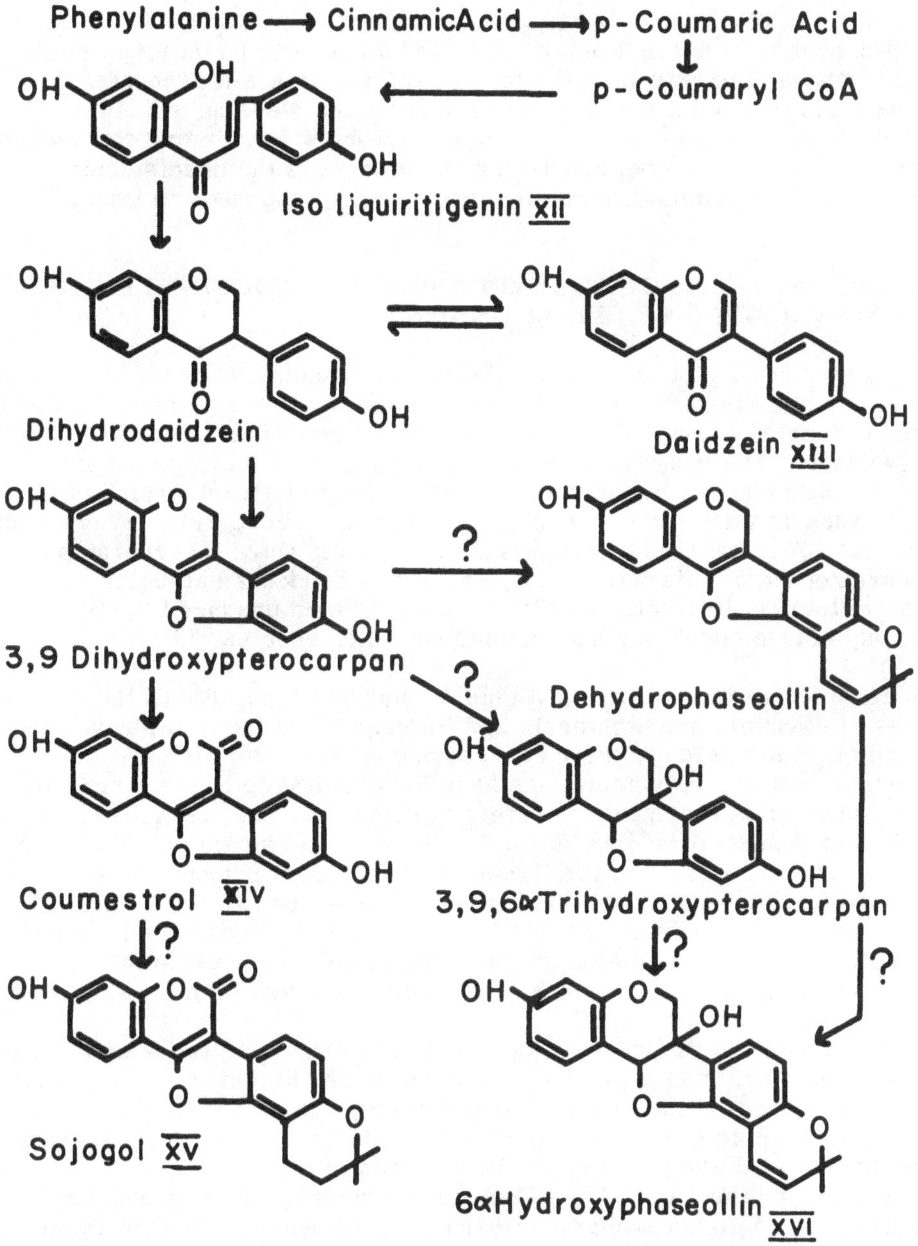

Figure 8. Proposed pathway of the synthesis of hydroxyphaseollin in infected soybean tissue.

(see 48). It should be noted in the scheme shown in Figure 8 that the aryl shift occurs between isoliquiritigenin and dihydrodaidzein. It is clear that, although the outline of the pathway of biosynthesis of these compounds is now known, few if any of the later steps have been characterised enzymically and thus the factors which control the accumulation of the various compounds are unknown as is the precise role played by the accumulating isoflavonoids in the responses to injury.

III. THE SYNTHESIS OF POLYMERIC POLYPHENOLS IN RESPONSE TO WOUNDING OF PLANT TISSUES

It is a common response of plant tissues on wounding to lay down a layer of a protective material over the cut surface which may form a barrier against loss of water and against invasion of the tissue by microorganisms. The material laid down is frequently either suberin or lignin. Suberin is a polymer of long chain fatty acids into which cinnamic acids such as ferulic and p-coumaric acids are incorporated by covalent linkages (97). Lignin is a polyphenolic polymer which incorporates phenylpropanoid units into a complex, highly condensed structure which has yet to be fully elucidated (31). Tissues which produced lignin on cutting include cucumber and various Brassica species (53).

Figure 9 shows the pathway of lignin formation which follows the normal route of cinnamic acid synthesis and its unique feature is the reduction of the cinnamic acids to their corresponding alcohols which are the precursors of the polymerisation reactions leading to lignin formation. The reduction occurs at three levels of substitution of the cinnamic acid moiety to produce hydroxycinnamyl, coniferyl and sinapyl alcohols. A major difference between the lignins from different sources is in the relative proportion of these monomeric units which are present in the polymer. There is evidence that the ratio of units having the hydroxycinnamyl, coniferyl and sinapyl ring substitution patterns in the resultant lignin is determined at the precursor stage (31).

The outline of the reduction system has been known for some time from experiments with radioactive lignin precursors (78) but it is only in the past few years that the enzymic systems responsible for the reduction of cinnamic acids to their alcohols have been described. Three enzymes are involved and the pathway has been shown to proceed via the CoA thioesters and the aldehydes. The first enzyme involved is hydroxycinnamate CoA ligase which catalyses the formation of the CoA thioesters from the cinnamic acids in the presence of CoA, ATP and Mg^{2+}. The enzyme exists in a number of isoenzymic forms. The first form which was studied (34, 92) had a high activity towards p-coumaric, ferulic, isoferulic and caffeic acids but had little or no activity towards cinnamic acid, sinapic acid and other methoxylated cinnamic acids. This iso-

Figure 9. Pathway of lignin formation in plants

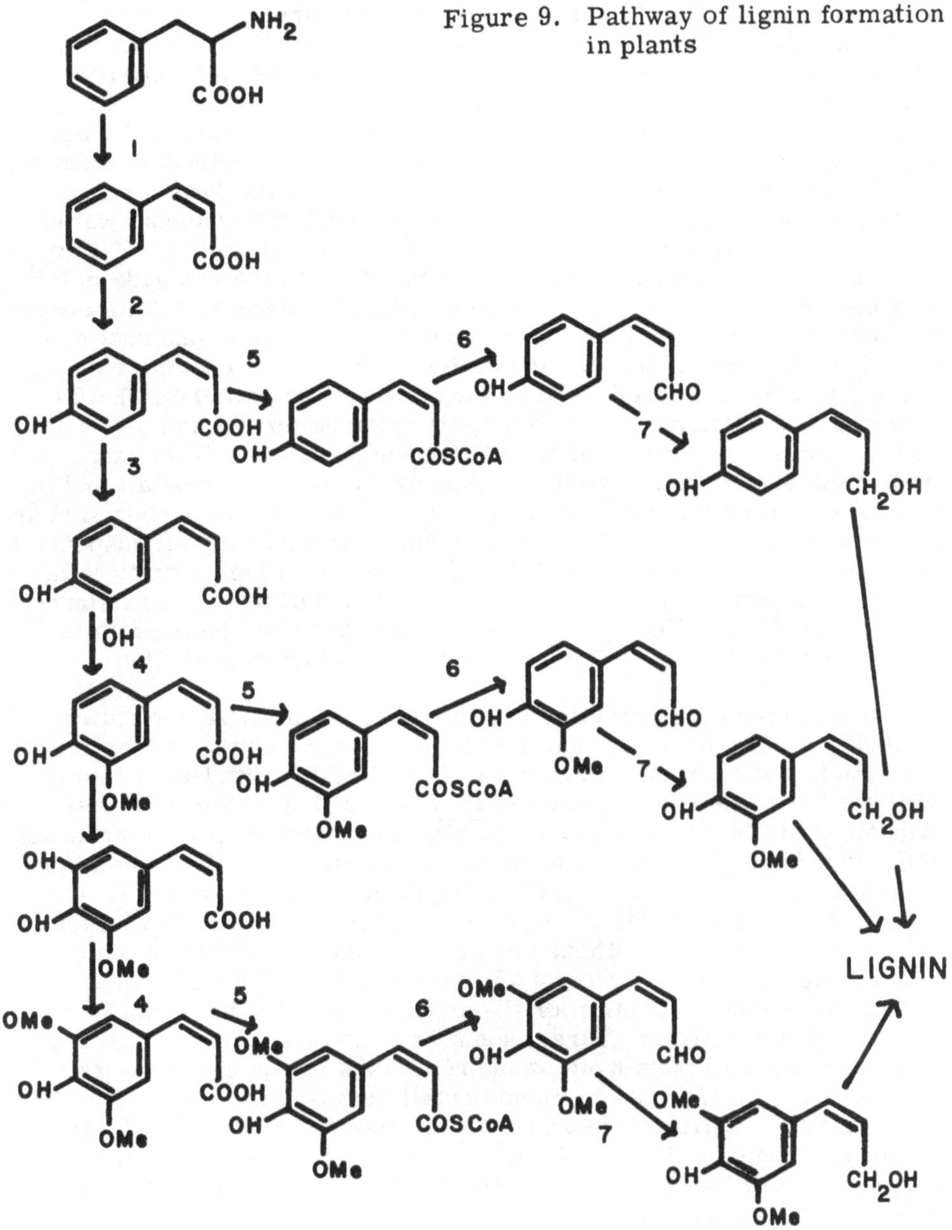

Enzymes involved:

1. PAL. 2. CAH. 3. p-coumarate hydroxylase. 4. OMT,
5. Hydroxycinnamate CoA ligase. 6. Hydroxycinnamyl CoA reductase.
7. Cinnamyl alcohol NADP oxidoreductase.

enzyme corresponds to form II in the nomenclature of Knobloch and Hahlbrock (56). These workers, in addition, described a form I which, although active towards p-coumaric acid, also shows high activity towards sinapic acid and other methoxylated cinnamic acids. Similarly, Ranjeva et al. (87) showed the presence of two isoenzymes of the ligase in Petunia tissue with similar properties to those described by Knobloch and Hahlbrock (56). Ligase II could be involved in the formation of hydroxycinnamyl and coniferyl residues for lignin biosynthesis while ligase I could be involved in the formation of sinapyl residues. The enzyme which reduces the hydroxycinnamyl CoA to the aldehyde is the hydroxycinnamyl CoA reductase which is NADPH specific. This enzyme has been purified from a number of sources including swede root disks (94), Forsythia cambium (33) and soybean (119). The enzyme from swede roots was resolved into two forms which separated on DEAE-cellulose and which both showed activity towards ferulyl and p-coumaryl CoA's. Gross and Kreiten (33) and Wengenmayer et al. (119) have studied the substrate specificity of the enzyme in greater detail and in both cases the major substrate is ferulyl CoA while both p-coumaryl and sinapyl CoA's are reduced at a slower rate. In soybean cell suspension cultures the hydroxycinnamyl CoA reductase shows high affinity towards ferulyl, p-coumaryl and 5-hydroxy ferulyl CoA but much less towards sinapyl CoA (119). The apparent molecular weight of the enzyme is 38,000 as judged from the elution volume from Sephadex G-100.

The final stage in the reduction is controlled by an NADP aromatic alcohol dehydrogenase (aromatic ADH) which reduces the hydroxycinnamyl aldehyde to the corresponding alcohol. An NADP dependent aromatic ADH has been described by Davies et al. (26) using potato tuber tissue. Mansell et al. (69) have described a cinnamyl alcohol specific aromatic ADH from Forsythia spp. which shows an absolute specificity both for the cinnamyl moiety and for NADP. Wyrambik and Grisebach (121) isolated two forms of $NADP^+$ dependent cinnamyl alcohol dehydrogenase from soybean extracts. These two isoenzymes have different molecular weights and differ by a factor of 75 in their affinities for coniferyl alcohol but are both specific for the cinnamyl moiety and do not react with benzyl derivatives. Three isoenzymes of NADP specific aromatic ADH were isolated from aged swede root disks in our earlier work (94) and these will be discussed in some detail later in the paper. The final stage in the formation of lignin is the polymerisation of the hydroxycinnamyl alcohols. The reaction is a dehydrogenative polymerisation studied in detail by Freudenburg (31) and it is suggested that in vivo it is catalysed by laccases and peroxidase (31).

We have studied in some detail the enzymic responses which follow the damage induced by cutting disks of the 'root' of Brassica napo-brassica (Swedish Turnip or Rutabaga). The development of the root of this species is such that, in the first year of growth, the bulk of the internal tissue consists of secondary xylem parenchyma which is largely unlignified. The root 'over winters' in this condition and during the following spring when the inflorescence develops the secondary xylem parenchyma undergoes very rapid lignification.

Figure 10 shows the distribution of two enzymes of phenylpropanoid metabolism, PAL and p-coumarate CoA ligase, through the various tissues of the root in the first year of growth. It can be seen that the enzymes are mainly found in the region corresponding to the cambial layer of the intact root and that they are only present in very low levels in the secondary xylem parenchyma. In the case of PAL, the cambial tissue is the only tissue in which activity was detected.

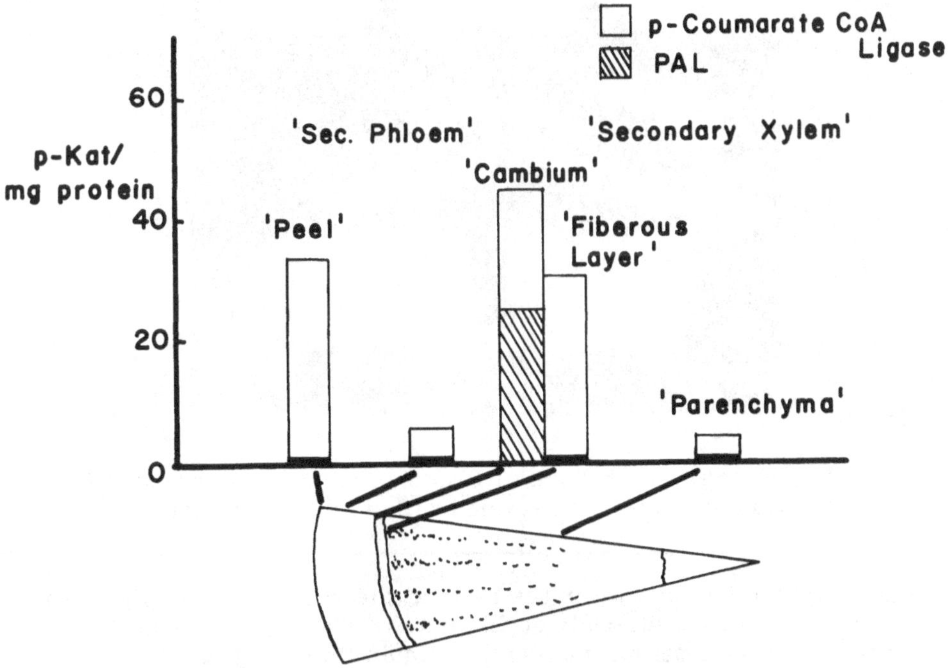

Figure 10. Distribution of two enzymes of phenylpropanoid synthesis in various tissues of Brassica napo-brassica 'root'.

When the secondary xylem parenchyma of first season's roots is cut into disks, and the disks are aged, a layer of lignin-like material forms in the cell walls of both the severed cells on the cut surface and of the next layer of unbroken cells. Fig. 11 shows a photograph of disks aged for various periods following wounding which have been stained with acid phloroglucinol. It can be seen that material staining with phloroglucinol is absent from disks immediately following cutting but increases steadily in the subsequent 24 hour period of ageing. This phloroglucinol

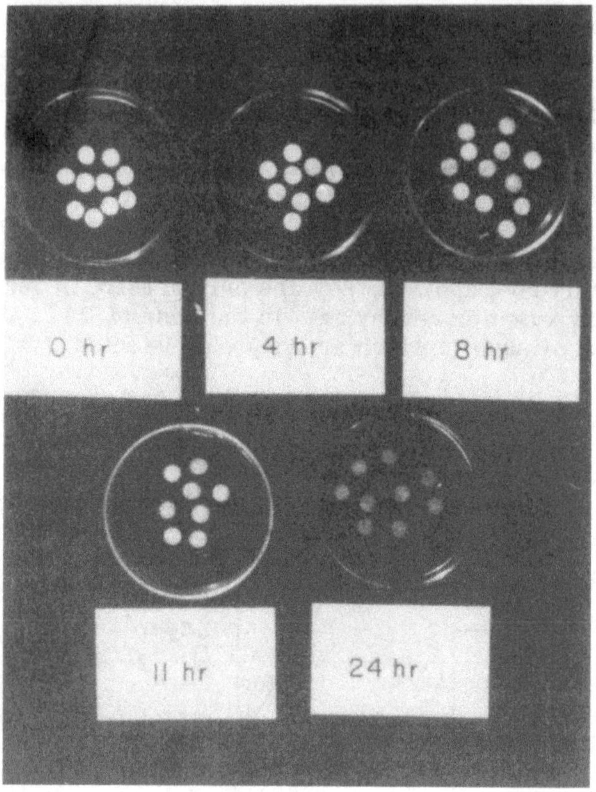

Figure 11. Photograph of disks of Brassica napo-brassica aged for various times and stained with acid phloroglucinol

positive material laid down in the cells at the surface of the disks was shown to absorb light strongly at 254 nm and subsequent chemical analysis suggests that this material is lignin (91).

In our work we have studied changes in the activity of a number of enzymes of both aromatic amino acid and of phenylpropanoid biosynthesis during the lignification that occurs when disks are aged. The effect of supplying low levels of exogenous ethylene (10 ppm) which accelerates the development of the lignin layer and enhances the activity of a number of enzymes involved in its synthesis was also studied. Figure 12 shows changes in the total activity of nine enzymes in disks which were aged in either air or in air containing 10 ppm ethylene. The responses of the enzymes fall into three groups. In one group are phenylalanine trans-aminase, coniferyl alcohol : NADP oxidoreductase and peroxidase which

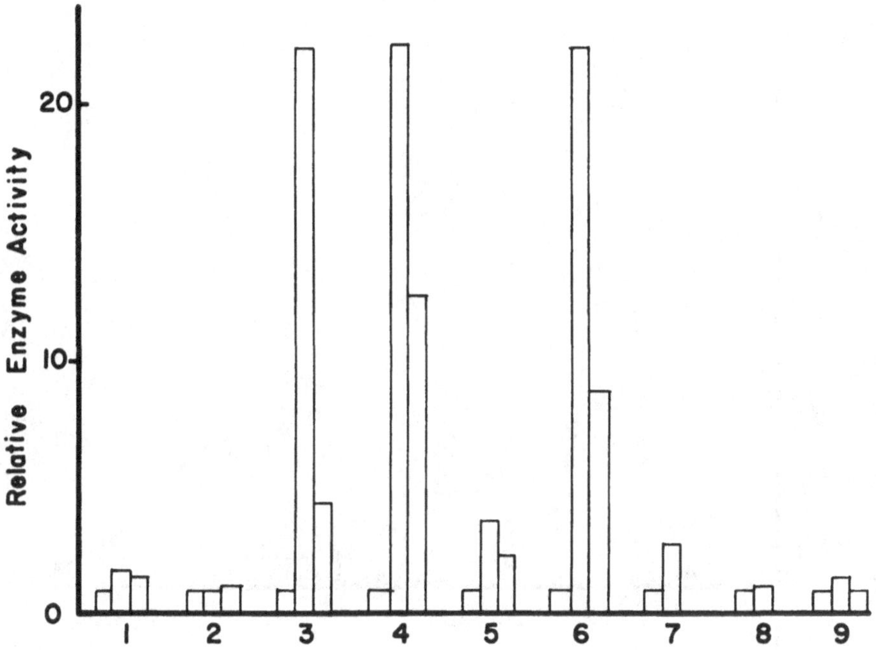

Figure 12.

Changes in the relative activities of nine enzymes of phenylpropanoid biosynthesis in Brassica napo-brassica root disks during ageing. First column in each group initial disks, second column disks aged in the presence of 10 ppm ethylene and third column disks aged in air. Group 1 shikimate dehydrogenase; 2. phenylalanine transaminase; 3. PAL; 4. CA 4H; 5. OMT; 6. Hydroxycinnamate CoA ligase; 7. Hydroxycinnamyl CoA reductase; 8. Cinnamyl alcohol : NADP oxidoreductase; 9. Peroxidase.
Data for group 7 & 8 for disks in air are not shown.

---

show little or no change in activity. In another, enzymes such as shikimate dehydrogenase, o-methyl transferase and hydroxycinnamyl CoA reductase increase two or three fold during ageing. In the third group are PAL, CAH and hydroxycinnamate CoA ligase which increase more than twenty fold on ageing. The enzymes of groups 1 and 2 are usually present in substantial amounts in initial disks while those in group 3 are only present at very low activity (90). Ethylene has very little effect on the activity of enzymes in groups 1 and 2 but stimulates PAL, CAH and ligase by 230-510%; the effect is greatest with PAL and somewhat less with CAH and ligase. The increase in activity of these

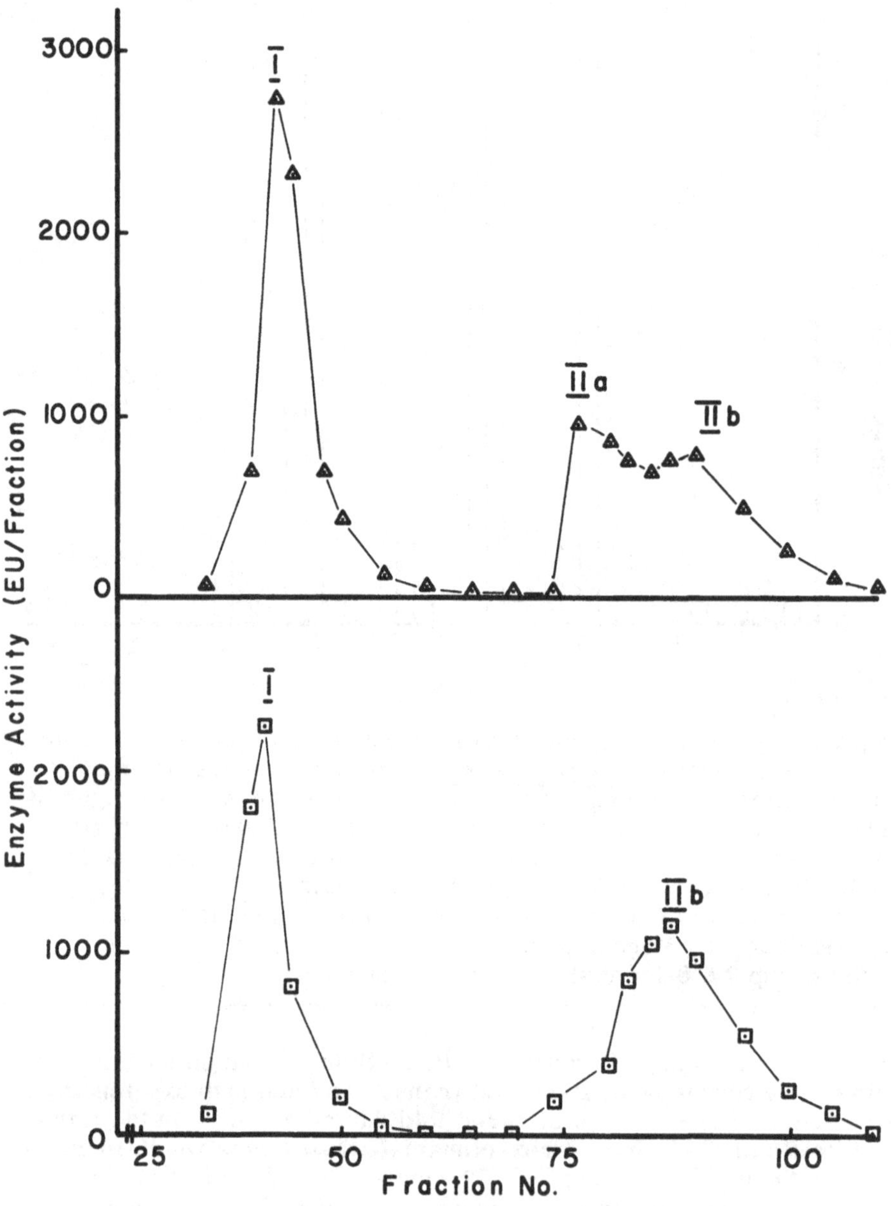

Figure 13. Separation of aromatic ADH's from extracts of aged disks of Brassica napo-brassica roots on DEAE-cellulose. Upper curve, disks aged for 24 hours. Lower curve, initial disks.

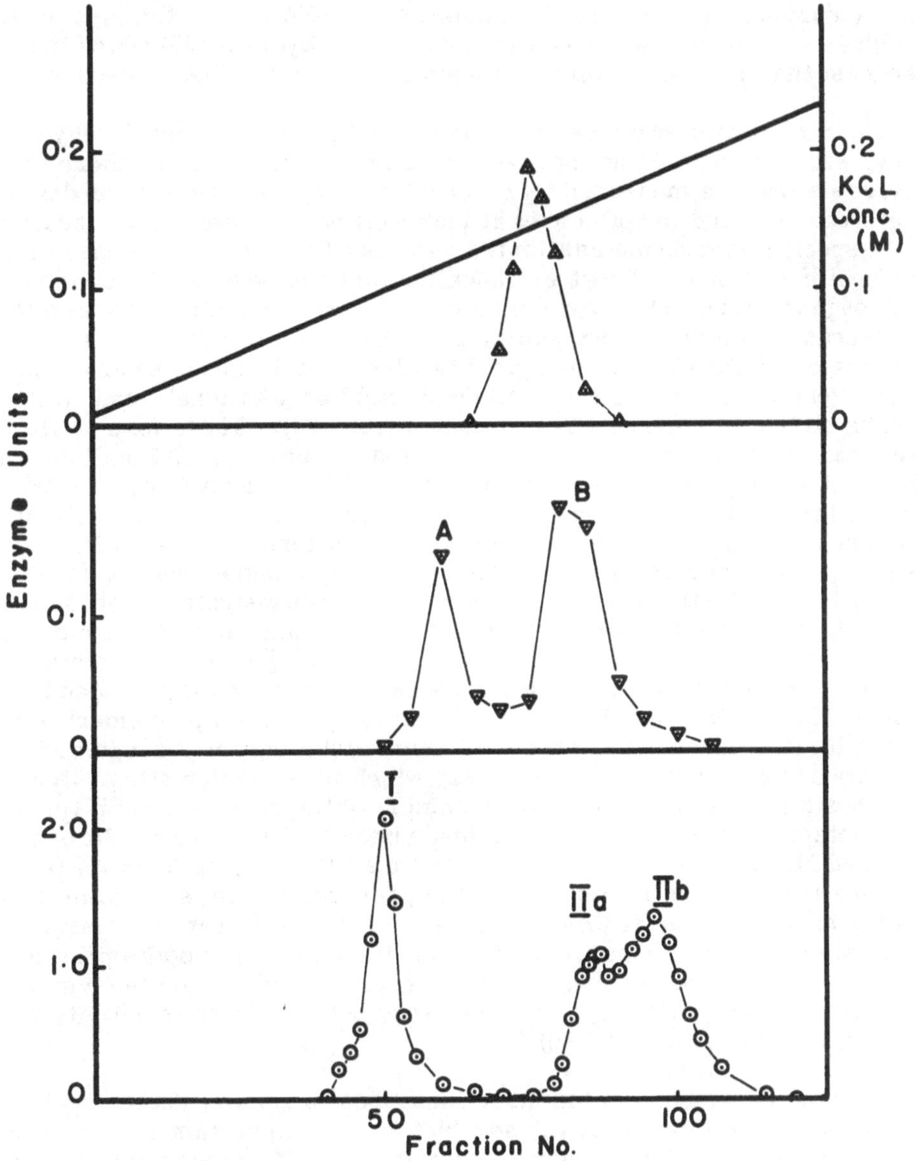

Figure 14. Separation of three enzymes, aromatic ADH (lower curve), hydroxycinnamyl CoA reductase (middle curve) and hydroxycinnamate CoA ligase (upper curve) from extracts of cambial tissue of the root of Brassica napo-brassica.

three enzymes is prevented by inhibitors of RNA and protein synthesis such as cycloheximide, puromycin and actinomycin D (90) which may suggest that de novo protein synthesis is involved in these responses.

In the case of two enzymes, peroxidase and aromatic alcohol dehydrogenase, showing little or no increase in activity during ageing, there is evidence that the measured total activities may obscure a great deal of heterogeneity and in both cases it may well be that there is an induction of a specific isoenzyme associated with lignification. In the case of aromatic ADH some direct evidence for this has been obtained since, although the total activity remains constant during ageing, there is the appearance of a new isoenzymic form. Extracts of initial disks separated on DEAE cellulose yield two isoenzymic forms of aromatic ADH (forms I and IIb) while aged disks yield an additional form (IIa) which elutes just before form IIb (see Figure 13). These forms, IIa and IIb, have been shown by repeated chromatography on DEAE cellulose to be distinct and are not inter-convertible. Table 3 shows the properties of the three forms of aromatic ADH. It can be seen that forms IIa and IIb are markedly different in properties from form I. Forms IIa and IIb show much higher affinities for the substrate, a higher degree of specificity of NADP and a higher sensitivity to substrate inhibition than form I. In addition, form I shows a relatively high activity with benzaldehyde as substrate while forms IIa and IIb are specific for the cinnamyl moiety. However, no major differences in properties were noted between forms IIa and IIb. Further evidence for the possible involvement of the form IIa in lignin biosynthesis has come from a chromatographic study of extracts of isolated cambial tissue which is the major site of lignin synthesis at this stage in the development of the root. Cambial tissue extracts separated on DEAE cellulose under the standard conditions yielded all three isoenzymes of aromatic ADH including form IIa (see Figure 14). Figure 14 also shows the presence of two isoenzymic forms of the hydroxycinnamyl CoA reductase in extracts of cambial tissue. Two similar isoenzymes were also found in extracts of both initial and aged disks. These two forms have very high affinities for their major substrate, ferulyl CoA and on ageing they both increase in activity by the same amount (i.e. 2 fold) (94).

Among the enzymes showing large increases in activity during ageing is the hydroxycinnamate CoA ligase which may be important in determining the composition of the precursor pool on which the polymerising enzymes act. The properties of the ligase of swede root disks were described in some of our earlier work (92). This is a type II ligase according to the definition of Knobloch and Grisebach (56) and can account for the activation of hydroxycinnamyl and ferulyl residues for lignin synthesis. It, however, is only feebly active towards sinapic acid. Although oxidation studies on purified fractions of swede lignin (unpublished data) show a small but definite content of syringyl lignin, a ligase isoenzyme similar to the ligase I of Knobloch and Grisebach (56) or the similar isoenzyme found by Ranjeva et al. (87) in Petunia tissue which are highly specific for sinapic acid, have not been found in the aged swede root disks in spite of an intensive search. However, a type I isoenzyme was found in this laboratory in extracts of etiolated pea shoots which have both ligase I and II (Wallis and Rhodes, unpublished data). In this connection, a comparison of the relative specificities of the ligases isolated from aged swede disks and from swede cambial tissue (see Figure 14) was made (Table 4). The two ligases have very similar specificities and are of type II with very little activity towards sinapic acid and other methoxylated cinnamic acids.

Table 3. Some properties of the various forms of the NADP aromatic ADH's of Brassica napo-brassica

	I	IIa	IIb
Coniferyl alcohol (μM)	550	4	2
Coniferyl aldehyde (μM)	140	NM	2
NAD activity as % of NADP activity	2.6	0.7	0.9
Activity with benzaldehyde (activity relative to coniferyl alcohol = 100)	32.8	0.7	0.1
Sensitivity to substrate inhibition	Low	High	High

NM = not measured

From these studies of changes in enzyme activity during lignification it seems that three enzymes, PAL, CAH and p-hydroxycinnamate CoA ligase, play an important role in the overall control of the pathway. This situation is analogous to that found in the synthesis of monomeric phenols such as chlorogenic acid where again the same three enzymes control the pathway and in which increases in these three activities are associated with the stimulation in the rate of synthesis of the end product. In the case of lignification, the ligase may have an additional controlling role in determining the relative concentrations of the substituted cinnamyl alcohols in the pool of substrates for the lignin polymerisation reactions. Such control by ligase may be particularly important when the ligase is present in separate isoenzymic forms, each specific for a particular substitution pattern in its cinnamic acid substrate.

Table 4. Substrate specificity of the hydroxycinnamyl CoA ligase from various tissues of Brassica napo-brassica

	Aged disks of secondary xylem parenchyma	Fresh tissue from cambial region
	pKat/mg protein (values in brackets activity relative to that with p-coumaric acid = 100)	
cinnamic acid	NA (0)	NA (0)
m-coumaric acid	58 (16)	5.2 (26)
o-coumaric acid	68 (18)	1.9 (9)
p-coumaric acid	368 (100)	20.1 (100)
m-methoxycinnamic acid	NA (0)	NA (0)
o-methoxycinnamic acid	NA (0)	NA (0)
p-methoxycinnamic acid	NA (0)	NA (0)
caffeic acid	250 (68)	8.8 (44)
ferulic acid	267 (72)	12.1 (60)
isoferulic acid	250 (68)	8.0 (40)
3,4-dimethoxycinnamic acid	NA (0)	NA (0)
2,4-dimethoxycinnamic acid	NA (0)	NA (0)
sinapic acid	7 (2)	0.9 (5)
3,4,5-trimethoxycinnamic acid	NA (0)	NA (0)

NA = not activated

The chemical nature of the lignin produced in response to wounding of Brassica roots was studied by subjecting lignin preparations from aged disks of secondary xylem parenchyma and from the fibrous region of the intact roots to alkaline CuO oxidation. The "wound lignin" was shown to be relatively rich in hydroxycinnamyl and deficient in sinapyl residues compared with the fibrous tissue of root. This agrees with the findings of other workers (4, 5) who found differences in the chemical composition of "wound" lignin compared to lignin found in other tissues of the plant. In both cases lignins with high contents of hydroxycinnamyl residues were indicated. It must be pointed out that with herbaceous tissues there is a danger of contamination of lignin preparations with proteins containing tyrosine which, on oxidation, can yield hydroxybenzaldehyde, the product also given by the hydroxycinnamyl residues of lignin. In our work, the lignin preparations were treated with proteolytic enzymes to reduce the contamination. It has been established, however, that these hydroxycinnamyl residues in the preparation are not solubilised by alkali treatment and are not analogous to the ester linked p-coumaric acid residues found in the lignin of monocotyledonous species (91). The evidence available does at least leave open the possibility that the lignin laid down in response to either wounding or infection, although closely related in structure and biosynthesis to native lignin, may in some respects differ from it.

The formation of lignin layers occurs in a number of tissues in response to either wounding or infection. Imaseki et al. (51) showed that a layer of lignin was formed when disks of sweet potato tissue were aged but in this case treatment with exogenous ethylene tended to inhibit lignification. Asada and Matsumoto (4) showed that a layer of lignin was laid down when the Japanese radish is infected with downy mildew fungus. These workers have studied extensively the nature of the product and its chemistry (4, 5). Ride (96) showed that lignin is synthesised after infection of wheat leaves with Botrytis cinerea and Mycosphaerella pinodes and that fungal growth was limited by the lignin layer. Subsequently, Maule and Ride (72) showed that this lignification was accompanied by increased activity of PAL and OMT in the lignifying tissue. Similarly, Vance and Sherwood (115) showed that leaf disks of reed canary grass inoculated with Helminthosporium avenae formed a lignin layer which was not penetrated by the fungal germ tubes. This lignification was associated with increased activity of PAL, p-coumarate CoA ligase and peroxidase in the lignifying zone.

The biosynthesis of lignin on the surface of a wounded tissue poses interesting problems in relation to the spatial orientation of the response, to the area in which the lignin precursors are synthesised and to their transport to the lignifying sites. In the case of disks such as those from Brassica roots, the lignin is often laid down in the walls of broken cells on the surface and here the synthesis of precursors must occur deeper in the disk and the products transported to the surface. The nature of the factors which restrict the response to the surface and prevent the whole tissue from lignifying is unknown. Factors such as O_2 supply may be important with only the surface tissue having an adequate supply of O_2. However, this seems unlikely since ageing the disks in pure O_2 does not induce more of the tissue to lignify. In the intact root, the secondary xylem parenchyma does not lignify during the first year of growth and it is probable that the hormonal regime in the root prevents it lignifying at this stage in its ontogeny. It is interesting that ethylene, which stimulates the activity of PAL and other enzymes in the pathway and accelerates lignification, does not alter the pattern of deposition of the lignin. It should, perhaps, be pointed out that many of the responses of disks following cutting are essentially surface effects (117) but the localisation of the lignification at the cut surface is particularly important in restricting the ingress of the pathogen. It seems likely that the orientation of the response depends on the activity of cells deeper in the tissue which, on injury, are damaged to some extent, possibly by rupture of some cell membranes, and the injury sets off a repair mechanism which leads to the activation of the enzymes of phenylpropanoid metabolism, to the formation of lignin precursors and the eventual deposition of lignin in the walls of cells on the surface.

SUMMARY

The wounding of plant tissues elicits a number of responses in the area of phenolic metabolism. These may include the oxidation of endogenous phenols or the production de novo of either monomeric or polymeric phenols. The first of these responses involves the breakdown of cell compartmentation and the mixing of cell contents of enzymes and their substrates which are normally held apart in vivo. The second type of response involves damage to certain cells which are stimulated to initiate repair processes. These processes include the activation of certain key enzymes of phenylpropanoid metabolism and this increase in activity leads to the accumulation of products of the pathway which may be either simple phenols such as chlorogenic acid or polymeric phenolic compounds such as lignin.

The role that these responses in phenolic metabolism play in the processes of repair that follow injury of plant tissues is unclear but they appear to have a general role in protecting the injured organ from infection. One of the most interesting questions is to try to define in biochemical terms the changes brought about by wounding and how such diverse factors as mechanical damage or viral infection can elicit certain common biochemical responses.

REFERENCES

1. Alibert, G., Ranjeva, R. : Recherches sur les enzymes catalysant la biosynthèse des acids phenoliques chez Quercus pedunculata (Ehrh). I. Formation des premiers termes du series cinnamique et benzöique. FEBS Lett. 19, 11-14 (1971).

2. Alibert, G., Ranjeva, R., Boudet, A. : Recherches sur les enzymes catalysant la formation des acides phenoliques chez Quercus pedunculata (Ehrh). II. Localisation interacellulaire de la

phenylalanine ammoniaque-lyase, de la cinnamate 4-hydroxylase et de la "Benzoate synthase". Biochim. Biophys. Acta 279, 282-289 (1972).

3. Alibert, G., Ranjeva, R., Boudet, A. : Recherches sur les enzymes catalysant la biosynthese des acides phenoliques chez Quercus pedunculata. III. Formation sequentielle, a partie de la phenylalanine, des acides cinnamique, p-coumarique et cafeique, par des organites cellulaires isolés. Physiol. Plant. 27, 240-243 (1972).

4. Asada, Y., Matsumoto, I. : Formation of lignin-like substance in the root tissue of Japanese Radish plant infected by Downy Mildew fungus. Ann. Phytopath. Soc. Japan 35, 160-167 (1969).

5. Asada, Y., Matsumoto, I. : The nature of lignin obtained from Downy Mildew-infected Japanese radish root. Phytopath. Z. 73, 208-214 (1972).

6. Bailey, J.A., Burden, R.S. : Biochemical changes and phytoalexin accumulation in Phaseolus vulgaris following cellular browning caused by tobacco necrosis virus. Physiol. Plant. Pathol. 3, 171-177 (1973).

7. Bartlett, D.J., Poulton, J.E., Butt, V.S. : Hydroxylation of p-coumaric acid by illuminated chloroplasts from spinach beet leaves. FEBS Lett. 23, 265-267 (1972).

8. Berlin, J., Dewick, P.M., Barz, W., Grisebach, H. : Biosynthesis of coumestrol in Phaseolus aureus. Phytochemistry 11, 168-1693 (1972).

9. Bilton, J.N., Debnam, J.R., Smith, I.M. : 6a-Hydroxypterocarpans from red clover. Phytochemistry 15, 1411-1412 (1976).

10. Birch, A.J., Donovan, F.W. : Studies in relation to biosynthesis. 1. Some possible routes to derivatives of orcinol and phloroglucinol. Aust. J. Chem. 6, 360 (1953).

11. Bonde, H.R., Millar, R.L., Ingham, J.L. : Induction and identification of sativan and vestitol as two phytoalexins from Lotus corniculatus. Phytochemistry 12, 2957-2959 (1973).

12. Brown, A.E., Swinburne, T.R. : Factors affecting the accumulation of benzoic acid in Bramley's seedling apples infected with Nectria galligena. Physiol. Plant. Pathol. 3, 91-99 (1973).

13. Buche, T., Sandermann, H. Jr. : Lipid dependence of plant microsomal cinnamic acid 4-hydroxylase. Arch. Biochem. Biophys. 158, 445-447 (1973).

14. Buhler, D.R., Mason, H.S. : Hydroxylation catalysed by peroxidase. Arch. Biochem. Biophys. 92, 424-437 (1961).

15. Burden, R.S., Bailey, J.A. : Structure of the phytoalexin from soybean. Phytochemistry 14, 1389-1390 (1975).

16. Camm, E. L., Towers, G. H. N. : Phenylalanine ammonia lyase. Phytochemistry 12, 961-973 (1973).

17. Camm, E. L., Towers, G. H. N. : Is cinnamic acid 4-hydroxylase of potato a light-induced enzyme? Can. J. Bot. 51, 824-825 (1973).

18. Chalutz, E. : Ethylene-induced phenylalanine ammonia-lyase activity in carrot roots. Plant Physiol. 51, 1033-1036 (1973).

19. Condon, P., Kuc, J. : Isolation of a fungitoxic compound from carrot root tissue inoculated with Ceratocystis fimbriata. Phytopath. 50, 267-270 (1960).

20. Coombs, J., Baldry, C., Bucke, C., Long, S. P. : o-Diphenol: oxygen oxidoreductase from leaves of sugar cane. Phytochemistry 13, 2703-2708 (1974).

21. Coxon, D. T., Curtis, R. F., Price, K. R., Levett, G. : Abnormal metabolites produced by Daucus carota roots stored under conditions of stress. Phytochemistry 12, 1881-1885 (1973).

22. Cruickshank, I. A. M., Perrin, D. R. : Isolation of a phytoalexin from Pisum sativum. Nature (Lond) 187, 799-800 (1960).

23. Cruickshank, I. A. M., Perrin, D. : Phytoalexins of the Leguminosae. Phaseollin from Phaseolus vulgaris. Life Sci. 2, 680-682 (1963).

24. Czichi, U., Kindl, H. : A model of closely assembled consecutive enzymes on membranes. Formation of hydroxycinnamic acids from L-phenylalanine on thylakoids of Dunaliella marina. Hoppe-Seylers Z. Physiol. Chem. 356, 475-485 (1975).

25. Czichi, U., Kindl, H. : Formation of p-coumaric and o-coumaric acid from L-phenylalanine by microsomal membrane fractions from potato : evidence of membrane-bound enzyme complexes. Planta (Berl.) 125, 115-125 (1975).

26. Davies, D. D., Ugochukwu, E. N., Patil, K. D., Towers, G. H. N. : Aromatic alcohol dehydrogenase from potato tubers. Phytochemistry 12, 531-536 (1973).

27. Dewick, P. M., Barz, W., Grisebach, H. : Biosynthesis of coumestrol in Phaseolus aureus. Phytochemistry 9, 775-783 (1970).

28. Ebel, J., Schaller-Hekeler, B., Knobloch, K., Wellman, E., Grisebach, H., Hahlbrock, K. : Co-ordinated changes in enzyme activities of phenylpropanoid metabolism during the growth of soybean cell suspension cultures. Biochim. Biophys. Acta 362, 417-424 (1974).

29. Finkle, B. J., Nelson, R. F. : Enzyme reactions with phenolic compounds : a meta-O-methyltransferase in plants. Biochim. Biophys. Acta 78, 747-749 (1963).

30. French, C.J., Vance, C.P., Towers, G.H.N. : Conversion of p-coumaric acid to p-hydroxybenzoic acid by cell free extracts of potato tubers and Polyporus hispidus. Phytochemistry 15, 564-566 (1976).

31. Freudenberg, K. : The constitution and biosynthesis of lignin. In 'The Constitution and Biosynthesis of Lignin' by Freudenberg, K. and Neish, A.C., Springer Verlag, Berlin, p. 47-122 (1968).

32. Gestetner, B., Conn, E.E. : The 2-hydroxylation of trans-cinnamic acid by chloroplasts from Melilotus alba Desr. Arch. Biochem. Biophys. 163, 617-624 (1974).

33. Gross, G.G., Kreiten, W. : Reduction of coenzvme A thioesters of cinnamic acids with an enzyme preparation from lignifying tissue of Forsythia. FEBS Lett. 54, 259-262 (1975).

34. Gross, G.G., Zenk, M.H. : Isolation and properties of hydroxycinnamate CoA ligase from lignifying tissue of Forsythia. Eur. J. Biochem. 42, 453-459 (1974).

35. Hadwiger, L.A. : The biosynthesis of pisatin. Phytochemistry 5, 523-525 (1966).

36. Hahlbrock, K., Grisebach, H. : Biosynthesis of flavonoids. In 'The Flavonoids', Harborne, J.B., Mabry, T.J., Mabry, H., eds. Chapman and Hall (London) p. 866-915 (1975).

37. Hahlbrock, K., Wellman, E. : Light dependent induction of enzymes related to phenylpropanoid metabolism in cell suspension cultures from parsley. Biochim. Biophys. Acta 304, 702-706 (1973).

38. Hanson, K.R., Havir, E.A. : Mechanism and properties of phenylalanine ammonia lyase from higher plants. Recent Adv. Phytochem. 4, 45-85 (1972).

39. Harborne, J.B. : Phenolic glycosides and their natural distribution. In 'Biochemistry of Phenolic Compounds' (Harborne, J.B., ed.) Academic Press, London, p. 129-170 (1974).

40. Harel, E., Mayer, A.M., Shain, Y. : Purification and multiplicity of catechol oxidase from apple chloroplasts. Phytochemistry 4, 783-790 (1965).

41. Harel, E., Mayer, A.M., Lehman, E. : Multiple forms of Vitis vinifera catechol oxidase. Phytochemistry 12, 2649-2654 (1973).

42. Havir, E.A., Hanson, K.R. : L-phenylalanine ammonia lyase (potato tubers). Meth. Enzymol. 17, 575-581 (1970).

43. Havir, E.A., Hanson, K.R. : L-phenylalanine ammonia lyase (maize and potato). Evidence that the enzyme is composed of four sub-units. Biochemistry 12, 1583-1591 (1973).

44. Havir, E.A., Hanson, K.R. : L-phenylalanine ammonia lyase (maize, potato and Rhodotorula glutinis). Studies of the prosthetic group with nitromethane. Biochemistry 14, 1620-1626 (1975).

45. Havir, E.A., Reid, P.D., Marsh, H.V. Jr. : L-phenylalanine ammonia lyase (maize). Evidence for a common catalytic site for L-phenylalanine and L-tyrosine. Plant Physiol. 48, 130-136 (1971).

46. Hearing, V.J., Ekel, T.M. :Mammalian tyrosinase. A comparison of tyrosine hydroxylation and melanin formation. Biochem. J. 157, 549-557 (1976).

47. Higuchi, T., Shimada, M., Ohashi, H. : Role of O-methyl-transferase in the lignification of bamboo. Agr. Biol. Chem. 31, 1459-1465 (1967).

48. Hijwegen, T. : Autonomous and induced pterocarpanoid formation in the Leguminosae. Phytochemistry 12, 375-380 (1973).

49. Hill, A.C.R., Rhodes, M.J.C. : The properties of cinnamic acid 4-hydroxylase of aged swede root disks. Phytochemistry 14, 2387-2391 (1975).

50. Hughes, J.C., Swain, T. : Scopolin production in potato tubers infected with Phytophthora infestans. Phytopathol. 50, 398-400 (1960).

51. Imaseki, H., Asahi, T., Uritani, I. : Investigations on the possible inducers of metabolic changes in injured plant tissue. In 'Biochemical Regulation in Diseased Plants or Injury', Hirai, T., ed. p. 189-201, Kyoritsu Printing Co., Tokyo (1968).

52. Jolley, R.L. Jr., Robb, D.A., Mason, H.S. : The multiple forms of mushroom tyrosinase. Association-dissociation phenomena. J. Biol. Chem. 244, 1593-1599 (1969).

53. Kahl, G. : Genetic and metabolic regulation in differentiating plant storage tissue cells. Bot. Rev. 39, 274-299 (1973).

54. Kanazawa, Y., Shichi, H., Uritani, I. : Biosynthesis of peroxidase in slices of black rot-infected sweet potato roots. Agr. Biol. Chem. 29, 840-847 (1965).

55. Keen, N.T., Zaki, A.I., Sims, J.J. : Biosynthesis of hydroxy-phaseollin and related isoflavonoids in disease-resistant soybean hypocotyls. Phytochemistry 11, 1031-1039 (1972).

56. Knobloch, K., Hahlbrock, K. : Isoenzymes of p-coumarate : CoA ligase from cell suspension cultures of Glycine max. Eur. J. Biochem. 52, 311-320 (1975).

57. Kojima, M., Uritani, I. : Elucidation of the structure of a possible intermediate in chlorogenic acid biosynthesis in sweet potato root tissue. Plant Cell Physiol. 13, 1075-1084 (1961).

58. Kojima, M., Uritani, I. : Studies on chlorogenic acid biosynthesis in sweet potato tissue with special reference to the isolation of a chlorogenic acid intermediate. Plant Physiol. 51, 768-771 (1973).

59. Kosuge, T. : The role of phenolics in host response to infection. Ann. Rev. Phytopath. 7, 195-222 (1969).

60. Koukol, J., Conn, E.E. : The metabolism of aromatic compounds in higher plants. IV. Purification and properties of the phenylalanine deaminase of Hordeum vulgare. J. Biol. Chem. 236, 2692-2698 (1961).

61. Kreuzaler, K., Hahlbrock, K. : Enzymic synthesis of an aromatic ring from acetate units. Partial purification and some properties of flavanone synthase from cell suspension cultures of Petroselinum hortense. Eur. J. Biochem. 56, 205-213 (1975).

62. Kuć, J. : Phytoalexins. Ann. Rev. Phytopath. 10, 207-232 (1972).

63. Lamb, C.J., Rubery, P.H. : A spectrophotometric assay for trans-cinnamic acid 4-hydroxylase activity. Analytical Biochem. 68, 554-561 (1975).

64. Lamb, C.J., Rubery, P.H. : Photocontrol of chlorogenic acid biosynthesis in potato tuber disks. Phytochemistry 15, 665-668 (1976).

65. Lamb, C.J., Rubery, P.H. : Differential effects of cycloheximide on the activity of PAL and CAH in light and dark-incubated potato tuber disks. Plant Sci. Letters 7, 33-37 (1976).

66. Lamb, C.J., Rubery, P.H. : Phenylalanine ammonia lyase and cinnamic acid 4-hydroxylase. Product repression of the level of enzyme activity in potato tuber disks. Planta (Berl.) 130, 283-290 (1976).

67. Levy, C.C., Zucker, M. : Cinnamyl and p-coumaryl esters as intermediates in the biosynthesis of chlorogenic acid. J. Biol. Chem. 235, 2418-2425 (1960).

68. MacDonald, I.R., Ellis, R.J. : Does cycloheximide inhibit protein synthesis specifically in plant tissues? Nature (Lond.) 222, 791-792 (1969).

69. Mansell, R.L., Gross, G.G., Stöckigt, J., Franke, H., Zenk, M.H. : Purification and properties of cinnamyl alcohol dehydrogenase from higher plants involved in lignin biosynthesis. Phytochemistry 13, 2427-2435 (1974).

70. Mason, H.S. : The chemistry of melanin. III. Mechanism of the oxidation of dihydroxyphenylalanine by tyrosinase. J. Biol. Chem. 172, 83-99 (1948).

71. Mason, H. S. : Comparative Biochemistry of the phenolase complex. Adv. Enzymol. XVI, 105-184 (1955).

72. Maule, A. J., Ride, J. P. : Ammonia-lyase and O-methyltransferase activities related to lignification in wheat leaves infected with Botrytis. Phytochemistry 15, 1661-1664 (1976).

73. Mueller, W. C., Beckman, C. H. : Ultrastructure of the phenol-storing cells in the roots of banana. Physiol. Plant Path. 4, 187-190 (1974).

74. Müller, K., Börger, H. : Experimentelle Untersuchungen uber die Phytophthora-resistenz der Kartoffel. Arb. Biol. Reichstanstat. Land-u. Forstwirtisch, Berlin 23, 189-231 (1940).

75. Munn, C. B., Drysdale, R. B. : Kievitone production and phenylalanine ammonia lyase activity in cowpea. Phytochemistry 14, 1303-1307 (1975).

76. Nagels, L., Parmentier, F. : Kinetic study of possible intermediates in the biosynthesis of chlorogenic acid in Cestrum poeppigii. Phytochemistry 15, 703-706 (1976).

77. Nari, J., Mouttet, C., Fouchier, F., Ricard, J. : Subunit interactions in enzyme catalysis. Kinetic analysis of subunit interactions in the enzyme L-phenylalanine ammonia lyase. Eur. J. Biochem. 41, 499-515 (1974).

78. Neish, A. C. : Monomeric intermediates in the biosynthesis of lignin. In 'Constitution and Biosynthesis of Lignin' by Freudenberg, K., Neish, A. C., Springer-Verlag, Berlin. p. 3-43 (1968).

79. Paxton, J., Goodchild, D. J., Cruickshank, I. A. M. : Phaseollin production by live bean endocarp. Physiol. Plant Pathol. 4, 167-171 (1974).

80. Perrin, D. R. : The structure of phaseolin (phaseollin). Tetrahedron Letters 1, 29-35 (1964).

81. Pierpoint, W. S. : Formation and behaviour of o-quinones in some processes of agricultural importance. Ann. Rpt. Rothamsted Expt. St. for 1970, Part 2, 199-218 (1970).

82. Potts, J. R. M., Weklych, R., Conn, E. E. : The 4-hydroxylation of cinnamic acid by sorghum microsomes and the requirement for cytochrome P450. J. Biol. Chem. 249, 5019-5026 (1974).

83. Poulton, J., Grisebach, H., Ebel, J., Schaller-Hekeler, B., Hahlbrock, K. : Two distinct S-adenosyl L-methionine 3:4 dihydric phenol 3-O-methyltransferases of phenylpropanoid metabolism in soybean cell suspension cultures. Arch. Biochem. Biophys. 173, 301-305 (1976).

84. Poulton, J., Hahlbrock, K., Grisebach, H. : Enzymic synthesis of lignin precursors. Purification and properties of the S-adenosyl L methionine : caffeic acid-3-O-methyl transferase from soybean cell suspension cultures. Arch. Biochem. Biophys. 176, 449-456 (1976).

85. Pridham, J.B., Saltmarsh, M.J. : The biosynthesis of phenolic glucosides in plants. Biochem. J. 87, 218-224 (1963).

86. Pueppke, S.G., VanEtten, H.D. : Accumulation of pisatin and three additional antifungal pterocarpans in Fusarium solani-infected tissues of Pisum sativum. Physiol. Plant Pathol. 8, 51-61 (1976).

87. Ranjeva, R., Boudet, A.M., Faggion, R. : Phenolic metabolism in Petunia tissues. IV. Properties of p-coumarate : Coenzyme A ligase isoenzymes. Biochimie 58, 1255-1262 (1976).

88. Raper, H.S. : The aerobic oxidases. Physiological Rev. 8, 245-282 (1928).

89. Rathmell, W.G. : Phenolic compounds and phenylalanine ammonia lyase activity in relation to phytoalexin biosynthesis in infected hypocotyls of Phaseolus vulgaris. Physiol. Plant Pathol. 3, 259-267 (1973).

90. Rhodes, M.J.C., Hill, A.C.R., Wooltorton, L.S.C. : Activity of enzymes involved in the lignin biosynthesis in swede root disks. Phytochemistry 15, 707-716 (1976).

91. Rhodes, M.J.C., Wooltorton, L.S.C. : Stimulation of phenolic acid and lignin biosynthesis in swede root tissue by ethylene. Phytochemistry 12, 107-118 (1973).

92. Rhodes, M.J.C., Wooltorton, L.S.C. : Formation of CoA esters of cinnamic acid derivatives by extracts of Brassica napobrassica root tissue. Phytochemistry 12, 2381-2387 (1973).

93. Rhodes, M.J.C., Wooltorton, L.S.C. : The p-coumarate CoA ligase of potato tubers. Phytochemistry 14, 2161-2164 (1975).

94. Rhodes, M.J.C., Wooltorton, L.S.C. : Enzymes involved in the reduction of ferulic acid to coniferyl alcohol during the ageing of disks of swede root tissue. Phytochemistry 14, 1235-1240 (1975).

95. Rhodes, M.J.C., Wooltorton, L.S.C. : The enzymic conversion of hydroxycinnamic acids to p-coumarylquinic acid and chlorogenic acid in tomato fruits. Phytochemistry 15, 947-951 (1976).

96. Ride, J.P. : Lignification in wounded wheat leaves in response to fungi and its possible role in resistance. Physiol. Plant Pathol. 5, 125-134 (1975).

97. Riley, R.G., Kolattukudy, P.E. : Evidence for covalently attached p-coumaric acid and ferulic acid in cutins and suberins. Plant Physiol. 56, 650-654 (1975).

98. Russell, D.W. : The metabolism of aromatic compounds in higher plants. X. Properties of the cinnamic acid 4-hydroxylase of pea seedlings and some aspects of its metabolic and developmental control. J. Biol. Chem. 246, 3870-3878 (1971).

99. Russell, D.W., Conn, E.E., Sutter, A., Grisebach, H. : Hydroxylation-induced migration and retention of tritium on conversion of [4-^{3}H]cinnamic acid to 4-hydroxycinnamic acid by an enzyme from pea seedlings. Biochim. Biophys. Acta 170, 210-213 (1968).

100. Sarkar, S.K., Phan, C.T. : Effect of ethylene on the qualitative and quantitative composition of the phenol content of carrot roots. Physiol. Plant 30, 72-76 (1974).

101. Sarkar, S.K., Phan, C.T. : Effect of ethylene on the phenylalanine ammonia lyase activity of carrot tissue. Physiol. Plant. 32, 318-321 (1974).

102. Sarkar, S.K., Phan, C.T. : The biosynthesis of 8-hydroxy 6-methoxy 3-methyl-3,4-dihydroisocoumarin and 5-hydroxy-7-methoxy-2-methylchromone in carrot root tissues treated with ethylene. Physiol. Plant 33, 108-112 (1975).

103. Shannon, L. : Physical and catalytic properties of peroxidase isoenzymes. In 'Biochemical Regulation in Diseased Plants or Injury' p. 181-188 (Hirai, T., ed.), Kjoritsu Printing Co. Ltd., Tokyo (1968).

104. Shimada, M., Kuroda, H., Higuchi, T. : Evidence for the formation of methoxyl groups of ferulic and sinapic acids in Bambusa by the same o-methyltransferase. Phytochemistry 12, 2873-2875 (1973).

105. Smith, D.A., VanEtten, H.D., Bateman, D.F. : Accumulation of phytoalexins in Phaseolus vulgaris hypocotyls following infection by Rhizoctonia solani. Physiol. Plant Pathol. 5, 51-64 (1975).

106. Sondheimer, E. : Chlorogenic acids and related depsides. Bot. Rev. 30, 667-712 (1964).

107. Stafford, H.A. : The metabolism of aromatic compounds. Ann. Rev. Plant Physiol. 25, 459-486 (1974).

108. Stafford, H.A. : Transformation of a large aggregation of hydroxycinnamate hydroxylase to lower molecular weight forms by sulphydryl agents in green leaves of sorghum. Plant Physiol. 55, 145-149 (1975).

109. Stafford, H.A. : Characteristics of a 4-hydroxycinnamate hydroxylase purified from Sorghum leaves. Plant Physiol. 57, 320-324 (1976).

110. Steck, W. : Metabolism of cinnamic acid in plants. Chlorogenic acid formation. Phytochemistry 7, 1711-1717 (1968).

111. Stockigt, J., Zenk, M.H. : Enzymatic synthesis of chlorogenic acid from caffeoyl coenzyme A and quinic acid. FEBS Lett. 42, 131-134 (1974).

112. Swinburne, T.R., Brown, A.E. : The biosynthesis of benzoic acid in Bramley's seedling apples infected by Nectria galligena Bres. Physiol. Plant Path. 6, 259-264 (1975).

113. Towers, G.H.N. : Enzymological aspects of flavonoid and lignin biosynthesis and degradation in plants. In 'MTP International Review of Science. Biochemistry Series I', Vol. 11 (Northcote, D.H. ed.) Butterworths, London, p. 247-276 (1974).

114. Uritani, I. : The biochemical basis of disease resistance induced by infection. Connecticut Agr. Exp. Sta. Bull. 663, 4-19 (1963).

115. Vance, C.P., Sherwood, R.T. : Regulation of lignin formation in reed canary grass in relation to disease resistance. Plant Physiol. 57, 915-919 (1976).

116. VanEtten, H.D. : Antifungal activity of pterocarpans and other selected isoflavonoids. Phytochemistry 15, 655-659 (1976).

117. Van Stevaninck, R.F.M. : The "washing" or "aging" phenomenon in plant tissues. Ann. Rev. Plant Physiol. 26, 237-258 (1975).

118. Vaughan, P.F.T., Butt, V.S. : The hydroxylation of p-coumaric acid by an enzyme from leaves of Spinach beet (Beta vulgaris L.) Biochem. J. 113, 109-115 (1969).

119. Wengenmayer, H., Ebel, J., Grisebach, H. : Enzymic synthesis of lignin precursors. Purification and properties of a cinnamoyl-CoA : NADP reductase from cell suspension cultures of soybean (Glycine max). Eur. J. Biochem. 65, 529-536 (1976).

120. Whiting, G.C., Coggins, R.A. : 4-p-coumarylquinic acid in apple fruits. Phytochemistry 14, 593 (1975).

121. Wyrambik, D., Grisebach, H. : Purification and properties of isoenzymes of cinnamyl alcohol dehydrogenase from soybean cell suspension cultures. Eur. J. Biochem. 59, 9-15 (1975).

122. Yahiel, V., Cotte-Martinon, M.G., Ducet, G. : Un cytochrome de type P_{450} dans la spadice d'Arum. Phytochemistry 13, 1649-1651 (1974).

123. Zenk, M. H. : In Proc. 2nd Meeting European Biochem. Soc. Vienna (Bullek, G., ed.) p. 45, Pergamon Press, Oxford (1965).

124. Zucker, M. : Induction of phenylalanine deaminase by light and its relation to chlorogenic acid synthesis in potato tuber tissue. Plant Physiol. 40, 779-784 (1965).

125. Zucker, M. : Sequential induction of phenylalanine ammonia lyase and a lyase-inactivating system in potato tuber disks. Plant Physiol. 43, 365-374 (1968).

126. Zucker, M. : Light and enzymes. Ann. Rev. Plant Physiol. 23, 133-156 (1972).

The Tissue Slice System as a Model for Studies of Host-Parasite Relationships

Ikuzo Uritani and Kazuko Ōba

I. INTRODUCTION

The slice system of plant storage tissues is valuable as a model for studying host-parasite relationships, and has provided many important findings. The system appears to have advantages over other systems using intact leaf, stem or root tissues.

This paper deals with the reasons for the usefulness of the storage tissues as research materials to study (i) biochemical and cytological alterations induced in host plants in response to the parasitic infection, (ii) roles of such alterations in the defense action (resistance) and the susceptibility of hosts, and (iii) parasite-recognition by hosts and its specificity.

On the other hand, the use of plant storage tissues has its own disadvantages. We should be cautious in extrapolating experimental results obtained with the tissue slices to the many kinds of interaction caused by the combination of various plants with various parasites. We will also discuss these aspects.

II. PLANT STORAGE TISSUES FOR STUDIES OF HOST-PARASITE RELATIONSHIPS

Biochemistry of Wounded Plant Tissues

There are various kinds of plant storage tissue (6, 7), usually swollen organs, e.g. roots such as sweet potato (Ipomoea batatas Lam.), sugar beet (Beta vulgaris L. var. saccharifera L.), carrot (Daucus carota L.), Yam (Dioscorea spp.) and parsnip (Pastinaca sativa L.); stems such as white potato or Irish potato (Solanum tuberosum L.) and Jerusalem artichoke (Helianthus tuberosus L.); hypocotyls such as radish (Raphanus sativus L.) and swede or rutabaga (Brassica napus L. var. napobrassica (L.) Pet.).

III. ADVANTAGES OF STORAGE TISSUES

1. Availability of Energy Sources in Large Amounts.

Generally speaking, plant storage tissues contain large amounts of polysaccharides (e.g. starch and inulin) and oligosaccharides (e.g. sucrose) in parenchymatous tissue cells. On the other hand, in response to the parasitic infection, plant tissues show various kinds of defense action involving the accumulation of phytoalexins in the infected region and the production of polyphenols in the non-infected adjacent tissue, which need a supply of ATP (23). The production of ATP is derived from the decomposition of reserve polysaccharides. Requirement for ATP is mainly due to the synthesis of enzymes involved in the production of phytoalexins, polyphenols and other abnormal metabolites (22, 23), as shown in Figure 1.

Furthermore, cellular membrane systems, such as mitochondria and microsomes are developed in response to the parasitic infection. ATP is also required in the biogenesis of these cellular membranes. Conceivably, the increased mitochondria and microsomes participate respectively in the supply of sufficient ATP and in the localization of some enzymes involved in the phytoalexin-and polyphenol-synthesizing enzyme systems (Figure 1). As shown in Figure 2, a vigorous increase in respiration occurs in the infected plant tissues. The overall reaction is obligatorily connected to the decomposition of the abundant polysaccharide and/or oligosaccharide reserves, yielding monosaccharides and related derivatives. It can be thus recognized that plant storage tissues are suitable for studying host-parasite relationships because of the potential supply of energy sources sustaining the defense action against the parasitic infection.

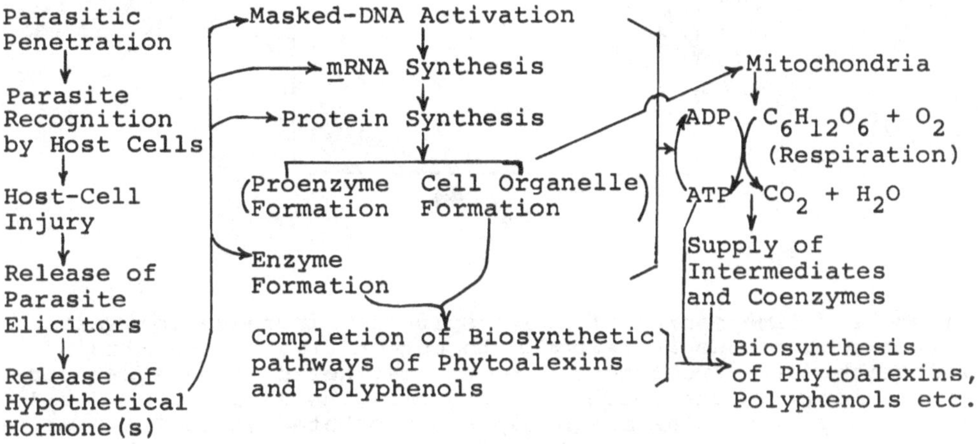

Figure 1. Possible mechanism of biosynthesis of abnormal metabolites in response to parasitic infection (22, 23).

2. Homogeneity of the Tissue Cells.

The majority of cells in plant storage tissues consist of parenchymatous cells which are extended and homogeneous. This anatomical feature is different from many other plant organs such as non-swollen leaves, stems and roots, which are composed of several different tissues, such as xylem, phloem and pith, and are often meristematic in part. It will be noted that any parts of plant storage tissues respond to the parasite attack in a relatively similar fashion, suffering from the parasite penetration to a more or less equal grade. As a consequence, parasite growth in the host tissues proceeds at relatively similar rate. Therefore, separation of the infected region, free from the non-infected tissue, can be quite easily manipulated, as modelled in Figure 3.

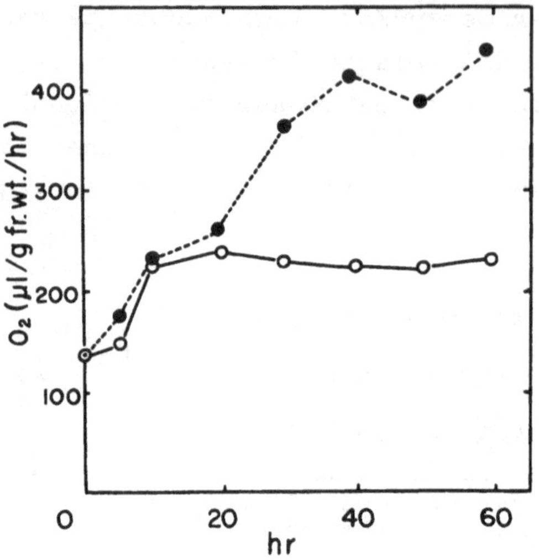

Figure 2. Time course of the respiratory increase in sweet potato root tissue in response to cut injury or parasitic infection (2). Sweet potato roots (cv. Norin 1) were cut perpendicularly in the thickness of 3 cm. The slices were divided into two groups. The one group was incubated at 29 ± 1°C without fungal inoculation, and called cut tissue. The other group was inoculated with the endoconidial suspension of _C. fimbriata_ on the cut-surface, then incubated at 29 ± 1°C, and called diseased tissue. Samples were taken at appropritate periods and discs (0.5 mm thick) were prepared from the healthy tissues in the 2 mm thickness adjacent to the cut-surface (cut-tissue) or to the infected region (diseased tissue). The respiratory rate of the slices were determined using a Warburg manometer. —○— ,--- ●---: cut tissue and diseased tissue, respectively.

3. Comparison of Metabolic Alterations between Parasite-infected and Cut-injured Tissues.

Regardless of the different sources of plant tissues, parasitic infection causes both biochemical and cytological alterations in the non-infected tissue adjacent to the infected region. However, somewhat similar alterations can be induced by the simple mechanical injury, such as cutting, although there are some parasite-specific phenomena. It is probable that the alterations specific to parasitic infection are largely related to defense action (resistance) or susceptibility. Therefore, it is feasible that the better understanding of the similarities and differences in metabolic alterations between parasite-infected response and cut-injured response may lead to the elucidation of the mechanism on either resistance or susceptibility in the host-parasite interactions. Obviously plant storage tissues are advantageous for such research purpose. For instance, sweet potato root tissues can be sliced laterally in several pieces, which may then be divided into two groups. One group may be inoculated on the cut surface by a spore suspension of a pathogenic fungus and incubated at an appropriate temperature and humidity; this group can be designated as "diseased tissue." The other group may be incubated at the same condition without inoculation, and referred to as "cut tissue." Subsequent time-course metabolic patterns can then be compared between both groups (14).

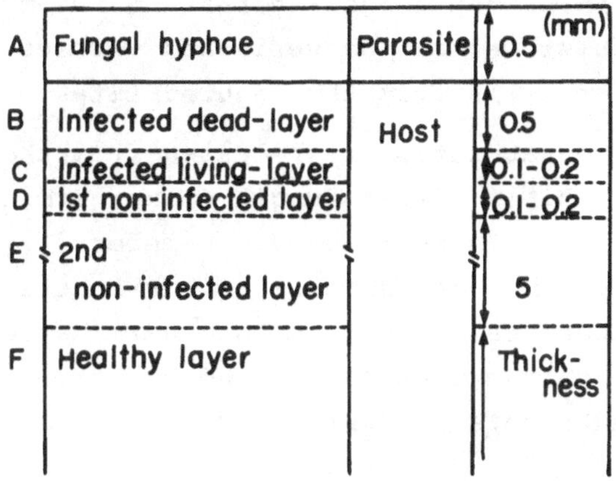

Figure 3. Model of host-parasite complex (22, 23). In the case of sweet potato root tissue infected by C. fimbriata for about 2 days, A: fungal hyphae (about 0.5 mm thick), B: infected dead-layer showing brown necrotic region and containing phytoalexins (about 0.5 mm thick), C: infected living-layer not-showing necrosis and containing phytoalexins, coumarins and polyphenols (about 0.1 to 0.2 mm thick), D: 1st non-infected layer containing phytoalexins, coumarins and polyphenols (about 0.1 to 0.2 mm thick), E: 2nd non-infected layer containing

polyphenols and showing increased respiration (about 5 mm thick), and F: Fresh layer not-suffering from the affect of infection severely. The average cell length of sweet potato parenchymatous tissue is 0.1 mm.

Here, it will be emphasized that large amounts of both groups can be prepared from plant storage tissues, whereas this can not be done so readily with other plant organs.

4. Facile Penetration of Parasites into Cut-injured Tissue.

Non-swollen tissues of leaf, stem and root are covered by waxes and cuticlae, deficient in nutrients for most parasites. Therefore, except for some particular combination of hosts and parasites, germination of many parasites such as fungi and subsequent penetration into host cells are prevented. Thus, before inoculating some fungal spores, one must mechanically injure the surface of such organs. On the other hand, in the case of plant storage tissues, upon their cut-injury, cytoplasmic substances are excreted from the injured cells, favoring the germination of the fungal spores and the mycelial penetration into host cells. Bacteria and viruses also easily penetrate into host cells from the injured sites.

Occasionally toxic metabolites are released from the injured cells, which may inhibit either fungal spore germination or parasite penetration. In such cases, therefore, it is necessary to thoroughly rinse the surface of cut-injured tissue with distilled water; reasonably uniform germination and penetration of fungi can thus be ensured, perhaps on account of the residual nutrients present in quite low amounts.

5. Easy application of Chemical Agents.

In order to investigate the matabolic alterations in diseased tissue, it is often necessary to administer some ^{14}C-labelled possible intermediates or metabolically active compounds to host-plant tissues before or after the parasite inoculation; in some cases they will be applied at the time of the parasite inoculation. Some valuable information may be obtained from such techniques.

We can readily apply a solution of a chemical agent to the cut surface, either directly or by applyting a filter paper disc moistedned with the chemical. If that is necessary to test the effect of a chemical agent, sliced tissue can be immersed in the test solution. If necessity, a vacumm-infiltration technique (usually, 30 to 40 mmHg for about 30 to 60 seconds, once or twice) can be also employed to replace the phase in the inter cellular-space with the chemical solution. The procedure is quite satisfactory, but care should be taken as sometimes the inner-cells of the host tissue will be mechanically damaged.

6. Long-term Storage of Plant Tissues.

Generally speaking, plant storage tissues exhibit little physiological and/or cytological activities as long as they are stored at a temperature low enough to repress sprouting; some organs can be kept fresh for as long as 7-12 months.

There are many storage organs of plants originating either from the temperate zone or from the tropical and subtropical zones, In the latter case, it is necessary to store them at 1 or 2°C higher than the transition points below which chilling injury occurs; ordinary storage temperature being 8-15°C, depending on the species, varieties and cultivars.

One has to be particularly careful to avoid chilling injury, as this causes irreversible deterioration in the physiological and cytological activities (7, 16, 26). Also alteration of the metabolic activity may occur, forming anomalous metabolites such as polyphenols (24). The biochemical mechanism of chilling injury and possible ways of recovering from the injury will be described elsewhere.

7. Potential Physiological and Cytological Activities of Plant Storage Tissues.

As described above, both physiological and cytological activities of plant storage tissues are not very high, remaining in the resting-state during the conditions of storage. Even if they are transferred to ambient temperatures, their activities are not still very high, showing a gradual increase as sprouting develops. However, when they are subjected to cut-injury or parasite inoculation, the metabolic activities are markedly enhanced. Obviously, such a consequential effect is suitable for examining the nature of metabolic alterations and cell-organelle formaiton in relation to the mechanisms of wounding, or host resistance and susceptibility.

These overall phenomena may be different from those occurring in other plant organs e.g. leaves, infected by the pathogenic agents. Upon infection or mechanical wounding, metabolic activities of leaf tissue are found to be enhanced slightly, in a similar way to that in the storage tissues, exhibiting a wound respiration type phenomenon. However, it should be noted that under such circumstances chloroplasts are structurally degraded, resulting in the deterioration of activities rather than activation.

8. High Moisture Content of Plant Storage Tissues.

The high moisture content of storage organs is important, enabling the plant tissue to respond to either cut injury or parasitic infection very rapidly. Furthermore, homogenates of tissues obtained by relatively gentle treatment can yield cell-organelles in an intact state.

Several vegetables and fruits such as cucumber, pumpkin, avocado, apple, pear, and banana contain expanded and rather homogeneous cells. However most fruits show the climacteric rise during the maturation stage, and it is difficult to keep them for such long periods as is possible with underground growing storage tissues. Vegetables such as cucumber and pumpkin are also difficult to store for longer periods. It will be noted also that the content of carbohydrates in vegetables and most fruits is not high.

IV. DISADVANTAGES OF STORAGE TISSUES

1. Low Levels of Nitrogenous Compounds.

In contrast to the high carbohydrate content in most storage organs whose origins are root, stem and hypocotyl, the content of nitrogenous compounds including proteins and amino acids in these tissues is much less compared to other organs such as leaves. Therefore if one could succeed in breeding storage plant tissues of high reserve protein levels, they would become useful research materials. As we have discussed above, it has been well established that, upon the parasitic infection of tissues, some enzymes involved in the production of secondary metabolites such as phytoalexins are synthesized de novo in host tissues (14, 22, 23). Clearly these synthetic reactions require a supply of free amino acids as well as some types of protein present in healthy tissues. Therefore, storage tissues containing the high content of nitrogenous compounds in addition to carbohydrates may facilitate the elevation of both physiological and cytological activities as a consequence of infection.

In this context, one can see that pea (Pisum sativum L.), bean (Phaseolus vulgaris L.) and soybean [Glycine max (L.) Merr.] seeds containing high contents of both proteins and carbohydrates are favorable materials for studying host-parasite relationships. However, maturating seeds or water-imbibed seeds after harvest are not in a resting state, but are metabolically active, and it is difficult to differentiate the physiological and cytological alterations in diseased tissue from those in the maturating or water-imbibed seed tissues.

2. Unsuitability of Plant Storage Tissues for Studies of Host-Virus Relationships.

Viruses can attack any kind of plant organs such as leaf, stem and root. However, virus multiplication can be observed usually in leaves instead of stems or roots, presumably because of the availability of nitrogenous compounds in leaves. Tuberous roots and tubers are generally unsuitable for the studies of host-virus relationships.

V. SIMILARITY OF METABOLIC ALTERATIONS BETWEEN MECHANICALLY CUT AND DISEASED TISSUES

Whether metabolic alterations induced by cut injury and parasitic infection of plant tissues are similar or dissimilar will be discussed by taking the case of sweet potato root tissues. The pathogenic agent of interest is _Ceratocystis fimbriata_, which causes black rot disease.

1. Respiratory Increase.

Upon the cut injury, there occurs a respiratory increase, commonly called wound respiration. The mechanism of wound respiration will be dealt in detail in other chapters of this book written by G. Laties and T. Asahi. It is recognized that wound respiration is induced by the enhanced synthesis of enzymes involved in the carbohydrate breakdown leading to pyruvate as well as by the mitochondrial biogenesis to enhance electron transport and oxidative phosphorylation. Furthermore, supply of monosaccharides derived from starch granules or sucrose (_e.g._ glucose) and the rapid synthesis of adenosine nucleotides such as ADP and ATP are presequisites in the process (Figure 1 and 2).

On the other hand, parasite penetration causes the injury of host cells, especially in the case of incompatible interactions between host and parasite. The infection also causes an increased respiration in diseased tissue apparently based on the same mechanisms as those in the mechanical injury in the initial stage of infection lasting for 1-day. As will be described in a later section, however, there exists the diseased-plant specific respiration occurring at the secondary stage.

Respiratory increases in wounded and diseased plants are comparatively in Figure. 2 (2).

One can thus understand that studies on the mechanisms of wound respiration in cut tissues are profitable in elucidating the nature of increased respiration in the pathogen-infected diseased tissues.

2. Formation of Polyphenols.

It is well established that both cut injury and parasite infection induce the production of various kinds of polyphenols (e.g. chlorogenic acid and isochlorogenic acid) in plant tissues, although the amounts synthesized are much higher in diseased tissue in comparison with those in cut tissue (Figure 4). However, technically it is more advantageous to use cut tissue than diseased plants to determine the biochemical mechanism of polyphenol production (20, 21, 24).

As described in other chapters by Rhodes and Kahl, preceding the polyphenol production, there occurs the de novo synthesis of various types of enzymes involved in polyphenol biosynthesis. (Figure 5). However, as a consequence of the degradative breakdown of enzymes at a later stage, polyphenol content tends to level off or maintain the plateau.

On the other hand, during the prolonged parasitic infection, polyphenol production in diseased tissue appears to continue for as long as the host plant tissue is alive, but the synthesis appears to terminate in parallel with the end of infection.

3. Formation of Peroxidase.

Fresh plant tissue contains peroxidase at much lower levels than in cut or diseased tissue. The enzyme molecules are newly synthesized following mechanical slicing after a lag period of about 1-day. The time-course pattern of peroxidase activity in diseased tissue is similar to that in cut tissue, the enzyme activity being generally higher in diseased tissue

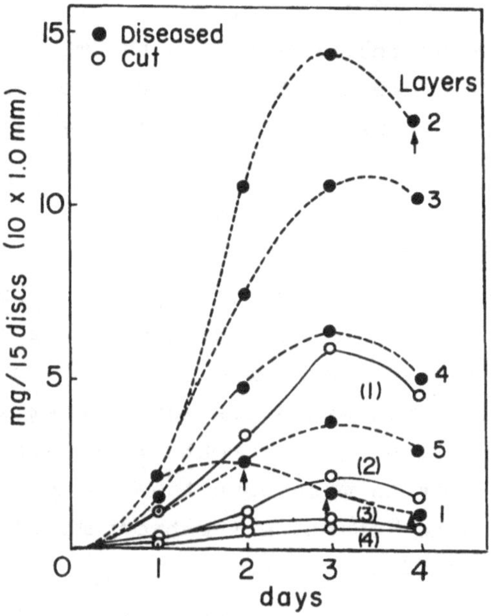

Figure 4. Production of polyphenols in sweet potato root tissue (about 3 cm thick) in response to cut injury or parasitic infection (22, 23). Cut tissue and diseased tissue (infected by C. fimbriata) were prepared in the similar method to that shown in the legend to Figure 2. Samples were taken at appropriate periods and discs (1 mm thick) were prepared from the cut surface toward the inner part, and the same layer discs were collected and subjected to the analysis. Polyphenol: mg (expressed as chlorogenic acid) per 15 discs (10 mm diameter and 1 mm thickness). ↑ : Shows the period when the discs were infected more than 50%.

(Figure 6) (8). Furthermore, one can note that the localization pattern of peroxidase activity from the outer tissue toward the inner portion is similar between the two differntly treated plant tissues, except that the enzyme activity in

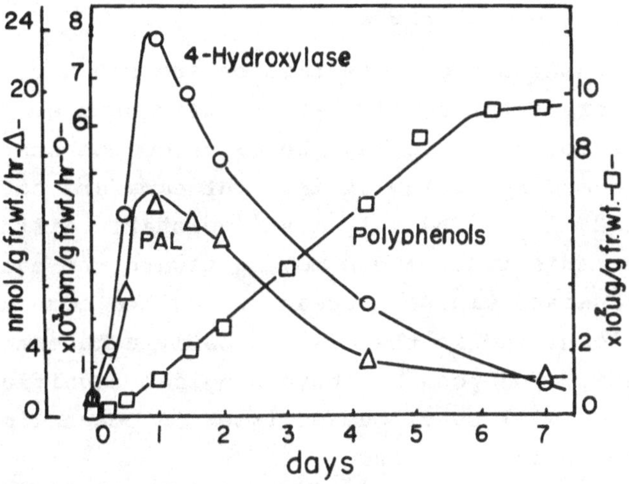

Figure 5. Time course of the development of phenylalanine ammonia-lyase (PAL) and trans-cinnamic acid 4-hydroxylase (4-hydroxylase) during incubation at 29 ± 1°C of discs (3 mm thick) of sweet potato root tissue (20). In addition to activities of the above two enzymes, the content of polyphenols (mostly chlorogenic acid and isochlorogenic acid) was determined.

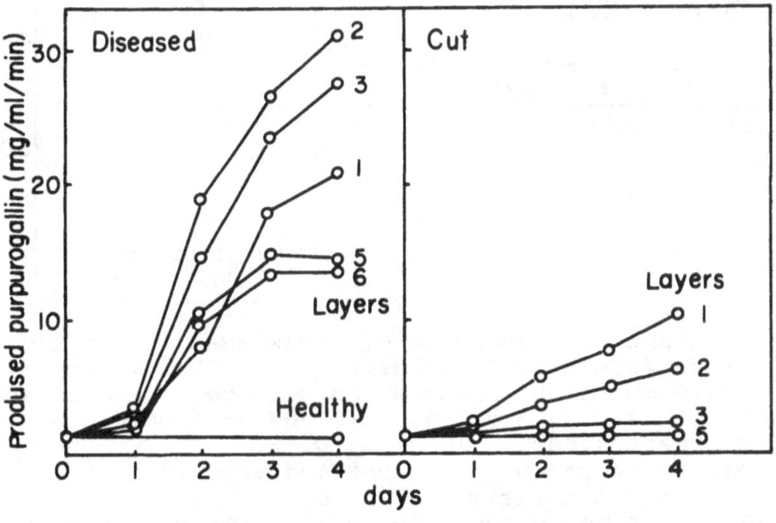

Figure 6. Time course of the development of peroxidase in sweet potato root tissue (about 3 cm thick) in response to cut-injury and parasitic infection (8). Cut tissue and diseased tissue (infected by *C. fimbriata*) were prepared in the similar method to that shown in the legend to Figure 2, and discs (1.0 mm thick) were prepared as shown in the legend to Figure 4. Peroxidase activity: purpurogallin (mg) produced from pyrogallol per min per ml of the incubation mixture.

diseased tissue is much higher.

The enzymes isolated from both tissues are shown to be composed of various isozyme forms, similar in their nature. From the analysis of zymogram patterns, the major components are shown to be anionic and neutral proteins, but some are cationic (Figure 7) (12, 18). The only distinguishable difference is that the two cationic isozymes moving toward the cathode are missing in diseased tissue. These are called component H_1 and H_2, and are localized in the cells closely adjacent to the cut tissue surface. The possible physiological significance of this phenomenon will be discussed later in connection with the lignin formaiton in cut tissue.

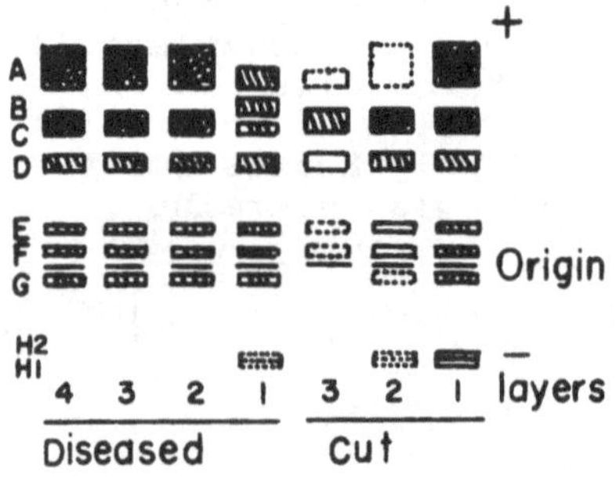

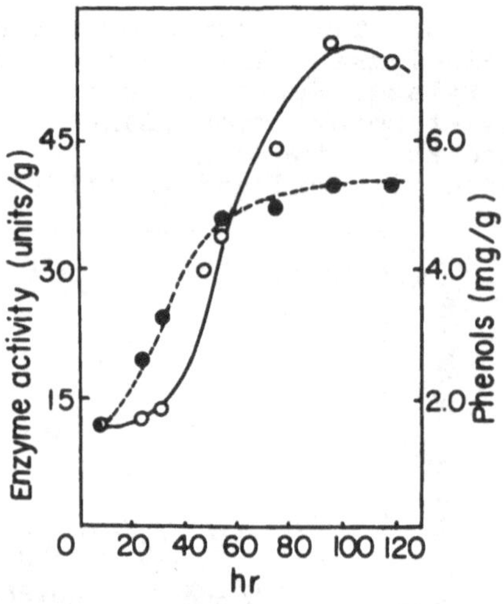

Figure 7. Starch-gel zymogram of peroxisase in cut and diseased tissues incubated for 2 days (12). Tissue incubation and disc (1.0 mm thick) preparation were performed as shown in the legends to Figure 2 and 4. (+) and (-): anodic and cathodic sides, respectively. A, B, C, D, E, F, G, H_1 and H_2: individual peroxidase components separated from each other by starch-gel electrophoresis. ▩ , ▧ , ▭ , ⬚ : degree of peroxidase staining in the order of intensity, strong to weak.

Figure 8. Changes in o-diphenol oxidase activity and polyphenol content in sweet potato root tissue (2 cm thick) during incubation (3). Sections in 2 mm thick were taken from the outer parts of the tissue and assayed for the enzyme activity and polyphenol content. The solid line shows o-diphenol oxidase activity in units per g of tissue and the dotted line indicates polyphenol content in mg per g of tissue.

4. Formation of Polyphenol Oxidase.

Polyphenol oxidase is synthesized *de novo* in both cut and diseased tissues with a time lag of more than 1 day, accompanying the initiation of peroxidase formation. Furthermore, the rates of enhancement of the enzyme activity exhibit no appreciable difference between cut and diseased tissues.

In contrast to peroxidase, however, the activity of polyphenol oxidase in fresh tissue is high, in spite of the fact that the activity in either cut tissue or diseased tissue is several-fold greater than in fresh tissue (Figure 8) (3). There are at least 3 isozymic forms of polyphenol oxidase in preparations isolated from fresh tissue and at least 6 isozymes in both cut and diseased tissues, among which the 3 isozymes appear to be the same as those in fresh extracts (4).

VI. DIFFERENCES IN METABOLIC ALTERATIONS BETWEEN CUT AND DISEASED TISSUES

1. Respiratory Increase at the Secondary Stage.

As seen in Figure 2, the time course pattern of the increased respiration shows a biphasic nature in diseased tissue (2). The respiratory activity in diseased tissue shows a secondary increase, responding to the developmental penetration of *C. fimbriata*. Thus, the preliminary experiment indicates that the increased respiration is partly due to the enhanced activity of the enzymes related to the breakdown of carbohydrate, and the electron transport system, such as D-glucose 6-phosphate dehydrogenase and cytochrome c oxidase. At the same time an increased production of ADP, resulting from the enhanced consumption of ATP, tightly linked to the synthesis of coumarins and phytoalexins appears to take part in the process (19) (see Figure 1).

2. Formation of Ethylene.

It is observed that small amounts of ethylene are synthesized during the initial 10 hr period after cutting, although amounts become negligible during the further incubation. However, there occurs a dramatic synthesis of ethylene in diseased tissue after about an 18-hr lag of period. Isotopic experiments have shown that, in response to the continuous injury of cells, either by the fungal penetration or by toxic chemicals, ethylene is formed from some unknown precursors, involving member(s) of the TCA cycle (5, 17). It is thus essential to separate the adjacent non-infected region of diseased tissue entirely free from the infected region for the biochemical studies on ethylene formation (see the chapter by Yang and Pratt).

3. Synthesis of Coumarins.

Simple cut injury of plant tissue does not cause the formation of coumarins. However, in response to the continuous stimulus such as fungal penetration and treatment by toxic chemicals, coumarins are formed in the injured part or in the tissue closely adjacent to the damaged part (Figure 9). It is shown that the synthesis occurs earlier than the phytoalexin formation.

In the case of the diseased sweet potato root tissue, it is known that umbelliferone, scopoletin and esculetin are formed. The glucosides of umbelliferone and scopoletin (i.e. skimin and scopolin respectively) are also shown to be synthesized (13). The mechanism for biosynthesis of coumarins is similar to that of phytoalexins, and coumarins are known to exhibit some antibiotic activity similar to phytoalexins. Their biosynthetic pathway is analogous to that of polyphenols such as chlorogenic acid, but is entirely different from that of the isoprenoid phytoalexins. In this context coumarins will not be included as a group of phytoalexins, although it appears likely they may also be classed as stress compounds (10).

4. Lignin Formation.

As described in the chapter by Rhodes, it is known that, upon cut injury of plant storage tissues of many species, lignin is formed in the limited cell layers (2 to 3 layers) of plant tissues just below the cut surface. This is exactly the case in sweet potato roots. The lignin formed will be referred to

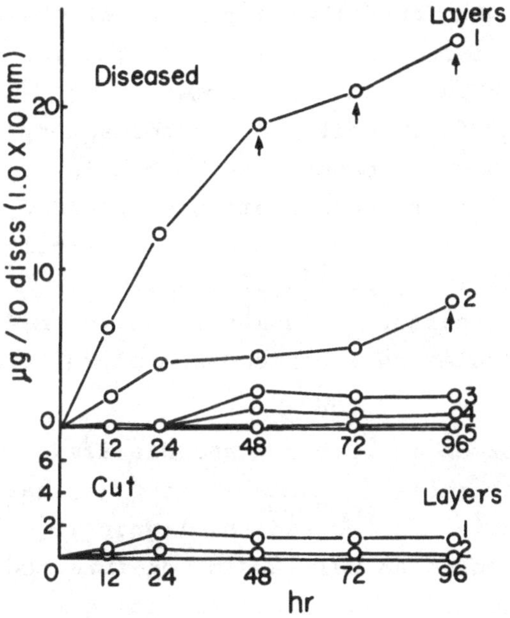

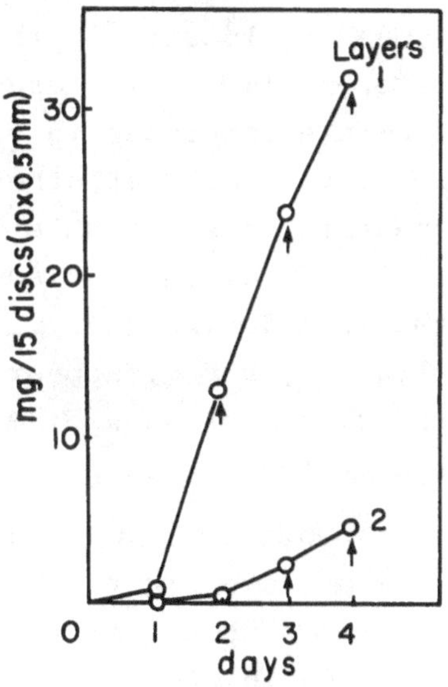

Figure 9. Production of umbelliferone (a coumarin) in sweet potato root tissue (about 3 cm thick) in response to cut injury and parasitic infection (13). Tissue incubation and disc (1 mm thick) preparation were performed as shown in the legends to Figure 2 and 4. Umbelliferone: µg per 10 discs (10 mm diameter and 1 mm thickness). ↑ : the same as shown in the legend to Figure 4.

Figure 10. Production of furano-terpenes in sweet potato root tissue (about 3 cm thick) in response to parasitic infection (22, 23). Tissue incubation and disc (0.5 mm thick) preparation were performed as shown in the legends to Figure 2 and 4. Furano-terpenes; mg (expressed as ipomeamarone) per 15 discs (10 mm diameter and 0.5 mm thickness). ↑ : the same as shown in the legend to Figure 4.

as "wound lignin." The structural skeltons of the wound lignin molecules in sweet potato appears to be the same as those in lignin present in the xylem of fresh sweet potato tissue, mainly of the feruroyl type.

On the other hand, when sweet potato root tissue is infected by Ceratocystis fimbriata, no lignin is formed in any part of diseased tissue. As described above, the cationic peroxidase isozymes, H_1 and H_2, are formed in the tissue closely adjacent to the cut surface (12). It was found that a portion of these isozymes are located in the microsomal fraction. Furthermore, two isozyme molecules are stable during the formation of lignin-like compounds in vitro from coniferyl alcohol and hydrogen peroxide. But other isozyme components, both anionic and neutral, are unstable during the peroxidation of coniferyl alcohol. In diseased tissue, isozymes H_1 and H_2 are scarcely synthesized during the initial 1-day period of the fungal inoculation, and disappear thereafter, presumably due to the degradative breakdown of molecules as a consequence of cellular damage by the fungal penetration (see Figure 7).

Is is known that lignin is formed in both cut and diseased tissues of some plants such as radish. Asada's group (1) has demonstrated that, in the fungi-infected radish, lignin is synthesized in the infected region as well as in non-infected tissue closely adjacent to the infected part. According to their results it appears that some specific peroxidase isozymes probably take part in the formation of lignin.

5. Phytoalexin Production.

Upon pathogenic infection, some antibiotic compounds are produced in both the infected region and the non-infected tissue closely adjacent to the infected region (14) (Figure 10). All these related compounds playing the role in the defense action of plants, due to their antibiotic activities are called "phytoalexins." Detailed discussion of phytoalexins will be treated by Kuč.

Phytoalexins are produced by a continuous stimulus such as

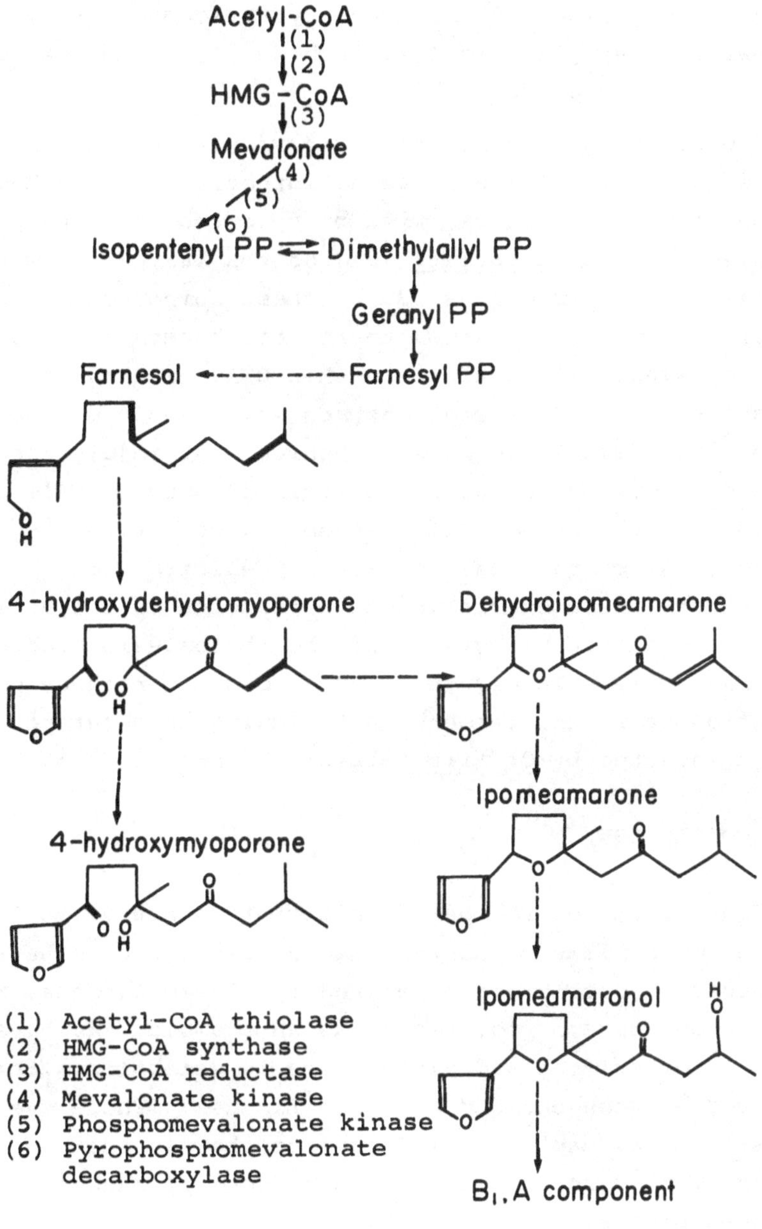

Figure 11. The possible pathway of furano-terpene biosynthesis (15, 24). B_1 and A components; structure-unknown components whose Rf values are smallest and largest among the terpenes, respectively.

fungal penetration and treatment by toxic chemicals, although ordinarily a single stimulus, such as the mechanical cut injury alone, does not produce them.

Here, production of sweet potato phytoalexins will be described only briefly. In response to infection, some unknown inducing agents (elicitors) will be released from the parasitic organism, i.e. C. fimbriata (9), or the larvae of sweet potato weevil, Cylas formicarius (25). These compounds may elicit the effect on plant tissues to release hormonal compound(s). They may eventually activate masked DNA, causing formation of mRNAs for the synthesis of enzymes involved in phytoalexin production. Even upon mere mechanical cut injury, some enzyme proteins involved in the phytoalexin production will be synthesized, although not all of them may be active. Hypothetical hormonal compound(s) released by elicitors may take part in the conversion of the inactive proteins to the active forms, completing the enzyme system of the phytoalexin production (14) (see Figure 1). In Figure 11, the possible pathway of the formation of furano-terpene phytoalexins in sweet potato root tissue infected by C. fimbriata are delineated (15, 24).

VII. CONCLUSION

The feasibility of the use of plant storage tissues for studying the mechanisms of host-parasite relationships have been discussed on both biochemical and cytological bases, mostly using sweet potato root tissue as an example. We have described the similarities and differences in metabolic alterations observed between cut and diseased tissues. These overall investigations may give us a clue to the elucidation of the mechanism of defense action (resistance) or susceptibility of diseased plants.

Finally one additional major advantage of using plant storage tissue in such study should be emphasized. During the entire course of pathogenic infection of storage tissues, parasitic organisms can be maintained without mutation. Otherwise some parasites mutate easily. In this case, we are also able to use callus cultures derived from plant storage tissues.

ACKNOWLEDGMENT: The authors express their thanks to Professor T. Akazawa, Nagoya University and Dr. T. Galliard, ARC Food Research Institute, Norwich for their helps in reviewing the manuscript.

REFERENCES

1. Asada, Y., Ohguchi, T., Matsumoto, I.: Biosynthesis of lignin in Japanese radish root infected by downy mildew fungus. In "Biochemistry and Cytology of Plant-Parasite Interaction", ed. K. Tomiyama et al., p. 200-212. Kodansha LTD., Tokyo and Elsevier Scientific Publishing Co., Amsterdam, 1976.
2. Greaksak, M., Asahi, T., Uritani, I.: Increase in mitochondrial activity in diseased sweet potato root tissue. Plant & Cell Physiol. 13, 1117-1121 (1972).
3. Hyodo, H., Uritani, I.: A study on increase in o-diphenol oxidase activity during incubation of sliced sweet potato tissue. Plant & Cell Physiol. 7, 137-144 (1966).
4. Hyodo, H., Uritani, I.: Properties of polyphenol oxidases produced in sweet potato tissue after wounding. Arch. Biochem. Biophys. 122, 299-309 (1967).
5. Imaseki, H., Uritani, I., Stahmann, M. A.: Production of ethylene by injured sweet potato root tissue. Plant & Cell Physiol. 9, 769-781 (1968).
6. Kahl, G. : Genetic and metabolic regulation in differentiating plant storage tissue cells. Bot. Rev. 39, 274-299 (1973).
7. Kahl, G. : Metabolism of plant storage tissue slices. Bot. Rev. 40, 263-314 (1974).
8. Kawashima, N., Uritani, I. : Occurrence of peroxidase in sweet potato infected by the black rot pathogen. Agr. Biol. Chem. 27, 409-417 (1963).
9. Kim, W. K., Uritani, I. : Fungal extracts that induce phytoalexins in sweet potato roots. Plant & Cell Physiol. 15, 1093-1098 (1974).
10. Kuč, J. A. : Phytoalexins. In "Physiolgical Plant Pathology", ed. R. Heitefuss and P. H. Williams, p. 632-652. Springer-Verlag, Berlin, Heidelberg, 1976.
11. Lyons, J. M. : Chilling injury in plants. Ann. Rev. Plant Physiol. 24, 445-466 (1973).
12. Matsuno, H., Uritani, I. : Physiological behavior of peroxidase isozymes in sweet potato root tissue injured by cutting or with black rot. Plant & Cell Physiol. 13,

1091-1101 (1972).
13. Minamikawa, T., Akazawa, T., Uritani, I. : Agr. Biol. Chem. 28, 230-233 (1964).
14. Ôba, K., Tatematsu, H., Yamashita, K., Uritani, I. : Induction of furano-terpene production and formation of the enzyme system from mevalonate to isopentenyl pyrophosphate in sweet potato root tissue injured by Ceratocystis fimbriata and by toxic chemicals. Plant physiol. 58, 51-56 (1976).
15. Oguni, I., Uritani, I. : Dehydroipomeamarone as an intermediate in the biosyntehsis of ipomeamarone, a phytoalexin from sweet potato root infected with Ceratocystis fimbriata. Plant Physiol. 53, 649-652 (1974).
16. Raison, J. K., Lyons, J., M. : Hibernation: alteration of mitochondrial membranes as a requisite for metabolism at low temperature. Proc. Nat. Acad. Sci., U.S.A. 68, 2092-2094 (1971).
17. Sakai, S., Imaseki, H., Uritani, I. : Biosyntehsis of ethylene in sweet potato root tissue. Plant & Cell Physiol. 11, 737-745 (1970)
18. Shannon, L. M., Imaseki, M., Uritani, I. : De nove synthesis of peroxidase isozymes in sweet potato slices. Plant Physiol. 47, 493-498 (1971).
19. Takamura, T., Uritani, I. : Changes in acid-soluble nucleotides in cut-injured sweet potato root tissue. Agr. Biol. Chem. 38, 1511-1515 (1973).
20. Tanaka, Y., Kojima, M., Uritani, I. : Properties, development and cellular-localization of cinnamic acid 4-hydroxylase in cut-injured sweet potato. Plant & Cell Physiol. 15, 843-854 (1974).
21. Tanaka, Y., Uritani, I. : Synthesis and turnover of phenylalanine ammonia-lyase in root tissue of sweet potato injured by cutting. Eur. J. Biochem. 73, 255-260 (1977).
22. Uritani, I. : Protein changes in diseased plants. Ann. Rev. Phytopathol. 9, 211-234 (1971).
23. Uritani, I. : Protein metabolism. In "Physiological Plant Pathology", ed. R. Heitefuss and P. H. Williams, p. 507-525. Springer-Verlag, Berlin, Heidelberg, 1976.
24. Uritani, I. : Biochemistry of host response to infection. Progress in Phytochem. 5, in press (1977).
25. Uritani, I., Saito, T., Honda, H., Kim, W. K. : Induction of furano-terpenoids in sweet potato roots by the larval components of the sweet potato weevils. Agr. Biol. Chem. 39, 1857-1862 (1975).
26. Yamaki, S., Uritani, I. : Mechanism of chilling injury in sweet potato XII. Temperature dependency of succinoxidase activity and lipid-protein interaction in mitochondria from healthy or chilling-stored tissue. Plant & Cell Physiol. 15, 669-680 (1974).

Control of Glycolysis in Plant Storage Tissue

David D. Davies

INTRODUCTION

The reactions of glycolysis are to be found in most text-books of biochemistry. However, recent studies on the glycolytic pathway in animals and yeast have demonstrated the anomeric specificity of the enzymes involved and produced a somewhat more complex metabolic pattern. Assuming that the anomeric specificities are the same in animals and plants allows us to draw the metabolic map of glycolysis shown in Figure 1.

Three of the four enzymes which possess anomerase activity have been demonstrated in plants. Aldose-1-epimerase which catalyses the anomerisation of glucose has been demonstrated in members of the Solanaceae, Cucurbitaceae and Liliaceae (10). Anomerase activity with glucose-6-phosphate and possibly with fructose-6-phosphate has been demonstrated for the phosphohexose isomerase from potato tubers and a glucose-6-phosphate epimerase which specifically catalyses the anomerisation of glucose-6-phosphate has also been demonstrated in the same tissue (79).

Anomerase activity associated with aldolase has not been demonstrated in plants. Rabbit aldolase appears to be specific for the β-anomer of fructose diphosphate, whereas yeast aldolase is anomerically unspecific due to its anomerase activity.

CONTROL SYSTEMS

Early considerations of the control of glycolysis assumed that a single control point or pacemaker determined the glycolytic flux. It is now clear that control is a property of the system - all reactions contributing to the overall rate. Kacser and Burns (40) and Heinrich and Rapoport (35) have provided theoretical treatments of metabolic control in a sequence of enzymes. The two treatments employ different symbols and definitions making comparisons difficult. However, the "sensitivity coefficient" (Z) defined by Kacser and Burns (40) is very similar to the coefficient of "Control Strength" (C) defined by Heinrich and Rapoport (35)

Biochemistry of Wounded Plant Tissues

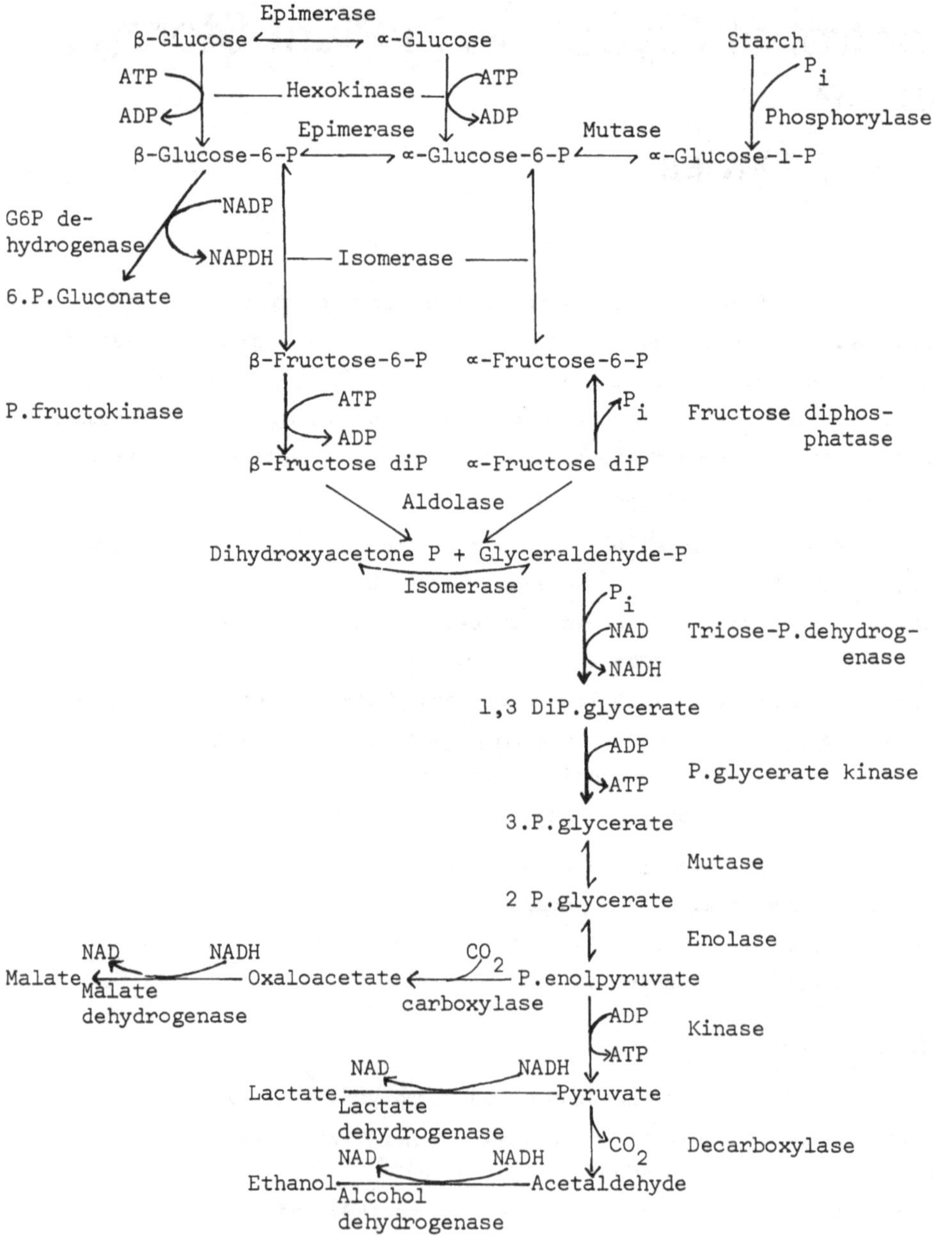

Figure 1: Reactions of Glycolysis

and Higgins (36).

The control strength of a particular reaction in a sequence is given by

$$C = (dF/F) / (dE/E) \quad ....(1)$$

where F is the flux through the system and E is the effective concentration of the enzyme whose sensitivity is being considered. The definition of control strength employed by Heinrich and Rapoport (35) involves changes in the velocity of the isolated enzyme under the influence of an effector. However, the change in rate, caused by a change in the effector is equivalent to a change in concentration of enzymes, so that the definitions are equivalent.

Equation (1) gives a simple experimental method for determining the control strength of an enzyme in a cell free system. If the glycolytic flux is measured and small increments of individual enzymes added, the expected changes are shown in Figure 2.

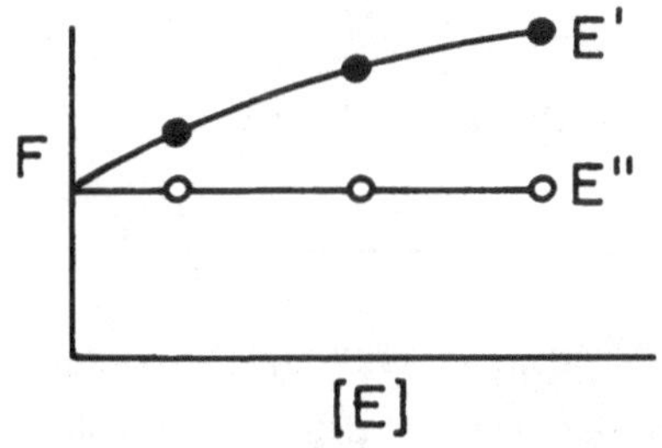

Figure 2: Relationship between flux through a multi-enzyme system and increments of a single enzyme.

Increasing amounts of E" have no effect on the flux and Z for this enzyme will be zero. The increase in flux on adding enzyme E' varies with the amount of E' added and response to a very small change in E' is needed. Unfortunately experiments of this type have not been performed with extracts from storage tissue nor for that matter with extracts from plants, although the glycolytic system from pea seeds developed by Turner (74) would seem most suitable.

An important extension of the concept of control strength is the summation theorem of Kacser and Burns (40), which states that the sum of all sensitivity coefficients in a metabolic sequence is equal to unity.

$$\Sigma Z_i = 1 \text{ ....(2)}$$

That is the control strengths of the individual reactions of glycolysis are distributed so that they are smaller than unity and only one reaction could approach full control, alternatively none of the enzymes can be considered as a "pacemaker" or "bottleneck."

Again in terms of glycolysis in storage tissue, we do not have sufficient data to use the summation theorem. However, glycolysis in red cells has been examined by Rapoport _et al_ (65) who concluded that the glycolytic flux was controlled only by the hexokinase and phosphofructokinase reactions; the control strength of the hexokinase reaction being _c_ 0.7 and the control strength of phosphofructokinase being _c_ 0.3 at pH 7.2. Nevertheless Rapoport _et al_ (65) argue that most changes in the flux produced by effectors are caused by interaction with phosphofructokinase. This apparent denial of the validity of the concept of control strength needs clarification.

Consider a metabolic sequence with feedback control

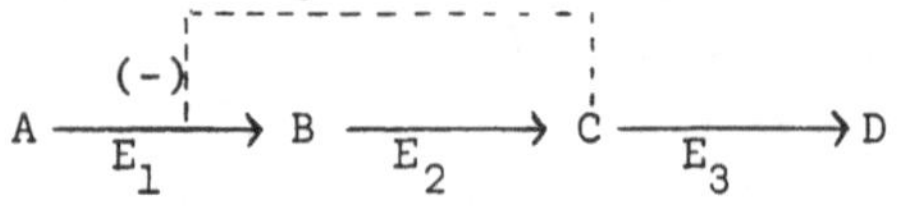

in which the reaction A → B is "controlled" by the feedback loop and C → D. To determine its control strength, E_3 is increased, producing a fall in C which is then counteracted by the de-inhibition of E_1. This has the effect of keeping the pool concentration constant over a wide range of concentrations of E_3. Since the pool is constant, the flux will vary directly with the concentration of E_3, i.e. the control strength of E_3 is close to unity and by the summation theory the control strength of the "controlled enzyme" E_1 is very low. This apparent paradox of a "controlled enzyme" exerting no control is resolved when we remember that E_1 responds to effectors. The effector strength χ is given by

$$\chi = (d\ v/v) \ / \ (d\ L/L) \text{ ....(3)}$$

where v is the velocity of the isolated enzyme and L is the concentration of an effector ligand. For any reaction there are as many effector strengths as there are metabolites and effectors which interact with the enzyme. If a particular effector strength is large, an effector can act on an enzyme with low control strength to produce large changes in flux. In this context it is worth noting that if the kinetics of inhibition are sigmoid the effector strength will be greater than if the kinetics of inhibition are competitive (Figure 3).

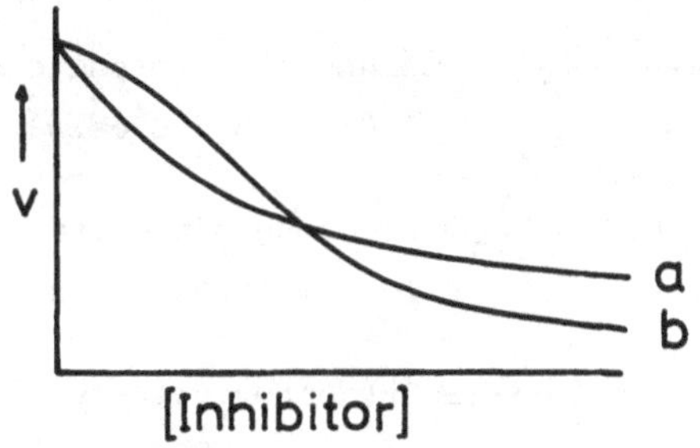

Figure 3: Relationship between enzyme activity and inhibitor concentration. (a) competitive inhibition (b) allosteric inhibition.

IDENTIFICATION OF CONTROL POINTS

Theoretical Considerations

Remembering that control is a system property, it is nevertheless possible to predict possible control points in a metabolic sequence based on considerations of irreversibility and order of reaction (19). In a linear sequence of enzyme catalysed reactions:

1. The first reactions showing zero order kinetics acts as the pacemaker for the whole process.
2. The rate of an irreversible reaction is unaffected by the concentration of its products, thus reactions subsequent to an irreversible reaction cannot affect the rate, but the first irreversible reaction of a sequence may be rate-determining.
3. The rate of an irreversible is of course affected by previous reactions; thus if a zero-order reaction precedes an irreversible reaction the overall rate is determined by the zero-order reaction.
4. Should an irreversible reaction precede a zero-order reaction a steady

state is impossible.

The application of these rules to the reactions of glycolysis is not simple. In theory all enzyme reactions are reversible, but when the free energy change is greater than 4-5 Keals the back reaction is so small that it may be neglected. The irreversible reactions of glycolysis are:

Glucose + ATP $\rightleftharpoons$ Glucose-6-phosphate + ADP
($\Delta G'$ = -4,600 cals)

Fructose-6-phosphate + ATP $\rightleftharpoons$ Fructose diphosphate + ADP
($\Delta G'$ = -4,200 cals)

1,3-Diphosphoglycerate + ADP $\rightleftharpoons$ 3Phosphoglycerate + ATP
($\Delta G'$ = -4,700 cals)

Phosphoenolpyruvate + ADP $\rightleftharpoons$ Pyruvate + ATP
($\Delta G'$ = -5,200 cals)

Three of these reactions were considered by Krebs in his Herter Lectures (51) as irreversible reactions and he identifies alternative reactions forming by-passes at these three points:

Glucose - 6-phosphate → Glucose + P_i

Fructose diphosphate → Fructose-6-phosphate + P_i

Pyruvate + ATP + CO_2) → Oxaloacetate + ADP)
Oxaloacetate + ATP) → Phosphoenolpyruvate + ADP + CO_2)

The possibility that these reactions play a role in the control of glycolysis is discussed later (see section on substrate cycling). The reaction catalysed by phosphoglycerate kinase was not discussed by Krebs and a by-pass of this reaction has not been identified, nevertheless the equilibrium constant of about 3 x 10^3 poses an unresolved problem in relation to reversibility.

As far as the control of glycolysis is concerned the rules suggest that it is the first irreversible reaction which is rate determining - suggesting that in glycolysis starting from glucose, hexokinase may be rate determining, whilst in glycolysis starting from starch, phosphofructokinase may be the rate determining step. However, the reactions of glycolysis contain a positive feedback loop in which the output (ATP) feeds back into the system.

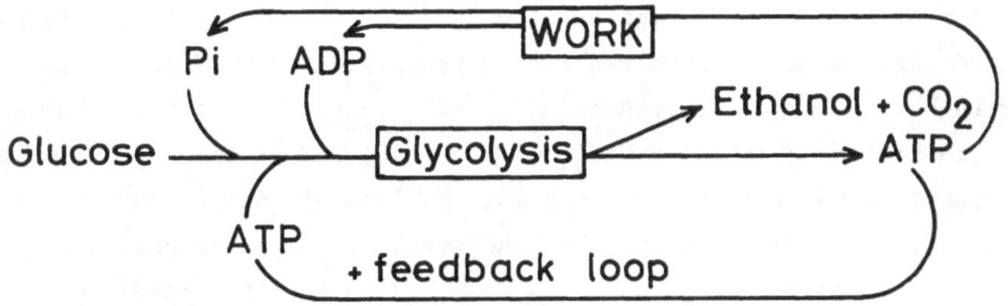

The effect of cycling is to convert a product into a substrate and consequently an irreversible reaction cannot determine the rate of cycling, by virtue of its irreversibility, but it does of course determine the direction of cycling.

Bücher and Rüssman (15) proposed a model for metabolic control based on a hydrodynamic analogy (Figure 4).

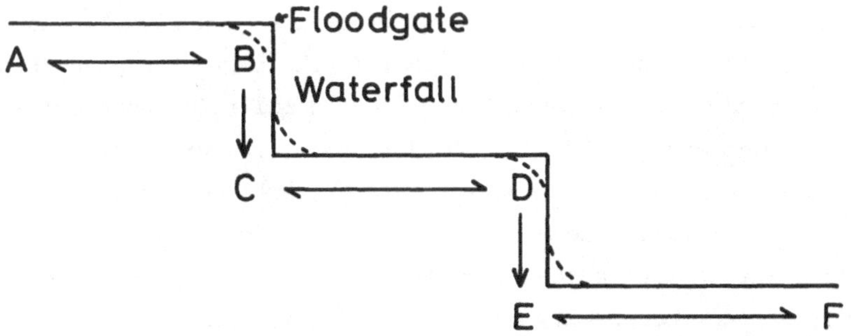

Figure 4: Schematic representation of a regulated waterway. The solid line represents the situation when the flood gates are raised and there is little flow through the system. The dashed line represents the increased flow when the flood gates are lowered.

The flow of water through the system is controlled by the flood gates. With the flood gates closed there is a negligible gradient across the reservoir and a slight flow of water over the dam. In metabolic terms the reaction A ⇌ B represents the reservoir and can be considered as a fast reaction close to equilibrium (ΔG small). The reaction B → C represents the waterfall and is a slow reaction far from equilibrium (ΔG large). The model can be extended to include the generation of useful work from the water flowing over the dam. Control points will tend to be associated with reactions involved in the production of useful work by coupled reactions, e.g. the production of ATP. However, the useful work function is not an essential feature of the model and in the absence of a coupled reaction to utilise the large ΔG, the loss of energy, say in the phosphofructokinase reaction, can be considered as the price paid for control. The essential feature of the model is to identify possible control points with reactions where ΔG is large.

Experimental Determination of Control Points

Krebs (52) has pointed out that"if the rate of flow changes in the opposite direction to changes in concentration of a substrate for a non-equilibrium enzyme then that enzyme must be regulatory to the metabolism of that substrate." It should, however, be noted that failure to observe the required changes does not exclude the possibility that a nonequilibrium reaction is regulatory. The proposition that if the flux increases the substrate concentration of the controlled reaction falls,is incorporated into the hydrodynamic analogy of Bücher and Rüssman (15) with the additional proposition that if the flux increases, the concentration of the product of the controlled reaction rises. The model thus predicts that if the flux through the system increases, the controlled reaction which is far from equilibrium will move towards equilibrium, i.e. the substrate will decrease and the product will increase. This gives an experimental method of detecting control points. If the concentrations of all glycolytic intermediates are measured before and after a change in flux, the mass action ratios of the steady state metabolite concentrations can be compared with the equilibrium constants for the reactions. The mass action ratios can be divided into two categories, equilibrium and non-equilibrium. The division is somewhat arbitrary but Rollerton (68) has suggested that if the value (mass action ratio/equilibrium constant) is less than 0.05, the reaction is a non equilibrium reaction and potentially important in control. If the value is greater than 0.2 the reaction

is an equilibrium reaction, not important in control (but see page
This leaves a number of reactions with values between 0.05 and 0.2 whose involvement in metabolic control must be assessed by the individuals judgment of what is significant. The decision on which reactions are important in the control of glycolysis is finally taken by seeing which reactions move towards equilibrium when the flux through the system is increased.

When this method is applied to discs of storage tissue, it is clear that using the criteria proposed by Rolleston (68), the kinases of glycolysis are non-equilibrium reactions (Table 1).

Table 1: Equilibrium and non-equilibrium reactions of glycolysis

		Potato tubers (41)		Carrot discs (aged) (32)	
Enzyme	K.equil.	Mass Action Ratio	Mass Action / K.equil.	Mass Action Ratio	Mass Action / K.equil.
Hexokinase	5.5×10^{3}	1.3×10^{-2}	6×10^{-4}	7.3×10^{-4}	1.3×10^{-7}
P.gluco-mutase	17	13	0.75	-	-
P.gluco-isomerase	0.47	0.21	0.45	0.22	0.96
P.fructo-kinase	1.2×10^{3}	0.196	1.6×10^{-4}	0.013	10^{-5}
Aldolase	1.2×10^{-4}	1.08×10^{-4}	0.9	3.3×10^{-5}	0.3
Triose-P. isomerase	22	1.5	0.07	4	0.2
Enolase	3	1.1	0.3	-	-
P.glycero-mutase + Enolase	2	-	-	0.27	0.14
Pyruvate kinase	1.6×10^{4}	1.28	8×10^{-3}	9.3	6×10^{-4}

When aged carrot discs were made anoxic, there was a significant change in the mass action ratio $\frac{\text{fructose diphosphate x ADP}}{\text{fructose-6-phosphate x ATP}}$ which moved towards equilibrium. On the basis of the criteria proposed by Bücher and Rüssman (15) this clearly identifies phosphofructokinase as a control point, but no other control point could be detected with any degree of certainty.

Before leaving this discussion, it is important to stress the difficulties and uncertainties connected with the determination of the cytoplasmic concentrations of metabolites. The extensive compartmentation of plant cells in general and the large storage vacuoles and food reserves of storage tissue in particular, makes it difficult to assess the relationship between metabolic content and concentration. The existence of separate pools of organic acids has been clearly demonstrated (56) and the distribution of inorganic phosphate between a non metabolic pool (95%) and a metabolic pool (5%) has been demonstrated in potato tissue (11). On the other hand it is generally assumed (albeit without evidence) that the glycolytic intermediates are restricted to the cytosol. Dennis and Green (27) have shown that proplasmids from castor bean endosperm contain a number of glycolytic enzymes. If this finding can be substantiated for storage tissue it casts doubt on the significance of the changes in metablite levels which have been determined before and after a change in glycolytic flux.

The Cross-over Plot

The most widely used method of interpreting changes in metabolite levels associated with changes in metabolic flux is by the use of cross-over plots. Metabolic intermediates are plotted on the abscissa in the sequence in which they are formed in the metabolic pathway and the concentration of each substrate, expressed as a percentage of the initial value is plotted on the ordinate. An example is shown in Figure 5 for the situation where the glycolytic flux in carrot discs is increased by transferring them from air to nitrogen.

The relationship between the cross-over plot and the Bücher-Rüssman model is seen by plotting the example given in Figure 4 as a cross-over plot.

The cross-over theorem (16) states that for the case where the flux through the system increases a (- +) pair in the sequence is defined as a

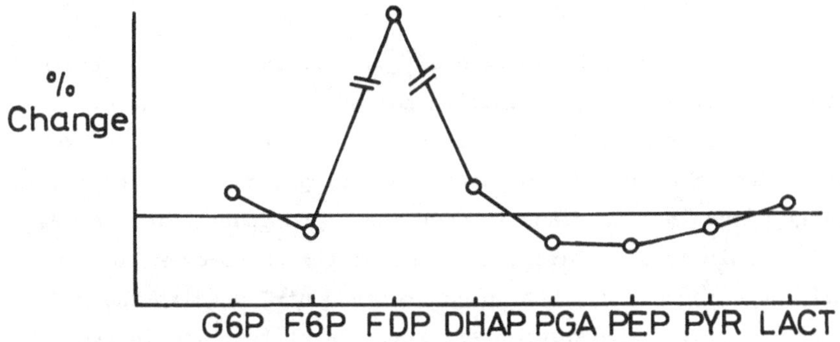

Figure 5: Cross-over plot of glycolytic intermediates following transfer of aged carrot discs from air to nitrogen (32). Intermediates were measured before transfer and 6 minutes after transfer from air to nigrogen.

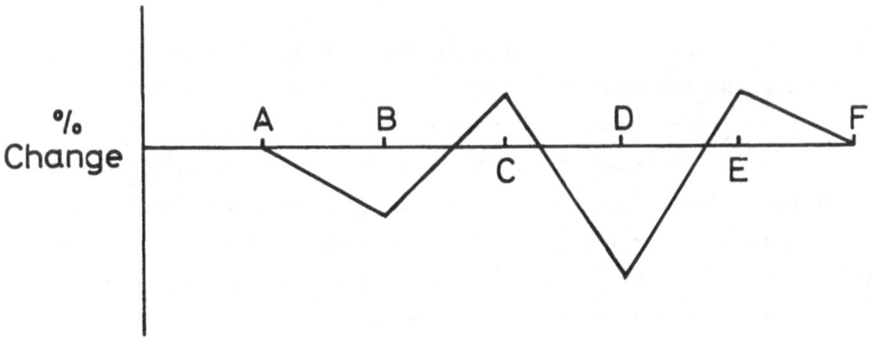

Figure 6: Cross-over plot of the model presented in Figure 4. The values are plotted for the condition that the flux through the system increases.

---

positive cross-over and a (+ -) pair as a negative or reverse cross-over. A positive cross-over is taken to indicate a site of interaction, that is a control point. However, Chance (16) argues that a negative cross-over does not indicate a control point, whereas Williamson (77) interprets a negative cross-over (in his terminology a reverse cross-over) as a control point.

In view of the widespread use of the cross-over plot to detect control points in glycolysis, it is important to recognise the limitations of the method, some of which are noted below:

1. Control points are usually located at quasi-irreversible reactions and such reactions are by definition not affected by products. Thus as noted by Rolleston (68) there is no logical requirement to use the concentration of a product to detect control points. However, it is the large percentage increase in fructose diphosphate which occurs when glycolysis is stimulated, which gives the cross-over plot such a dramatic appearance and which no doubt has contributed to the wide usage of the plot. The special significance of the rise in fructose diphosphate during increased glycolysis is discussed on page

2. If a reaction involving a cofactor is close to equilibrium a false cross-over may be obtained. Consider the system

$$A \rightleftharpoons B \underset{\text{slow}}{\overset{x \quad y}{\rightleftharpoons}} C \rightleftharpoons D \underset{\text{fast}}{\overset{y \quad x}{\rightleftharpoons}} E$$

where the reaction $B + x \rightleftharpoons C + y$ is far from equilibrium and rate determining and the reaction $D + y \rightleftharpoons E + x$ is very close to equilibrium. If the flux through the system increases, the concentrations of B and x will decrease and y will increase. Consequently the concentration of D will decrease and E will increase to maintain the equilibrium, thereby producing a false crossover between B and C. With reference to the reactions of glycolysis the reaction $B + x \rightleftharpoons C + y$ could be considered as the phosphofructokinase reaction and $D + y \rightleftharpoons E + x$ could be considered as the phosphoglycerate kinase reaction, or as the pyruvate kinase reaction. Thus if a positive cross between fructose-6-phosphate and fructose-diphosphate is followed by a positive cross-over between 13 diphosphoglycerate and 3 phosphoglycerate or between phosphoenolpyruvate and pyruvate, the interpretation of the second cross-over as a control point should be treated with caution.

3. The cross-over theorem is valid only if the intermediates in the sequence are conserved and only the ratio substrate/product varied. In most applications of the theorem to glycolysis, the intermediates are not conserved so that the interpretation of the cross-over plots is difficult and the <u>firm</u> assertion that a cross-over identifies a control point is illogical.

4. The cross-over theorem as stated by Chance et al (17) and the proof offered by Holmes (37) considers changes from one steady state to another. Many studies (probably the majority) concern the transition from one steady state to another and in these circumstances the cross-over plot must be interpreted with caution since it lacks a theoretical basis. In some plant studies (34) the cross-over plot is based on steady state values but in most cases (2, 18, 32, 33, 49, 57), the plot is used to study transients.

There is nothing wrong with studying transients, indeed Barker et al (8) stated that "Any metabolic change which occurs after 15 minutes in anoxia is probably not closely related to the development of the Pasteur Effect." The investigator must, however, recognise that control may be passing from one point to another during the transition and employ a somewhat empirical approach. For example, the first cross-over in Figure 5 is consistent with an increased rate at the phosphofructokinase step, the negative cross-over which occurs between dihydroxyacetone phosphate and 3-phosphoglycerate could be considered as a pseudo-cross-over by applying the formal cross-over theorem, but recognising that we are dealing with transients it seems reasonable to state that 3-phosphoglycerate is being removed more rapidly than it is being formed and thus although phosphofructokinase may have been rate determining in the steady state before transfer to nitrogen, the transfer speeds up this reaction and a constraint then appears in the system somewhere between dihydroxyacetone phosphate and 3-phosphoglycerate.

Enzyme Activities

If an enzyme is sufficiently active to maintain its reactants close to equilibrium and therefore to catalyse the reverse reaction at almost the same rate as the forward reaction in spite of the flux through the pathway, it cannot be limiting that flux. Thus it appears possible that by determining the activities of the glycolytic enzymes we could predict which enzymes are likely to be control points. Some of the technical difficulties associated with isolating enzymes from plants have been discussed by ap Rees (3) and some of the problems associated with enzyme assays have been noted by Davies (22). The results of a number of investigations are shown in Table 2.

Table 2: Activity of some glycolytic enzymes in plants

A. Potato tubers: Activity as n moles/min/mg protein (43, 44)

B. Potato discs (aged): Activity as n moles/min/mg protein (43,44)

C. Potato tubers: Activity as n moles/min/gm fresh weight
Glucose equivalent to O_2 uptake = 1.4 n moles/min/gm fresh weight (76)

D. Carrot discs (fresh): Activity as n moles/min/gm fresh weight
Glucose equivalent to O_2 uptake = 8.4 n moles/min/gm fresh weight (67)

E. Carrot discs (ages): Activity as n moles/min/gm fresh weight
Glucose equivalent to O_2 uptake = 24 n moles/min/gm fresh weight (67)

F. Maize roots: Activity as n moles/min/gm fresh weight (67)

G. Pea cotyledons: Activity as n moles/min/gm fresh weight (45)

Enzyme	A	B	C	D	E	F	G
P.Glucomutase	180	100	200	-	-	-	-
Hexokinase	8	7.8	40	17	67	700	5000
P.glucoisomerase	120	110	480	-	-	3000	15000
P.fructokinase	-	-	-	69	39	300	600
Aldolase	3	3.5	-	-	-	-	-
Triose P.isomerase	280	260	-	-	-	-	-
Triose P.dehydrogenase	-	-	-	-	-	3000	3000
P.glycerate kinase	-	-	-	-	-	40000	90000
P.glycerate mutase	18	37	-	-	-	-	-
Enolase	180	100	-	-	-	8000	35000
Pyruvate kinase	10	16	-	-	-	1000	8000
Glucose-6-p.dehydrog-enase	35	65	160	29	174	-	-

A tentative conclusion from this data is that phosphofructokinase has the lowest activity and may therefore be a rate controlling enzyme. However, we note that the maximum glycolytic flux through the tissue (calculated by assuming that all the CO_2 produced by the tissue is derived from glucose) is very much lower than the activity of phosphofructokinase. Thus the conclusion that phosphofructokinase may be a rate determining enzyme is at best tentative. Ricardo and ap Rees (67) have suggested that the decline in activity of phosphofructokinase and the rise in glucose dehydrogenase contribute directly to the relative increase in the activity of the pentose phosphate pathway during the first 24 hours of ageing. This conclusion must also at best be tentative since the activities of most enzymes of glycolysis and the pentose phosphate pathway appear to exceed the flux through these pathways and the causal relationship require that the activity of the enzymes be rate limiting.

In stating that most of the enzymes are in excess, relative to the flux through the pathway, we meet a semantic difficulty in interpreting what is meant by excess. Control is a system property, all enzymes contribute to the overall rate and its control. The effect of having relatively active enzymes in a metabolic sequence is that the substrates and products of these reactions will be close to equilibrium so that there will be little loss of free energy at these steps and the system will be buffered with minimal fluctuations in levels of intermediates. It seems likely that evolutionary pressure would result in an effective balance between the activities of the enzymes in a sequence and such pressure would tend to remove any "excess" activity.

There is also some uncertainty about the amount of activity which an enzyme may possess under *in vivo* conditions. There is a large gap in our knowledge between enzymology and metabolic or physiological biochemistry. The enzymologist is interested in the mechanism of enzyme action; and to study a reaction such as $A + B \rightleftharpoons P + Q$ he derives a rate law. To analyse the situation he measures initial rates, that is he puts P and Q equal to zero; such a treatment simplifies the analysis and enables the enzymologist to test proposed mechanisms but it makes the kinetic data of limited value to the physiologist. The activity of an enzyme may be large when the concentrations of its products are low, but the activity of the enzyme may be low when the substrates and products of the enzyme are present in concentrations close to their equilibrium values. Purich and Fromm (64) have considered the kinetic consequences of the substrates and products being present and provided computer solutions for some examples. The information which the physiologist requires about enzyme activities is the rate at physiological concentrations of substrates, products and effectors and at the pH of the cytoplasm rather than the pH optimum of the enzyme (20). We also need to know how the rate changes when the concentrations of metabolites and effectors change. This is a formidable task since to establish the complete rate law for a single enzyme with 8 reactants and effectors would require 1,679,616 assays (71).

THE GLYCOLYTIC FLUX IN STORAGE TISSUE

An analysis of the control of the glycolytic flux must be seen against a background of uncertainty concerning the relative contribution of the major metabolic pathways to the overall rate of respiration in storage tissue. The metabolic changes associated with the ageing process have been thoroughly reviewed (41, 42) and are discussed elsewhere in this volume. However, it is clear that the information is qualitative rather than quantitative and the glycolytic flux in storage tissue is simply not known. To determine control points it is necessary to change the flux through the system and the minimum requirement for analysis is that we know whether the flux has increased or decreased. The model proposed by Lynen (55) for the glycolytic flux in air and nitrogen (Figure 7) fits available data for storage tissue but analysis requires quantitative information.

Oxidative phosphorylation. Phosphorylation associated with glycolysis.

Dephosphorylation.

Figure 7: The control of glycolysis by the phosphate cycle

Consider the case of a freshly cut storage disc being transferred from air to nitrogen. We can measure the CO_2 output before and after transfer to nitrogen - in the case of carrot discs the CO_2 output remains relatively constant. The problem is how to relate this to the glycolytic flux? The conversion of glucose to ethanol and CO_2 produces only 2 mol CO_2/mol of glucose, compared with 6 mol CO_2/mol of glucose produced under aerobic conditions. Consequently it is possible that glycolysis is stimulated by anoxia without producing an overall increase in CO_2. It is, however, a particularly tentative conclusion on which to base an analysis of rate control through all the metabolites of glycolysis. In the case of "aged" slices anoxia produces a significant increase in CO_2 production and it is reasonable to assume that the glycolytic flux has increased. This conclusion can be strengthened by determining the production of $^{14}CO_2$ from glucose-3,4-^{14}C in air and nitrogen and such experiments indicate that glycolysis is stimulated by anoxia (33). Additional methods which are available include the release of ^3H as ^3H_2O from specifically labelled glucose (see Katz and Rognstad (46)).

THE END PRODUCTS OF GLYCOLYSIS IN STORAGE TISSUE

Under aerobic conditions the reactions of glycolysis convert glucose into pyruvate which then enters the Krebs cycle. Relatively small amounts may be converted into other products which can be end products under anaerobic conditions. These compounds include, malate, lactate and ethanol.

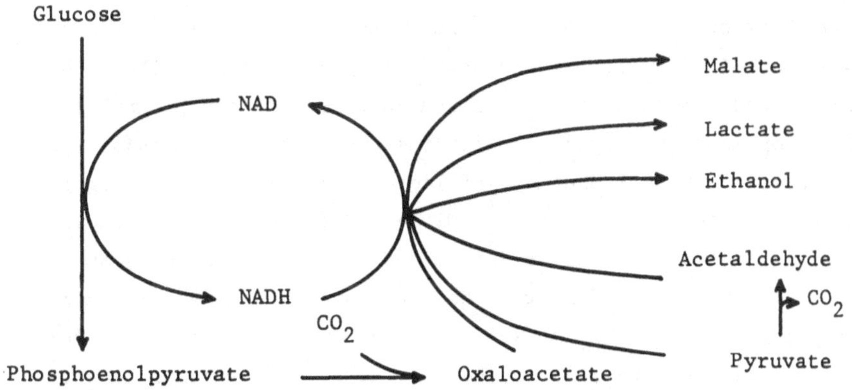

There is some uncertainty concerning which products are formed under anaerobic conditions. Following the discovery of phosphoenolpyruvic carboxylase, Mazelis and Vennesland (59) suggested that malate may be an important end product of anaerobic metabolism in plants. More recently it has been suggested (58) that plants which are intolerant of flooding suffer an acceleration of glycolysis and produce ethanol under anaerobic conditions, whereas tolerant species avoid the acceleration of glycolysis and undergo a metabolic switch from ethanol to malate production.

In the case of carrot tissue ethanol is clearly the major product of glycolysis (39) but the situation in potato tubers is less certain. A number of investigators (45) have shown that alcohol is formed when potato tubers are made anaerobic and Link and Barron (39) have shown that the enzymes pyruvic decarboxylase and alcohol dehydrogenase are present. However, Barker and El Saif (6) have shown that with the variety "King Edward," lactic acid is produced immediately on transfer to nitrogen, but the production of ethanol is delayed for 8 - 10 days, although this delay is reduced to about 24 hours when the tubers are stored at 1° before being transferred to nitrogen (7). The lack of alcohol production in King Edward potatoes appears to be due to the lack of pyruvic decarboxylase (38). Pyruvic decarboxylase could not be isolated from potatoes stored in air, but activity was demonstrated after the potatoes had been stored in nitrogen for some days. The explanation of these differing results is not known.

Factors controlling the partitioning of pyruvate between ethanol and lactate have been examined using cell free preparations from pea seeds and parsnip roots (24). The results were interpreted to mean that under aerobic conditions the pH of the cytoplasm is approximately neutral and pyruvic decarboxylase is inactive. With the onset of anaerobiosis, glycolysis leads to an accumulation of lactate with a corresponding fall in pH, which activates pyruvic decarboxylase. The partitioning of pyruvate between lactate and ethanol is determined by the kinetic constants of pyruvate decarboxylase and lactate dehydrogenase. The pH optimum of lactate dehydrogenase is alkaline whilst the pH optimum of pyruvate decarboxylase is acid (Figure 10), so that as lactic acid accumulates it tends to cut off its own synthesis and pyruvate is shunted towards ethanol

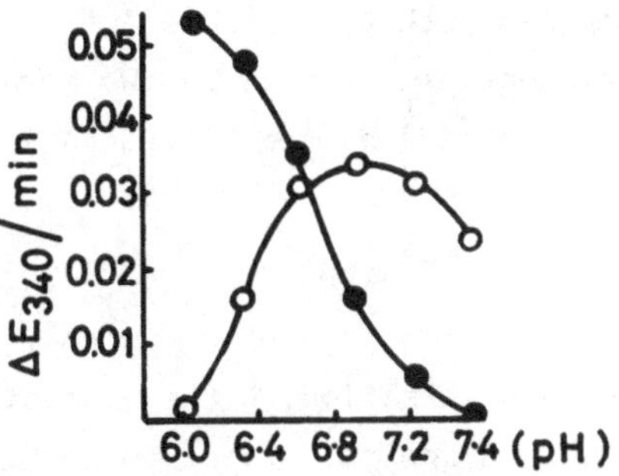

Figure 8: Effect of pH on the activity of pyruvate decarboxylase and lactate dehydrogenase in an extract of parsnip root (66)

production. Thus the two enzymes function as a pH stat and the possibility of excess production of lactate is further controlled by the inhibition of lactate dehydrogenase by ATP (23).

METABOLIC CONTROL DURING THE AGEING OF DISCS

During the ageing of the discs there is a complex change in metabolism (41, 42). Everson and Rowan (31) observed a correlation between the concentration of fructose diphosphate and those phosphates and the rate of respiration of carrot discs and concluded that a reaction in the pathway prior to the production of fructose diphosphate appeared to regulate the respiratory rate (Figure 9).

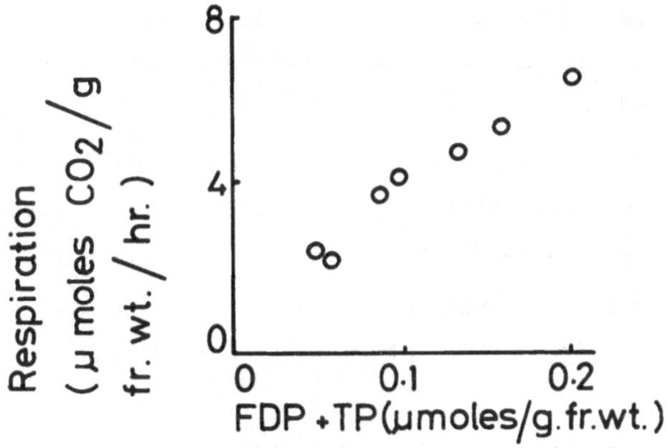

Figure 9: Correlation between rate of respiration and concentration of fructose diphosphate and triose phosphate in carrot discs washed at 21° (31).

This correlation can be compared with the correlation between fructose diphosphate and the rate of glucose utilisation by two strains of E.coli (28). The two quantities are related by the Hill equation:

$$v = \frac{Vmax\ (A)^N}{(A_{0.5})^N + A^N} \qquad .... (4)$$

where v is the rate of glucose utilisation, A the concentration of fructose-diphosphate and N the Hill number calculated to be 2.29 (Figure 10).

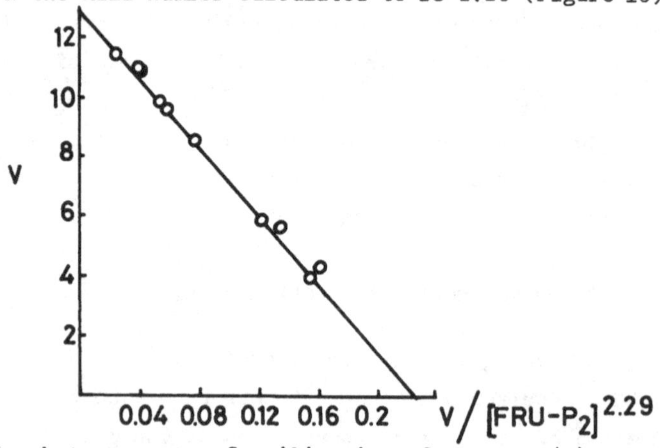

Figure 10: Correlation between rate of utilisation of glucose (v) and the concentration of fructose diphosphate, during exponential growth of two strains of E.coli (28).

Dihydroxyacetone phosphate and phosphoglyceraldehyde covary with fructose diphosphate in E.coli (54) so that it is not possible to say which of the three metabolites is an effector.

The cross-over theorem has been employed by Adams and Rowan (2) to examine glycolytic control of respiration during ageing of carrot root tissue and their conclusions are presented in Figure 11.

In considering these results it should be remembered that the use of the cross-over theorem to identify control points in a time-dependent process has no theoretical basis.

The ageing process which occurs on washing discs of storage tissue is a complex process involving protein synthesis and protein degradation.

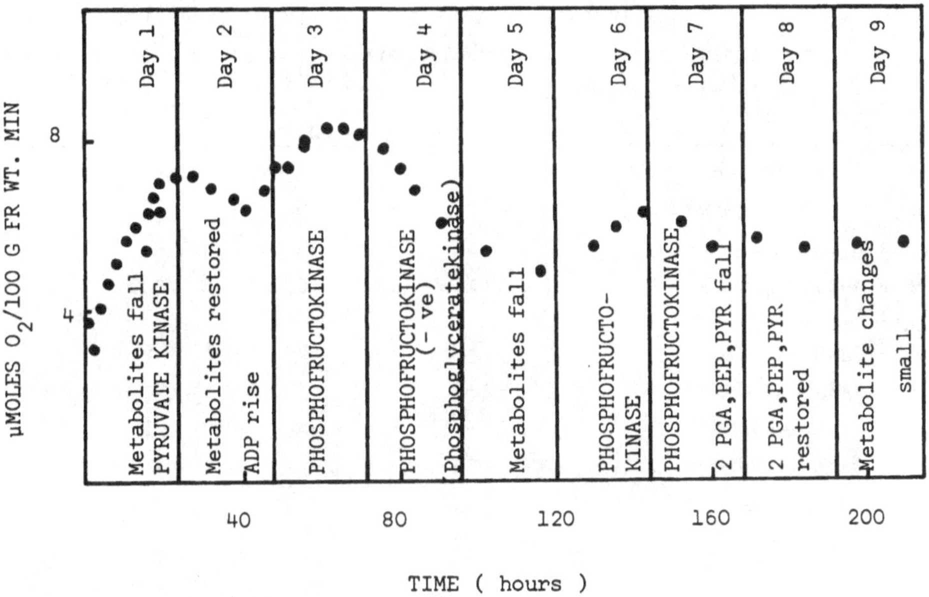

Figure 11: Sequence of major regulator reactions (indicated in upper case) and changes in metabolites in carrot slices washed for 220 hr in distilled water.

During this period some glycolytic enzymes increase whilst others decrease as seen in Table 3. Against this background of changing enzyme levels it is unlikely that a simple analysis of enzyme substrate levels will permit the identification of metabolic control points.

Changes in carbohydrate metabolism associated with the ageing of discs of

Table 3: Effect of washing for 24 hrs on the activity of various enzymes in storage tissue. The results are expressed as a percentage of the initial activity.

	A	B	C	D	E	F
Phosphoglucomutase	-	-	55	-	-	-
Alcohol dehydrogenase	225	225	-	98	-	-
Lactate dehydrogenase	115	118	-	-	-	-
G-6-P dehydrogenase	168	178	186	160	117	600
Gluconate-6-P dehydrogenase	-	-	161	120	400	150
Phosphoglucoisomerase	-	-	92	-	-	-
Phosphofructokinase	-	-	-	-	-	56
Isocitrate dehydrogenase	200	390	-	-	180	-
Malate dehydrogenase	70	86	-	72	-	-
Shikimate dehydrogenase	565	620	-	-	-	-
Hexokinase	107	112	97	-	-	394
Glutamate-OAA transaminase	190	186	-	97	-	-
Aspart.phenylpyruvate transaminase	210	180	-	-	-	-
PAL	3500	1940	-	-	-	-
Malic enzyme	80	160	125	-	64	-
Aldolase	138	137	110	-	-	-
Triose phosphate isomerase	-	-	93	-	-	-
Phosphoglyceromutase	-	-	205	-	-	-
Enolase	-	-	56	-	-	-
Pyruvate kinase	-	-	160	-	-	-
Glutamate-pyruvate transaminase	-	-	80	-	-	-
Ribonuclease	388	394	-	-	-	-
Acid phosphatase	101	83	-	-	-	-
Peroxidase	130	130	-	-	-	-
Chlorogenic acid oxidase	210	260	-	-	-	-
Catalase	61	62	-	-	-	-

A. Potato discs in light (69) C. Potato discs (41) E. Beetroot discs (66)
B. Potato discs in dark (69) D. Swede discs (69) F. Carrot discs (67)

Jerusalem artichoke have been attributed to changes in the allosteric properties of phosphofructokinase which was reported to lose its sensitivity to allosteric inhibition by ATP after washing the discs for 18 hrs (12). Loss of allosteric inhibition by ATP was also noted in the ageing process of Brussel sprouts (26). However, Ashihara _et al_ (4) reported no loss of allosteric response to ATP during the ageing of carrot discs and they also failed to confirm the findings of Black and Wedding (58) for Jerusalem artichoke.

The remainder of this article will concentrate on the control of glycolysis in fresh and aged tissue.

EVIDENCE THAT PHOSPHOFRUCTOKINASE IS A CONTROL POINT OF GLYCOLYSIS IN STORAGE TISSUE

The following evidence taken from Faiz-ur-Rahman et al (32, 33) points to the controlling influence of phosphofructokinase in the glycolytic pathway of aged carrot discs.

1. ΔG for phosphofructokinase is large and the mass action ratio is far from equilibrium.
2. After short periods of anoxia, there is a small decrease in the concentration of fructose-6-phosphate and a large increase in the concentration of fructose diphosphate, giving a positive cross-over and a movement of the mass action ratio towards equilibrium.
3. The enzyme is the least active of all the glycolytic enzymes. Whilst this evidence is consistent with phosphofructokinase being a control point, it does not rule out other enzymes in the sequence also having a controlling role. The evidence is qualitative and does not enable us to estimate the control strength of the reaction. The evidence that phosphofructokinase has a controlling role in fresh slices is not strong. The enzyme has low activity and the mass action ratio is far from equilibrium, but when fresh carrot slices are transferred to nitrogen, the concentrations of the metabolites remain unchanged and no conclusions can be drawn.

Evidence that Pyruvate Kinase is a Control Point of Glycolysis in Storage Tissue

As an alternative to the phosphofructokinase activation/inhibition theory of the control of glycolysis, Barker _et al_ (8, 9) have proposed that the glycolytic enzymes are located in a particulate structure which is more permeable to ADP than ATP. Under nitrogen, the concentration of ADP in the cytoplasm increases, it then passes into the particle where it activates pyruvate kinase to produce ATP which subsequently stimulates the production of fructose diphosphate. The evidence for this proposal was the demonstration that on transfer to nitrogen the concentration of phosphoenolpyruvate fell before the concentration of fructose diphosphate rose and in some tissues the level of fructose diphosphate did not increase. The data obtained with aged carrot discs also shows a rapid fall in the level of phosphoenolpyruvate on transfer from air to nitrogen (32, 33), which is consistent with pyruvate kinase acting as a control point, though in this case the increase in fructose diphosphate is not a _consequence_ of the increased activity of pyruvate kinase since the fall in phosphoenolpyruvate does not precede the rise in fructose diphosphate. Adams and Rowan (2) argue that pyruvate kinase is the control point of glycolysis in fresh discs of carrot and offer two pieces of evidence:

1. A positive cross-over is observed between phosphoenolpyruvate and pyruvate after washing the discs for 24 hours.
2. Pyruvate - the product of the kinase stimulates the respiration of fresh carrot discs.

The lack of a theoretical basis for applying the cross-over theorem to a time dependent process has already been discussed. The stimulation of respiration by pyruvate can hardly be taken as evidence for pyruvate kinase acting as a control point since if control is located at _any_ point in the glycolytic sequence it will limit the production of pyruvate.

THE MECHANISM OF CONTROL

The rate of glycolysis is closely connected with the rate of phosphorylation (Figure 7) and the controlling influence of adenine nucleotides and inorganic phosphate has been discussed in relation to glycolysis in plants (1, 34). Atkinson (5) has developed the concept of energy charge

$$\text{Energy charge} = \frac{[ATP] + 0.5\ [ADP]}{[ATP] + [ADP] + [AMP]}$$

as a central controlling influence on metabolism. Some glycolytic enzymes from plants respond to energy charge in the manner proposed by Atkinson (63) but since phosphofructokinase from plants does not respond to AMP it is unlikely that energy charge plays the dominant role in the control of plant glycolysis which has been proposed for animals.

THE CONTROL OF PHOSPHOFRUTOKINASE

The current theory of the control of glycolysis in animals (62) proposes that the phosphofructokinase is normally inhibited by ATP. When the level of ATP drops slightly, as under anaerobic conditions, phosphofructokinase is activated, but the activator is AMP rather than ATP. The concentration of AMP in the cell is small compared with the concentration of ATP and ATP and all three nucleotides are maintained in equilibrium by adenylase kinase.

$$2ADP \longleftrightarrow ATP + AMP$$

$$K.equi = \frac{(ATP)\ (AMP)}{(ADP)^2}$$

A small percentage decrease in ATP will produce a large percentage increase in AMP, producing an amplification system to activate phosphofructokinase.

All plant phosphofructokinase so far isolated are slightly inhibited by AMP rather than being activated. Consequently an alternative mechanism is required and it appears likely that the proposal that phosphoenolpyruvate controls the activity of phosphofructokinase (48) is correct. The enzyme from pea seeds is inhibited 50% by 1.3μM phosphoenolpyruvate, which also induces a number of other changes. Thus in the presence of phosphoenolpyruvate the Michaelis-Menten kinetics with respect to fructose-6-phosphate are converted into sigmoid kinetics and the sensitivity of the enzyme to changes in pH is increased. The kinetics of phosphofructokinase are complex - the activity of the plant enzyme is modified by at least ten metabolites and as discussed by Turner and Turner (75) these properties coupled with the strategic position of the enzyme point to an important and involved role for phosphofructokinase in the control of glycolysis.

With regard to the role of phosphofructokinase in storage tissue, the enzyme has been isolated from carrots (25) and shown to be inhibited by ATP, ADP, AMP and citrate, to be activated by P_i and to show sigmoid kinetics with respect to fructose diphosphate. These kinetics were interpreted as being consistent with the model of control proposed for animals. However, as noted by Adams and Rowan (2), unless the concentration of ATP in the cytoplasm is at least 50-fold higher than the concentration expressed per unit fresh weight, ATP cannot be a negative effector in carrot. Faiz-ur Rahman et al (32) in analysing their data on the Pasteur effect in carrots concluded that phosphoenolpyruvate was an important negative effector for phosphofructokinase. Since phosphoenolpyruvate is the substrate for pyruvate kinase control could pass from one site to the other. However, the level of phosphoenolpyruvate could be affected by other reactions and in particular by phosphoenolpyruvate carboxylase (see page

Furthermore to consider the control of phosphofructokinase solely in terms of phosphoenolpyruvate would be an oversimplification. Citrate has a strong inhibitory effect on the enzyme and this metabolite could well be involved in controlling the activity of glycolysis relative to the activity of the Krebs cycle.

Substrate Cycling in the Control of Phosphofructokinase

Newsholme and Crabtree (61) proposed that the control of glycolysis at the level of phosphofructokinase is achieved by a substrate cycle.

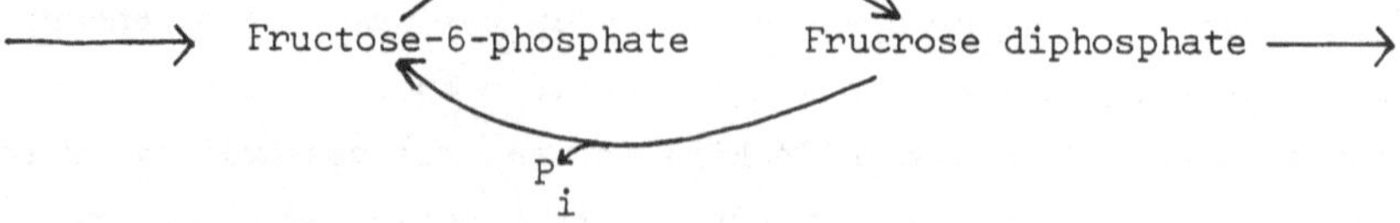

To illustrate how this system might work, consider three cases where the ratio of kinase to phosphatase is different but the flux through the system is constant (10 units). The effect of stimulating the kinase by 10% is shown below.

Before Stimulation | After Simulation of kinase by 10% | % Increase in Flux

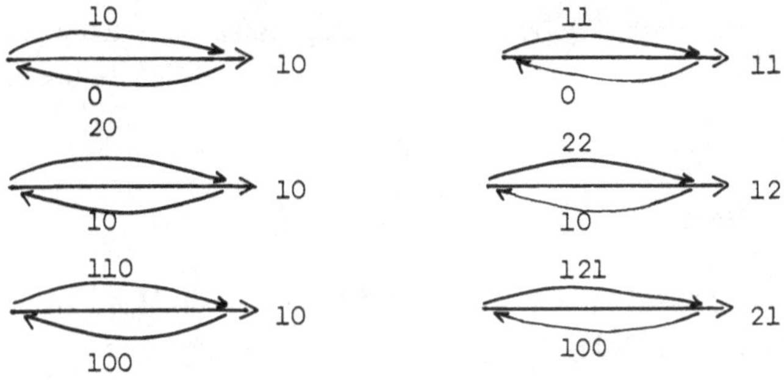

% Increase in Flux: 10, 20, 210

If the effector which stimulated the kinase by 10% were to also inhibit the phosphatase by 10% - the amplification would be even greater. Similar cycles could also operate at the other irreversible reactions of glycolysis.

Evidence that these substrate cycles occur in animal tissues has been reviewed (46) but there is no evidence one way or the other for plants. It is clear that if these substrate cycles occur to any extent, there will be a significant production of heat. This could be considered the price to be paid for metabolic control. However, in some cases the physiological significance of substrate cycling could be the production of heat as in Arum (60) or in the flight muscle of some insects (13).

The Relationship between Aldolase and Phosphofructokinase

The rapid increase in activity of phosphofructokinase which occurs øn transfer to nitrogen may be attributed to a positive feedback loop since fructose diphosphate stimulates its own synthesis.

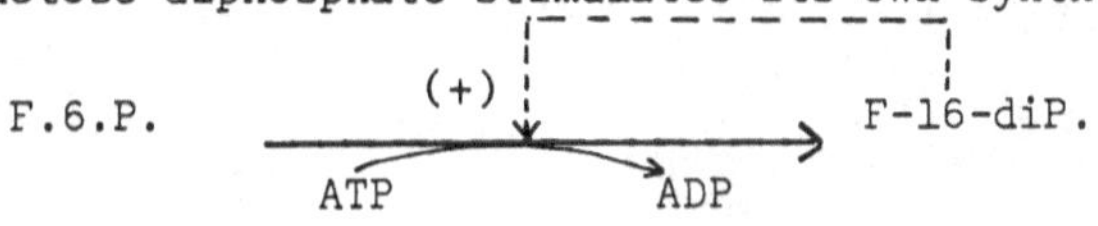

If the activity of the next enzyme in the sequence (aldolase) was high with respect to the activity of phosphofructokinase, the fructose diphosphate would be rapidly removed and thus the positive feedback effect would be reduced. This is an example of a system property; control does not involve a single master reaction, the control strength of an enzyme depends not only on its own kinetic properties but on the kinetic constants of other enzymes in the sequence. In the case of phosphofructokinase it can be shown experimentally that increasing amounts of aldolase reduce the overall rate of conversion of fructose-6-phosphate to trioses (30).

THE CONTROL OF PYRUVATE KINASE

The control properties of pyruvate kinases from liver and muscle have been extensively described (47, 72). These properties include sigmoid kinetics with respect to phosphoenolpyruvate and activation by fructose diphosphate. Pyruvate kinase from cotton seeds (29) and carrots (73) shows Michaelis-Menten kinetics with respect to phosphoenolpyruvate and does not respond to fructose diphosphate. On the other hand citrate and ATP inhibit the enzyme from both sources whilst the cotton seed enzyme is activated by fumarate and AMP, and inhibited by malate. Whilst these responses are consistent with a regulatory role for pyruvate kinase, the lack of response to fructose diphosphate is perplexing. However, the reported stimulation by AMP (29) can be regarded as equivalent. Thus if the level of ATP falls slightly, it will be reflected by a large percent change in the concentration of AMP due to the adenylate kinase equilibrium. This activation of pyruvate kinase by AMP is reinforced by the inhibition produced by ATP. In the steady state the activity of pyruvate kinase is controlled by the balance of ATP/AMP, during anoxia this ratio falls and pyruvate kinase would then be activated. It should, however, be noted that AMP activation of pyruvate kinase has not been reported for the enzyme from carrots.

The Relationship between Pyruvate Kinase and Phosphoenolpyruvate Carboxylase

Phosphoenolpyruvate carboxylase from aetiolated maize (78) and potato tubers (14) exhibits allosteric responses to a number of metabolites, consistent with its proposed role in a metabolic pH-stat (20, 21). The activation of the carboxylase by fructose diphosphate has consequences for the control of glycolysis. The increase in fructose diphosphate which occurs in anoxia would activate the carboxylase and thus lower the concentration of phosphoenolpyruvate, which was inhibiting phosphofructokinase, thereby leading to a further increase in fructose diphosphate. However, the associated decrease in phosphoenolpyruvate reduced the flux through pyruvate kinase. The competition between the kinase and the carboxylase is thus a potentially important controlling factor, which is currently under investigation in this laboratory.

ANOMERIC SPECIFICITY IN RELATION TO METABOLIC CONTROL

The allosteric sites of hexokinase,phosphofructokinase and pyruvate kinase do not appear specific for the anomers of the hexose phosphates, suggesting that anomeric specificity is not involved in feedback and feedforward mechanisms of control.

The half time for the non-enzymic interconversion of some hexoses is shown in Table 4.

Table 4: Rates of non-enzymic anomerisation

Sugar	Anomerisation half-life $\alpha \rightarrow B$
D-glucose	130 mins.
D-glucose-6-phosphate	15 secs.
D-fructose-6-phosphate	0.4 secs.
D-fructose-1,6-diphosphate	0.09 secs.

These relatively slow rates could conceivably be rate determining in glycolysis and raise the possibility that the anomerases may play a role in the control of flux. Two possibilities will be considered.

The entry of sugar molecules into the glycolytic or pentose phosphate pathways involves the partitioning of glucose-6-phosphate between glucose-6-phosphate dehydrogenase (β-specific) and phosphohexoseisomerase (Fig.12).

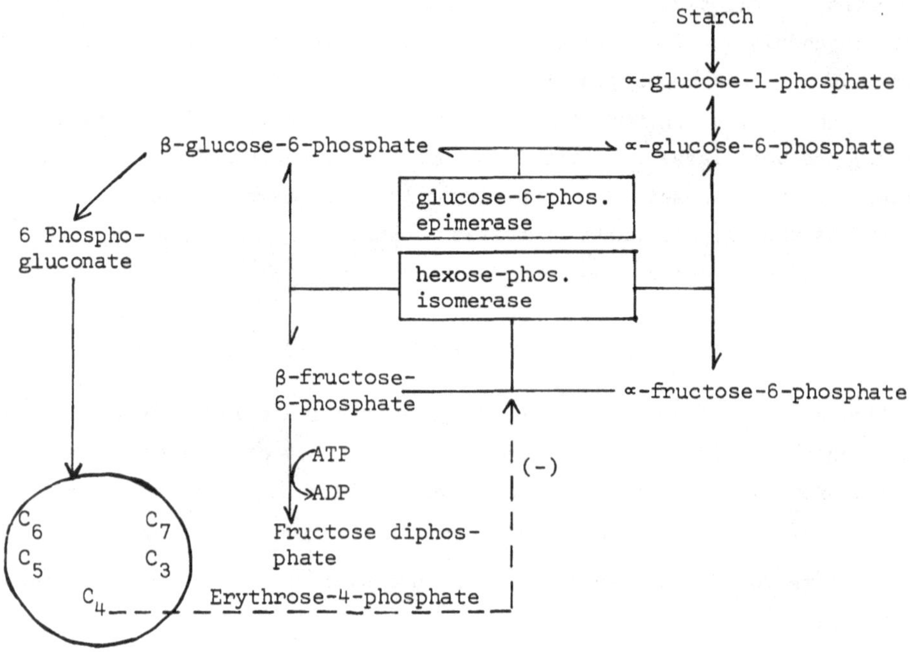

Figure 12: Anomerases associated with the partitioning of glucose between glycolysis and the pentose phosphate pathway.

D-glucose entering the pathway as the α-isomer can be converted to fructose-6-phosphate without a change of configuration, whereas the conversion to 6 phosphogluconate requires anomerisation. Thus the partitioning could be affected by the rate of anomerisation which is subject to control at the isomerase level by the products of the pentose phosphate pathway - especially erythrose-4-phosphate (70). This possibility could be tested by comparing the ratio

$$\frac{^{14}CO_2 \text{ produced from hexose-1-}^{14}C}{^{14}CO_2 \text{ produced from hexose-6-}^{14}C}$$

obtained with ^{14}C-labelled glucose and a ^{14}C hexose such as galactose which enters by way of α-glucose-6-phosphate. If the anomerisation step is fast relative to glycolysis, both sugars should give the same ratio. Such experiments have not been performed with plants but a related experiment performed with red cells indicates that anomerisation is very fast and does not contribute to partitioning.

A second possible involvement of anomerisation lies in the substrate cycling discussed on page . Phosphofructokinase and fructose diphosphatase can function together to produce an effective ATPase, but the enzymes have reciprocal activator/inhibitor responses which control the futile cycling. The fact that the two enzymes prefer opposite anomers may also affect the rate of futile cycling.

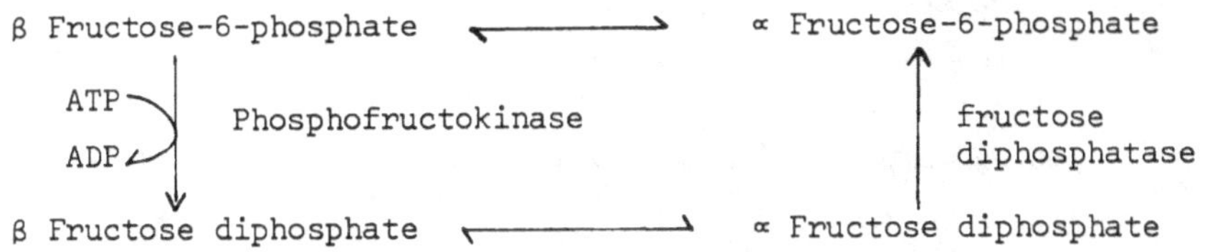

If the rate of anomerisation is slow then it could limit the amount of cycling or putting it in another way, is the rate of anomerisation sufficient to allow significant cycling? An approximate calculation based on the half lives of anomerisation given in Table 4 and the concentrations of fructose diphosphate and fructose-6-phosphate compared with the glycolytic flux in carrot tissue suggests that anomerisation is fast enough to allow extensive futile cycling. Thus in carrot discs the glycolytic flux is of the order 1μmole/100gm/min and the rate of anomerisation of fructose 6-phosphate is <u>c</u> 400μmoles/100gm/min and of fructose diphosphate <u>c</u> 800 μmoles/100gm/min.

CONCLUSIONS

The recognition that control is a system property, removes the all or none effect associated with earlier considerations of metabolic control. The uncertainties associated with the theoretical basis of the methods used to identify control points as well as the practical difficulties involved make an objective assessment of control particularly difficult. Nevertheless there is now a census of opinion indicating that glycolytic control is dominated by the interaction between phosphofructokinase and pyruvate kinase - although other control points are no doubt involved to some extent.

The molecular basis on which control is achieved is equally uncertain. The enzymes respond to numerous allosteric effectors and it is easy to produce complicated schemes of interaction between enzymes, substrates and effectors. It is more difficult to determine if these schemes operate in the plant as well as in the mind of the investigator. However, if only for a point of further discussion, some of the proposed interactions are shown below.

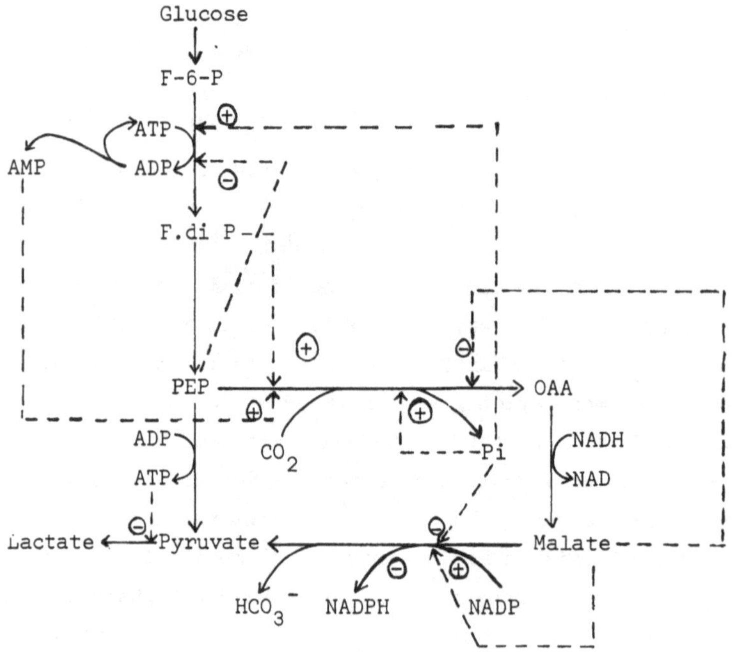

Figure 13: Interactions involved in the control of glycolysis

ACKNOWLEDGEMENT

The author is grateful to the Science Research Council for supporting this work.

REFERENCES

1 Adams, P. B. Effect of adenine nucleotides on levels of glycolytic intermediates during the development of induced respiration in carrot root slices. Plant Physiol. 45, 500-503 (1970).

2 Adams, P. B., Rowan, K. S. Glycolytic control of respiration during ageing of carrot root tissue. Plant Physiol. 45, 490-494 (1970).

3 ap Rees, T. Pathways of carbohydrate breakdown in higher plants. In: M.T.P. Int. Rev. Sci. Biochem. V. 11. Ed. Kornberg, H. L. & Phillips, D. C. 89-127 (1974).

4 Ashihara, H., Komamine, A., Shimokoriyma, M. Phosphofructokinase of cultured and aged carrot root slices. Phytochem. 11, 2717-2721 (1972).

5 Atkinson, D. E. The energy charge of the adenylase pool as a regulatory parameter. Interaction with feedback modifiers. Biochemistry 7, 4030-4034 (1968).

6 Barker, J., El Saifi, A. F. Studies in the respiratory and carbohydrate metabolism of plant tissues II. Inter-relationship between the rates of production of carbon dioxide, lactic acid and of alcohol in potato tubers under anaerobic conditions. Proc. Roy. Soc. London 140B, 508-522 (1952).

7 Barker, J., Mapson, L. W. Studies in the respiratory and carbohydrate metabolism of plant tissues. XII. Further studies of the formation of CO_2 and the changes in lactate, alcohol, sucrose, pyruvate and αketo-glutarate in potato tubers in nitrogen and in the air following anaerobic conditions.Proc.Roy.Soc.London 157B, 383-402 (1963).

8 Barker, J., Khan, M. A. A., Solomos, T. Studies in the respiratory and carbohydrate metabolism of plant tissues XXI. The mechanism of the Pasteur effect in peas. New Phytologist 66, 577-596, (1967).

9 Barker, J., Khan, M. A. A. Studies in the respiratory and carbohydrate metabolism of plant tissues. XXII. Pasteur effect in potatoes and apples. New Phytologist 67, 205-212 (1968).

10 Bentley, J. M., Fishman, P. H., Pentchev, P. G. Studies on mutarotases 1. Purification and properties of mutarotase from higher plants. J. Biol. Chem. 242, 4263 (1967).

11 Bielski, R. L., Laties, G. G. Turnover rates of phosphate esters in fresh and aged slices of potato tuber tissue. Plant Physiol. 38, 586-594 (1963).

12 Black, M. I., Wedding, R. T. Effects of storing and ageing on properties of phosphofructokinase from Jerusalem artichoke tubers. Plant Physiol. 43, 2066-2069 (1968).

13 Bloxham, D. P. Metabolic flux through phosphofructokinase and fructose 1-6, diphosphate. Int. J. Biochem. 5, 429-435 (1974).

14 Bonugli, K., Davies, D. D. The regulation of phosphoenolpyruvate carboxylase in relation to a metabolic pH-stat. Planta, 133, 281-287 (1977).

15 Bücher, T., Rüssman, W. Equilibrium and nonequilibrium in the glycolysis system. Angew. Chemie Int. Ed. 3, 426-439 (1964).

16 Chance, B. The respiratory chain as a model for metabolic control in multi-enzyme systems. In: Control of Energy Metabolism, ed. Chance, B. Estabrook, R. W., Williamson, J. R. Acad. Press (1965).

17 Chance, B., Holmes, W. F., Higgins, J. J. Connelly, C. M. Localisation of interaction sites in multicomponent transfer systems: Theorems derived from analogues. Nature 182, 1190-1193 (1958).

18 Chapman, E. A., Graham, D. The effect of light on the tricarboxylic acid cycle in green leaves II. Intermediary metabolism and the location of control points. Plant Physiol. 53, 886-892 (1974).

19 Davies, D. D. Intermediary metabolism in plants. Pub. Cambridge Univ. Press (1961).

20 Davies, D. D. Control of and by pH. In: Rate control of biological processes. Ed. Davies, D. D. Cambridge Univ. Press 513-529 (1973).

21 Davies, D. D. Metabolic control in higher plants. In: Biosynthesis and its control in plants. Ed. Milborrow, B. V. Acad. Press 1-20 (1973)

22 Davies, D. D. In: Symposium of the Phytochem. Soc. Ed. Smith, H. (1977) In Press.

23 Davies, D. D., Davies, S. Purification and properties of L(+) lactate dehydrogenase from potato tubers. Biochem. J. 129, 831-839 (1972).

24 Davies, D. D., Grego, S., Kenworthy, P. The control of the production of lactate and ethanol by higher plants. Planta 118, 297-310 (1974).

25 Dennis, D. T., Coutate, T. P. Phosphofructokinase, a regulatory enzyme in plants. Biochem. Biophys. Res. Commun. 25, 187-191 (1966).

26 Dennis, D. T., Coutate, T. P. The regulatory properties of a plant phosphofructokinase during leaf development. Biochim. Biophys. Acta 148, 129-137 (1967).

27 Dennis, D. T., Green, T. R. Soluble and particulate glycolysis in developing castor bean endosperm. Biochem. Biophys. Res. Commun. 64, 970-975 (1975).

28 Dietzler, D. N., Leckie, M. P., Bergstein, P. E., Sughrue, M. J. J. Evidence for the coordinate control of glycogen synthesis, glucose utilisation and glycolysis in Escherichia coli. 1. Quantitative covariance of the rate of glucose utilisation and the cellular level of fructose 1,6-diphosphate during exponential growth and nutrient limitation. Biol. Chem. 250, 7188-7193 (1975).

29 Duggleby, R. G., Dennis, D. T. Characterisation and regulatory properties of pyruvate kinase from cotton seeds. Arch. Biochem. Biophys. 155, 270-277 (1973).

30 El-Badry, A. M., Otani, A., Mansour, T. E. J. Studies on heart phosphofructokinase. Role of fructose 1,6-diphosphate in enzyme activity. Biol. Chem. 248, 557-563 (1973).

31 Everson, R. G., Rowan, K. Phosphate metabolism and induced respiration in washed carrot slices. Plant Physiol. 40, 1247-1250 (1965).

32 Faiz-ur-Rahman, A. T. M., Trewavas, A. J., Davies, D. D. The Pasteur effect in carrot root tissue. Planta 118, 195-210 (1974).

33 Faiz-ur-Rahman, A. T. M., Davies, D. D., Trewavas, A. J. The control of glycolysis in aged slices of carrot root tissue. Planta 118, 211-224 (1974).

34 Givan, C. V. Short-term changes in hexose phosphates and ATP in intact cells of Acer pseudo platanus L. subjected to anoxia. Plant Physiol. 43, 948-952 (1968).

35 Heinrich, R., Rapoport, T. A. A linear steady-state treatment of enzymatic chains. General properties, control and effector strength. Eur. J. Biochem. 42, 89-95 (1974).

36 Higgins, J. Dynamics and control in cellular reactions. In: Control of energy metabolism. Ed. Chance, B., Estabrook, R. W., Williamson, J. R. Pub. Acad. Press 13-46 (1965).

37 Holmes, W. F. Locating sites of interactions between external chemicals and a sequence of chemical reactions. Trans. Farad. Soc. 55, 1122-1126 (1959).

38 James, W. O. Respiration of higher plants. Adv. Enzymol. 18, 281-318 (1957).

39 James, W. O., Ritchie, A. F. The anaerobic respiration of carrot tissue. Proc. Roy. Soc. London 143B , 302-310 (1955).

40 Kacser, H., Burns, J. A. The control of flux. In: Rate control of biological processes. Ed. Davies, D. D. Pub. Cambridge Univ. Press 65-104 (1973).

41 Kahl, G. Genetic and metabolic regulation in differentiating plant storage tissue cells. Bot. Rev. 39, 274-299 (1973).

42 Kahl, G. Metabolism in plant storage tissue. Bot. Rev. 40, 263-314 (1974).

43 Kahl, G., Lange, H., Rosenstock, G. Substrate levels, enzymatic activities and genetic regulation after derepression in plant storage tissue. Z. Naturforschg. 24b, 911-918 (1969).

44 Kahl, G., Lange, H., Rosenstock, G. Regulation of glycolysis by the synthesis and degradation of enzymes. Z. Naturforschg. 24b, 1544-1549 (1969).

45 Khavkin, E. E., Zeleneva, I. V. Glycolytic enzyme pattern and constant proportion group in plant cells as related to their developmental and functional state. FEBS Lett. 21, 269-272 (1972).

46 Katz, J., Rognstad, R. Futile cycles in the metabolism of glucose. Curr. Top. in Cell Regul. 10, 237-289 (1976).

47 Kayne, F. L. Pyruvate kinase. Enzymes 8, 353-382 Pub. Acad. Press (1973).

48 Kelly, G. J., Turner, J. F. The regulation of pea-seed phosphofructokinase by phosphoenolpyruvate. Biochem. J. 115, 481-487 (1969).

49 Kobr, M. J., Beevers, H. Gluconeogenesis in the castor bean endosperm 1. Changes in glycolytic intermediates. Plant Physiol. 47, 48-52 (1971).

50 Kostytschew. Pflanzenatmung. Pub. Springer, Berlin (1924).

51 Krebs, H. A. Considerations concerning the pathways of synthesis in living matter. Synthesis of glycogen from non-carbohydrate precursors. John Hopkins Hosp. Bull. 95, 19-33 (1954).

52 Krebs, H. A. Control of metabolic processes. Endeavour 16, 125-132 (1957).

53 Link, G. K. K., Barron, E. S. G., Klein, R. M., Michel, B. E. The metabolism of potato slices. Arch. Biochem. Biophys. 28, 377-398 (1951).

54 Lowry, D. H., Passoneau, J. V., Hasselberger, F. X., Schultz, D. W. Effect of ischemia on known substrates - cofactors of the glycolytic pathway in brain. J. Biol. Chem. 239, 18-30 (1964).

55 Lynen, F., Hartman, G., Netter, K. R., Schuegraf, A. Phosphate turnover and Pasteur effect. In: Regulation of cell metabolism. Ed. Woolstenholme, G. E. W., O'Connor, C. M. Pub. Churchill, London, 256-273 (1959).

56 MacLennan, D. H., Beevers, H., Harley, J. H. "Compartmentation" of acids in plant tissues. Biochem. J. 89, 316-327 (1963).

57 Macnicol, P. K. Metabolic regulation in the senescing tobacco leaf II. Changes in glycolytic metabolite levels in the detached leaf. Plant Physiol. 51, 798-801 (1973).

58 McManmion, M., Crawford, R. M. M. A metabolic theory of flooding tolerance: The significance of enzyme distribution and behaviour. New Phytologist 70, 299-306 (1971).

59 Mazelis, M., Vennesland, B. Carbon dioxide fixation into oxalacetate in higher plants. Plant Physiol. 32, 591-600 (1957).

60 Meeuse, B. J. D. Thermogenic respiration in Aroids. Ann. Rev. Plant Physiol. 26, 117-126 (1975).

61 Newsholme, E. A., Crabtree, B. Metabolic aspects of enzyme activity regulation. In: Rate control of biological processes. Ed. Davies, D. D. Pub. Cambridge Univ. Press 429-460 (1973).

62 Newsholme, E. A., Start, C. Regulation in Metabolism. Pub. Wiley, London, (1973).

63 Pacold, I., Anderson, L. Energy charge control of the Calvin cycle enzyme 3-phosphoglyceric acid kinase. Biochem. Biophys. Res. Comm. 51, 139-143 (1973).

64 Purich, D. L., Fromm, H. J. A possible role for kinetic reaction mechanism dependent substrate and product effects in enzyme regulation. Curr. Top. Cell Regul. 6, 131-166 (1972).

65 Rapoport, T. A., Heinrich, R., Jacobasch, G., Rapoport, S. A linear steady-state treatment of enzymatic chains. A mathematical model of glycolysis of human erythrocytes. Eur. J. Biochem. 42, 107-120 (1974).

66 Reed, D. J., Kolattukudy, Metabolism of red beet slices I. Effects of washing. PLant Physiol. 41, 653-660 (1966).

67 Ricardo, C. P. P., ap Rees, T. Activities of key enzymes of carbohydrate oxidation in discs of carrot storage tissue. Phytochem. 11, 623-626 (1972).

68 Rolleston, F. S. A theoretical background to the use of measured concentrations of intermediates in study of the control of intermediary metabolism. Curr. Top. Cell Regul. 5, 47-75 (1972).

69 Sacher, J. A., Towers, G. H. N., Davies, D. D. Effect of light and ageing on enzymes, particularly phenylalanine ammonia lyase, in discs of storage tissue. Phytochem. 11, 2383-2391 (1972).

70 Salas, M., Vinuela, E., Sols, A. J.Spontaneous & enzymatically catalysed anomerisation of glucose 6-phosphate and anomeric specificity of related enzymes. Biol. Chem. 240, 561 (1965).

71 Savageau, M. A. The behaviour of intact biochemical control systems. Curr. Top. Cell Regul. 6, 63-130 (1972).

72 Seubert, W., Schoner, W. The regulation of pyruvate kinase. Curr. Top. Cell. Regul. 3, 237-267 (1971).

73 Tomlinson, J. D., Turner, J. F. Pyruvate kinase of higher plants. Biochim. Biophys. Acta 329, 128-139 (1972).

74 Turner, J. F. The biosynthesis of sucrose. Biochem. J. 67, 450 (1957).

75 Turner, J. F., Turner, D. H. The regulation of carbohydrate metabolism. Ann. Rev. Plant Physiol. 26, 159-186 (1975).

76 Verleur, J. D. The role of changes in enzyme activities in the regulation of carbohydrate dissimilation in potato tissue after wounding or infection with Gibberella zeae. Acta. Bot. Neerl. 18, 353-366 (1969).

77 Williamson, J. R. Metabolic control in the perfused rat heart. In: Control of energy metabolism. Ed. Chance, B., Estabrook, R. W., Williamson, J. R. Acad. Press 333-346 (1965).

78 Wong, K. P., Davies, D. D. Regulation of phosphoenolpyruvate carboxylase of Zea mays by metabolites. Biochem. J. 131, 451-458, (1972).

79 Wurster, B., Hess, B. Glucose-6-phosphate-1-epimerase from bakers yeast. A new enzyme. FEBS Lett. 23, 341-345 (1972).

Induction and Degradation of Enzymes in Aging Plant Storage Tissues

Günter Kahl

INTRODUCTION

The actual number of molecules of a particular enzyme per cell is a function of its rate of synthesis and degradation.These profound parameters thus regulate the actual enzyme pool of the cell,whereas sophisticated mechanisms of activation and inactivation determine the number of active enzyme molecules within the cellular population of a distinct catalyst.Over the past two decades a vast number of experiments were dedicated to enzyme synthesis,so that the syntheic machinery is now well known.Likewise a picture of various activation and inactivation mechanisms is emerging,strongly indicating a very elaborate and complex regulation at this level.Usually activation-inactivation mechanisms serve as a tool for changing enzyme activities rapidly - as compared to the relatively time-consuming synthesis of a protein - and consequently represent extremely potent means of the cell to adapt its metabolism economically to a changing environment.The underlying mechanisms on one hand are more or less physical,on the other chemical modifications of the respective enzyme protein.

Although the cell has also to get rid of an enzyme molecule after a round of catalyses,we do not know too much of the responsible mechanisms,in fact we do not know anything in wounded plant storage tissue.Mammalian tissues such as small intestines and skeletal muscles firstly inactivate the enzymes no longer in use by group-specific "inactivases",which recognize free coenzyme-binding sites of their substrate proteins,cleave the latter into smaller oligopeptides,which in turn are subject to unspecific proteolytic de-

Biochemistry of Wounded Plant Tissues

gradation (60-63).Although it is highly unlikely,that there exists a specific "inactivase" for each other enzyme in a cell,it is nevertheless evident,that in a limited number of enzyme systems this is actually the case (11,13,28,30,59).

Since all these different mechanisms are likely to work also in wounded plant storage tissues,it seems necessary to firstly review some of the physical and chemical activation-inactivation mechanisms shortly and then discuss the experiments concerned with enzyme induction and degradation in the tissue slice model system.

PHYSICO-CHEMICAL MODIFICATIONS AND ACTIVITY OF ENZYME PROTEINS.

Activation and inactivation through association and dissociation of subunits and enzymes.

An essential structural element of enzymes with quartary structure is the subunit,which - isolated - is catalytically inactive.After association with another subunit or subunits,however,the enzyme becomes active and gains regulativity.It is possible,that each subunit owns an active site,but only in combination with others is able to bind and convert substrates.The best known example represents the alkaline phosphatase of E.coli,the two inactive identical subunits of which are dimerized by Zn^{++} and thereby activated.This process is fully reversible.A more elaborate mechanism has been evolved for the aspartate transcarbamylase of E.coli.Even under physiological conditions it seems possible,that only the six catalytic subunits are structurally organized into an active enzyme structure.However,this enzyme can't be regulated allosterically (see below),unless the six regulatory subunits are also fixed to the core.

Specific aggregation of whole enzyme molecules to high molecular aggregates has been shown for a number of different enzymes and definitely alter the properties of the individual molecules drastically. The glutamate dehydrogenase of the liver consists of several identical subunits,which altogether have a molecular weight of 280 000 daltons.In the presence of ADP and distinct amino acids the enzymes -

if highly concentrated - form complexes with each other and a particle weight of 2,2 million daltons as well as L-glutamic dehydrogenase activity.In the presence of NADH,GTP,both effectors together or after dilution the different enzymes dissociate from the complex. It is not known with certainty whether these mechanisms have any importance in vivo.

Activation and inactivation of enzymes through allosteric mechanisms.

Rapid switch-off and switch-on of metabolic chains is exerted through small molecular weight metabolites ("effectors"),which bind to enzymes of strategic importance ("regulative" or "allosteric" enzymes) and alter the environment of the active center.Such enzymes usually possess quartary structure.Activators are positive effectors,which after binding to the protein at distinct allosteric sites distort the peptide chains around the active center to only about few Ångstroms,so that the substrate can be bound more rapidly and accurately.So the activator induces a higher complementarity between active center and substrate molecule ("induced fit").Negative effectors ("inactivators") on the contrary induce a conformation of the substrate pocket,which does not allow proper binding of the substrate. One of the best known regulatory enzyme,the aspartate transcarbamylase (ATCase),catalyses the condensation of carbamylphosphate and aspartate to carbamylaspartate and feeds the sequence for the synthesis of uridine mono - and triphosphate and hence cytidine-5'-triphosphate.This enzyme is effectively inhibited by the end-product of the sequence,CTP,in a so-called "feed-back"-inhibition.CTP pronouncedly lowers the affinity of the enzyme towards its substrate. Binding of the effector at distinct sites of the regulatory subunits (R-chains) induces a conformational change of these polypeptides, which again is transmitted to the neighbouring catalytic subunits (C-chains) and forces them to change their shape ("allos stéros"). Result:inactivation of the enzyme.This process is totally reversible,if the concentration of CTP at the enzyme's surface drops below a certain level and allows a sophisticated control of the synthesis of certain intermediates (26,27).

Two important hypotheses try to explain the mechanism of protein-protein relationships in regulatory enzymes.According to the "non-or-all" model of Monod,Wyman and Changeux strict cooperativity is characteristic for these proteins,i.e.all the subunits have either one or the other conformation ("relaxed" or R-form:catalytic subunit active,regulatory subunit non-inhibited,since occupied by effector; and "tense" or T-form:catalytic subunit inactive,regulatory one inhibited).The immediate transformation of one into another state is defined by a delicate interplay between substrate and effector(s). Somewhat to the contrary,the "induced-fit"-model of Koshland allows a sequential transformation of individual subunits,i.e.the enzyme may acquire a whole series of different conformative states and hence of different catalytic capabilities (68,69,90).

Activation and inactivation of enzymes through enzymatic modification.

Enzymatic modification of enzymes,i.e.the catalyzed transfer of specific groups onto proteins where they are fixed covalently,in all cases so far known lead to profound changes either of activity of the enzyme or its regulation.Such chemical modifications are carbamoylations (transfer of a carbamyl group from carbamylphosphate to distinct groups of the enzyme),phosphorylations,acylations,oxidations and reductions of thiol groups (which are necessary for structure and function of the enzyme) and adenylations (transfer of adenylate residues).The effect of such modifications may be the activation of previously not or only weakly active enzymes,more frequently,however, the inactivation of previously active enzymes (table 1).These modifications enable the cell to extremely rapid reactions upon a changing intra- or extracellular parameter and it seems more and more likely,that procaryotic and all eucaryotic cells acquired the genes for modifying enzymes.It is evident that chemical modifications of enzymes are energy-saving (as compared to enzyme synthesis) and allow the persistence of partial activities in multifunctional enzymes, as has been demonstrated with glutamine synthetase of E.coli.The 12 subunits of this enzyme may be sequentially modified by enzymatic attachment of adenylyl moieties from ATP to the phenolic hydroxyl group of a tyrosyl residue (65,112,126).This is accompanied by constant lowering of the activity.The whole process is reversible and under delicate and multifactorial control (35).

Table 1: Chemical modification of different enzymes.

Chemical modification	enzyme	effect
carbamylation	glutamic dehydrogenase	inactivation
phosphorylation	phosphoglucomutase	inactivation
	glycogen phosphorylase	activation
	phosphorylase b kinase	activation
	phosphorylase b	activation
	glycogen synthetase	inactivation
	pyruvate dehydrogenase complex	inactivation
	fructose-1,6-diphosphatase	inactivation
acylation	phosphofructokinase	loss of allosteric properties
	glutamine synthetase	inactivation
SH-SS(oxidation)	nitrate reductase	inactivation
	pyruvate lyase	inactivation
adenylylation	glutamine synthetase	inactivation
ADP-ribosylation	translocase	inactivation

Activation and inactivation of enzymes through limited proteolysis.

A series of enzymes is stored as inactive precursors after synthesis ("zymogens,proenzymes"),but may rapidly be converted to the active form by a specific,irreversible,limited proteolysis.Many well-known digestive enzymes are present in such a proenzymic state.The pro-

enzyme trypsinogen as one representative is only weakly active shortly after its synthesis. The proteolytic cleavage of a hexapeptide at the N-terminal end of the molecule, however, induces a change in conformation of the polypeptide, which in turn allows the formation of a highly effective active site.
This relatively "primitive" process of activation has the advantage, that the inactive precursor may be stored inert within the cell and may be activated at once by proteases, once accessible. Such activations seem to be widespread in nature (i.e. chitin synthetases in yeast; ß-amylases in aleurone cells of certain Gramineae; all digestive enzymes in mammals). Inactivations by limited proteolysis, on the other hand, seem to initiate the degradation of enzyme molecules, as has been demonstrated to be the case in mammals and is suggested to be likely for some enzymes in wounded plant storage tissues.
The highly superficial and by no means complete survey readily demonstrates the need for respective experiments in the promising plant storage tissue system.

THE INDUCTION OF ENZYMES IN SLICED PLANT STORAGE TISSUE.

Wounding of whatever kind of plant storage tissue induces specific changes in the activity of most enzymes. The great majority of all these proteins increases in activity, some enzymes do not react at all and some decrease their activity (table 2).

Increase in activity may be due to activation of previously inactive enzymes or the de novo synthesis of the respective proteins. Both mechanisms seem to operate in wounded tissues and although activation stricto sensu is not related to induction, i.e. the de novo synthesis of a protein as induced by one or several external or internal factors, an example for an activated enzyme (ribonuclease) precedes the discussion of some truely induced enzymes (peroxidase; phenylalanine ammonia lyase, glucose-6-phosphate dehydrogenase).

organ	enzyme	effect	mechanism	reference
white potato tuber	glucose-6-phosphate-and 6-phosphogluconate dehydrogenase isozymes	increase	de novo synthesis	46
	phosphoglucomutase isozymes	decrease	degradation?	50,53
	sucrose synthetase	increase	"activation"	75
	sucrose phosphorylase	no change	-	75
	hexokinase	no change	-	47
	phosphoglucoisomerase	no change	-	47
	aldolase	small increase	-	47
	triosephosphateisomerase	small decrease	-	47
	phosphoglyceromutase	increase	?	47
	enolase	decrease	degradation?	47
	pyruvate kinase	increase	synthesis?	47
	glutamate-pyruvate-transaminase	decrease	degradation?	47
	"malic enzyme"	increase	synthesis?	47
	malate dehydrogenase	no change	-	132
	phenylalanine ammonia lyase	increase	de novo synthesis	37,110
	fatty acid synthetase	increase	synthesis?	127
	RNA polymerase isozymes	increase	?	54,55
	peroxidase isoenzymes	increase	?	14,32
	RNAase	increase	activation	98
sweet potato root	glucose-6-phosphate-and 6-phosphogluconate dehydrogenase isoenzymes	increase	synthesis?	91,92
	alkaline invertase	no change	-	80,81
	acid invertase	increase	synthesis?	80,81
	pyruvate decarboxylase	decrease	?	94
	3-deoxy-D-arabino-heptulosonic acid 7-phosphate synthetase	no change	-	86
	5-dehydroquinate hydrolyase	increase	synthesis?	66,85
	shikimate:NADP oxidoreductase	increase	synthesis?	66,85
	trans-cinnamic acid 4-hydroxylase	increase	synthesis	18,87
	phenylalanine ammonia lyase	increase	de novo synthesis	19,88,121
	tyrosine ammonia lyase	increase	synthesis	48
	HMG-CoA-reductase	decrease/increase	?	116
	NADPH-cytochrome c-reductase	decrease/increase	?	117,125
	NADH-cytochrome c-reductase	increase	?	125
	acid phosphatase	decrease	?	117
	peroxidase isoenzymes	increase	de novo synthesis	57,64,111
	polyphenol oxidase isozymes	increase	de novo synthesis	39,40,41
carrot root	ascorbate oxidase	increase	?	97
	acid invertase	increase	?	107
	alkaline invertase	no change	-	107
	hexokinase	increase	?	106
	pyruvate kinase	increase	?	1
Jerusalem artichoke	acid invertase	increase	?	21
	ascorbate oxidase	increase	?	21
	cinnamic acid hydroxylase	increase	?	10
	NADPH-cytochrome P-450 reductase	increase	?	10
sugar beet	acid invertase	increase	synthesis?	7,15
	RNA polymerase	increase	?	20
red beet	glucose-6-phosphate-and 6-phosphogluconate dehydrogenase	increase	?	103
dandelion and chicory root	fructosan hydrolase	increase	?	24

Table 2: Wounding-induced changes of enzymatic activities in different plant storage tissues.

Activation of ribonuclease (E.C.3.1.4.23)

The injury of potato leaves and potato tuber parenchyma as in other plant organs leads to an increase in RNA hydrolyzing activity around the wound.Usually the hydrolases reach their maximum of activity two days after wounding and afterwards level off.At least three different enzymes seem to contribute to this RNA degradation:RNAse,an acid phosphomonoesterase and a phosphodiesterase.The diesterase activity - in contrast to both other enzymes - does not change appreciably after wounding.

Experiments designed to elucidate the mechanism of the increase in RNAse activity lead to contradictory results.On one hand cycloheximide,although at fairly high concentrations (50 µg/ml) drastically inhibits the development of RNAse activity after wounding potato tuber tissue,whereas actinomycin D (30 µg/ml) only marginally affects the same process.On the other hand density labelling experiments using D_2O and isopycnic density gradient centrifugation of the RNAse do not reveal any shift in buoyant density of the enzyme.This is true for the RNAse from wounded leaves and wounded tuber tissue.From these experiments alone it cannot be decided whether there is synthesis or activation of the enzyme especially since insufficient labelling of newly formed enzymes cannot be strictly excluded:the wound-healing tissue was incubated with 50% (v/v) D_2O only (98).So the enzyme was purified and used as an antigen for the production of antibodies (100).Employing the immuno-diffusion technique of Ouchterlony,it could be demonstrated clearly,that the protein content of RNAse did not change after wounding and consequently did not parallel the increase in activity of the enzyme (table 3).

Although in view of the stable buoyant density of the enzyme after deuteration of the wounded tissue and the fact,that there is no increase in enzyme mass,it is certainly tempting to suggest wound-induced enzyme activation.However,wounding triggers variations in the isoenzyme pattern and the newly appearing variants may possess different antigenic properties to those of the antigens from resting tuber tissue (which was used as antigen source) and hence escape detection in the immunoassay (Fig 1).So the data may be interpreted as

Table 3: Changes in RNAse activity and RNAse protein content after wounding in potato tuber tissue (adapted from 100).

Time after wounding (hrs)	RNAse activity (OD 260 nm)	Mean diameter of precipitin rings (mm)
0	3.1	6.61
2	3.4	6.48
16	11.5	6.41
24	26.9	6.40
48	8.4	3.29

indication that the RNAses present in quiescent potato tuber tissue do not increase in mass,but leave it an open question,whether or not the increase in activity is due to an activation of exactly these or a de novo synthesis of other isozymes.

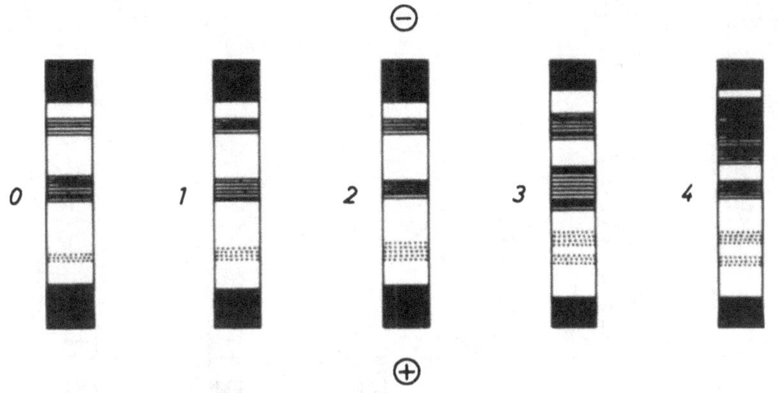

Fig.1: Wounding-induced variations of the RNAase isozyme pattern in white potato tuber tissue (Courtesy of C.Gräf,Department of Biology,University Frankfurt,GFR).

The good agreement of the immunoprecipitation and the density labeling data,however,strongly point to this enzyme as a possible candidate for future research on enzyme activation.

De novo synthesis of peroxidase isozymes (E.C.1.11.1.7).

It has long been known,that wounding of white potato and sweet potato parenchymatic tissue results in the rapid appearance of peroxidatic activity at the wound sites.Frequently a correlation between this activity and the ability of the tissue to survive a phytopathogenic attack has been constructed.Inspite of the fact,however,that up to now no natural substrates for these enzymes are known with certainty and hence no distinct physiological role can be dedicated to peroxidases,they nevertheless represent excellent markers for different processes,i.e."activation"of resting cells,cellular differentiation and others.

Resting potato tuber tissue contains some 8-11 peroxidase isoenzymes,mostly anionic forms (14,131,132).The increase in total peroxidase activity after wounding the tissue reflects the activation or synthesis of nearly all these isozymes and appearance of three new molecular variants (Fig.2).

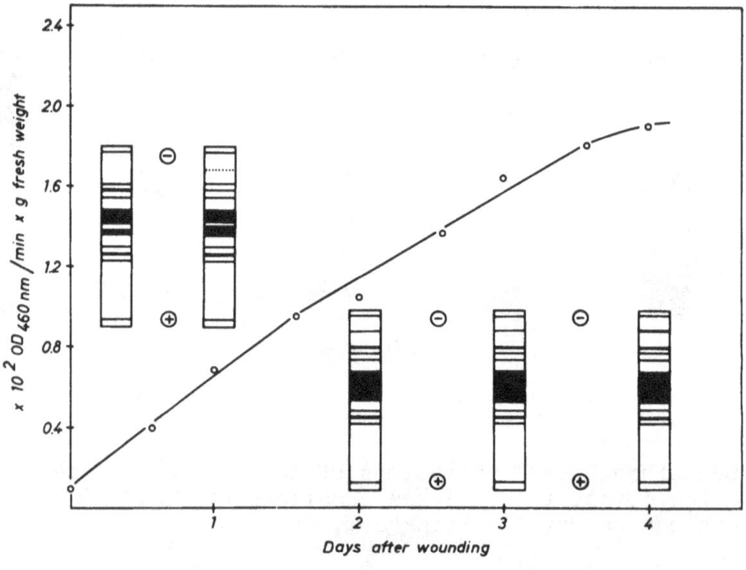

Fig.2.: Wounding-induced increase of peroxidase activity and the pattern of peroxidase isoenzymes in white potato tuber tissue, modified after 132.

The different isoperoxidases are not uniformly distributed in the wounded tissue,but instead arranged in a defined spatial pattern,indicating the read-out of different peroxidase genes in different cell types.So the suberization zone - comprising mostly the mitotically inactive,suberin synthesizing cell layer beneath the wound-surface - owns three main and three minor peroxidase isozymes,which again are in appreciably lower activity in the cell division zone together with some 8 other peroxidases.The less active tissue in the interior again exhibits another set of peroxidases,only partially identical to that of the two other tissue zones (14).This pattern seems to hold true also for wounded sweet potato tissue,where about 1 mm from the wound surface a cationic peroxidase is to be found,which seems to catalyze coniferyl-type-lignin synthesis.This particular isoenzyme is formed prior to lignin formation and degraded before completion of the lignin layer at the cut surface (125).

A rise in peroxidase activity is also characteristic for wounded as well as infected sweet potato tissue (64).This process may be stimulated pronouncedly by the application of ethylene,although this compound does not change the quality of the peroxidase pattern (43, 44,45).The rise in peroxidase activity can be prevented by various metabolic inhibitors such as Antimycin A and 2,4-DNP,puromycin,cycloheximide,canavanine,Blasticidin S and actinomycin D,although neither substance has any influence on the enzymes themselves (6,25,57). This fact already suggests tentatively the de novo synthesis of these enzymes.However,interpretation of experiments with inhibitors is difficult and presents many pitfalls to the experimenter (23).Other proof for a de novo synthesis,however,comes from incorporation studies using ^{59}Fe.This ion is integrated into b-type haems of both the soluble and microsomal fractions of sweet potato root tissue cells and this process is greatly enhanced by wounding and interpreted as slicing-induced stimulation of - at least partly - peroxidase synthesis (58).The wound-induced de novo synthesis of at least some of the peroxidase isoenzymes could be demonstrated unequivocally using ^{14}C - leucine (111).Slices of sweet potato root tissue were soaked in ^{14}C - leucine and incubated in an atmosphere containing 1 μl ethylene per liter.After extraction and separation of the three main peroxidase fractions the incorporated radioactivity and the activity of the enzymes were determined (Fig.3).As already mentioned,peroxidase acti-

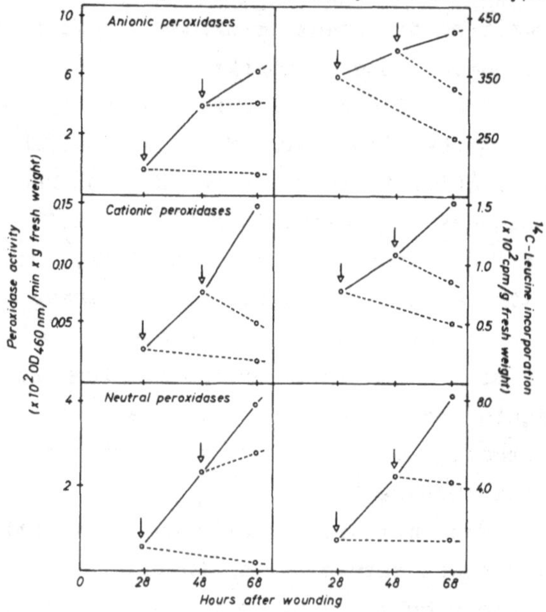

Fig.3: The effect of blasticidin S on the development of peroxidase activity (left panel) and ^{14}C-leucine incorporation into peroxidases (right panel) of wounded sweet potato root tissues. Arrows indicate the time of application of 1 µg/ml of blasticidin S for one minute (Modified after 111).

vity in the anionic,cationic and neutral fractions as well increased dramatically after wounding (left column in Fig.3).At the same time ^{14}C-leucine incorporation into all three peroxidatic fractions was enhanced (right column in Fig.3).If protein synthesis is blocked by Blasticidin S,then both the rise in peroxidase activity (dotted line, left panel) and the ^{14}C-labeling of peroxidases are shut off immediately (dotted line,right panel).This again can be demonstrated with the four major anionic peroxidases of sweet potato root tissue,separable by chromatography (57).So radioactivity of ^{14}C-leucine can be found in all four anionic species with different specific activity,probably reflecting a different extent of synthesis of the different enzymes (Fig.4).This can be reproduced by gel electrophoretic experiments,in which most - though not all - ^{14}C-leucine was found in

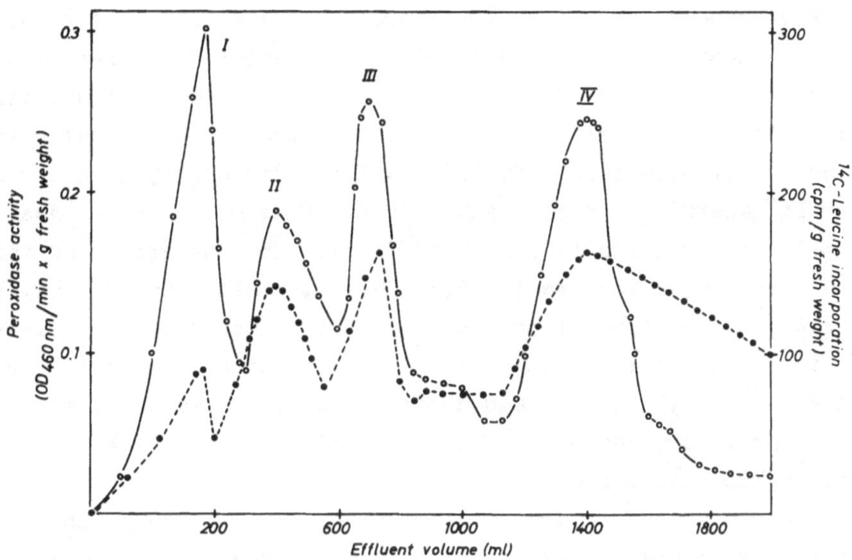

Fig.4: Separation of the major anionic peroxidases of wounded sweet potato root tissue by DEAE-cellulose chromatography and measurement of the incorporated ^{14}C-leucine label (modified after 111).

the peroxidase bands.The same holds true for neutral peroxidases.

It should be stressed at that point,that the mere demonstration of incorporated label in what is designed "peroxidase" activities is certainly insufficient to prove de novo synthesis of those enzymes. Coincidence of label together with highly purified peroxidase - purified at least to electrophoretical homogeneity as well as immunological prove of association of radioactivity with peroxidatic antigen would serve as better indicators.However,taken together with the inhibitor experiments,the present results very strongly suggest that the increase in peroxidase activity after wounding sweet potato root tissue in fact results from de novo synthesis of the enzymes.

De novo synthesis of glucose-6-phosphate dehydrogenase (E.C.1.1.1.49)

The extensive synthesis of nucleic acids of all kinds as well as coenzymes and fatty acids in wounded plant storage tissues presupposes the action of the pentose phosphate shunt (48).This pathway is initiated by two dehydrogenases,glucose-6-phosphate- and 6-phosphogluconate dehydrogenase, both of which increase in activity as a consequence of wounding. This holds true for red beet (103), sweet potato (91), chicory, swede (106, 110), white potato tissue (46) and a series of other tissues as well, i.e. Kohlrabi tuber, radish hypocotyl, carrot root and Scorzonera tissue (56). In fact, the frequently observed increase in general dehydrogenase activity after wounding as measured with tetrazolium salts is a good indication for the action of the pentose cycle dehydrogenases (34, 36, 37, 52). Usually the activity rises to a peak at the first or second day after wounding and then levels off. This pattern agrees well with $^{14}CO_2$-evolution from administered glucose-1-^{14}C (4, 5, 67, 108) and demonstrates that the shunt contributes to glucose breakdown and CO_2 output along with glycolysis and the tricarboxylic acid cycle. In white potato tuber tissue 5 isoenzymes of glucose-6-phosphate dehydrogenase (G-6-PDH) contribute to the overall activity and it seems that wounding enhances the action of all these catalysts (Fig. 5). The pattern of 6-phosphogluconate dehydrogenase

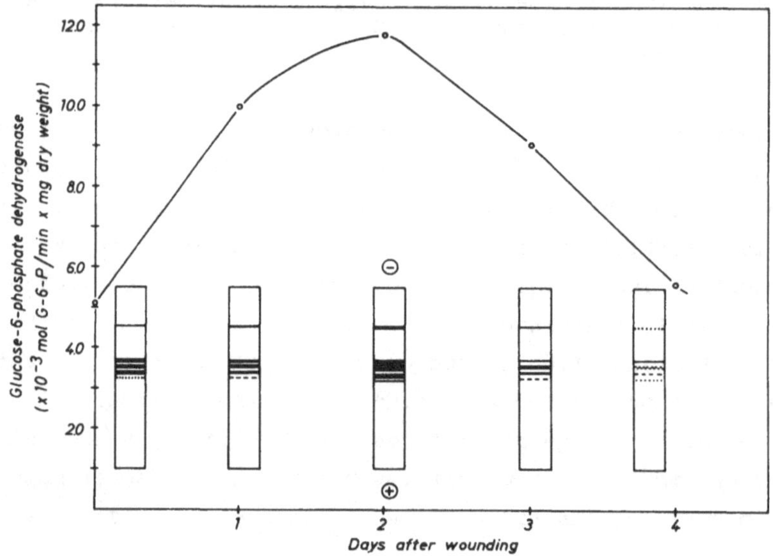

Fig. 5: Wounding-induced increase of glucose-6-phosphate dehydrogenase activity and the pattern of its isoenzymes in white potato tuber tissue.

(6-PGDH) isoenzymes does not respond as uniformly to slicing. Six of eight isoenzymes increase their activity considerably and determine the extent of overall activation of 6-PGDH,whereas two isozymes seem to be degraded at the same time.Density labeling methods have not proven beyond doubt,that the increase in activity of 6-PGDH is due to de novo synthesis as suggested by inhibitor experiments (46,50).As in potato tuber slices preferential activation of one of the tree isozymes of 6-PGDH in wounded sweet potato tuber tissue can be detected as soon as eight hours after injury (91).

Inhibitors of transcription and translation prevent the increases of both dehydrogenases after slicing potato tuber tissue (50),a rather vague indication,but by no means a proof for de novo synthesis of these enzymes.More direct proof for such an enzyme synthesis can be obtained by deuteration of newly formed enzymes and separation of labelled and unlabelled enzyme generations by isopycnic equilibrium centrifugation in CsCl.For these experiments one portion of the chopped tuber tissue was dipped in H_2O,another one in 99,8% D_2O (110) for 30 minutes and subsequently aged under comparable conditions.After this period the enzyme was extracted and layered on top of a CsCl gradient and centrifuged to equilibrium (46).Careful examination of the gradients revealed,that the buoyant density of G-6-PDH increased as a result of the treatment with D_2O (Fig.6).Since there is no reason to believe that this density shift represents the deuteration of a possible carbohydrate moiety of the enzyme (3,102),it can be concluded that G-6-PDH isoenzymes are synthesized de novo as a result of wounding the tissue.Again,it should be stressed,that the good accordance of inhibitor and density labelling experiments favors this view,but neither result alone is sufficient evidence for slicing-induced enzyme formation.

De novo synthesis of phenylalanine ammonialyase (E.C.4.3.1.5)

Wounding plant storage tissues leads to the rapid and intense accumulation of a series of phenolic compounds,of which shikimic acid, caffeic acid and chlorogenic acid are the most prominent ones (71,129). Polyphenol production proceeds up to two to three days after slicing the tissue and then decreases again (89).Two enzymes of strategic importance in phenol synthesis,phenylalanine ammonia lyase (PAL) and

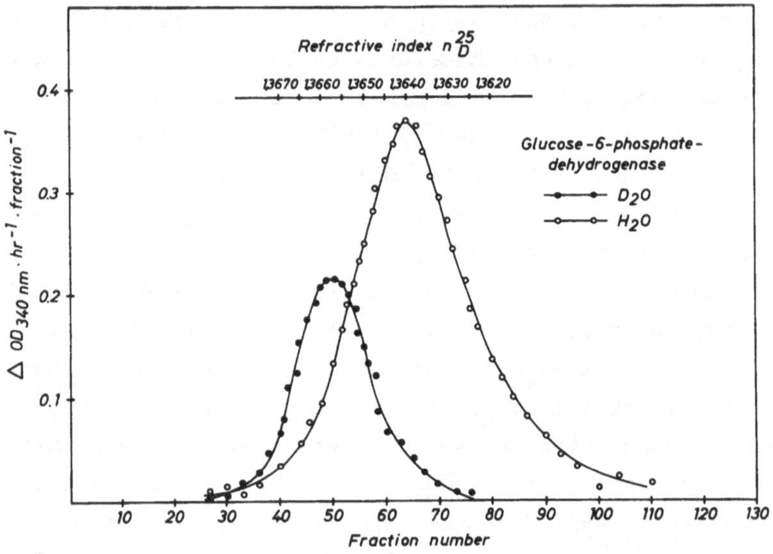

Fig.6: Equilibrium distribution in CsCl gradients of glucose-6-phosphate dehydrogenase of white potato tuber tissue after 28 hours of woundhealing in D_2O respectively H_2O (46).

trans-cinnamic acid 4-hydroxylase both are scarcely present both in intact white potato and sweet potato tuber tissue (118,137).Wounding, however,induces an increase in activity of both enzymes after a lag phase of about 3-6 hours and leads to maximal activity 1-2 days thereafter.This development is characteristic also for swede and Jerusalem artichoke slices (93,104,105).In aged sweet potato tissue two PAL isoenzymes with different Michaelis constants and activation energies exist,both sensitive to allosteric control by t-cinnamic acid and p-coumaric acid as well as caffeic acid,which again suggests a regulatory role of these compounds,in particular t-cinnamic acid (33,88,89).

The appearance of PAL activity after slicing appears to be due to de novo synthesis of the enzyme(s).In fact the most advanced and different techniques altogether point to newly occurring formation of PAL proteins in wounded tissues.

1) A whole series of different inhibitors of different metabolic steps (ethionine,benzimidazol,6-propylthiouracil,p-fluorphenylalanine,puromycin,antimycin A,2,4-DNP,actinomycin D,blasticidin S and cycloheximide;see 48) as well as metabolites (i.e.phenylalanine and cinnamic

acid,see 74,134) effectively interfere with PAL appearance.Minute amounts of cycloheximide inhibit very rapidly and irreversibly.Incubation of potato tuber disks with cycloheximide for only the first 30 minutes after wounding prevents enzyme formation just as if the tissues were exposed continously for 24 hours (see also 42).

2) Density labelling experiments convincingly demonstrate,that PAL activity increases as a consequence of de novo synthesis of the enzymes (110).

3) More direct proof for a de novo synthesis of PAL proteins after wounding sweet potato root tissue comes from immunochemical experiments (119,121).PAL from injured tissues was purified to near electrophoretical homogeneity (120) and antiserum against the purified protein was generated through immunization of a white rabbit,in turn purified and used as precipitant in quantitative immunoprecipitation tests. Antigens were prepared from intact sweet potato roots and root tissue after different times of aging.In the Ouchterlony double diffusion test there was no precipitin band formed with antiserum and antigen from resting root tissue,indicating,that PAL proteins may be totally absent or in very low concentrations only.Slicing this tissue,however, leads to the appearance of antigenic material,the concentration of which increases up to the second day of woundhealing and afterwards decreases again (119).This development can be affirmed by SDS polyacrylamide gel electrophoresis of the PAL subunits from immunoprecipitates:again the precipitated material as measured with Coomassie brilliant blue dye follows the pattern (Fig.7).These results strongly suggest that the enzyme protein is synthesized de novo as a consequence of wounding sweet potato root tissue.

This view is even more strongly supported by elegant experiments using radioactively labelled amino-acids (121).Wounded tissue was incubated together with ^{14}C-leucine and the label incorporated into PAL was assayed by measuring the radioactivity of the corresponding immunoprecipitate as well as that of the subunits after SDS-polyacrylamide gel electrophoresis.These experiments convincingly demonstrate,that during a lag period of about 3-4 hours neither PAL activity increases nor the introduction of label into both the immunoprecipitate and enzyme subunits.However,after this lag phase incorporation of label into

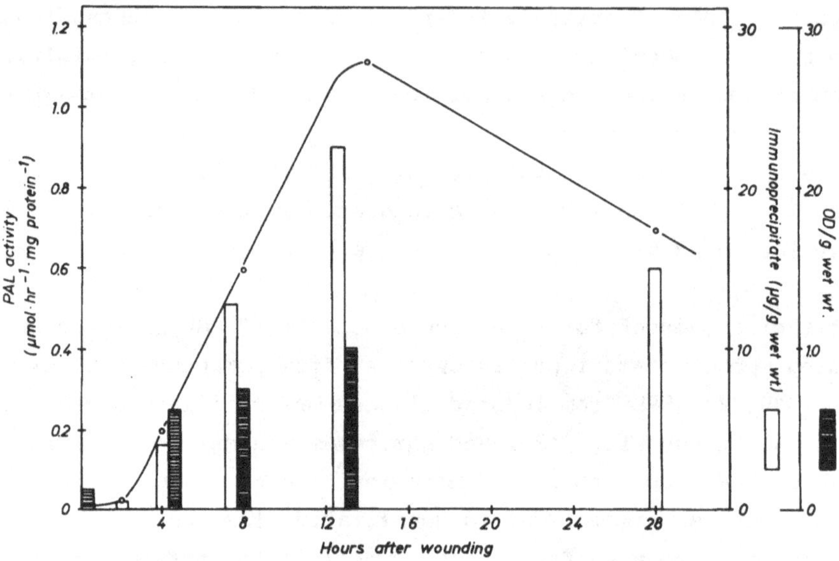

Fig.7: Wound-induced changes in PAL activity and PAL content,measured as protein content in the immunoprecipitate (open bars) or as optical density of the stained subunit bands of the enzyme from immunoprecipitates (striped bars).Modified after 119 and 121.

the subunits of the enzyme increases pronouncedly,reaches a maximum about 6 hours after wounding and subsequently is maintained at a constant level inspite of the fact that PAL activity already begins to decrease (Fig.8).

Taken altogether,these data make up the best known and most thoroughly investigated example of a true de novo synthesis of an enzyme after wounding plant storage tissues.This in turn points to the importance of transcriptional and/or translational processes set into motion after injury of resting and differentiated storage tissue cells.

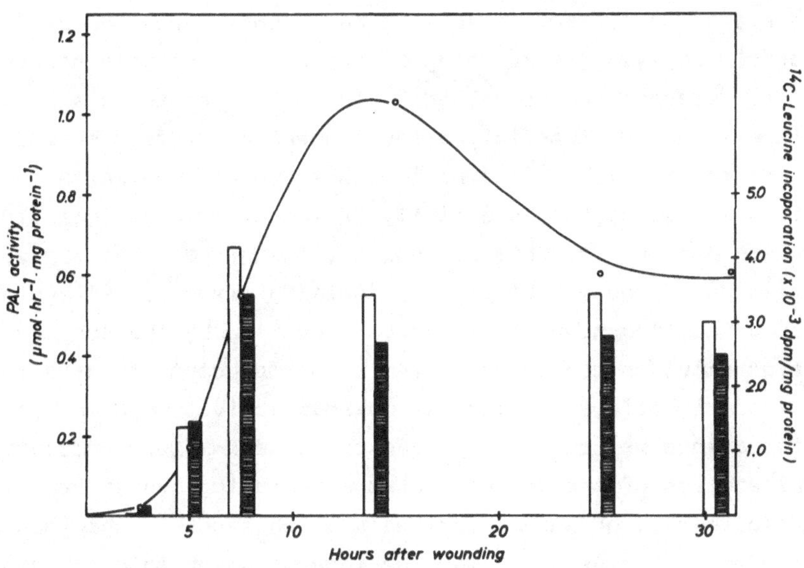

Fig.8: Wound-induced changes in PAL activity and PAL synthesis, measured as rate of incorporation of ^{14}C-leucine label into the immunoprecipitate (open bars) or into the enzyme subunit (striped bars). Modified after 121.

THE DEGRADATION OF ENZYMES IN SLICED PLANT STORAGE TISSUE.

Turnover of phenylalanine ammonia lyase (E.C.4.3.1.5)

It is obvious that the amount of a particular enzyme in a eukaryotic cell is regulated by its rate of synthesis and degradation and there is reason to believe that both processes work side by side in the cell at any time. However, their relative activity may and in some well established cases actually do vary in response to external or cellular signals. This holds true for wounded storage tissues, too. Generally, there seems to be an induction or activation of specific degrading systems after the onset of de novo syntheses of a particular enzyme. We do not know anything about the nature of these degrading principles, except that they must exist, as some experiments tentatively suggest.

As already discussed, PAL proteins are synthesized de novo after slicing sweet potato root tissue and as a consequence PAL activity rises considerably.After reaching a maximum the activity decreases again.This decrease can be halted by incubation of the tissue slices in cycloheximide,pointing to the involvement of protein synthesis in protein degradation.This is true for sweet potato and white potato tuber tissue.However,the sensitivity of the protein synthesizing machinery against the drug has appreciably changed at that time.So a concentration of the inhibitor,which totally prevents PAL-synthesis, does not affect inactivation,which is prevented by the tenfold concentration only.The degrading system in white potato tuber slices appears shortly before PAL reaches maximal activity and may well be of proteinaceous nature (135,137).As the data in Fig.8 indicate,it is an accelleration of the activity of the degrading system rather than a decrease in rate of enzyme synthesis which leads to the loss of activity from the tissue: ^{14}C-leucine incorporation into subunits does not change appreciably in times of a decrease in activity.So it may be that the PAL degrading system does not operate in intact tissue but is induced some 10-15 hours after wounding and by its appearance regulates the amount of active enzyme within the cell as suggested by late Milton Zucker (Fig.9).

Degradation of phosphoglucomutase(E.C.2.7.5.1)

In wounded white potato tuber tissue most,if not all carbon atoms - used for the synthesis of many different products - are derived from storage starch,which is rapidly degraded via glycolysis and pentose phosphate shunt.Starch corrosion is visible under the microscope about 10 hours after wounding,first in the phellogen and sub-phellogen layer and then towards the interior of the tissue (48,114).After only two days of wound-healing most of the starch in the phellogen and phellem cells is degraded and with completion of the periderm after about 8-10 days,almost no starch is visible in these cells.As a consequence of this process,all glycolytic intermediates increase in concentration (47,50).However,in contrast to the uniform accumulation of intermediates,the pattern of the involved enzymes is quite different.Some of the glycolytic enzymes are synthesized,some are de-

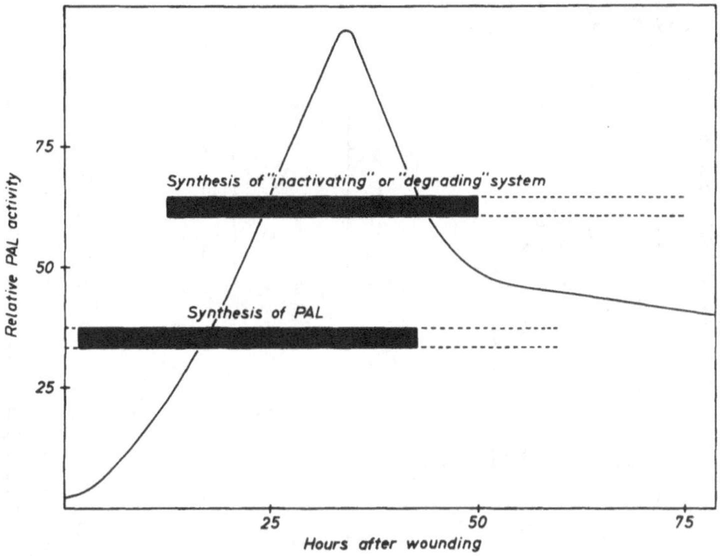

Fig.9: Schematic representation of the sequential induction of PAL and the system inactivating or degrading the enzyme.Modified after late Dr.Zucker (137).

graded or inactivated and some do not respond at all to the perturbation caused by wounding.So the enzymes in a metabolic chain acting one after the other are regulated independently from one another (47).

One of the most active catalysts in dormant potato tuber tissue, phosphoglucomutase (PGM),is rapidly degraded after wounding and the degradation process affects all three isoenzymes to a comparable extent (Fig.10;see 53).It should be stressed,that the term "degradation" in this context is used as circumscription of the disappearance of the isoenzymes and does not mean,that a degradation of the proteins has been proved.

The degradative principle is definitely different from simple phenol inactivation of the enzyme,which in turn has to be taken into account during extraction procedures,since the phenol content of the tissue rises considerably after wounding (51).As judged from inhibitor studies,the degradation process is dependent on oxidative metabolism,translation and possibly on transcription (Fig.11).

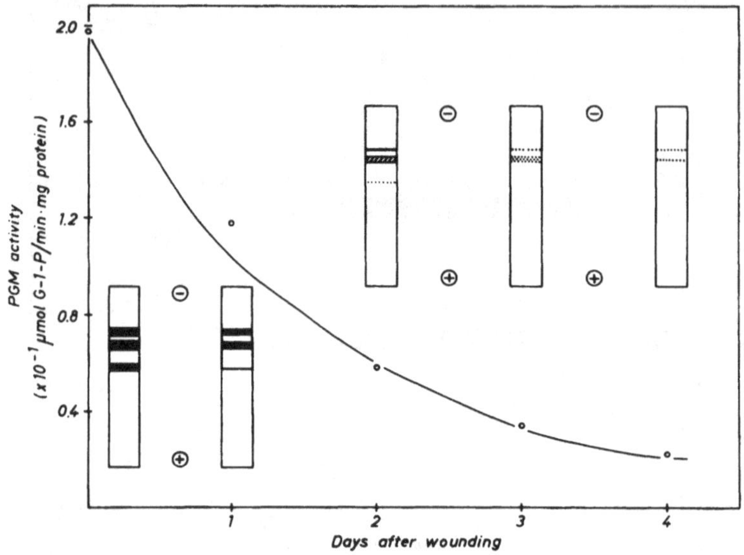

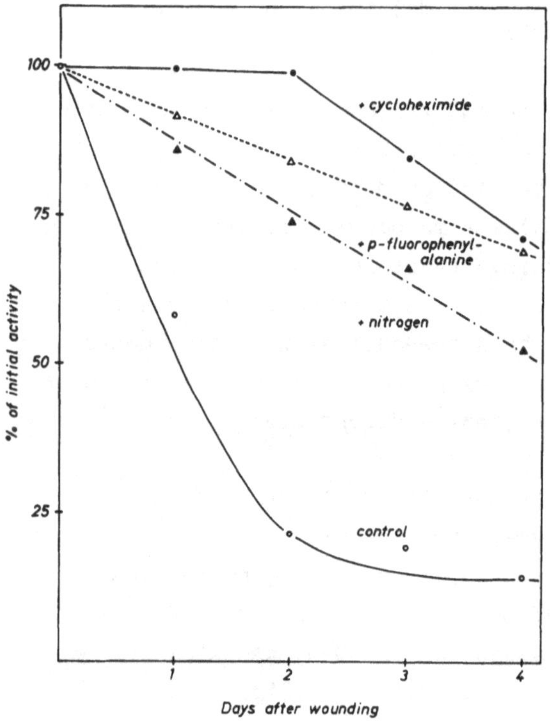

Fig. 10 (above): Wound-induced decrease in activity of white potato tuber phosphoglucomutase and the isoenzyme pattern of this enzyme.

Fig. 11 (left): Inhibition of wound-induced degradation of phosphoglucomutase in a nitrogen atmosphere and by inhibitors of protein synthesis (cycloheximide: 20 µg/ml; p-fluorophenylalanine (100 µg/ml).

In summary,the synthesis of proteins is a necessary prerequisite for PGM destruction.One is forced to think of the synthesis of either specific or unspecific proteases.Enzyme-specific proteases (i.e.19) and enzyme group-specific proteases (i.e.61-63) are well-known as degradative principles in eukaryotic cells.So,the working hypothesis had to be tested,that after wounding either specific or unspecific proteases are induced,which inactivate or degrade the PGM isoenzymes. In fact,trypsin-like and chymotrypsin-like proteases increase in activity after wounding white potato tuber tissue (56).Consequently we started to prove or disprove the following concept:in resting tuber tissue the PGM isoenzymes are physically separated from proteases or proteases are practically absent,whereas wounding of the tissue either bring together proteases and protein substrates or induces the formation of new proteases.In any event,the proteases in some way must recognize,that certain proteins are to be degraded,others not. This could be achieved by the fact,that proteins,if associated with protective effectors (i.e.substrates,cofactors),are not susceptible to proteolytic breakdown.After dissociation of the effector,the conformation changes to a highly susceptible one:the protein is recognized as substrate and degraded (see discussion).

Whatever the degradation process may be,no active fragments of PGM isoenzymes can be found using various methods (49).Consequently the molecular weight of the active enzymes remains constantly at about 59 000 $\pm$ 4 000 daltons.All the different methods,i.e.gel filtration or gel electrophoresis for all their virtues are not suitable to detect small variations in size of the protein molecules.So it may well be,that only short stretches of the peptide chain are split off, which are not active as phosphotransferase and cannot be detected.

Since proteases are prominent candidates for PGM destruction,the latter enzyme was subjected to proteolysis in vitro in order to simulate possible in vivo mechanisms.Expectedly the three PGM isoenzymes decrease in activity,when subjected to massive tryptolysis,chymotryptolysis or pronatic digestion (49).So the in vitro degradation of the purified enzyme by various proteases strikingly mimics the degradative process in vivo (Fig.12).

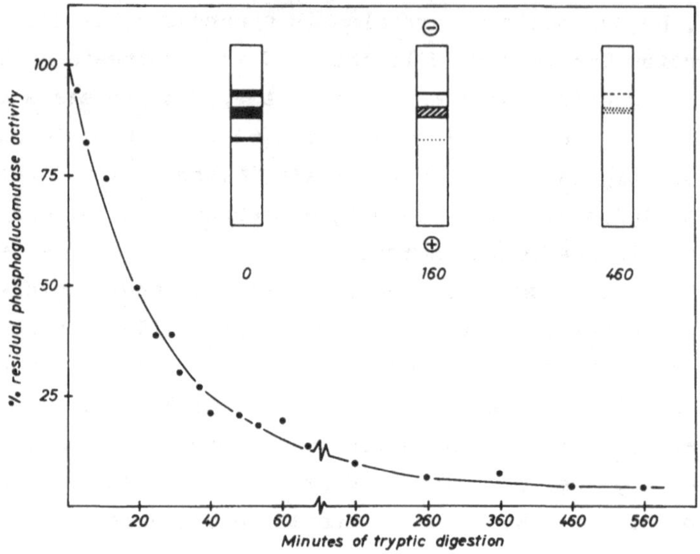

Fig. 12: Digestion of partially purified white potato phosphoglucomutase by 1.25% bovine trypsin (E.C.3.4.21.4) and electrophoretic pattern of PGM isoenzymes after limited tryptolysis.

The tryptolysis of the purified phosphoglucomutase in vitro can be largely prevented by the presence of some effectors, all glycolytic intermediates. Of a series of different substances tested, only glucose-1,6-diphosphate (the "co-substrate" of PGM), fructose-1-phosphate and fructose-1,6-diphosphate protect the enzyme from being degraded, inspite of the fact, that they activate the protease to a considerable extent.

Evidently the mutase binds the effectors, a prerequisite for this binding is at least one phosphate group attached to the fructose molecule or two phosphate residues in the glucose molecule. This attachment stabilizes the molecule in a distinct conformation, which is not susceptible or less susceptible to tryptic breakdown (Fig. 13 A). Additional evidence for this conclusion comes from denaturation experiments, in which unfolding of the mutase by urea can be prevented by the addition of, say, fructose-1,6-diphosphate (Fig. 13 B). However, a fact deserves serious consideration: the effective concentrations of the effectors range in between 10 and 70 mM, in first view certainly exceeding the intracellular concentrations. Again, it should be

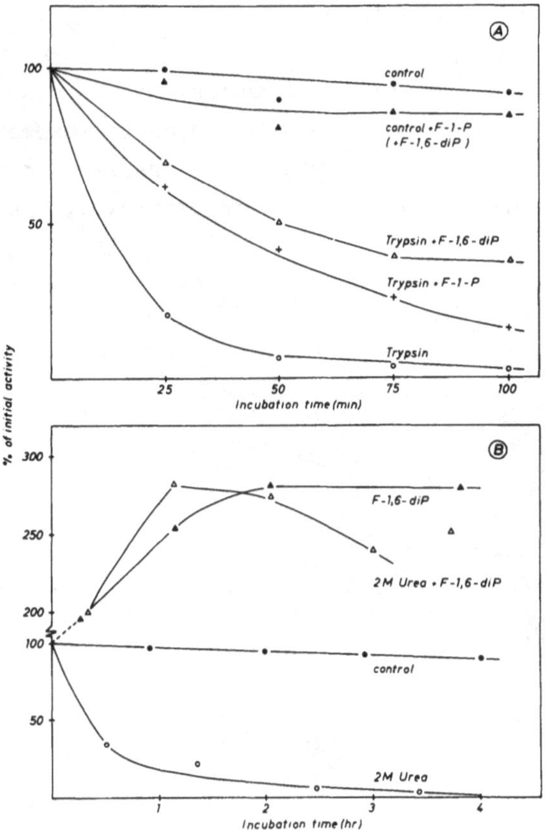

Fig. 13: Effectors of PGM delay tryptolysis and prevent denaturation of the partially purified enzyme from potato tubers.
A) Digestion of PGM by 1.25% bovine trypsin ± 15 mM fructose-1-phosphate (F-1-P) and 70 mM fructose-1,6-diphosphate (F-1,6-diP).
B) Denaturation of PGM by 2 M Urea ± 70 mM F-1,6-diP.

stressed that the in vitro conditions are highly artificial and the protease concentrations extremely high. However, intermediates may well accumulate locally to very high concentrations, by far exceeding 70 mM. Taken altogether, we are left with the preliminary conclusion that protective effectors for all three PGM isoenzymes do exist, that they prevent or delay tryptic degradation of the enzymes and that the mutase protein is available for destruction, if the effectors dissociate from the molecule.

MECHANISMS OF DEGRADATION OF ENZYMES.

Proteins are steadily synthesized and degraded in vivo,as is known for many eukaryotic systems and is confirmed by the experiments discussed in this chapter.The induction process occurs side by side with activation of some enzymes and some of its regulatory aspects are summarized in Fig.14:

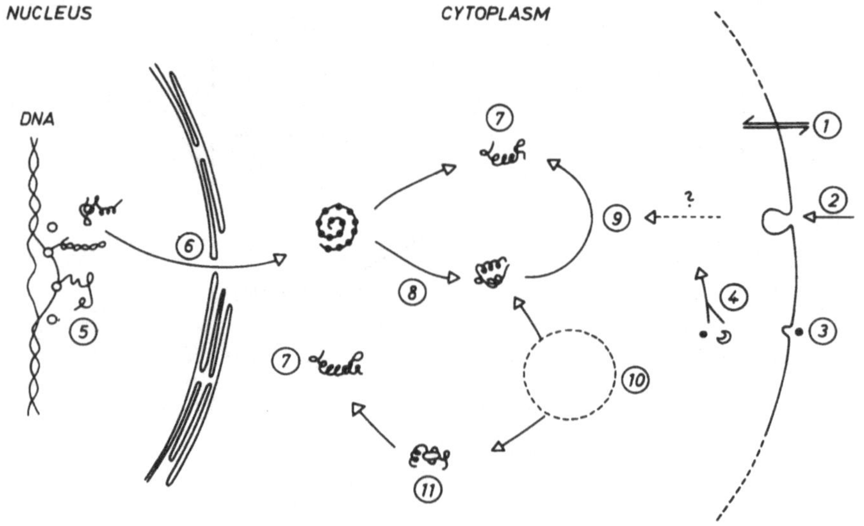

Fig.14: Induction and activation processes.Wounding induces changes in membrane permeability,which may be transmitted via symplasmic continua to cells far away from the wound site and which may facilitate extensive ion fluxes or the release of membrane bound factors (1).Products of wounded cells may be ingested pinocytotically (2) or hormones may be bound by specific receptors of the plasmamembrane (3) or cytoplasmic receptors (4).It is totally unknown, which of these factors (or others) are how transferred to the genome.However,one consequence of wounding is the activation of genes (5),the de-novo synthesis of peroxidase-,PAL- and G-6-PDH-messengers,which after their transport into the cytoplasm (6) are translated by polysomes either to yield the active enzyme (7) or an inactive precursor (8),which in turn may be activated by limited proteolysis (9).It has been shown,that lysosome-like particles are labilized after wounding in potato parenchyma cells (10),thus giving rise to secretion of proteases,which - hypothetically - may destroy inhibitor proteins (11) and consequently activate "latent" enzymes.

The degradation process obviously is specific for each protein, it concerns different proteins with different speed and extent. Now, how does the cell know, that a distinct protein should be degraded? There is no doubt, that there is not a specific degrading enzyme for each enzyme, a specific degrading enzyme for each degrading enzyme and so forth. On the contrary, the cell works with a limited repertoire of degrading enzymes (proteases), which are relatively unspecific (i.e. split the peptide bonds in the vicinity of distinct amino acid side chains) and by no means are able to discriminate between different enzymes. How then is specificity introduced into this system? There is much evidence, that it is the protein conformation, which signals the beginning of degradation. According to this concept all proteins acquire basically two conformations: one, which is not and the other, which is susceptible to proteolytic breakdown. This relationship between the structure of a protein and the speed of its degradation can be proven:

1. Abnormal proteins are more rapidly degraded. This concerns proteins, which are not readily synthesized, proteins with abnormal amino acid residues in the peptide sequence, proteins coded for by mutated gens, proteins, which denature spontaneously as well as proteins, which arise by errors in the transcriptional or translational process (28,29). It is completely unknown, how the degradative system recognizes such proteins; however, all these proteins possess certainly other conformations than their "normal" counterparts.
2. Proteins with a high molecular weight are degraded more rapidly than small molecules. Bigger protein molecules evidently are more susceptible to proteolytic attack, because they own more protease-sensitive sites. Moreover, these molecules are less stable, they denature more easily even under in vivo conditions and it seems, that bigger proteins contain errors in the peptide sequence more frequently than smaller ones (18).
3. Ligands bind to proteins, induce conformational changes and protect thereby against proteolysis. Most of the cellular proteins are by no means rigid in structure. On the contrary, they are highly flexible and easily change their overall conformation after the

attachment of small molecules (2,70). It is known, that proteins continously "breath" (i.e. they unfold partially and refold again) and that this process may be drastically influenced by ligands (22). The presence of such effectors in turn alters the sensitivity of the proteins towards proteases (16). This is certainly the case with potato PGM, which is protected by fructose-1- and fructose-1,6-diphosphate and glucose-1,6-diphosphate as well. However, there are numerous other reports with various living systems and enzymes, which altogether add credence to the concept of ligand-induced insensitivity of proteins towards proteolytic attack. Only few such examples will be mentioned. The binding of the substrate stabilizes many proteins and renders them resistant to proteolysis. Glucose protects hexokinase, substrate analogues induce resistance against carboxypeptidase in aldolase and other ligands reduce the sensitivity of serum albumin (78), aspartate transcarbamylase (83), nuclease (122) and ribonuclease (79) towards proteases. Moreover, the binding of coenzymes stabilizes the non-susceptible configuration. Serin-dehydratase and tyrosine aminotransferase are protected from proteolysis by their coenzyme (12). In fact, proteins are more stable, which bind their cofactors (coenzymes) very tightly (76,77). This fact becomes understandable in view of the existence of enzyme group-specific proteases which selectively inactivate pyridoxal- or NAD^+-dependent enzymes (61,62). Even the de-stabilizing effect of ligand binding is known: binding of oxygen by hemoglobin and of glucose by glycogen phosphorylase accelerates proteolytic breakdown (31,133).

4. <u>Proteins are not attacked by proteases, if attached to cellular structures.</u> As long as proteins are associated with other proteins, RNA, DNA or membranes, they are only poor substrates for proteases. For that reason the proteins of ribosomes (95), membranes (123) and some of the chromatin proteins (8) are relatively stable. As soon as they dissociate from the complex, however, they rapidly become substrates for proteases. The microsomal NADPH-cytochrome-C-reductase in rat liver cells is only stable if attached to the membranes (72,73). Thus, the half-life time of a particular enzyme is also dependent on the half-life of cellular complexes, i.e. membranes, ribosomes, to which it is bound.

For reasons of simplicity it may be assumed that a protein degrading system is working continuously in the living cell.A distinct protein becomes a substrate for degradation as soon as its conformation is changed from a non-susceptible to a susceptible state.This event may even occur spontaneously or may be the consequence of the dissociation of protective effectors from the protein or the detachment of the protein from membranes or other protective structures. The susceptible conformation certainly is characterized by denaturation of part of the polypeptide chain.Proteases can only bind protein substrates which in the region of the critical amino acids are unfolded,that is denatured (115).In principle this touches the question,how the denaturation process is achieved.First of all,the difference in free energy between the native conformation of a globular protein and its fully unfolded state is extraordinarily small.This means that proteins may denature rather easily and frequently within the living cell (96),this process being partly reversible ("pulsing").Then the interaction between the protein and its environment steadily changes and a sudden change of,say,hydratation may favor partial denaturation.Actually hydrophobic residues,which stabilize the interior of globular proteins,are exposed to the environment during the denaturation process,thereby changing the overall charge of the protein.Moreover,the hydrophobic forces may be responsible for the increased affinity of denatured proteins for lipids,which is of importance for the association of such protein molecules with lysosomal membranes and hence their susceptibility to proteolysis (17).

Proteolysis by unspecific, group-specific or enzyme-specific proteases.

Once a distinct enzyme is partly denatured,it becomes substrate for various proteases,which may be unspecific or may identify distinct groups of enzymes by the conformation of their coenzyme binding site or may even be specific for one particular enzyme:

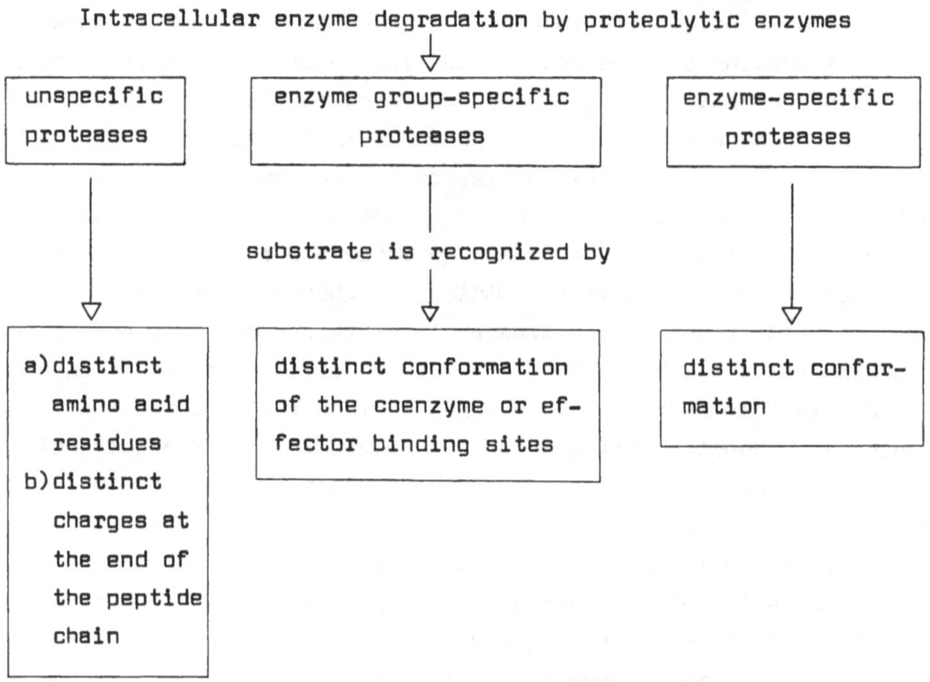

Whereas the unspecific proteases, mostly characterized as "trypsin"- or "chymotrypsin-like" proteases in plant tissues are catalysts well-known from animal organs and deserve no further comment here, the higher cell evidently has evolved group-specific proteases. These enzymes recognize the coenzyme-binding site of certain enzymes and split the apoprotein at this very site. In nearly all parts of the small intestine of rats, less in skeletal muscle, "inactivases" are present, which degrade pyridoxal-dependent enzymes lacking "their" coenzyme. This degradation is specific, since enzymes, which do not use pyridoxal phosphate for catalysis, are not attacked. Addition of the coenzyme protects the enzymes from being destroyed. The NAD^+-dependent dehydrogenases in turn possess their own, specific inactivases and again are stabilized by binding of the coenzyme. The same holds true for a series of FAD-containing enzymes (61-63).

The inactivases of the intestines of rats exist both in latent and active forms. Inactivation of the inactivases is achieved by certain inhibitors, the concentration of which decreases during vitamine, that is coenzyme deficiency.At the same time the inactivases become activated and the enzyme substrates are consequently degraded (63). Certainly each higher cell contains such group-specific proteases side by side with more enzyme-specific proteases,which at least in some cases may operate within the cell,both pro- and eukaryotic ones. As briefly discussed,insulin is split in the skeletal muscle of rats by a protease,which hydrolyzes only 9 peptide bonds.This protease is highly specific for insulin and unable to bind proinsulin (19).In higher plant tissues such enzyme-specific proteases seem to exist, although in no case conclusive proof has been presented (i.e.38,130, 136).

Protease inhibitors.

Most,if not all higher cells possess another instrument for the regulation of protein degradation,the protease inhibitors.These inhibitors,altogether proteins of comparably low molecular weight (5000 - 50 000 daltons) bind to the active center of proteases in steric complementarity,but are nevertheless not split and hence inhibit the activity of the enzymes competitively (124).Some of the protease inhibitors are polyvalent,they inhibit different proteases.

It is not known,how the dissociation of such an enzyme-inhibitor-complex is brought about.Evidently some proteolytic steps are necessary for this process,which possibly occurs in the living cell ("temporary inhibition",124).And the actual importance of the inhibitors for cellular metabolism is by no means clear.Their participation in various processes of mammals are known (blood clotting,fibrinolysis,inflammation,cancer),but their role in higher plant cells,which are rich in these inhibitor proteins,is obscure.Potato tubers contain more than 13 different protease inhibitors (9,84),which are highly active towards microbial or animal proteases,but do not show appreciable interference with plant enzymes (109).The concentration of inhibitors fluctuate during developmental processes,but again no unequivocal proof for their role in regulation of proteolysis is at hand (82). Possibly these proteins serve a storage role (109).On the other hand

it becomes more and more evident,that these inhibitors may function as protection against invading plant pathogens,which are known to secrete proteases in order to penetrate into host tissues.Actually plant organs (leaves) react upon invading microbes or insects as well as mechanical wounding with the extensive synthesis of large amounts of protease inhibitors (113).
It is obvious that the whole process of enzyme degradation in general and enzyme degradation in wounded plant storage cells in particular deserves much future work in order to understand better the importance of the various regulatory processes summarized in Fig 15:

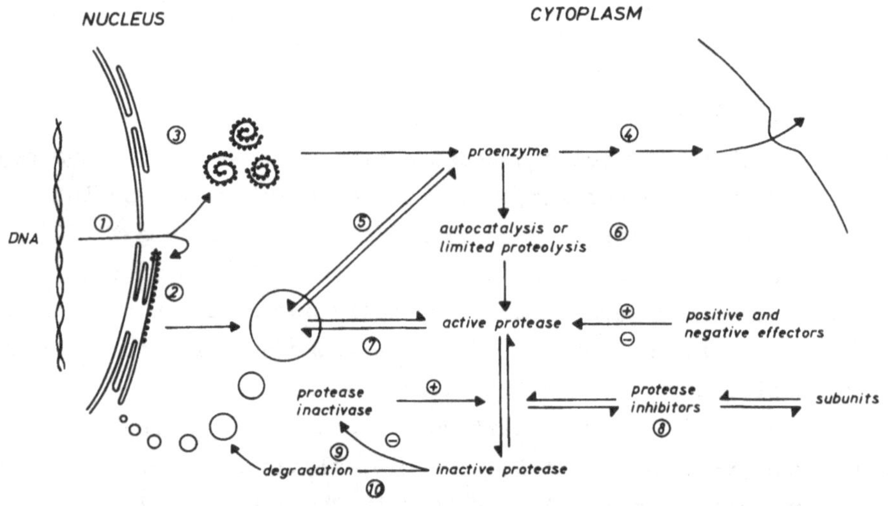

Fig.15: Regulatory factors of the intracellular protein degradation. The synthesis of mRNAs for proteases (1) may well be one of the limiting steps during the formation of an intracellular proteolytic system.However,once the mRNAs are transported into the cytoplasm and translated (2,3),the resulting enzymes may be secreted as proenzymes (4),packed into lysosomes and kept there latent (5) or be activated through autocatalysis or limited proteolysis (6).Possibly lysosomes steadily release small amounts of proteases into the cytoplasm (7;see 99,101).The activity of proteases in turn is regulated by a series of different protease inhibitors,some of which are inhibitory only,if associated in a multimeric complex (8).Protease-specific inactivases ("protease inactivase") destroy the proteases,but are themselves subject to a feed-back inhibition by inactivated proteases (9).The total degradation of the proteases (10) is followed by the re-use of the resulting amino acids for syntheses.

ACKNOWLEDGMENTS

Part of the work of the author was supported by Stiftung Volkswagenwerk (Hannover) and Deutsche Forschungsgemeinschaft (Bad Godesberg). Professor Dr.I.Uritani (Nagoya University,Japan) generously permitted to reproduce work on peroxidase and phenylalanine ammonia lyase synthesis and turnover.The assistance of Mrs.S.Kost and Mrs.C. Gräf during preparation of this manuscript is gratefully acknowledged.

REFERENCES

1. Adams,P.B., Rowan,K.S.: Glycolytic control of respiration during aging of carrot root tissue.
Plant Physiol. 45, 490-494 (1970).

2. Anfinsen,C.B.: Principles that govern the folding of protein chains.
Science 181, 223-230 (1972).

3. Anstine,W., Jacobson,J.V., Scandalios,J.G., Varner,J.E.: Deuterium oxide as a density label of peroxidases in germinating barley embryos.
Plant Physiol. 45, 148-152 (1970).

4. ApRees,T., Beevers,H.: Pathways of glucose dissimilation in carrot slices.
Plant Physiol. 35, 830-838 (1960).

5. ApRees,T., Beevers,H.: Pentose phosphate pathway as a major component of induced respiration of carrot and potato slices.
Plant Physiol. 35, 839-847 (1960).

6. Asahi,T., Majima,R.: Effect of antibiotics on biogenesis of mitochondria during aging of sliced sweet potato tissue.
Plant Cell Physiol. 10, 317-323 (1969).

7. Bacon,J.S.D., McDonald,J.R., Knight,A.H.: The development of invertase activity in slices of the root of Beta vulgaris L. washed under aseptic conditions.
Biochem. J. 94, 175-182 (1965).

8. Bartley,J., Chalkley,R.: Further studies of a thymus nucleohistone-associated protease.
J. Biol. Chem. 245, 4286-4292 (1970).

9. Belitz,H.D., Kaiser,K.P., Santarius,K.: Trypsin and α-chymotrypsin inhibitors from potatoes : isolation and some properties. Biochem. Biophys. Res. Comm. 42, 420-427 (1971).

10. Benveniste,J., Salaun,J., Durst,F.: Wounding-induced cinnamic acid hydroxylase in Jerusalem artichoke tuber. Phytochem. 16, 69-73 (1977).

11. Blobel,G., Potter,V.R.: Relation of ribonuclease and ribonuclease inhibitor to the isolation of polysomes from rat liver. Proc. Natl. Acad. Sci., USA 55, 1283-1288 (1966).

12. Bond,J.S.: A comparison of the proteolytic susceptibility of several rat liver enzymes. Biochem. Biophys. Res. Comm. 43, 333-339 (1971).

13. Bonsignore,A., De Flora,A., Mangiarotti,A.M., Lorenzoni,J., Alema,S.: A new hepatic protein inactivating glucose-6-phosphate dehydrogenase. Biochem. J. 106, 147-154 (1968).

14. Borchert,R.: Isoperoxidases as markers of the wound-induced differentiation pattern in potato tuber. Develop. Biol. 36, 391-399 (1974).

15. Cherry,J.H.: Regulation of invertase in washed sugar beet tissue. In: Biochemistry and Physiology of Plant Growth Substances, p. 417-431 (F.Wightman, G.Setterfield, eds.),Ottawa,Runge Press (1968).

16. Citri,N.: Conformational adaptability in enzymes. Adv. Enzymol. 37, 397-648 (1973).

17. Dean,R.T.: Concerning a possible mechanism for selective capture of cytoplasmic proteins by lysosomes. Biochem. Biophys. Res. Comm. 67, 604-609 (1975).

18. Dice,J.F., Dehlinger,P.J., Schimke,R.T.: Studies on the correlation between size and relative degradation rate of soluble proteins. J. Biol. Chem. 248, 4220-4228 (1973).

19. Duckworth,W.C., Heinemann,M.A., Kitabchi,A.E.: Purification of insulin-specific protease by affinity chromatography. Proc. Natl. Acad. Sci. USA 69, 3698-3702 (1972).

20. Duda,C.T., Cherry,J.H.: Chromatin- and nuclei- directed ribonucleic acid synthesis in sugar beet root. Plant Physiol. 47, 262-268 (1971).

21. Edelman,J., Hall,M.A.: Enzyme formation in higher plant tissues. Development of invertase and ascorbate-oxidase activities in mature storage tissue of Helianthus tuberosus L.
Biochem. J. 95, 403-410 (1965).

22. Englander,S.W., Downer,N.W., Teitelbaum,H.: Hydrogen exchange.
Ann. Rev. Biochem. 41, 903-924 (1972).

23. Filner,P., Wray,J.L., Varner,J.E.: Enzyme induction in higher plants.
Science 165, 358-367 (1969).

24. Flood,A.E., Rutherford,P.P., Weston,E.W.: The response of chicory root tissue disks to treatment with 2,4- and 3,5-dichlorophenoxyacetic acids.
Phytochem. 9, 2431-2437 (1970).

25. Gahagan,H.E., Holm,R.E., Abeles,F.B.: Effect of ethylene on peroxidase activity.
Physiol. Plant. 21, 1270-1279 (1968).

26. Gerhart,J.C., Schachman,H.K.: Distinct subunits for the regulation and catalytic activity of aspartate transcarbamylase.
Biochemistry 4, 1054-1062 (1965).

27. Gerhart,J.C., Schachman,H.K., Changeux,J.-P.: Allosteric interactions in aspartate transcarbamylase. I. Binding of specific ligands to the native enzyme and its isolated subunits.
Biochemistry 7, 531-538 (1968).

28. Goldberg,A.L., Dice,J.F.: Intracellular protein degradation in mammalian and bacterial cells.
Ann. Rev. Biochem. 43, 835-869 (1974).

29. Goldberg,A.L., St.John,A.C.: Intracellular protein degradation in mammalian and bacterial cells: Part 2.
Ann. Rev. Biochem. 45, 747-803 (1976).

30. Gancedo,C., Holzer,H.: Enzymatic inactivation of glutamine synthetase in Enterobacteriaceae.
Eur. J. Biochem. 4, 190-192 (1968).

31. Graves,D.J., Mann,S.A.S., Philip,G., Oliveira,R.J.: A probe into the catalytic activity and subunit assembly of glycogen phosphorylase.
J. Biol. Chem. 243, 6090-6098 (1968).

32. Gruess,J.: Abhandlungen über Enzymwirkungen.
Z. Pflanzenkrankheiten 17/18, 65 resp.202 (1907/1908).

33. Havir,E.A., Hanson,K.R.: Purification and properties of the L-phenylalanine ammonia lyase (E.C.4.3.1.5) from potato tubers. Fed. Proc. 25, 790 (1966).

34. Herrmann,J.: Atmungssysteme in der Kartoffelknolle. Die Nahrung 2, 1011-1035 (1958).

35. Holzer,H., Duntze,W.: Metabolic regulation by chemical modification of enzymes. Ann. Rev. Biochem. 40, 345-374 (1971).

36. Honda,K., Oda,Y.: Enzymatic study on browning and respiration of potato. I. Respiration of potato slice. J. Agric. Chem. Soc. Japan 29, 32-36 (1955).

37. Honda,K., Oda,Y.: Respiration and carbohydrate metabolism of potato tuber. Part I. Bull. Agr. Chem. Soc. Japan 24, 269-273 (1960).

38. Huffaker,R.C., Peterson,L.W.: Protein turnover in plants and possible means of its regulation. Ann. Rev. Plant Physiol. 25, 363-392 (1974).

39. Hyodo,H., Uritani,I.: The inhibitory effect of some antibiotics on increase in o-diphenol oxidase activity during incubation of sliced sweet potato tissue. Agr. Biol. Chem. 30, 1083-1086 (1966).

40. Hyodo,H., Uritani,I.: A study on increase in o-diphenol oxydase activity during incubation of sliced sweet potato tissue. Plant Cell Physiol. 7, 137-144 (1966).

41. Hyodo,H., Uritani,I.: Properties of polyphenol oxidases produced in sweet potato tissue after wounding. Arch. Biochem. Biophys. 122, 299-309 (1967).

42. Hyodo,H.: Wound-induced phenylalanine ammonia-lyase in potato tuber tissue. J. Biochem. 79, 277-282 (1976).

43. Imaseki,H.: Induction of peroxidase activity by ethylene in sweet potato. Plant Physiol. 46, 172-174 (1970).

44. Imaseki,H., Asaki,T., Uritani,I.: Investigations on the possible inducers of metabolic changes in injured plant tissues. In: Biochemical Regulation in Diseased Plants and Injury; T. Hirai, J.Hidaka,I.Uritani, eds. The Phytopathological Society of Japan, Tokyo, p. 189-201 (1968).

45. Imaseki,H., Uchiyama,M., Uritani,I.: Effect of ethylene on the inductive increase in metabolic activities in sliced sweet potato roots. Agr. Biol. Chem. 32, 387-389 (1968).

46. Kahl,G.: De novo synthesis of glucose-6-phosphate-(E.C.1.1.1.49) and 6-phosphogluconate dehydrogenase (E.C.1.1.1.44) in plant storage tissue slices.
Z. Naturforschung 29c, 700-704 (1974).

47. Kahl,G.: Genetic and metabolic regulation in differentiating plant storage tissue cells.
Bot. Rev. 39, 274-299 (1973).

48. Kahl,G.: Metabolism in plant storage tissue slices.
Bot. Rev. 40, 263-314 (1974).

49. Kahl,G., Gaul,E.: In vivo and in vitro degradation of white potato phosphoglucomutase (E.C.2.7.5.1).
Z. Pflanzenphysiol. 75, 217-228 (1975).

50. Kahl,G., Lange,H., Rosenstock,G.: Substratspiegel,Enzymaktivität und genetische Regulation nach Derepression in pflanzlichen Speichergeweben.
Z. f. Naturforschung 24b, 911-918 (1969).

51. Kahl,G., Müller,M.: Phenol-induced inactivation of phosphoglucomutase (E.C.2.7.5.1) in extracts from different organs of the potato plant.
Biochem. Physiol. Pflanzen 169, 281-288 (1976).

52. Kahl,G., Rosenstock,G., Lange,H.: Topographische Aspekte der Stoffwechselaktivität bei dereprimiertem Knollenparenchym der Kartoffelpflanze.
Biol. Zentralbl. 89, 765-776 (1970).

53. Kahl,G., Stegemann,H.: Enzyme degradation in higher plants: phosphoglucomutase.
FEBS-Letters 32, 325-329 (1973).

54. Kahl,G., Wechselberger,M.: Activation of chromatin-bound DNA-dependent RNA polymerase (E.C.2.7.7.6) in plant storage tissue slices.
Z. Naturforschung 32c, 229-235 (1977).

55. Kahl,G., Wielgat,B.: Regulation of transcriptional activity in wounded potato tuber tissues.
Physiol. Vég. 14, 725-738 (1976).

56. Kahl,G., unpublished.

57. Kanazawa,Y., Shichi,H., Uritani,I.: Biosynthesis of peroxidases in sliced or black rot-infected sweet potato roots.
Agr. Biol. Chem. 29, 840-847 (1965).

58. Kanazawa,Y., Asaki,T., Uritani,I.: Change in b-type haem content in relation to peroxidase biosynthesis in injured sweet potato roots.
Plant Cell Physiol. 8, 249-262 (1967).

59. Katsunuma,T., Schott,E., Elsässer,S., Holzer,H.: Purification and properties of tryptophan-synthase-inactivating enzymes from yeast.
Eur. J. Biochem. 27, 520-526 (1972).

60. Katunuma,N.: New intracellular proteases in various organs and their regulation,in: Metabolic interconversion of enzymes, Third Int. Symp., Seattle, Springer-Verlag Berlin-Heidelberg-New York, p. 313-324 (1974).

61. Katunuma,N., Kito,K., Kominami,E.: A new enzyme that specifically inactivates apo-protein of NAD-dependent dehydrogenases.
Biochem. Biophys. Res. Comm. 45, 76-81 (1971).

62. Katunuma,N., Kominama,E., Kominami,S.: A new enzyme that specifically inactivates apo-protein pyridoxal enzymes.
Biochem. Biophys. Res. Comm. 45, 70-75 (1971).

63. Katunuma,N.: Enzyme degradation and its regulation by group-specific proteases in various organs of rats.
in: Current Topics in Cellular Regulation (Horecker,B.L., Stadtman,E.R., eds.) 7, 175-203 (1973).

64. Kawashima,N., Uritani,I.: Occurrence of peroxidases in sweet potato infected by the black rot pathogen.
Agr. Biol. Chem. 27, 409-417 (1963).

65. Kingdon,H.S., Shapiro,B.M., Stadtman,E.R.: Regulation of glutamine synthetase.
VIII. ATP: glutamine synthetase adenylyltransferase,an enzyme that catalyzes alterations in the regulatory properties of glutamine synthetase.
Proc. Natl. Acad. Sci. USA 58, 1703-1710 (1967).

66. Kojima,M., Minamikawa,T., Uritani,I.: Some properties of skikimate:NADP oxidoreductase produced in sweet potato root tissue after slicing.
Plant Cell Physiol. 10, 245-257 (1969).

67. Kollattukudy,P.E., Reed,D.J.: Metabolism of red beet slices.
II. Effects of aging.
Plant Physiol. 41, 661-669 (1966).

68. Koshland,D.E.,Jr.: Conformational aspects of enzyme regulation.
in: Current Topics in Cellular Regulation 1, 1-27 (1969).

69. Koshland,D.E.Jr., Nemethy,G., Filmer,D.: Comparison of experimental binding data and theoretical models in proteins containing subunits.
Biochemistry 5, 365-385 (1966).

70. Koshland,D.E.Jr., Neet,K.E.: The catalytic and regulatory properties of enzymes.
Ann. Rev. Biochem. 37, 359-410 (1968).

71. Kosuge,T.: The role of phenolics in host response to infection.
Ann. Rev. Phytopathol. 7, 195-222 (1969).

72. Kuriyama,Y., Omura,T.: J.Biochem. (Japan) 69, 659-675 (1971).

73. Kuriyama,Y., Omura,T., Siekevitz,P., Palade,G.E.: Effects of phenobarbital on the synthesis and degradation of the protein components of rat liver microsomal membranes.
J. Biol. Chem. 244, 2017-2026 (1969).

74. Lamb,C.J., Rubery,P.H.: Phenylalanine ammonia-lyase and cinnamic acid 4-hydroxylase : product repression of the level of enzyme activity in potato tuber discs.
Planta 130, 283-290 (1976).

75. Lavintman,N., Cardini,C.E.: Changes in sucrose synthetase activities in aging potato tuber slices.
Plant Physiol. 43, 434-436 (1968).

76. Litwack,G., Rosenfeld,S.: Coenzyme dissociation,a possible determinant of short half-life of inducible enzymes in mammalian liver.
Biochem. Biophys. Res. Comm. 52, 181-188 (1973).

77. Litwack,G., Singer,S.: Biochem. Actions Horm. 2, 113-126 (1972).

78. Markus,G.: Protein substrate conformation and proteolysis.
Proc. Natl. Acad. Sci. USA 54, 253-258 (1965).

79. Markus,G., Barnard,E.A., Castellani,B.A., Saunders,D.: Ligand-induced conformational changes in ribonuclease.
J. Biol. Chem. 243, 4070-4076 (1968).

80. Matsushita,K., Uritani,I.: Change in invertase activity of sweet potato in response to wounding and purification and properties of its invertases.
Plant Physiol. 54, 60-66 (1974).

81. Matsushita,K., Uritani,I.: Effects of cycloheximide,actinomycin D and ethylene on the increase and subsequent decrease in acid invertase activity in wounded sweet potato.
Plant Cell Physiol. 16, 203-210 (1975).

82. Mayer,A.M., Shain,Y.: Control of seed germination. Ann. Rev. Plant Physiol. 25, 167-193 (1974).

83. McClintock,D.K., Markus,G.: Conformational changes in aspartate transcarbamylase. I. Proteolysis of the intact enzyme. J. Biol. Chem. 243, 2855-2862 (1968).

84. Melville,J.C., Ryan,C.A.: Chymotrypsin inhibitor I from potatoes. Large scale preparation and characterization of its subunit components. J. Biol. Chem. 247, 3445-3453 (1972).

85. Minamikawa,T., Kojima,M., Uritani,I.: Dehydroquinate hydro-lyase and skikimate:NADP oxidoreductase in sliced root of sweet potato. Arch. Biochem. Biophys. 117, 194-195 (1966).

86. Minamikawa,T., Uritani,I.: 3-deoxy-D-arabino-heptulosonic acid 7-phosphate synthase in sweet potato roots. J. Biochem. 61, 367-372 (1967).

87. Minamikawa,T., Uritani,I.: Phenylalanine ammonia lyase in sliced sweet potato roots.Effect of antibiotics on the enzyme formation and its relation to the polyphenol biosynthesis. Agr. Biol. Chem. 29, 1021-1026 (1965).

88. Minamikawa,T., Uritani,I.: Phenylalanine ammonia-lyase in sliced sweet potato roots. J. Biochem. 57, 678-688 (1965).

89. Minamikawa,T., Uritani,I.: Phenylalanine deaminase and tyrosine deaminase in sliced or black-rot infected sweet potato roots. Arch. Biochem. Biophys. 108, 573-574 (1964).

90. Monod,J., Wyman,J., Changeux,J.P.: On the nature of allosteric interactions : a plausible model. J. Mol. Biol. 12, 88-118 (1965).

91. Muto,S., Asahi,T., Uritani,I.: Increase in the dehydrogenase activities of the pentose phosphate pathway in sweet potato root tissue after slicing. Agr. Biol. Chem. 33, 176-189 (1969).

92. Muto,S., Uritani,I.: Glucose-6-phosphate dehydrogenase from sweet potato. Plant Cell Physiol. 11, 767-776 (1970).

93. Nitsch,C., Nitsch,J.P.: Effet de la lumière sur l'induction de la phénylalanine-déaminase dans les tissus de tubercule d'Helianthus tuberosus L. Compt. Rend. Acad. Sci. Paris 262, 1102-1105 (1966).

94. Oba,K., Uritani,I.: Purification and characterization of pyruvate decarboxylase from sweet potato roots.
J. Biochem. 77, 1205-1213 (1975).

95. Östner,U., Hultin,T.: The use of proteolytic enzymes in the study of ribosomal structure.
Biochim. Biophys. Acta 154, 376-387 (1968).

96. Pace,C.N.: Crit. Rev. Biochem. 1, 1-57 (1975).

97. Palmer,J.M.: The effect of some plant growth substances on the induction of enzymatic activities in thin slices of plant tubers.
In: Biochemistry and Physiology of Plant Growth Substances, p. 401-415 (F. Wightman, G.Setterfield, eds.),Ottawa,Runge Press (1968).

98. Pitt,D., Galpin,M.: Increase in ribonuclease activity following mechanical damage to leaf and tuber tissues of Solanum tuberosum L.
Planta 101, 317-332 (1971).

99. Pitt,D., Coombes,C.: The disruption of lysosome-like particles of Solanum tuberosum cells during infection by Phytophthora erythroseptica.
J. Gen. Microbiol. 53, 197-202 (1968).

100. Pitt,D.: Purification of a ribonuclease from potato tubers and its use as an antigen in the immunochemical assay of this protein following tuber damage.
Planta 101, 333-351 (1971).

101. Pitt,D., Coombes,C.: Release of hydrolytic enzymes from cytoplasmic particles of Solanum tuber tissues during infection by tuber-rotting fungi.
J. Gen. Microbiol. 56, 321-329 (1969).

102. Quail,P.H., Varner,J.E.: Combined gradient-gel electrophoresis procedures for determining buoyant densities or sedimentation coefficients of all multiple forms of an enzyme simultaneously.
Anal. Biochem. 39, 344-355 (1971).

103. Reed,D.J., Kolattukudy,P.E.: Metabolism of red beet slices. 1. Effect of washing.
Plant Physiol. 41, 653-660 (1966).

104. Rhodes,M.J.C., Wooltorton,L.S.C.: The effect of ethylene on the respiration and on the activity of phenylalanine ammonia lyase in swede and Parsnip root tissue.
Phytochem. 10, 1989-1997 (1971).

105. Rhodes,M.J.C., Wooltorton,L.S.C.: Stimulation of phenolic acid and lignin biosynthesis in swede root tissue by ethylene. Phytochem. 12, 107-118 (1973).

106. Ricardo,C.P.P., ApRees,T.: Activities of key enzymes of carbohydrate oxidation in disks of carrot storage tissue. Phytochem. 11, 623-626 (1972).

107. Ricardo,C.P.P., ApRees,T.: Invertase activity during the development of carrot roots. Phytochem. 9, 239-247 (1970).

108. Romberger,J.A., Norton,G.: Changing respiratory pathways in potato tuber slices. Plant Physiol. 36, 20-29 (1961).

109. Ryan,C.A.: Proteolytic enzymes and their inhibitors in plants. Ann. Rev. Plant Physiol. 24, 173-196 (1973).

110. Sacher,J.A., Towers,G.H.N., Davies,D.D.: Effect of light and ageing on enzymes,particularly phenylalanine ammonia lyase, in discs of storage tissue. Phytochem. 11, 2383-2391 (1972).

111. Shannon,L.M., Uritani,I., Imaseki,H.: De novo synthesis of peroxidase isozymes in sweet potato slices. Plant Physiol. 47, 493-498 (1971).

112. Shapiro,B.M., Stadtman,E.R.: 5'-Adenylyl-o-tyrosine.The novel phosphodiester residue of adenylylated glutamine synthetase from E. coli. J.Biol.Chem. 243, 3769-3771 (1968).

113. Shumway,L.K., Yang,V.V., Ryan,C.A.: Evidence for the presence of proteinase inhibitor I in vacuolar protein bodies in plant cells. Planta 129, 161-165 (1976).

114. Steward,F.C., Wright,R., Berry,W.E.: The absorption and accumulation of solutes by living plant cells. Protoplasma 16, 576-611 (1952).

115. Stroud,R.M.: A family of protein-cutting proteins. Sci. Am. 231, 74-88 (1974).

116. Suzuki,H., Oba,K., Uritani,I.: The occurrence and some properties of 3-hydroxy-3-methylglutaryl coenzyme A reductase in sweet potato roots infected by Ceratocystis fimbriata. Physiol. Plant Path. 7, 265-276 (1975).

117. Suzuki,H., Uritani,I.: Subcellular localization of 3-hydroxy-3-methylglutaryl coenzyme A reductase and other membrane-bound enzymes in sweet potato roots.
Plant Cell Physiol. 17, 691-700 (1976).

118. Tanaka,Y., Kojima,M., Uritani,I.: Properties,development and cellular localization of cinnamic acid 4-hydroxylase in cut-injured sweet potato.
Plant Cell Physiol. 15, 843-854 (1974).

119. Tanaka,Y., Uritani,I.: Immunochemical studies on fluctuation of phenylalanine ammonia-lyase activity in sweet potato in response to cut injury.
J. Biochem. 79, 217-219 (1976).

120. Tanaka,Y., Uritani,I.: Purification and properties of phenylalanine ammonia lyase in cut-injured sweet potato.
J. Biochem. (Japan) 81, 963-970 (1977).

121. Tanaka,Y.,Uritani,I.: Synthesis and turnover of phenylalanine ammonia-lyase in root tissue of sweet potato injured by cutting.
Eur. J. Biochem. 73, 255-260 (1977).

122. Taniuchi,H., Morávek,L., Anfinsen,C.B.: Ligand-induced resistance of Staphylococcal nuclease and nuclease-T to proteolysis by subtilisin,α-chymotrypsin and thermolysin.
J. Biol. Chem. 244, 4600-4606 (1969).

123. Taylor,J.M.: Drug Metab. Disposition 1, 84-115 (1973).

124. Tritz,H., Tschesche,H.: Proteinase Inhibitors.
Proc. Int. Res. Conf. Munich, Walter de Gruyter,Berlin, New York (1971).

125. Uritani,I.: Protein metabolism.
In: Encyclopedia of Plant Physiology, NS 4, p. 509-525 (1976).

126. Wulff,K., Mecke,D., Holzer,H.: Mechanism of the enzyme inactivation of glutamine synthetase from E. coli.
Biochem. Biophys. Res. Comm. 28, 740-745 (1967).

127. Willemot,C., Stumpf,P.K.: Fat metabolism in higher plants. XXXIV. Development of fatty acid synthesis as a function of protein synthesis in aging potato tuber slices.
Plant Physiol. 42, 391-397 (1967).

128. Wallace,W.: A nitrate reductase inactivating enzyme from maize root.
Plant Physiol. 52, 197-201 (1973).

129. Yoshida,S.: Biosynthesis and conversion of aromatic amino acids in plants.
Ann. Rev. Plant Physiol. 20, 41-62 (1969).

130. Zielke,H.R., Filner,P.: Synthesis and turnover of nitrate reductase induced by nitrate in cultured tobacco cells.
J. Biol. Chem. 246, 1772-1779 (1971).

131. Zimmermann,H.J., Rosenstock,G.: Proteingehalt,Proteinmuster, Peroxydase- und Malatdehydrogenase-Isoenzymmuster während der Entwicklung und Lagerung der Knollen von Solanum tuberosum L..
Biochem. Physiol. Pflanzen 169, 321-336 (1976).

132. Zimmermann,H., Rosenstock,G.: Proteingehalt,Proteinmuster,Peroxydase- und Malatdehydrogenase-Isoenzymmuster beim Speicherparenchym von Solanum tuberosum L. nach Verwundung.
Biochem. Physiol. Pflanzen 169, 487-499 (1976).

133. Zito,R., Antonini,E., Wyman,I.: The effect of oxygenation on the rate of digestion of human hemoglobins by carboxypeptidases.
J. Biol. Chem. 239, 1804-1808 (1964).

134. Zucker,M.: Induction of phenylalanine deaminase by light and its relation to chlorogenic acid synthesis in potato tuber tissue.
Plant Physiol. 40, 779-784 (1965).

135. Zucker,M.: Sequential induction of phenylalanine ammonia lyase and a lyase-inactivating system.
Fed. Proc. 26, 454 (1967).

136. Zucker,M.: Light and enzymes.
Ann. Rev. Plant Physiol. 23, 133-156 (1972).

137. Zucker,M.: Sequential induction of phenylalanine ammonia lyase and a lyase-inactivating system in potato tuber disks.
Plant Physiol. 43, 365-374 (1968).

Biogenesis of Cell Organelles in Wounded Plant Storage Tissue Cells

Tadashi Asahi

I. INTRODUCTION

Biochemical and cytological studies in the last decade have revealed that metabolic changes induced in higher plant cells in response to alterations in physiological and environmental conditions are occasionally accompanied by induced biogenesis, development or degradation of cell organelles. For instance, the rise of respiration during germination of seeds is closely related to either the structural development of mitochondrial organelles with little or no internal structure to fully developed and active mitochondria, or to the biogenesis of mitochondria, that is to an increase in the number of mitochondria (for review see 47). Extensive development of endoplasmic reticulum and Golgi apparatus has also been observed during seed germination (for review see 47). An increase in respiratory activity per cell during cell enlargement also seems to be connected with the development of the internal structure of mitochondria (44,60,65,76). A marked increase and a subsequent decrease in the rate of the conversion of lipid to sugar in germinating oil seeds are linked to the biogenesis and the degradation, respectively, of the glyoxysomes, which carry the enzymes for the glyoxylate cycle and the β-oxidation of fatty acids (for reviews see 12,19,67). Peroxisomes, which contain some enzymes involved in photorespiration and amino acid metabolism (the so called glycolate pathway), are formed in illuminated leaves in the same pattern as chloroplast development (for reviews see 12,19,67). Thus,

Biochemistry of Wounded Plant Tissues

it is a frequent occurrence in higher plant cells that metabolic regulation is connected with the biogenesis, development or degradation of cell organelles.

As mentioned in other chapters in this monograph, various metabolic changes are induced in plant storage tissues in response to wounding. Much biochemical and cytological evidence has been presented to show that some of these changes are closely related to induced biogenesis or development of cell organelles. This chapter describes biochemical studies on the biogenesis and development of cell organelles in wounded plant storage tissues with special regard to the biogenesis of mitochondria.

II. BIOGENESIS AND DEVELOPMENT OF CELL ORGANELLES IN WOUNDED PLANT STORAGE TISSUE CELLS

1. Biogenesis of Mitochondria

It is well known that when slices of plant storage tissues are aerobically aged, the respiratory activity of the slices increases markedly. Much information and many reviews on the widespread occurrence and the process of the respiratory increase are available (for reviews see 7,35,36,40,68,70). It has been emphasized in the literature that the increase is brought about by induced activation of catabolism, including glycolysis, the pentose phosphate pathway, the tricarboxylic acid cycle, and the oxidation of fatty acids, as well as of anabolism which utilizes ATP. In Figure 1 is shown the rise of respiration in wounded sweet potato root tissue. The tissue respiration increases after a lag phase of less than 8 hours, and reaches a plateau (about double the original activity) after about one day during aging of the sliced tissue. The increased respiration is resistant or insensitive to cyanide inhibition, just as it has been well established that cyanide-insensitive respiration develops in wounded plant storage tissues (for reviews see 7,23,27).

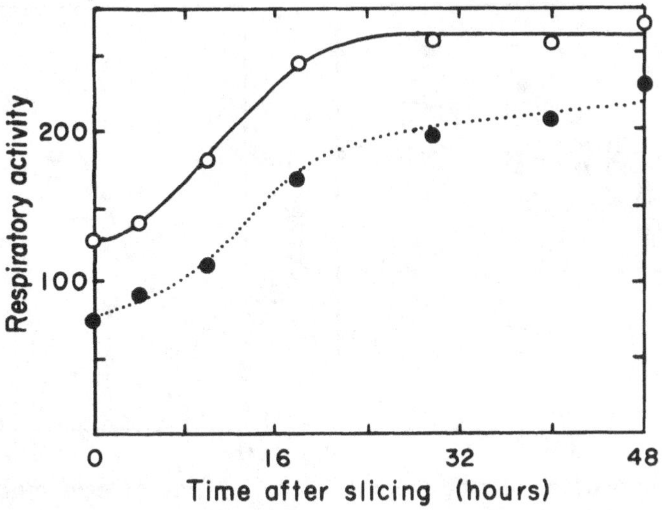

Figure 1. Respiratory increase in wounded sweet potato root tissue. Tissue slices (about 3 cm in thickness) were incubated at 30°C under moist conditions for the indicated times, then disks (1 mm in thickness and 10 mm in diameter) were taken from the second mm (layer between 1 and 2 mm) below the wounded surface and assayed for their respiratory activity. The activity is expressed as µl O_2 consumed per hour per g of tissue. o ; without cyanide. ● ; with 10^{-3}M cyanide.

The rise of tissue respiration in wounded plant storage tissues seems to be related to an induction of active biogenesis of mitochondria, namely a rapid increase in the number of mitochondria, without accompanying cell division (2,3,11, 18,41,74). In potato tuber tissue, biogenesis was shown to occur most markedly in the layers with cambial activity just below the wounded surface (73).

Mitochondria isolated from wounded plant storage tissues are much more active with respect to respiration with succinate as the substrate than those from fresh tissues on the basis of fresh tissue weight (22,52,74). Moreover, the former mitochondria show resistance to inhibition by cyanide and antimycin A, whereas the latter are highly sensitive to these inhibitors in the same manner as the slices themselves (24,27,

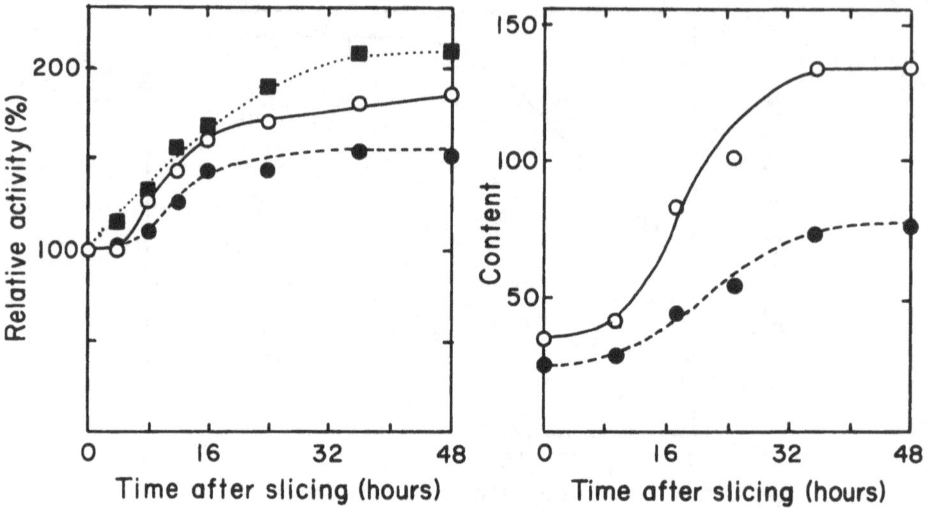

Figure 2 (Left). Increases in activities of mitochondrial enzymes in wounded sweet potato root tissue (Seto, T., Asahi, T. :unpublished data). Crude mitochondrial fraction was prepared from slices (3 mm in thickness) aged for the indicated times and was assayed for enzymatic activities. ○ ; cytochrome oxidase activity. ● ; succinate dehydrogenase activity. ■ ; succinate-cytochrome c reductase activity. The activities are expressed on a relative basis per unit fresh weight.

Figure 3 (Right). Increases in content of mitochondrial membrane-protein and -phospholipid in wounded sweet potato root tissue (51). Mitochondrial membrane fragments were prepared from slices (3 mm in thickness) aged for the indicated times as described previously (56) and were assayed for protein and phospholipid content. The contents are expressed as μg per g of tissue. ○ ; protein. ● ; phospholipid.

52). Accordingly, the situation in isolated mitochondria resembles that in tissues with respect to the activity and properties of respiration.

As shown in Figures 2 and 3, cytochrome oxidase and succinate dehydrogenase activities and mitochondrial membrane-protein and -phospholipid content in sweet potato root tissue increase after a lag phase of about 8 hours, and reach plateaus after about one day during aging of sliced tissue. It is now evident that a rapid and net synthesis of mitochondrial membrane takes place in wounded sweet potato root tissue. When crude mitochondrial fraction is assayed for the number of Janus green B-stained particles with a light microscope, the number is observed to increase in the early stages of aging of sweet potato root tissue slices (3), suggesting that the number of mitochondria increases without accompanying cell division in the wounded tissue. It seems likely that the number almost doubles during the first day after slicing (4). Consequently, it is inferred that the increase in the number of mitochondria per cell accounts for the increase in the prevalence of mitochondrial membrane in the wounded tissue.

There seems to be a correlation between the length of the lag phase and the time taken to attain a plateau, and the rise of tissue respiration and the biogenesis of mitochondria. This suggests that biogenesis of mitochondria is a factor in the rise of tissue respiration. It should be noted that the extent of the increase of tissue respiration, mitochondrial function or mitochondrial components during aging of sliced plant storage tissues seems to change as the organs are stored. However, it seems likely that there is a good correlation between the course and magnitude of the respiratory increase and mitochondrial biogenesis provided that organs stored for the same period after harvest are used for experiments.

2. Development of Endoplasmic Reticulum

The prevalence of endoplasmic reticulum also changes in wounded plant storage tissues. Ellis and MacDonald (17) reported that the crude microsomal fraction prepared from red beetroot tissue by differential centrifugation became more active in protein synthesis *in vitro* and rich in RNA content during aging of sliced tissue. Cytological observations with Jerusalem artichoke showed increases in the amount and extent of the endoplasmic reticulum without marked changes in the population of the dictyosomes and the microtubules, and a significant increase in the number of ribosomes, of which the majority seemed associated with the endoplasmic reticulum, during aging of slices (20). BenAbdelkader (8,9) has shown increases in protein and phospholipid content and antimycin A-insensitive NADH-cytochrome c reductase activity in crude microsomal fractions with an accompanying increase in polyribosomes bound to endoplasmic reticulum during aging of potato tuber tissue slices. The increases were shown to be inhibited by cycloheximide (8).

The development of endoplasmic reticulum induced in response to wounding of tissues seems rather complicated in some plant storage tissues. It was suggested from cytological studies that slicing and washing of red beetroot tissue resulted first in reduction of the lamellar endoplasmic reticulum to a number of small cytoplasmic vesicles, followed by the reconstitution of lamellae and an increase in amount of lamellae with slice aging (31). This situation was confirmed biochemically by Rungie and Wiskich (54), who showed that protein content and some reductase activities in crude microsomal fractions first decrease after slicing, and thereafter increased during aging of turnip, swede and beet storage slices.

During aging of sweet potato root tissue slices, however, microsomal NADPH-cytochrome c reductase and antimycin A-insensitive NADH-cytochrome c reductase activities increase without initial decline and reach maxima after 2 days (66). Thereafter, antimycin A-insensitive NADH-cytochrome c reductase activity begins to decrease (66). Interestingly, NADPH-cytochrome c reductase activity in the crude microsomal fraction begins to decrease after 3 days with a concomitant increase in this activity in the supernatant fraction (66). Therefore, it is likely that in sweet potato root tissue, the endoplasmic reticulum first develops, then degrades during aging of sliced tissue.

Thus, it is concluded that endoplasmic reticulum, especially rough-surfaced endoplasmic reticulum, develops during aging of plant storage tissue slices, at least in some stages. The development must be closely related to an activation of protein synthesis in wounded plant storage tissues. It may also be connected with induction of polyphenol and lipid syntheses in response to wounding of tissues. However, plant microsomal fractions prepared by differential centrifugation contain many kinds of membranes other than endoplasmic reticulum, e.g. mitochondrial membranes (broken mitochondria), plasma membranes, Golgi membranes, and tonoplast membranes. Therefore, detailed biochemical studies with membrane fractions purified by density gradient centrifugation and other techniques are required for elucidating how endoplasmic reticulum develops in wounded plant storage tissues.

3. Other Cell Organelles

Little biochemical information is available as to whether or not the population of cell organelles other than mitochondria and endoplasmic reticulum changes in wounded plant storage tissues. First, there are biochemical data suggesting that the Golgi apparatus develops in wounded plant storage tissues. The synthesis of extensin, a hydroxyproline-rich glycoprotein present in the cell walls, is greatly enhanced during aging of sliced carrot storage tissue (13). Since the synthesis takes place in the Golgi apparatus (21), it is probable that the organelle develops in the wounded tissue. However, there is no direct evidence for the prospect that the Golgi apparatus increases in wounded plant storage tissues, probably because of difficulties in isolating Golgi apparatus in a pure state as well as in a good yield from plant storage tissues.

Secondly, it seems that the population of microbodies increases in wounded plant storage tissues. Huang and Beevers (25,26) have isolated and characterized microbodies from potato tuber and carrot root tissues. The microbodies differ from either peroxisomes or glyoxysomes with respect to their enzyme constituents, and are designated nonspecialized microbodies. Their biochemical function is still unknown. It has been shown that catalase activity, of which a part is recovered in particulate fractions, increases in wounded sweet potato root tissue (37). It seems likely that the increase occurs the second day after slicing (37), and accordingly lags behind the biogenesis of mitochondria and the development of endoplasmic reticulum. Moreover, when a crude mitochondrial fraction from sweet potato root tissue is centrifuged on a linear Urografin density gradient, a protein peak carrying no cytochrome oxidase is found at a position denser than the mitochondrial peak in the case of wounded tissue, but not in the case of fresh tissue (57), suggesting that the microbodies develop in wounded sweet potato root tissue. Recently, Van der Plas et al. (69) reported that pelletable nitrogen almost doubled during the aging of potato tuber tissue slices for one day without increase in cytochrome oxidase activity. They proposed that part of the increase might be due to the development of peroxisomes.

Some cytological evidence is available as to changes in the morphology of nuclei in wounded plant storage tissues. In particular, remarkable morphological changes have been observed in the nucleoli, such as the formation of granules and vacuoles, which have been proposed to be related to active synthesis of ribosomal RNA (6,20,32,33,35,53). So far, however, there is no biochemical work on nuclei of wounded plant storage tissues in connection with their morphological changes.

Jackman and Steveninck (31) proposed that the initial decline in lamellar endoplasmic reticulum in wounded red beetroot tissue was connected with production of lysosomal organelles, but there is no biochemical evidence for their proposal.

III. BIOCHEMICAL STUDIES ON THE BIOGENESIS OF MITOCHONDRIA IN WOUNDED PLANT STORAGE TISSUE CELLS

1. Biochemical Mechanism of Mitochondrial Biogenesis in Wounded Plant Tissue Cells

An increase in the content of mitochondrial protein during aging of sliced sweet potato root tissue was shown to be inhibited by either cycloheximide or chloramphenicol, when the antibiotics were applied to slices just after slicing (56). The result indicates that both mitochondrial (chloramphenicol sensitive) and cytoplasmic (cycloheximide sensitive) protein-synthesizing systems cooperate in the biogenesis of mitochondria. The requirement for the cooperation of both protein-synthesizing systems for mitochondrial biogenesis has been well established with yeast and *Neurospora crassa* (for reviews see 5,58). It was found, however, that it was hard to show an effect of chloramphenicol on the increase in the content of mitochondrial protein in wounded sweet potato root tissue, when it was applied to slices which had been aged for about 10 hours (56). On the other hand, cycloheximide strongly inhibits the increase even when applied to slices pre-aged for about 10 hours (56). Therefore, it is proposed that the mitochondrial protein-synthesizing system is required only in the early stages of aging for the biogenesis of mitochondria. On the other hand, the synthesis of mitochondrial protein on cytoplasmic ribosomes is much more important in the biogenesis of mitochondria in the later stages of aging. This situation is rather unique, since the need for the simultaneous cooperation of both protein-synthesizing systems for mitochondrial biogenesis has been emphasized with microorganisms and animal tissues (for reviews see 5,58).

As seen in Figure 2, succinate-cytochrome *c* reductase activity appears to increase without a lag phase during aging of sliced sweet potato root tissue, whereas cytochrome oxidase and succinate dehydrogenase activities increase after a lag phase. Some other differences in the course of mitochondrial enzyme activities with time have been observed during aging of plant storage tissue slices (55,75).

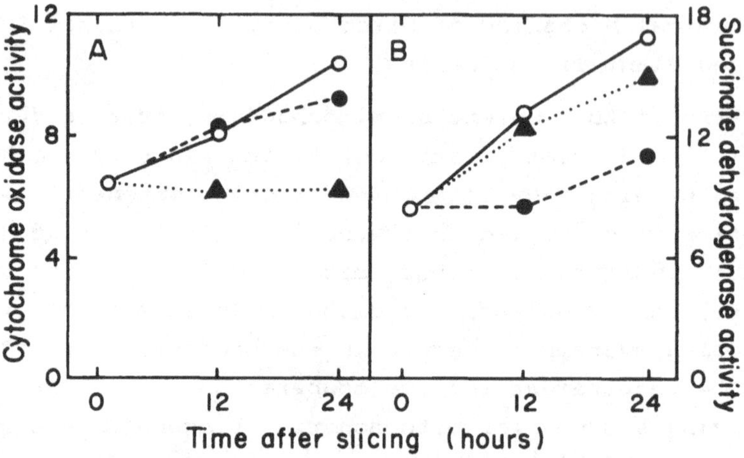

Figure 4. Effects of cycloheximide and chloramphenicol on increases in cytochrome oxidase and succinate dehydrogenase activities in wounded sweet potato root tissue (Seto, T., Maejima, M., Asahi, T. :unpublished data). Slices (3 mm in thickness and 18 mm in diameter) were incubated in water (○), 10^{-5} M cycloheximide (●) or 6 x 10^{-3} M chloramphenicol (▲) at room temperature for 1 hour, then under moist conditions. A; cytochrome oxidase activity expressed as decrease in absorbance of reduced cytochrome c at 550 nm per minute per g of tissue. B; succinate dehydrogenase activity expressed as decrease in absorbance of 2,6-dichlorophenolindophenol at 600 nm per minute per g of tissue.

Figure 4 shows the effects of cycloheximide and chloramphenicol on increases in cytochrome oxidase and succinate dehydrogenase activities in wounded sweet potato root tissue when applied to the slices just after slicing. The increase in succinate dehydrogenase activity is inhibited by cycloheximide but not by chloramphenicol. This is consistent with a generally accepted proposal that succinate dehydrogenase protein is synthesized on cytoplasmic ribosomes (45,46,72). Under the same conditions, the increase in cytochrome oxidase activity is inhibited by chloramphenicol but not by cycloheximide. Much biochemical evidence with yeast, <u>Neurospora crassa</u> (for review see 58) and animal tissues (39) has been presented which indicates that cytochrome oxidase is composed of six or seven subunits, some of which are synthesized on mitochondrial ribosomes and the others on cytoplasmic ribosomes. Therefore, it seems likely that the increase in cytochrome oxidase activity in wounded sweet potato root tissue is achieved by assembly of cytoplasmically synthesized subunits preexisting in intact root tissue with subunits newly synthesized on the mitochondrial ribosomes after wounding of the tissue. This possibility is now under examination in my laboratory.

In conclusion, there is something about the mechanism of mitochondrial biogenesis in wounded plant storage tissues, which distinguishes it from the mechanism in other organisms. Detailed biochemical studies on the distinctive features of biogenesis of mitochondria in wounded plant storage tissues are expected to find new information on the mechanism of the biogenesis of mitochondria. Such studies would help to elucidate how mitochondrial and cytoplasmic protein-synthesizing systems interact with each other in the course of mitochondrial biogenesis, and how membrane-bound enzymes are assembled in the membrane.

2. Differences in Biochemical Properties between Mitochondria from Fresh and Wounded Plant Storage Tissue Cells

Studies with sucrose density gradient centrifugation of the crude mitochondrial fraction have suggested that mitochondria from wounded plant storage tissues are different in their density and/or size from those from intact or fresh tissues (51,52). Moreover, as seen in Figure 3, the relative content of phospholipid to protein in the mitochondrial membrane becomes low as sliced sweet potato root tissue is aged. Therefore, mitochondria in wounded plant storage tissues seem quite to differ from those in fresh tissues.

Protein in mitochondrial membranes from wounded sweet potato root tissue was found to be relatively insoluble in solutions of deoxycholate and cholate compared with that from fresh tissue (Figure 5). The solubility of the protein seems to

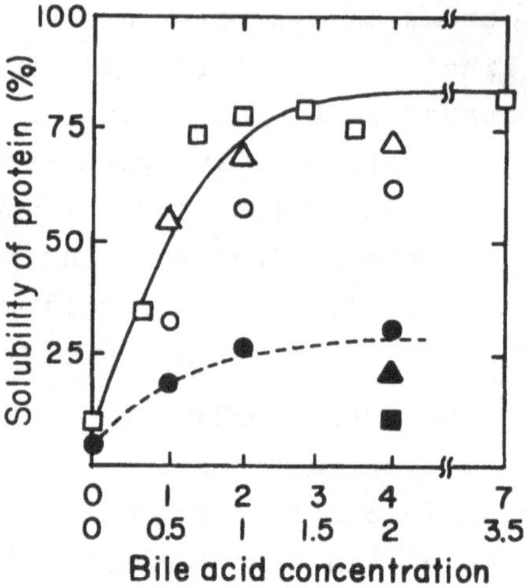

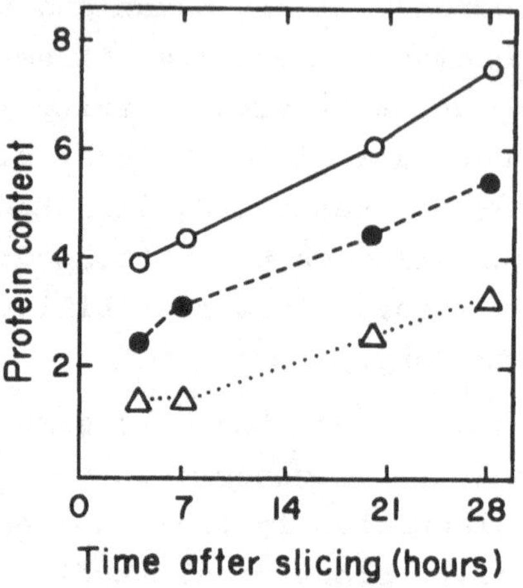

Figure 5 (Left). Solubilization of mitochondrial membrane-protein from sweet potato root tissue by deoxycholate and cholate (51). The upper and lower numerals on the abscissa indicate the concentrations of deoxycholate and cholate, respectively, in mg per mg of protein. Open symbols; membrane suspensions from fresh tissue at the concentrations of 1.8 (□), 2.9 (○), and 3.7 (△) mg of protein per ml. Solid symbols; membrane suspensions from slices aged for 2 days at the concentrations of 2.0 (■), 2.9 (●), and 3.3 (▲) mg of protein per ml.

Figure 6 (Right). Increase in mitochondrial membrane-protein in wounded sweet potato root tissue (56). ○ ; whole mitochondrial protein. ● ; mitochondrial membrane-protein. △; mitochondrial membrane-protein insoluble in deoxycholate (4 mg per mg protein) and cholate (2 mg per mg protein) solution. Protein contents are expressed as mg of protein per 50 g of tissue.

depend upon the relative amount of detergent to protein, and not upon the absolute concentrations. Detergent-insoluble, but not detergent-soluble protein, in mitochondrial membrane increases during aging of sliced tissue (Figure 6). Experiments with ^{14}C-leucine showed that the amino acid was actively incorporated into detergent-insoluble mitochondrial membrane-protein but not into detergent-soluble protein in wounded sweet potato root tissue (56). On the other hand, about 80% of mitochondrial membrane-phospholipid is solubilized by deoxycholate and cholate in the cases of both fresh and wounded sweet potato root tissues (Table I). Whatever concentrations of the detergents are used, the relative proportion of phospholipid to protein in both detergent-soluble and -insoluble fractions is the same as that in the untreated membrane in the case of fresh tissue. This is consistent with a proposal that the detergents reduce membranes to stable lipoprotein complexes (64). However, in the case of wounded tissue, the protein and phospholipid are not coequally solubilized, but rather the phospholipid is selectively solubilized, while the protein remains insoluble. Sodium dodecylsulfate-polyacrylamide gel electrophoresis of mitochondrial membrane-protein solubilized by heating the membrane in the presence of sodium dodecylsulfate and mercaptoethanol shows no significant differences in polypeptide composition between mitochondrial membranes from fresh and wounded tissues (51). There is also no difference in sugar content between them (51). So far, the only difference detected between them is the relative proportion of

Table I. Solubilization of Mitochondrial Membrane from Sweet Potato Root Tissue with Deoxycholate and Cholate (51)

Concentration of detergents[a]	Detergent-soluble			Detergent-insoluble		
	Protein (μg/g tissue)	PL[b] (μg/g tissue)	PL/ protein (μg/μg)	Protein (μg/g tissue)	PL[b] (μg/g tissue)	PL/ protein (μg/μg)
Fresh tissue						
Untreated membrane	----	----	----	28.7	12.8	0.45
0.75 ; 0.325	20.4	8.3	0.41	9.1	4.2	0.46
1.00 ; 0.50	23.5	10.0	0.43	7.0	2.8	0.40
4.00 ; 2.00	25.2	13.1	0.52	4.6	1.1	0.24
Wounded tissue						
Untreated membrane	----	----	----	75.8	24.6	0.32
1.00 ; 0.50	20.3	14.8	0.73	50.1	7.8	0.16
4.00 ; 2.00	26.3	21.9	0.82	42.8	2.2	0.05

[a]The concentrations of deoxycholate and cholate are shown by the numerals before and after the semicolons, respectively.
[b]PL, phospholipid.

phospholipid to protein. Therefore, it is proposed that the insolubility of protein in mitochondrial membrane from wounded tissue in detergent solution is caused by a low phospholipid content and not by a difference in protein composition from that of membrane from fresh tissue. The lowering of phospholipid content in membrane may result in replacement of the interaction between protein and phospholipid by that between proteins, which may bring about an insolubility of the protein in detergent solutions. Recently, Arrigoni et al. (1) reported that mitochondrial protein from wounded Jersalem artichoke tuber tissue contained much more hydroxyproline than that from the fresh tissue. Therefore, there is a possibility that a high content of hydroxyproline in mitochondrial membrane-protein causes the insolubility of the protein in detergent solutions. However, it is doubtful whether hydroxyproline content in mitochondrial membrane-protein increases in wounded sweet potato root tissue or not, because of no significant differences in polypeptide composition between mitochondrial membranes from fresh and wounded tissues.

When mitochondrial membrane fragments are centrifuged to equilibrium on a sucrose density gradient, two bands of membrane fragments are obtained from wounded sweet potato root tissue but only one band from the fresh tissue (Figure 7). The lighter band from the wounded tissue is identical to the single band from the fresh tissue with respect to density, the proportion of phospholipid to protein, and the solubility of the protein in deoxycholate and cholate solution (Table II). The yield of the lighter band from the wounded tissue is about the same as that of the single band from the fresh tissue on the basis of tissue fresh weight. Therefore, it is very likely that the lighter band represents mitochondrial membrane pre-existing in the fresh tissue. The denser band from the wounded tissue is quite different from the other band. The band contains a very low amount of phospholipid relative to protein, and a large amount of protein insoluble in

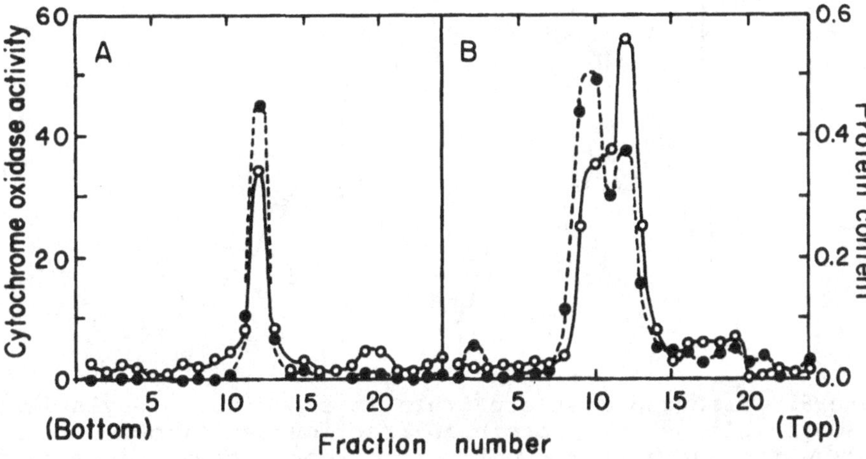

Figure 7. Distribution of cytochrome oxidase activity and protein after equilibrium density centrifugation of disrupted mitochondrial membrane from sweet potato root tissue (51). The membrane fragments from 85 g of fresh tissue (A) or from slices aged for 2 days (B) were centrifuged on a sucrose density gradient (32.5-65%, w/v) until equilibrium density was reached. Then the gradient was divided into 0.5 ml-fractions. O ; cytochrome oxidase activity expressed as decrease in absorbance of reduced cytochrome c at 550 nm per minute per fraction. ● ;protein content in mg per fraction.

Table II. Properties of Fractionated Mitochondrial Membrane Fragments from Sweet Potato Root Tissue (51)

Membrane fragments	Density (g/cm^3)	Solubility of protein in deoxycholate-cholate solution (%)	Phospholipid content (mg/mg protein)
Band from fresh tissue	1.185	70	0.51
Lighter band from wounded tissue	1.185	71	0.51
Denser band from wounded tissue	1.207	21	0.21

deoxycholate and cholate solution (Table II). However, there is a good correspondence of polypeptides between the lighter and denser bands in scans of sodium dodecylsulfate-polyacrylamide gel electorphoretic patterns (Figure 8). It is now

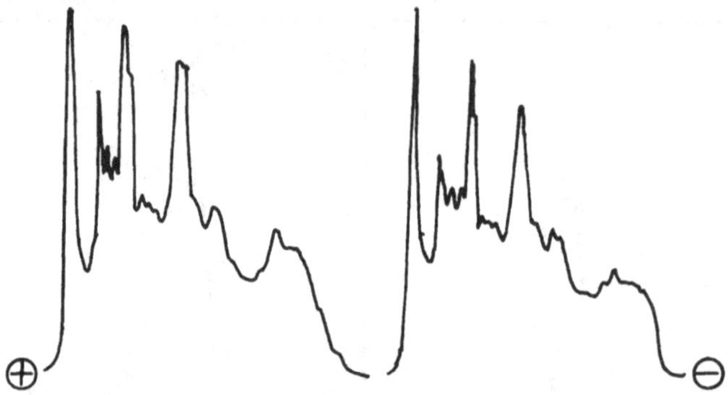

Figure 8. Electrophoretic pattern in a sodium dodecylsulfate-polyacrylamide gel of protein in fractionated mitochondrial membrane fragments from sweet potato root tissue slices aged for 2 days (51). The membrane fragments were solubilized by heating in the presence of sodium dodecylsulfate and mercapto-ethanol, then subjected to the electrophoresis. The gels were stained with Coomassie brilliant blue and scanned at 570 nm with a Joyce-Loebl Chromatoscan. Left; solubilized protein of the lighter band. Right; solubilized protein of the denser band.

evident that whether or not mitochondrial membrane-protein is soluble in detergent solution does not depend upon differences in protein composition. Rather, the differences observed between mitochondrial membranes from fresh and wounded tissues are due to the presence of the phospholipid depleted denser mitochondrial membrane fraction in wounded tissue.

Thus, it is likely that the denser band is mitochondrial membrane formed newly in response to wounding of the tissue. That is, a phospholipid-poor mitochondrial membrane is formed in wounded sweet potato root tissue. The synthesis of phospholipid has been shown to be activated in plant storage tissues in response to wounding (8,10). Nevertheless, the synthesis may not be sufficiently rapid to form normal mitochondrial membrane. It has been established that the major intracellular site of phospholipid synthesis in plant cells is the endoplasmic reticulum, as in animal cells (38,42, 43,49,50) and that phospholipid synthesized in the endoplasmic reticulum is transferred to mitochondria through specific carrier proteins (34,48). Therefore, it is probable that in wounded plant storage tissues, newly synthesized phospholipid is preferentially utilized for the development of the endoplasmic reticulum, and consequently the supply of phospholipid to the mitochondria is limited.

3. Relationship between the Biogenesis of Mitochondria and the Development of Cyanide-Insensitive Respiration in Wounded Plant Storage Tissue Cells

As shown in Table III, the reduction of cytochrome c with NADH

Table III. Oxidation of NADH in the Presence of Cytochrome c and Sodium Azide by Mitochondrial Membrane Fragments from Sweet Potato Root Tissue (51)

Membrane fragments	Oxidation of NADH[a] (μmoles/min/mg protein)			Reduction of cytochrome c (μmoles/min/mg protein)		
	Control	+Ant. A[b]	Inhibition (%)	Control	+Ant. A[b]	Inhibition (%)
Membrane from fresh tissue[c]	142	34	76	231	47	80
Membrane from wounded tissue[c]	107	67	37	191	30	84
Membrane from fresh tissue[d]	27.2	1.3	95	50.4	0.9	98
Lighter membrane from wounded tissue[d]	43.4	2.0	95	80.0	1.4	98
Denser membrane from wounded tissue[d]	20.1	12.1	40	17.4	1.2	93

[a]The amount of NADH oxidized was calculated from the decrease in absorbance at 340 nm (for details see 51).
[b]Ant. A, antimycin A.
[c]The membrane fragments before fractionation by equilibrium density centrifugation.
[d]The membrane fragments after fractionation by equilibrium density centrifugation.

by all fractions after equilibrium density centrifugation of mitochondrial membrane fragments from sweet potato root tissue is almost completely inhibited by antimycin A, indicating that the bands are of the mitochondrial inner membrane and free from either the mitochondrial outer membrane or the microsomes. The oxidation of NADH in the presence of cytochrome c and sodium azide by the membrane fraction from fresh tissue and the lighter membrane fraction from the wounded tissue is also sensitive to inhibition by antimycin A. However, the oxidation by the denser membrane fraction from wounded tissue is highly resistant to inhibition. The result indicates that a cyanide-insensitive respiratory chain is operating in the denser membrane fragments from wounded tissue but not in the other membrane fragments.

Preliminary experiments in my laboratory have shown that when phospholipid is supplied *in vitro* to mitochondrial membrane fragments from wounded sweet potato root tissue, electron flow through the cyanide-insensitive pathway is greatly depressed. As shown in Table IV, when the membrane fragments (which contain both the lighter and denser membrane fragments) were pre-incubated with soybean phospholipid, their NADH oxidase activity in the presence of sodium azide and antimycin A is markedly repressed. On the other hand, the activity in the presence of sodium salicylhydroxamate, which is known to be a specific inhibitor for cyanide-insensitive respiration (59), is not affected by the supplement of phospholipid. Moreover, when mitochondrial membrane fragments from wounded tissue are pre-incubated with lecithin micelles, the oxidation of NADH in the presence of cytochrome c and sodium azide by the membrane fragments becomes more sensitive to inhibition by antimycin A (Figure 9). The rate of oxidation in the absence of antimycin A is not affected by the pre-incubation. Therefore, it is likely that electron flow through the cyanide-insensitive pathway becomes slow as the mitochondrial inner membrane become rich in phospholipid.

Table IV. Properties of NADH Oxidase of Mitochondrial Membrane Fragments from Wounded Sweet Potato Root Tissue (Otsuka, K., Iwatsuki, N., Asahi, T.: unpublished data)

Mitochondrial membrane fragments	NADH oxidase activity		
	Control[a]	SHAM[b]	NaN_3+ant.A[c]
	(nmoles NADH oxidized/minute/tube)		
Untreated fragments	4.50	1.63	2.35
Fragments pre-incubated			
without PL[d]	3.42	0.83	2.44
with PL[d]	2.24	0.78	1.08

[a]None of inhibitors was present in the assay mixture.
[b]Sodium salicylhydroxamate was added to the assay mixture at a final concentration of 2 mM.
[c]NaN_3 and antimycin A were added to the assay mixture at final concentrations of 2 mM and 0.15 μM, respectively.
[d]Mitochondrial membrane fragments were pre-incubated in a buffer in the presence or absence of soybean phospholipid (PL) at 0°C for 1 hour before assay.

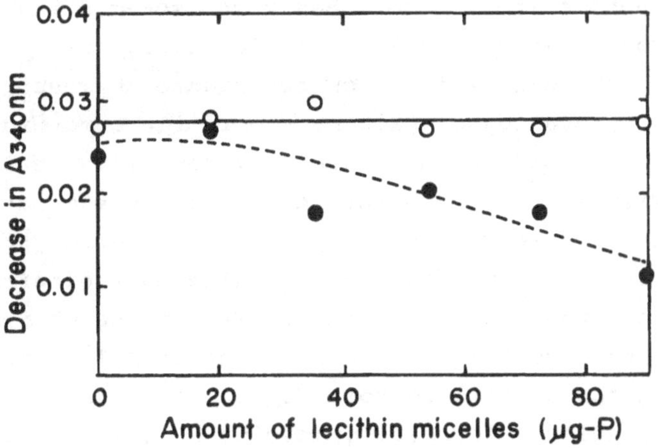

Figure 9. Effect of lecithin micelles on the sensitivity of NADH oxidation in the presence of cytochrome c to antimycin A (Otsuka, K., Iwatsuki, N., Asahi, T.: unpublished data). Mitochondrial membrane fragments (27 μg of protein) from sweet potato root tissue slices aged for 2 days were pre-incubated with the indicated amounts of lecithin micelles, and were then assayed for decrease in absorbance at 340 nm in an assay system containing NADH, cytochrome c, and sodium azide in the presence or absence of antimycin A. ○ ; without antimycin A. ● ; with antimycin A.

It is now inferred that the development of cyanide-insensitive respiration in wounded plant storage tissues is attributable to the formation of phospholipid-poor mitochondrial inner membrane. The cyanide-insensitive respiration has been observed in aged plant tissues such as older leaves, and in rather particular plant tissues such as petals and spadices (for a review see 7). It may be reasonable to postulate that in such tissues too, the rate of phospholipid synthesis is not rapid enough to form normal mitochondrial membranes during turnover of mitochondrial membranes.

Many investigators have focused their attention on the electron transport pathway of the cyanide-insensitive respiration. According to recent knowledge, a flavoprotein and a non-heme iron protein are involved in the alternate electron transport pathway, which is believed to branch from a flavoprotein or coenzyme Q. But neither the flavoprotein nor the non-heme iron protein has been isolated and characterized (for a review see 27). It is likely that skunk cabbage mitochondria which show resistance to respiratory inhibition by cyanide contain much more flavoprotein than do mung bean mitochondria which are very sensitive to inhibition (for a review see 27). Therefore, it may be proposed that the sensitivity of respiration to cyanide is inversely related to the prevalence of components of the alternate electron transport pathway. However, there seems to be no difference in either the cytochrome content of the normal respiratory chain or in the overall ferrosulfoprotein content (which presumably includes ferrosulfoproteins of the alternate path) between mitochondria from fresh and wounded plant storage tissues (14,15). Further, as previously indicated, there are no significant differences in polypeptide composition between the lighter and denser mitochondrial inner membranes from wounded sweet potato root tissue (Figure 8). Moreover, cycloheximide, but not chloramphenicol, prevents the development of the cyanide-insensitive respiration in wounded potato tuber tissue (16). In fact, chloramphenicol induces cyanide-insensitive respiration in rice coleoptiles (71) and *Neurospora crassa* (18). These phenomena are quite hard to interpret if one presupposes that the cyanide-insensitive respiration develops in response to the active synthesis of proteins involved in the alternate electron transport pathway.

It has been proposed in recent papers (16,51,61,62,63) that the cyanide-insensitive respiration may develop when there is an alteration of the structure or composition of components other than the proteins involved in the alternate pathway of the mitochondrial inner membrane. The author would propose that whether or not the cyanide-insensitive respiration occurs in cells may be dominated by the phospholipid level in the mitochondrial inner membrane rather than by the presence or absence of the proteins involved in the alternate pathway. The strengthening of the interaction among proteins in the mitochondrial inner membrane, or the lessening of membrane fluidity because of a low content of phospholipid may cause the operation of the cyanide-insensitive pathway, which is unfavorable with respect to the efficiency of oxidative phosphorylation.

IV. MECHANISM OF INDUCTION OF BIOGENESIS AND DEVELOPMENT OF CELL ORGANELLES IN WOUNDED PLANT STORAGE TISSUE CELLS

The findings that the biogenesis and development of cell organelles are induced in plant storage tissues in response to wounding raise the question of how wounding leads to the induction of biogenesis and development. Many investigators have turned their attention to a similar question with respect to the induced synthesis of many cytosol enzymes in wounded plant storage tissues. The problem of organelle induction and biogenesis is very difficult to separate from the latter problem. However, clues must be sought if there is to be an understanding of the control mechanisms of cell organelle formation.

Ethylene has been shown to stimulate the respiratory rise in wounded sweet potato root tissue (28), which in turn produces ethylene (30). Solomos and Laties (62) reported an induction of a sharp increase in respiration of intact potato tubers by treating them with either ethylene or cyanide. They proposed that the increase was due to the activation or alteration of the link between the cyanide-insensitive and -sensitive electron transport pathways. According to preliminary experiments in my laboratory, treatment of sliced sweet potato root tissue with ethylene results in a further enhancement of the increase in cytochrome oxidase activity during aging of the slices (Asahi, T., Majima, R., Uritani, I.: unpublished data). However, the increase in the number of mitochondrial particles per slice during aging is not enhanced by the treatment. Therefore, it seems that the development of the internal structure of mitochondria, but not the increase in the number of mitochondria, in response to wounding is enhanced by ethylene. A similar phenomenon has been observed with sliced sweet potato root tissue infected with Ceratocystis fimbriata (3), in which ethylene is actively produced (29). Accordingly, it is proposed that an enhancement of the respiratory increase in wounded sweet potato root tissue by infection with fungi is related to the development of the internal structure of the mitochondria, which is engendered by ethylene.

Thus, ethylene seems to act as a trigger for the development of the inner mitochondrial membrane, but not for the biogenesis of mitochondria in wounded plant storage tissues. Further biochemical studies are required for elucidatin how ethylene affects on the formation of cell organelles.

V. CONCLUSION

There is no doubt that drastic changes in the intracellular structure are induced in plant storage tissues in response to wounding. Biochemical evidence has been presented to show that the number and/or the internal structure of cell organelles including mitochondria and endoplasmic reticulum increase in wounded plant storage tissues. The population of Golgi bodies and microbodies also seems to increase in these tissues. The changes seem closely related to certain metabolic changes, including the rise of tissue respiration and the activation of protein synthesis in the cells. The author would emphasize that in general, much more attention should be directed to the changes in the number and the internal structure of cell organelles in connection with metabolic control in higher plant cells.

It is evident that a marked net synthesis of mitochondrial and endoplasmic reticulum membranes occurs without accompanying cell division in wounded plant storage tissues. This leads us to use these tissues as a tool for biochemical studies on the mechanism of membrane biogenesis. It is rarely in organisms other than higher plants and perhaps in yeast that a net increase in the mitochondrial inner membrane takes place without an increase in cell number. Moreover, chloramphenicol and cycloheximide have unique effects upon increases in some mitochondrial components and enzyme activities in wounded plant storage tissues. Therefore, biochemical studies on events in wounded plant storage tissues seem to have the advantage of providing knowledge on the formation of the mitochondrial inner membrane. It is very likely that an increase in the number of mitochondria accounts for the net increase in the mitochondrial inner membrane in wounded plant storage tissues. It is most uncommon for the number of mitochondria to increase without accompanying cell division. Therefore, the tissues should be used for studies on the mechanism and the control of mitochondrial biogenesis.

The mitochondrial inner membrane from wounded plant storage tissues quite differs in biochemical and biophysical properties from that from fresh tissues. Detailed studies on the differences have led us to the proposition that the cyanide-insensitive respiration develops when a phospholipid-poor mitochondrial inner membrane is being formed. Probably, the occurrence of the alternate pathway in older or particular plant tissues is due to a limited supply of phospholipid to mitochondria during turnover of the inner membrane. Although more evidence is needed for the proposal, it should be noted that this is the first proposition pertaining to the physiological meaning of the cyanide-insensitive respiration. Detailed biochemical studies on the biogenesis and development of cell organelles in wounded plant storage tissues will provide new concepts in the field of plant physiology as well as new information on the mechanism of membrane biogenesis.

ACKNOWLEDGEMENT: The author expresses his thanks to Professor G. G. Laties for his editorial help in the preparation of the manuscript.

REFERENCES

1. Arrigoni, O., DeSantis, A., Arrigoni-Liso, R., Calabrese, G.: The increase of hydroxyproline-containing proteins in Jerusalem artichoke mitochondria during the development of cyanide-insensitive respiration. Biochem. Biophys. Res. Commun. 74, 1637-1641 (1977).
2. Asahi, T., Honda, Y., Uritani, I.: Increase of mitochondrial content in sweet potato root tissue after wounding. Arch. Biochem. Biophys. 113, 498-499 (1966).
3. Asahi, T., Honda, Y., Uritani, I.: Increase of mitochondrial fraction in sweet potato root tissue after wounding or infection with Ceratocystis fimbriata. Plant Physiol. 41, 1179-1184 (1966).
4. Asahi, T., Majima, R.: Effect of antibiotics on biogenesis of mitochondria during aging of sliced sweet potato root tissue. Plant & Cell Physiol. 10, 317-323 (1969).
5. Ashwell, M., Work, T.S.: The biogenesis of mitochondria. Ann. Rev. Biochem. 39, 251-290 (1970)
6. Barckhausen, R., Rosenstock, G.: Feinstrukturelle Untersuchungen zur Vorkommen von "Riesenvakuolen" in Nucleoli verletzer Parenchymzellen aus der Knolle von Solanum tuberosum L. Biochem. Physiol. Pflanz. 164, 327-337 (1973)
7. Beevers, H.: "Respiratory Mechanisms in Plants". Harper Row Co., Evanston, Ill., 1961.
8. BenAbdelkader, A.: Influence de la "survie" (ageing) sur la biosynthèse des phospholipides dans le "microsomes" de tubercule de Pomme de terre (Solanum tuberosum L.). C. R.

Acad. Sci. Paris, Ser. D. 268, 2406-2409 (1969).
9. BenAbdelkader, A., Auderset, G.: Formation de polyribosomes dans des tranches de tubercules de Pomme de terre (Solanum tuberosum L.) maintenues en "Survie". C. R. Acad Sci. Paris, Ser. D. 274, 1311-1314 (1972).
10. BenAbdelkader, A., Mazliak, P.: Infuence de la "survie" (ageing) sur la biosynthèse des phospholipides dans les cellules entières ou les mitochondries du tubercule de Pomme de terre (Solanum tuberosum L.). C. R. Acad. Sci. Paris, Ser. D. 267, 609-612 (1968).
11. BenAbdelkader, A., Mazliak, P., Castersson, A.M.: Biogenese des lipides mitochondriaux au cours de la "survie" (ageing) de disques de parenchyme de tubercule de Pomme de terre. Phytochemistry 8, 1121-1133 (1969).
12. Breidenbach, R.W.: Microbodies. In "Plant Biochemistry", 3rd Ed., ed. J. Bonner and J.E. Varner, p91-114. Academic Press, New York, N. Y., 1976.
13. Chrispeels, M.J., Sadava, D., Cho, Y.P.: Enhancement of extensin biosynthesis in ageing disks of carrot storage tissue. J. Exp. Bot. 25, 1157-1166 (1974).
14. Dizengremel, P.: La voie d'oxydation insensible au cyanure dans les mitochondries de tranches de Pomme de terre (Solanum tuberosum L.) maintenues en survie. Physiol. Vég. 13, 39-54 (1975).
15. Dizengremel, P., Chauveau, M., Lance, C.: Résistance au cyanure et teneur en ferrosulfoprotéines des mitochondries végétales. C. R. Acad. Sci. Paris, Ser. D. 277, 239-242 (1973).
16. Dizengremel, P., Lance, C.: Control of changes in mitochondrial activities during aging of potato slices. Plant Physiol. 58, 147-151 (1976).
17. Ellis, R.J., MacDonald, I.R.: Activation of protein synthesis by microsomes from aging beet disks. Plant Physiol. 42, 1297-1302 (1967).
18. Edwards, D.L., Rosenberg, E., Maroney, P.A.: Induction of cyanide-insensitive respiration in Neurospora crassa. J. Biol. Chem. 249, 3551-3556 (1974).
19. Frederick, S.E., Gruber, R.J., Newcomb, E.H.: Plant microbodies. Protoplasma 84, 1-29 (1975).
20. Fowke, L.C., Setterfield, G.: Cytological responses in Jerusalem artichoke tuber slices during aging and subsequent auxin treatment. In "Biochemistry and Physiology of Plant Growth Substances", ed. F. Wightman and G. Setterfield, p581-602. The Runge Press Ltd., Ottawa, Canada, 1968.
21. Gardiner, M., Chrispeels, M.J.: Involvement of the Golgi apparatus in the synthesis and secretion of hydroxyproline-rich cell wall glycoproteins. Plant Physiol. 55, 536-541 (1975).
22. Greksak, M., Asahi, T., Uritani, I.: Increase in mitochondrial activity in diseased sweet potato root tissue. Plant & Cell Physiol. 13, 1117-1121 (1972).
23. Hackett, D.P.: Respiratory mechanisms and control in

higher plant tissues. In "Control Mechanism in Respiration and Fermentation", ed. B. Wright, p105-127. The Ronald Press Co., New York, N. Y., 1963.
24. Hackett, D.P., Haas, D.W., Griffiths, S.K., Niederpruem, D.J.: Studies on the development of cyanide resistant respiration in potato tuber tissue. Plant Physiol. 35, 8-19 (1960).
25. Huang, A.H.C., Beevers, H.: Isolation of microbodies from plant tissues. Plant Physiol. 48, 637-641 (1971).
26. Huang, A.H.C., Beevers, H.: Localization of enzymes within microbodies. J. Cell Biol. 58, 379-389 (1973).
27. Ikuma, H.: Electron transport in plant respiration. Ann. Rev. Plant Physiol. 23, 419-436 (1972).
28. Imaseki, H., Asahi, T., Uritani, I.: Investigations on the possible inducers of metabolic changes in injured plant tissues. In "Biochemical Regulation in Diseased Plants or Injury", ed. T. Hirai, J. Hidaka and I. Uritani, p189-201. Phytopathol. Soc. Japan, Tokyo, 1968.
29. Imaseki, H., Teranishi, T., Uritani, I.: Production of ethylene by sweet potato roots infected by black rot fungus. Plant & Cell Physiol. 9, 769-781 (1968).
30. Imaseki, H., Uritani, I., Stahmann, M.A.: Production of ethylene by injured sweet potato root tissue. Plant & Cell Physiol. 9, 757-768 (1968).
31. Jackman, M.E., Van Steveninck, R.F.M.: Changes in the endoplasmic reticulum of beetroot slices during aging. Aust. J. Biol. Sci. 20, 1063-1068 (1967).
32. Jordan, E.G., Chapman, J.M.: Ultrastructural changes in the nucleoli of Jerusalem artichoke (Helianthus tuberosus) tuber discs. J. Exp. Bot. 22, 672-634 (1971).
33. Jordan, E.G., Chapman, J.M.: Nucleolar and nuclear envelope ultrastructure in relation to cell activity in discs of carrot root (Daucus carota L.). J. Exp. Bot. 24, 197-209 (1973).
34. Kader, J.C.: Proteins and the intracellular exchange of lipids. I. Stimulation of phospholipid exchange between mitochondria and microsomal fractions by proteins isolated from potato tuber. Biochim. Biophys. Acta 380, 31-44 (1975).
35. Kahl, G.: Genetic and metabolic regulation in differentiating plant storage tissue cells. Bot. Rev. 39, 274-299 (1973).
36. Kahl, G.: Metabolism in plant storage tissue slices. Bot. Rev. 40, 263-314 (1974).
37. Kanazawa, Y., Asahi, T., Uritani, I.: Changes in b-type haem content in relation to peroxidase biosynthesis in injured sweet potato roots. Plant & Cell Physiol. 8, 249-262 (1967).
38. Kates, M., Marshall, M.O.: Biosynthesis of phosphoglycerides in plants. In "Recent Advances in the Chemistry and Biochemistry of Plant Lipids", ed. T. Galliard and E.I. Mercer, p115-201. Academic Press, London, 1975.
39. Koch, G.: Synthesis of the mitochondrial inner membrane

in cultured Xenopus laevis oocytes. J. Biol. Chem. 251, 6097-6107 (1976).
40. Laties, G.G.: Control of respiratory quality and magnitude during development. In "Control Mechanisms in Respiration and Fermentation", ed. B. Wright, p129-155. The Ronald Press Co., New York, N.Y., 1963.
41. Lee, S.G., Chasson, R.M.: Aging and mitochondrial development in potato tuber tissue. Physiol. Plant. 19, 199-206 (1966).
42. Lord, J.M., Kagawa, T., Beevers, H.: Intracellular distribution of enzymes of the cytidine diphosphate choline pathway in castor bean endosperm. Proc. Nat. Acad. Sci. U. S. A. 69, 2429-2432 (1972).
43. Lord, J.M., Kagawa, T., Moore, T.S., Beevers, H.: Endoplasmic reticulum as the site of lecithin formation in castor bean endosperm. J. Cell Biol. 57, 659-667 (1973).
44. Lund, H.A., Vatter, A.E., Hanson, J.B.: Biochemical and cytological changes accompanying growth and differentiation in the root of Zea mays. J. Biophys. Biochem. Cytol. 4, 87-98 (1958).
45. Mahler, H.R., Perlman, P.S.: Mitochondriogenesis analyzed by blocks on mitochondrial translation and transcription. Biochemistry 10, 2979-2990 (1971).
46. Mahler, H.R., Perlman, P.S., Mehratra, B.D.: Mitochondrial specification of the respiratory chain. In "Autonomy and Biogenesis of Mitochondria and Chloroplasts", ed. N.K. Boardman, A.W. Linnane and R.M. Smillie, p492-511. North-Holland Publishing Co., Amsterdam, 1971.
47. Mayer, A.M., Shain, Y.: Control of seed germination. Ann. Rev. Plant Physiol. 25, 167-193 (1974).
48. Mazliak, P., Douady, D., Demandre, C., Kader, J.C.: Exchange processes between organelles involved in membrane lipid biosynthesis. In "Recent Advances in the Chemistry and Biochemistry of Plant Lipids", ed. T. Galliard and E.I. Mercer, p301-318. Academic Press, London, 1975.
49. Moore, T.S., Lord, J.M., Kagawa, T., Beevers, H.: Enzymes of phospholipid metabolism in the endoplasmic reticulum of castor bean endosperm. Plant Physiol. 52, 50-53 (1973).
50. Moore, T.S.: Phosphatidylcholine synthesis in castor bean endosperm. Plant Physiol. 57, 383-386 (1976).
51. Nakamura, K., Asahi, T.: Changes in properties of the inner mitochondrial membrane during mitochondrial biogenesis in aging sweet potato tissue slices in relation to the development of cyanide-insensitive respiration. Arch. Biochem. Biophys. 174, 393-401 (1976).
52. Nakano, M., Asahi, T.: Biochemical studies on mitochondria formed during aging of sliced potato tuber tissue. Plant & Cell Physiol. 11, 499-502 (1970).
53. Rose, R.J., Setterfield, R.J., Fowke, L.C.: Activation of nucleoli in tuber slices and the function of nucleolar vacuoles. Exp. Cell Res. 71, 1-16 (1972).
54. Rungie, J.M., Wiskich, J.T.: Changes in microsomal electron transport of plant storage tissues induced by

slicing and aging. Aust. J. Biol. Sci. 25, 103-113 (1972).
55. Sakano, K., Asahi, T.: Biochemical studies on biogenesis of mitochondria in wounded sweet potato root tissue. I. Time course analysis of increase in mitochondrial enzymes. Plant & Cell Physiol. 12, 417-426 (1971).
56. Sakano, K., Asahi, T.: Biochemical studies on biogenesis of mitochondria in wounded sweet potato root tissue. II. Active synthesis of membrane-bound protein of mitochondria. Plant & Cell Physiol. 12, 427-436 (1971).
57. Sakano, K., Hirai, M., Asahi, T.: Induced synthesis of mitochondrial RNA in sweet potato root after wounding. Agr. Biol. Chem. 35, 953-955 (1971).
58. Schatz, G., Mason, T.L.: The biogenesis of mitochondrial proteins. Ann. Rev. Biochem. 43, 51-87 (1974).
59. Schonbaum, G.R., Bonner, W.D., Storey, B.T., Bahr, J.T.: Specific inhibition of the cyanide-insensitive respiratory pathway in plant mitochondria by hydroxamic acids. Plant Physiol. 47, 124-128 (1971).
60. Simon, E.M., Chapman, J.A.: The development of mitochondria in Arum spadix. J. Exp. Bot. 12, 414-420 (1961).
61. Solomos, T., Laties, G.G.: Similarities between the actions of ethylene and cyanide in initiating the climacteric and ripening of avocados. Plant Physiol. 54, 506-511 (1974).
62. Solomos, T., Laties, G.G.: The mechanism of ethylene and cyanide action in triggering the rise in respiration in potato tubers. Plant Physiol. 55, 73-78 (1975).
63. Solomos, T., Laties, G.G.: Effects of cyanide and ethylene on the respiration of cyanide-sensitive and cyanide-resistant plant tissues. Plant Physiol. 58, 47-50 (1976).
64. Steck, T.L., Fox, C.F.: Membrane proteins. In "Membrane Molecular Biology", ed. C.F. Fox and A.D. Keith, p27-75. Sinauer Associates Inc., Stamford, Conn., 1972.
65. Tanaka, Y., Asahi, T.: Changes in mitochondrial activity during enlargement of mung bean root cells. Plant & Cell Physiol. 14, 965-971 (1973).
66. Tanaka, Y., Kojima, M., Uritani, I.: Properties, development and cellular-localization of cinnamic acid 4-hydroxylase in cut injured sweet potato. Plant & Cell Physiol. 15, 843-854 (1974).
67. Tolbert, N.E.: Microbodies-peroxisomes and glyoxysomes. Ann. Rev. Plant Physiol. 22, 45-74 (1971).
68. Uritani, I.: Protein metabolism. In "Encyclodia of Plant Physiology, Vol.4., Physiological Plant Pathology", ed. R. Heitefuss and P.H. Williams, p509-525. Springer-Verlag, Berlin, 1976.
69. Van der Plas, L.H.W., Wagner, M.J., Verleur, J.D.: Changes in respiratory pathway of potato tuber (Solanum tuberosum L.) after various times of storage of the tuber. Comparison of wounded and non-wounded tissues. Z. Pflanzenphysiol. 79, 218-236 (1976).

70. Van Steveninck, P.F.M.: The "washing" or "aging" phenomenon in plant tissues. Ann. Rev. Plant Physiol. 26, 237-258 (1975).
71. Vartapetian, B.B., Maalov, A.I., Andreeva, I.N.: Biochemical and cytological studies of mitochondrial development in chloramphenicol-induced rice coleoptiles. Biochim. Biophys. Acta 411, 357-368 (1975).
72. Vary, M.J., Edwards, C.L., Stewart, P.R.: The biogenesis of mitochondria. IX. Formation of the soluble mitochondrial enzymes, malate dehydrogenase and fumarase in Saccharomyces cerevisiae. Arch. Biochem. Biophys. 130, 235-243 (1969).
73. Verleur, J.D.: Observations on the induction of mitochondrial particles in potato tuber tissue after wounding. Z. Pflanzenphysiol. 61, 299-309 (1969).
74. Verleur, J.D., Uritani, I.: Respiratory activity of the mitochondrial fractions isolated from healthy potato tubers and from tuber tissue incubated after cutting or infection with Ceratocystis fimbriata. Plant Physiol. 40, 1008-1012 (1965).
75. Verleur, J.D., van der Velde, H.H., Sminia, T.: The heterogeneity of the mitochondrial fraction in relation to the increase of mitochondrial content in wounded and diseased potato tuber tissue. Z. Pflanzenphysiol. 62, 352-361 (1970).
76. Whaley, W.G., Mollenhauer, H.H., Leech, J.H.: The ultrastructure of the meristematic cell. Am. J. Bot. 47, 401-449 (1960).

The Development and Control of Respiratory Pathways in Slices of Plant Storage Organs

George G. Laties

INTRODUCTION

A significant stumbling block, albeit often unrecognized, to our understanding of the respiratory behavior of bulky storage organ slices has been the tacit presumption that fresh slice respiration is representative of that of the parent organ. Thus the immediate respiratory rise in freshly cut slices, the wound respiration, has been viewed primarily as a quantitative phenomenon, as has to a considerable degree the subsequent wound-induced, or induced, respiration which characterizes aged slices (66). In the latter case, however, the frequent advent of cyanide resistance with ageing has called attention to the likelihood of attendant qualitative alterations. As we shall see, fresh slice respiration is often anomalous, while the respiration of aged slices may be qualitatively more nearly akin to that of the parent tissue. The respiration rate of fresh slices is in most cases 3 to 5 times that of the intact organ, while ageing elicits another comparable increase. Thus the respiration rate of an aged slice may be 25 times that of the intact organ. Interest centers on the nature and genesis of both the instant wound respiration in fresh slices, and the time-dependent induced respiration in aged slices.

Although slices from different plant materials do not invariably behave identically, as will be demonstrated, it is nevertheless useful to consider potato slices as a prototype, both because of their extensive use, and because generalizations drawn from potato slice studies may serve as a frame of reference and comparison. One may begin therefore with the oft-stated contention that the tricarboxylic acid cycle (TCAC), and

Biochemistry of Wounded Plant Tissues

glycolysis as well, is inoperative in fresh potato slices (63). The deduction has been made on the basis of the fact that fresh slice respiration is predominantly malonate resistant, at the same time that fresh slices fail to evolve radioactive CO_2 from ^{14}C-labelled glucose or from any labelled glycolytic product or component of the tricarboxylic acid cycle (90, 63). With ageing, malonate-resistant, cyanide sensitive fresh slices become malonate sensitive and cyanide resistant (62, 63, 90). It is difficult to imagine how a higher plant tissue putatively bereft of glycolytic and TCAC activity survives — taking into account the normal dependence of oxidative phosphorylation on the operation of the TCAC. If the fresh potato slice were in fact representative of the parent organ the means of survival of potato tubers would be a vexing question. However, extensive studies with intact tubers support if not prove that tuber respiration is conventional — comprising glycolytic and TCAC metabolism as major elements (18). Parenthetically it is appropriate to ask whether fresh potato slices are unique in their anomalous characteristics. On the basis of studies with carrot slices apRees and Beevers (7) have asserted that fresh slices do in fact display TCAC activity, and that much of the induced rise reflects the build-up of the pentose phosphate pathway (PPP). The difference in question between potato and carrot is more apparent than real. Fresh carrot slices do show a modicum of TCAC activity while potato slices show none. However in both cases the rise in TCAC activity is extensive, and even in carrot dwarfs the initial activity (7). In both cases as well, PPP activity is augmented with ageing. As will be further substantiated below, the absence of conventional metabolism may be a feature of fresh slices of many storage organs, and ageing redresses the failing (6).

FRESH SLICE RESPIRATION

To return to the original problem, it is to be noted that whereas the respiration of fresh potato slices is markedly cyanide sensitive, that of intact tubers is not only insensitive, but frequently stimulated from 5 to 8 fold (28, 38, 96)! Thus there is little question that fresh slice respiration differs from tuber respiration. Since fresh slices have proven unwilling or unable to metabolize any of the conventional glycolytic or respiratory substrates, resort has been made to mass-spectrophotometric analysis of the respiratory CO_2 in an effort to deduce the nature of the endogenous substrate. The ratio of ^{13}C to ^{12}C differs in various classes of compounds in plant cells (44, 93) as the result of natural mass isotopic selectivity attending certain biosynthetic events. Providentially, potato lipid comprises a distinctly lower ratio of $^{13}C/^{12}C$ than does potato protein or carbohydrate. Accordingly when the $^{13}C/^{12}C$ ratio of slice respiratory CO_2 is compared with the $^{13}C/^{12}C$ ratio of the CO_2 derived from the combustion of different substrate classes in potato, it becomes evident that tuber respiratory CO_2 is carbohydrate in origin, whereas fresh slice CO_2 is predominantly lipid in origin. In the course of a day the endogenous substrate of aged slices once again becomes carbohydrate (Figs. 1, 2). If the lipid-derived CO_2 were released by conventional means — viz. by β oxidation of fatty acid followed by the complete oxidation of acetyl CoA — involvement of the TCAC would be indicated. Yet as noted above TCAC activity is absent in fresh potato slices. Thus another means of carbon release from lipid in fresh slices must be sought.

A clue to the answer to this problem is to be found in the response of fresh slices to phosphorylation uncouplers which raise the fresh slice rate from 2 to 3 times. Chicory slices make the point most dramatically (59, 60), albeit potato slices act much the same way (69). Whereas the uncoupler educed increment is totally malonate (and cyanide) sensitive, there is still no TCAC activity. Malonate does not change the RQ, nor does malate restore the uninhibited respiration rate, as it does when aged slices are inhibited with malonate (59). The picture which emerges is that of a fumarate-succinate electron shuttle linked to substrate dehydrogenase (NAD-linked fatty acid α oxidation) on one side, and to the electron transport chain by way of succinic dehydrogenase on the other, much as suggested by Szent-Gyorgyi almost 40 years ago (5). The significant point is that malonate inhibition is coextensive with imidazole inhibition (69), imidazole being an inhibitor of fatty acid α oxidation (77). Studies with carboxyl-labelled long chain fatty acids have reaffirmed the prevalence of α oxidation in fresh slices (68, 69). Fresh slice respiration per se is inhibited by imidazole by no more than 20 to

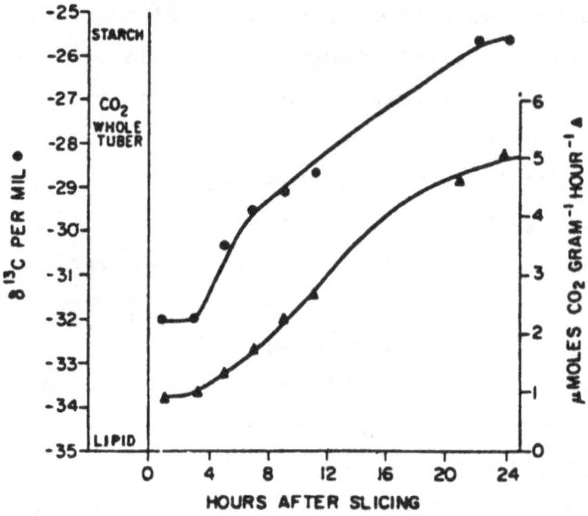

Fig. 1 The $^{13}C/^{12}C$ Ratio of Respiratory Carbon Dioxide of Potato Slices as a Function of Slice Age

$$\delta^{13}C \text{ per ml} = \frac{(^{13}C/^{12}C)\text{ sample} - (^{13}C/^{12}C)\text{ standard}}{(^{13}C/^{12}C)\text{ standard}} \times 1000$$

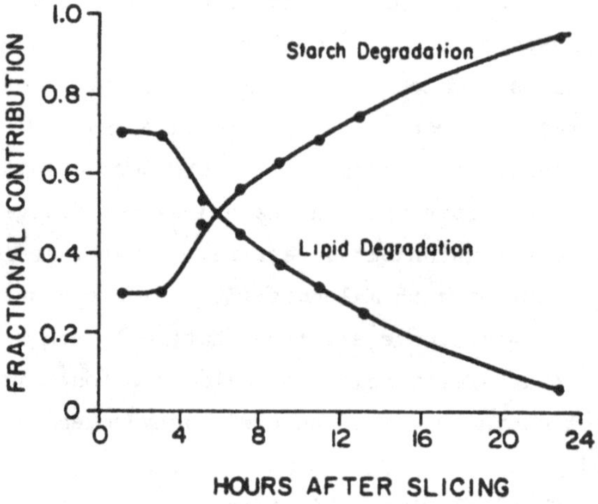

Fig. 2 The Relative Contribution of Starch and Lipid Respectively to Potato Slice Respiration as a Function of Time

30 per cent. Nevertheless mass spectrophotometric measurements suggest that a much larger percent of fresh slice respiration entails lipid oxidation (44). It may be well to point out at this juncture that there is a fraction of storage organ slice respiration in general about which we are totally ignorant. Whereas cyanide inhibits the bulk of fresh slice respiration, the small cyanide-resistant residue is not touched by inhibitors of the CN-resistant, or alternate, path (see below). Aged slices in the presence of both CN and CLAM (Cl-benzhydroxamic acid) or SHAM (salicylhydroxamic acid) — inhibitors of the CN-resistant path (92) — also manifest an inhibitor-resistant residue of unknown nature. This truly residual respiration has consistently been swept under the rug and must soon bear examination. The monoxygenases, specifically cytochrome P450 monoxygenases, have but recently begun to receive attention in plant tissues, and in all probability contribute to the overall oxygen utilization by tissue slices. It is especially noteworthy that hydroxylating monoxygenase activity rises with time in aging slices (19, 30).

The Significance of Fresh Slice Respiration

It is evident that the preparation of fresh slices from potato tubers immediately alters the respiratory metabolism — the substrate changing from carbohydrate to lipid, and the respiration changing from CN resistant to CN sensitive. What is the signal which elicits the change? What is the source of lipid? The answer to the latter question was suggested by ultrastructural studies carried out by Van Steveninck and Jackman (43, 104) on red beet roots which showed that in fresh slices the integrity of cellular membranes — in mitochondria and ER alike — is grossly impaired, whereas in aged slices membranes appear inviolate. Potato tuber tissue is quite low in lipid, and accordingly we have come to the view that the source of lipid serving fresh slice respiration comprises breakdown products of cellular membranes. Lipid acyl hydrolases (as well as lipoxygenase) are rife in potato (35, 40, 41), and slicing either activates these degradative enzymes or makes membrane lipids suddenly susceptible to lipid acyl hydrolase attack. In minutes at 0°

slicing causes the loss of 35 percent of membrane phospholipid and 50 percent of the cellular galactolipid (101). In thin slices (1 mm) phospholipid regeneration begins soon after cutting and proceeds with ageing. In thick slices (8 mm) by contrast, phospholipid degradation continues with time (Fig. 3). A basic maxim of slice respiratory physiology has been that the development of the induced respiration is inversely related to slice thickness (61, 62, 66). In the relation of phospholipid biosynthesis to slice thickness we perceive a close correlate to the respiratory behavior, and there is reason to believe the similarity is highly meaningful. Free fatty acids may be markedly inhibitory to electron transport and oxidative phosphorylation (31, 107) as well as to

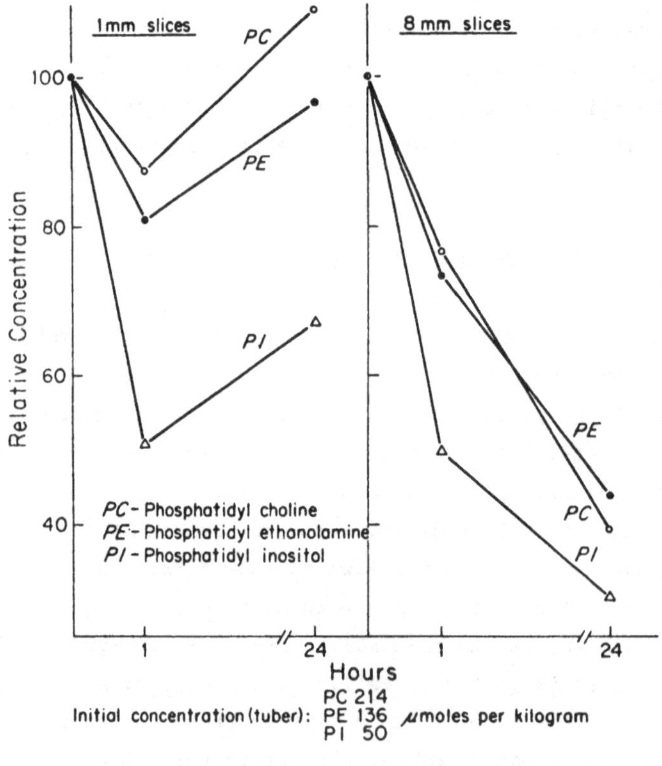

Fig. 3 Phospholipid Levels in Potato Slices as a Function of Time from Cutting

glycolysis (87). Presumably it is free fatty acids which inhibit conventional metabolism — glycolysis and the TCAC — in fresh potato slices. In this view the first hours after slicing include a period of scavenger metabolism in which lipid breakdown products are degraded — as well as more positive events which seemingly underlie the initiation and course of the induced respiration.

What type of signal in response to cutting evokes membrane degradation in slices to a depth of at least 4 mm from each surface, in minutes, at low as well as at room temperature? The manipulation of neither tonicity, pH, pCO_2, pO_2 nor the provision of a selection of macromolecular addenda affects the changes which attend slicing. Might the signal be electrical, in the nature of an action spike or variation potential such as described by Pickard in plant tissues (86)? The symplasm may well simulate an axon in terms of a continuum able to propagate an electrical signal — and slicing would certainly be an adequate affront to initiate such a signal. The consequence of membrane depolarization may be to allow cell wall calcium access to the cytoplasm, with an attendant stimulation of phospholipase activity (41), or a distortion of membrane conformation leading to an enhanced susceptibility of membrane phospholipid to enzymatic degradation. Since both are fanciful concepts there is little harm in introducing a third — the possibility of a self-perpetuating wave of free radical generation by way of hydroperoxides of membrane unsaturated fatty acids [lipoxygenase is plentiful in potato tissue (35)]. Where recovery — i.e. phospholipid regeneration — is concerned, the basis of the inverse relationship between thickness and biosynthetic activity may well be the same as that which governs the development of the induced respiration (61, 73, 74).

Cyanide Resistance in Relation to Membrane Integrity

A survey of a variety of bulky storage organs has found them to fall into two categories with respect to the effect of cutting on cyanide resistance. In all cases the respiration of the parent organ is CN resistant

(97). In approximately half the cases, however, CN resistance is lost immediately on cutting, while in the rest, CN resistance survives cutting (Table I). Where CN resistance persists in fresh slices there is no measurable lipid breakdown, whereas where CN resistance is lost, lipid degradation is readily demonstrable (Theologis and Laties, unpublished)! Assuming the seat of CN resistance to be in the mitochondria (94, 97), it may be deduced that some degree of mitochondrial membrane alteration leads to the loss of the CN resistant, or alternate path, activity. Interestingly, whether fresh slices are CN sensitive (e.g. white potato) or CN resistant (e.g. sweet potato) the respiration rate rises with ageing. Whereas the induced increment in initially sensitive slices is largely CN resistant, the increment in initially resistant slices is predominantly CN sensitive (Table I).

The foregoing reciprocal relationship is not as remarkable as it seems. Table II makes it clear that in almost every case the respiration of uncoupled fresh slices exceeds that of aged slices. Accordingly, it cannot be overemphasized that with few exceptions the electron transport capacity needed to sustain the respiration of aged slices <u>is there from the outset</u>. In initially cyanide-sensitive slices it is obvious that the existing cytochrome path is more than adequate to carry the full electron transport traffic of the aged slice. In initially cyanide-resistant slices the answer is the same. Titration of uncoupled (as well as coupled) CN-resistant fresh slices with an inhibitor of the alternate path (see below) indicates no contribution of the alternate path in the absence of cyanide. In short, the cytochrome path can sustain rates in excess of aged-slice rates. Accordingly, we regard substrate mobilization as a major consequence of slice aging. Where the alternate path is repaired or reestablished with aging, as in potato, there is perforce a corresponding increase in cyanide-resistant respiration in the presence of cyanide. On the other hand, where the alternate path remains more or less intact with aging, or even suffers some curtailment (Table I), and where electron transport in the fresh slice is already adequate to saturate the alternate path in the presence of cyanide, substrate mobilization reasonably leads to an apparent increase in cytochrome-mediated respiration, since it is

Table I The Effect of Cyanide on the Respiration of Fresh and Aged Slices

	FRESH			AGED			Relative Increase		
CN-Sensitive Slices	Control	CN	CN Resistance	Control	CN	CN Resistance	Control	CN	Δ CN / Δ Control
	µl O_2/g/hr		percent	µl O_2/g/hr		percent	Ratio: fresh/aged		percent
Potato	35	10	28	130	130	100	3.7	13	136
Turnip	74	3	4	113	78	69	1.5	26	194
Rutabaga	25	10	13	117	62	53	1.6	6.2	123
Radish	33	< 1	0	70	33	47	2.1	33	89
Beet	58	17	29	138	40	29	2.4	2.4	27
CN-Resistant Slices									
Sweet Potato	110	82	75	195	64	32	1.8	0.78	-20
Red Sweet Potato	82	149	181	138	134	97	1.7	0.9	-26
Parsnip (xylem)	67	97	144	150	54	36	2.2	0.56	-40
Parsnip (phloem)	106	140	132	219	50	22	2.1	0.36	-79
Carrot	100	72	80	131	35	26	1.3	0.49	-19
Banana	63	66	104	105	62	59	1.7	0.94	- 6
Avocado	115	138	120	147	163	110	1.3	1.2	78

Slices, 1 mm thick. Aged, 24 hrs. CN, 0.1 mM.

Table II Effect of Uncoupler on the Respiration of Fresh Slices

CN-Sensitive Slices	Intact Organ	Fresh	Fresh + Uncoupler	Stimulation	Aged	Aged/ Uncoupled Fresh
	µl CO_2/g/hr	µl O_2/g/hr		percent	µl O_2/g/hr	percent
Potato	8	35	128	365	135	105
Turnip	12	74	190	256	113	59
Rutabaga	10	75	203	270	117	58
Beet	14	58	199	343	138	69
CN-Resistant Slices						
Red Sweet Potato	20	82	195	237	138	70
Parsnip	21	105	281	267	170	60
Carrot	12	112	245	218	142	58
Banana	10	63	184	292	102	55
Avocado	40	115	235	204	147	62

Uncoupler: Cl ccp (carbonyl-cyanide Cl-phenylhydrazone), 10^{-5}M

the cytochrome path whose full capacity is unfulfilled in fresh slices.

The proposition has been put forward that membrane lipid degradation is the basis of the loss of CN resistance in a variety of fresh slices, and that the free fatty acids arising from lipid breakdown inhibit the conventional glycolytic and respiratory paths. Accordingly it was expected that CN-resistant fresh slices, which show no discernible lipid breakdown, might resemble aged slices in their ability to metabolize glucose and TCAC components. The general observation that the uncoupler-evoked increment in CN-resistant fresh slices is indifferent to imidazole, an inhibitor of fatty acid α oxidation, whereas that in CN-sensitive slices is inhibited by imidazole, added to the expectation. Nevertheless Table III shows fresh CN-resistant slices no better able to oxidize exogenous glucose than CN-sensitive slices. In each case the capacity for glucose oxidation blooms with ageing, as does the observable capacity for oxidation of exogenous organic acids (not shown). Whereas the disparity in glucose oxidation between fresh and aged slices markedly exceeds the disparity in glucose uptake, an impairment of substrate absorption cannot be ruled out with certainty as the cause of diminished utilization since nothing is known of the relative compartmentation in space and time of substrate taken up by fresh and aged slices respectively.

Nevertheless, the evidence in fresh potato slices indicates not only that glucose is absorbed (63, 90) but that organic acids are absorbed as well (63). Since glucose uptake is almost three times as great in CN-resistant red sweet potato slices as in CN-sensitive white potato slices (Table III) it is problematical to attribute the inability of CN-resistant slices to utilize exogenous glucose exclusively to an impairment of absorption.

At the risk of further beclouding the uncertain relationship between metabolic behavior and absorption, it is to be noted that the utilization of exogenous glucose in CN-treated aged potato slices is quite low in the face of considerable glucose uptake, at the same time respiration is unaffected by cyanide (67). Reenforcing the foregoing anomaly is the

Table III Glucose (^{14}C-UL) Oxidation by Fresh and Aged Slices of Various Tissues

CN-Sensitive Slices	$^{14}CO_2$ EVOLUTION			UPTAKE		
	Fresh	Aged	Ratio Aged/Fresh	Fresh	Aged	Ratio Aged/Fresh
	dpm × 10^{-4}/3 g/hr			% initial radioactivity		
Potato	0.24	120	500	6	94	16
Turnip	1.8	128	71	15	93	6
Red Beet	19.1	150	8	38	95	2.5
CN-Resistant Slices						
Red Sweet Potato	0.43	108	251	16	95	6
Parsnip	0.44	91	207	12	86	7
Avocado	0.71	47	66	2	97	48
Carrot	0.56	60	107	10	83	8

Initial radioactivity of external solution: 18.2 × 10^6 dpm/15 mls

observation that the δ ^{13}C of the respiratory CO_2 of aged potato slices seemingly drops from the carbohydrate value to the lipid value in response to cyanide (45).

Fatty acid biosynthesis underlies developmental changes with ageing in CN-resistant slices even though there is no obvious evidence of lipid breakdown therein. Specifically, cerulenin, an inhibitor of fatty acid synthesis [(83), see below], prevents fresh CN-resistant sweet potato slices from developing the marked capacity for the oxidation of exogenous glucose and TCAC organic acids that ageing elicits in CN-sensitive and CN-resistant slices alike (Table III). Fatty acid synthesis may represent a derepressive response to slicing whether or not slicing causes lipid breakdown. Whereas it is an open question whether the effectiveness of cerulenin bespeaks previous lipid degradation, it is possible that as a result of slicing a localized level of free fatty acid, too low to impair the alternate path and too low to note by our analyses, may yet be involved in the obvious derangement of metabolism and/or absorption in CN-resistant slices.

CN-resistant slices aged in cerulenin for 24 hours continue to resemble fresh slices insofar as the induced respiration rise is suppressed and they remain unable to oxidize exogenous substrates, whether glucose or organic acids (Theologis and Laties, unpublished). In the absence of measurable lipid breakdown (e.g. in sweet potato slices) one may well question the nature of the substrate which sustains slice respiration under the circumstances. If carbohydrate is the prime suspect, as surely it must be, the inability of cerulenin-aged slices to oxidize glucose may point to glucose activation (i.e. phosphorylation) as a limiting event in such tissue. Phosphorylative activity is meager in fresh tissue (20, 72). Phosphorylase-mediated starch mobilization, linked to PPP metabolism, may allow endogenous carbohydrate oxidation where the oxidation of exogenous glucose is precluded. The foregoing suggestion is inadequate, however, since it doesn't explain the inability of fresh slices to metabolize organic acids. In any event, to establish the scope of the effect of cutting on CN-resistant slices it will be necessary to carry out the same type of experiments that were done with potato slices — experiments which involve determination of inhibitor sensitivities (90), phosphorylation capacity (20, 72), label distribution among respiratory intermediates (63), and the $\delta\ ^{13}C$ of respiratory CO_2 (44).

Lest the focus be placed on free fatty acids as the exclusive explanation for the absence of significant TCAC activity in fresh slices, it is well to recall that fresh potato slices contain levels of γ hydroxy α ketoglutaric acid which inhibit aconitase, isocitrate dehydrogenase and α ketoglutaric dehydrogenase alike (65, 84, 85) and that the block in the TCAC in fresh slices has been shown to be between citrate and α ketoglutarate (63). The levels of γ hydroxy α ketoglutaric acid and of glyoxylate drop rapidly on slicing, and at the same time there is an actinomycin-insensitive rise in TCAC activity which follows a hyperbolic time course for some 3 hours (84). The rise is perceptible by measurements involving $^{14}CO_2$ release from methine-carbon labelled succinate (84) or glucose (90), and is not perceived manometrically. Withal it may provide enough ATP generation to initiate the biosynthetic events which are central to the development of the induced respiration (see below). In this connection it is noteworthy that malonate totally inhibits the development of the induced respiration (90), and inhibits suberization, a concomitant of ageing under conventional conditions, when applied within the first 6 to 7 hours after slicing (R. Borchert, personal communication).

Regulatory Aspects of the Fresh Slice Respiration

The respiration of intact bulky storage organ slices is normally well below capacity as evidenced by two phenomena in particular. First, fresh slice respiration is 3 to 5 times tuber respiration, and the cyanide-sensitive fresh slice respiration is in turn raised 2 to 3 times by uncouplers. Thus the cytochrome path at the very least is extensively damped in the intact tuber. Secondly, we know that cyanide (as well as ethylene, and low levels — less than 1 percent — of CO) raises tuber respiration many-fold (28, 38, 96). Thus we may deduce that the cyanide-resistant electron transport path is also severely limited in the intact organ.

The rise in fresh slices which attends cutting is essentially instantaneous, and cannot be ascribed to a spurt of biosynthesis. This is so whether the slices become CN-sensitive or remain CN-resistant. The most obvious candidates for control (e.g. pO_2, pCO_2, pH, tonicity) have been rejected. That is to say, the respiration of fresh potato slices, for example, is not reduced to the tuber rate by raising the pCO_2 or dropping the pO_2 within reasonable limits, or by cutting slices into isotonic solution or by adjusting the external pH. Whether one or another regulatory volatile is involved (62, 80) remains an open question. The problem is compounded by the heterogeneity of fresh slice respiration. The respiration rate of fresh slices is well above that of the parent organ whether fresh slice respiration is predominantly CN-sensitive or CN-resistant (Table II). Whereas a case has been made for the contribution of anomalous fatty acid oxidation to the respiration of fresh potato slices, the extent to which lipid oxidation contributes to the respiration of other CN-sensitive fresh slices remains to be rigorously established. In general, imidazole modestly inhibits coupled fresh slice respiration and extensively inhibits uncoupled fresh slice respiration in CN-sensitive slices, much as in potato. It has yet to be established whether, or to what degree, lipid oxidation is involved in the respiration of CN-resistant fresh slices. CN-resistant slices show no lipid breakdown, and their respiration is unaffected by imidazole, whether coupled or uncoupled.

Contrary to the view engendered by oxygen electrode measurements of respiratory control in isolated mitochondria, tissue respiration rates are probably controlled by substrate availability. Whereas electron transport per se has recently been shown to be controlled by the ratio of ADP to ATP (56), or more rigorously, by the adenylate energy-charge [the ratio ATP + $\frac{1}{2}$ ADP/AMP + ADP + ATP, a number which ranges from zero to 1; see Atkinson (13)], there is essentially no control at values below 0.9 (Bishop and Atkinson, personal communication), and most observed values are lower (26). However variations in energy-charge near 0.85 profoundly affect the regulatory enzymes of glycolysis (12), among others, and the fact that tissue respiration rates normally represent but a fraction of maximal capacity (12, 95, 96) suggests that substrate mobilization controls the respiration rate in tissues and organs, and that the usual stimulatory effect of uncouplers on tissue respiration reflects a conventional Pasteur effect. The response of carrot slice respiration to pyruvate and to adenine nucleotides strengthens this view (1).

In spite of the foregoing generalization I wish to suggest that uncoupler stimulation of fresh CN-sensitive slices [viz. potato (69), chicory (59)] represents a special and anomalous case. In the latter instance it is not so much a mere quantitative change which is at issue, but rather the elicitation of a large qualitatively distinct component by the addition of uncoupler. Thus, the low level of fresh potato slice respiration is inhibited some 20 percent at best by imidazole, while the two to three hundred percent increment due to uncoupler is fully inhibited by imidazole (69). The Pasteur effect is presumably not at issue here since as indicated, slicing evokes imidazole-sensitive α oxidation of fatty acids in fresh potato slices, and lipid is the predominant substrate (44). Whereas fatty acids have long been known to uncouple oxidative phosphorylation, they also serve to inhibit the ADP/ATP translocator (107). To the extent that the latter phenomenon is seemingly more sensitive (107), low levels of free fatty acid may inhibit electron transport in a manner which is the equivalent of inhibition by oligomycin. Under such circumstances uncoupler will bring about a direct stimulation of respiration on electron transport per se, and it is proposed that such is the case in fresh CN-sensitive slices. It follows that electron transport linked to fatty acid α oxidation is mediated by at least the terminal part of the cytochrome path, and the cyanide sensitivity of both coupled and uncoupled fresh potato slice respiration bears out the presumption.

The response to uncoupler in CN-resistant slices, wherein there is no evident lipid breakdown, may reflect a different mode of action, since CN stimulates fresh red sweet potato and parsnip slices, as does uncoupler (Tables I, II). As noted, uncoupled fresh CN-resistant slices are resistant to imidazole, and are seemingly not involved in fatty acid α oxidation in any event. Stimulation by CN speaks to a Pasteur effect (although an alternative possibility is broached with respect to the CN-stimulated peroxidation of NADPH by phagocytosing leukocytes; see below), and if a Pasteur effect is at issue in CN-resistant slices we face the paradox, extensively discussed above, of their apparent inability to oxidize exogenous glucose.

We are left with the question of why the respiration of intact bulky storage organs is so low. The capacity for very much higher rates is clearly there, as the response to ethylene and cyanide, for example, shows (28, 94, 95, 96). Oxygen tension is not at issue, since the oxygen tension in the center of a good sized potato tuber is still 15 percent or higher (25), and elevation of the oxygen tension externally [except in the presence of ethylene (29)] does not raise the respiration rate. Endogenous volatiles (62, see 80) have yet to be ruled out as a regulatory element, even though a convincing case for the involvement of such has yet to be made. Where fresh slice respiration demonstrably differs from tuber respiration qualitatively as well as quantitatively as in potato, the prospect of a triggering or stimulatory response to cutting must be considered as against a simple deinhibition. In those cases where the qualitative relationship between fresh slice and parent organ respiration is uncertain (e.g. sweet potato) there is still reason to suspect a complicated transition in metabolism between the intact organ and the fresh slice, since ageing seemingly influences such slices with respect to the manifestation of conventional respiratory metabolism (Table III).

THE INDUCED RESPIRATION

Biochemical Aspects

The respiratory phenomenon most endearing to slice devotees is the induced respiration which develops in slices in the hours or days after cutting. There is little question that the induced respiration depends upon one or more biosynthetic events following cutting. Specifically, the inhibition of RNA synthesis or protein synthesis prevents the development of the induced respiration. Since a restriction so fundamental as that of RNA or protein synthesis may be expected to exert a multitude of effects, we wished to know which of the many biosynthetic events following slicing are requisite to the induced rise and which are merely incidental. In 1963 Click and Hackett (29) demonstrated that with respect to the development of the induced respiration in potato slices, which proceeds through a period of at least 24 hours, the process is inhibited by actinomycin, an inhibitor of RNA transcription, or puromycin, an inhibitor of protein synthesis, only when the latter are given in the first 8 to 10 hours (Fig. 4). It is in the first 8 to 10 hours that new ribosomes are made (91) and presumably critical new messages as well, as evidenced by enhanced polysome formation (49). We adapted the concept of time-restricted inhibition by actinomycin as a tool in separating the causative biosynthetic events of the first hours from the consequential. Since from what has been mentioned above there seems every reason to believe that membrane biosynthesis plays a crucial role in the genesis and course of the induced respiration, enzymes and products of phospholipid metabolism were scrutinized in particular.

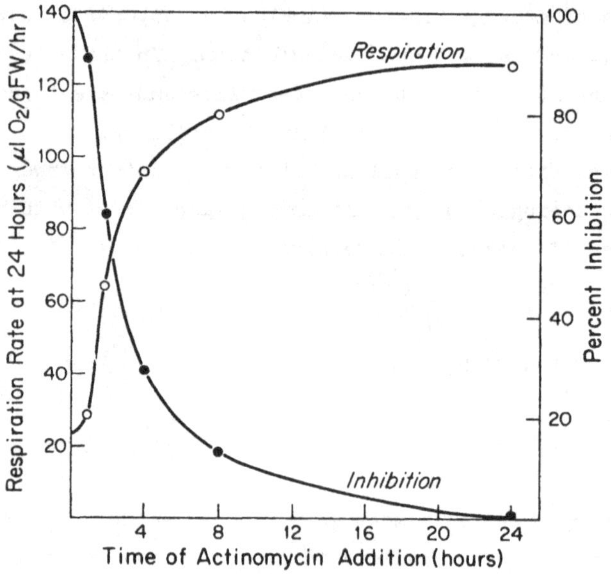

Fig. 4 The Effect of Time of Actinomycin Addition on the Development of the Induced Respiration in 24 Hours. [Adapted from Click and Hackett (36)]

The influence of actinomycin on choline incorporation into phospholipid was found to be time-restricted just as the effect of actinomycin on the development of the induced respiration (105, cf 29). That is actinomycin influences choline incorporation only when given in the first 8 to 10 hours (Fig. 5). Similarly cerulenin, an antibiotic produced by the fungus Cephalosporium caerulens, which inhibits the first step in fatty acid synthesis (83), suppresses respiratory development (and choline incorporation) in potato slices in a time-restricted way, as does dimethylaminoethanol (DME), an inactive analogue of choline [Figs. 6, 7, see (106)]. In the latter instance, whereas DME replaces choline, the lecithin analogue which is formed fails to substitute for lecithin per se in the synthesis of membranes which sustain the developed respiration and evoke CN resistance. Finally, it is of particular significance not only that the enzymes of lecithin biosynthesis, viz. phosphoryl choline cytidyl transferase, phosphoryl choline glyceride transferase, and phosphatidyl phos-

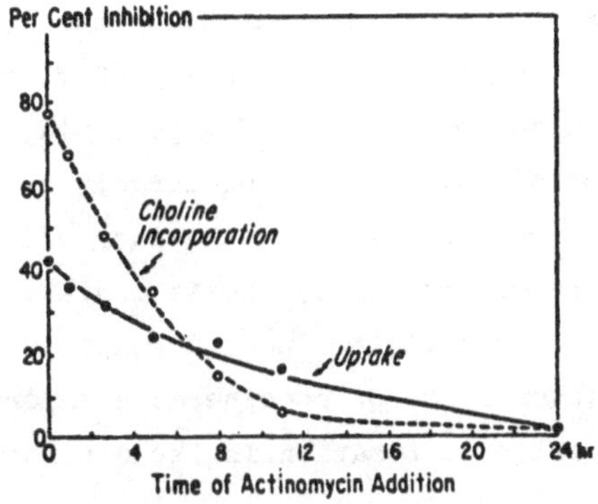

Fig. 5 Time-Restricted Effects of Actinomycin on Choline Incorporation

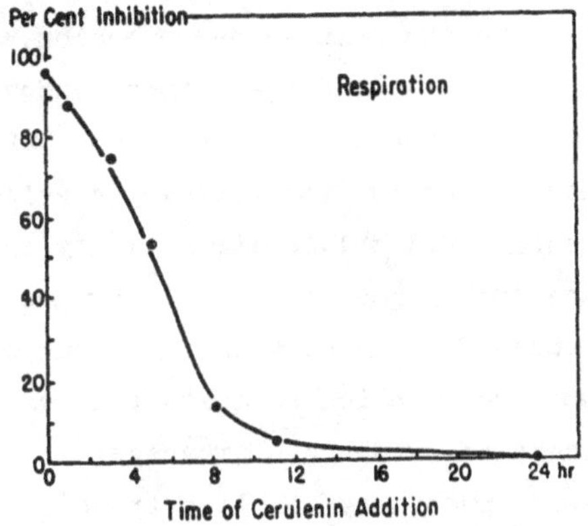

Fig. 6 Inhibition by Cerulenin of Development of Induced Respiration

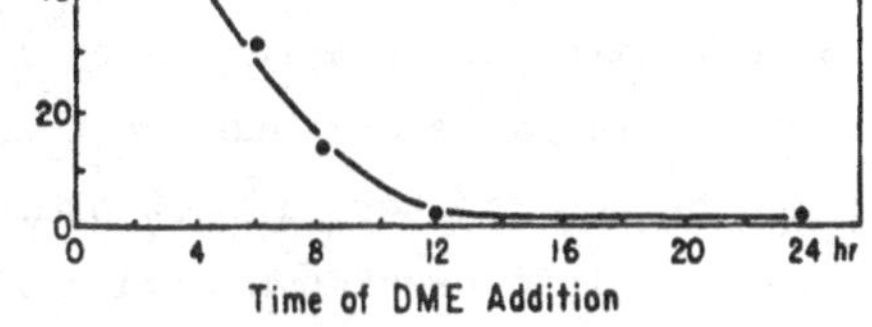

Fig. 7 Inhibition by DME of Development of Induced Respiration

phatase rise with time after cutting, but that their enhancement is curtailed by actinomycin in a time-restricted way (Fig. 8). By contrast, the activity of a selection of enzymes of glycolysis, the pentose cycle, and the TCAC remains steady with time, and largely unaffected by actinomycin through a 24 hour period (Fig. 9). Neither chloramphenicol nor erythromycin, inhibitors of protein synthesis mediated by mitochondrial ribosomes, were found to inhibit the induced respiratory rise in potato slices (33, 105), an observation of considerable importance in evaluating the role of mitochondrial proliferation in the development of the induced respiration.

It is difficult to estimate the role of derepression, if such occurs, in the multitude of events which attend respiratory induction. Actinomycin inhibition does not perforce bespeak derepression, but rather, simply transcription. In this connection there is ample evidence of the marked activation of DNA-dependent RNA polymerase activity in potato slices and slices of other storage organs [see (49)]. If new messages are not unmasked on slicing the need for RNA synthesis in connection with the development of the induced respiration suggests that existing messages must be destroyed. There is evidence of message turnover following slicing, since in the case of phosphoryl choline transferase, for example, the enzyme level in slices aged 24 hours in actinomycin is lower than the enzyme level in fresh slices. On the other hand once certain messages have been resynthesized they must be quite stable in view of the absence of any effect of actinomycin presented after 8 to 10 hours (105). The evidence for de novo synthesis of enzymes following slicing is another matter, and unquestionably occurs, since cycloheximide, for example, readily inhibits the build-up of a variety of enzymes in ageing slices (47, 48). The ultimate criterion for derepression is new DNA template activity, such as is demonstrable in potato buds when dormancy is broken (102), and such as has been demonstrated in sugar beet root slices with ageing (34). Should derepression be contributory if not exclusively causative, it is worth pondering whether lipid breakdown products serve as derepressors in the initiation of lipid — and ultimately, membrane — synthesizing enzymes.

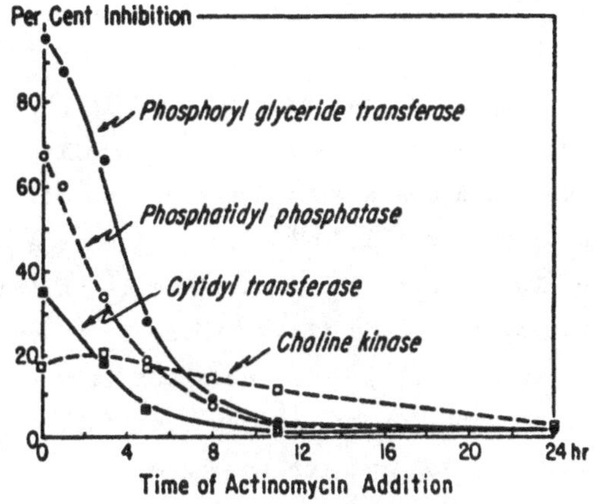

Fig. 8 Time-Restricted Inhibition by Actinomycin of the Increase in Lecithin Synthesizing Enzymes

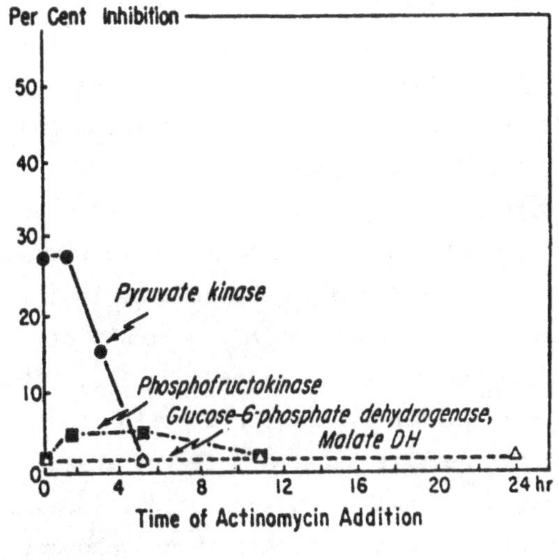

Fig. 9 The Effect of Actinomycin during Ageing on Respiratory Enzymes in Potato Slices

Is Mitochondrial Proliferation a Factor in Wound Induced Respiration?

A large body of experimentation has been offered in support of the proposition that the induced respiration is due to a proliferation of mitochondria. The bulk of the work deals with sweet potato slices, albeit two reports focus on white potato (71, 82). The evidence adduced is summarized in this volume (10). Whether or not there is an increase in mitochondrial number with ageing — and the increase in a day in sweet potato is at best less than twofold (11) — the question is whether a modest increase in number can account for the almost qualitative change in behavior between fresh and aged slices, and for the extensive (ca. 25 fold) disparity in respiratory rate between aged slices and parent organ. As already suggested it may be a snare to compare the respiration rates of aged and fresh slices on the premise that the predominant change is quantitative, notwithstanding the change in CN sensitivity. The most significant datum which must be addressed is that in both white and sweet potato slices the maximal cytochrome oxidase potential is present at all times and need only be expressed. Thus when the maximal cytochrome oxidase activity of fresh potato slices is estimated by a measure of the CN-susceptible oxygen uptake in the presence of uncoupler, the value turns out to be almost exactly the same as the maximal cytochrome oxidase activity in aged slices as measured by CN susceptibility in the presence of uncoupler and a hydroxamic acid inhibitor of the CN-resistant electron transport path (Theologis and Laties, unpublished). The same is true in sweet potatoes. Thus there is no need for an increase in the amount of cytochrome oxidase with ageing to explain the change in slice behavior. Since the evidence indicates that there is little if any contribution to the total respiration of aged slices by the alternate path in the absence of cyanide (see below) Table II makes essentially the same point. Consistent with this expectation we find little or no increase in mitochondrial cytochrome oxidase with ageing in white potato (105) in accordance with the observations of Hackett et al. (36) and Van der Plas et al. (103); cf Dizengremel and Lance (33). There is often some increase in "mitochondrial" pellet volume with ageing which speaks to the urgency of establishing pellet composition, and in recognizing the effect of tissue age on the sedimentability of mitochondria from the homogenate under given conditions of centrifugation (11, 103).

It is difficult to envisage a derepressive change in response to cutting which elicits a program for a totally new population of mitochondria (e.g. a CN-resistant population) coexisting with the original population. When a tissue changes from CN sensitive — viz. fresh potato slices — to wholly CN resistant — viz. aged potato slices, where is the contribution of the CN-sensitive population (apart from the fact that potato tuber respiration is CN resistant)? How is the transition from malonate resistance in slices to malonate sensitivity (90), of phosphorylative inadequacy to phosphorylative competence (20, 72) envisaged in response to a modest increase in mitochondrial number? The point need not be belabored. The transition of slice behavior is more in keeping with a change in the characteristics or capabilities of the total mitochondrial population, as well as in regulatory controls which may govern glycolytic and mitochondrial behavior.

We have suggested the intussusception of new mitochondrial membrane components into the entire mitochondrial population during ageing (105, 106, see 46, 78). In the absence of synchronization there may well be mitochondrial heterogeneity during the ageing process. In this context the work of Nakamura and Asahi (81) seems particularly apt insofar as they find a bile-salt insoluble mitochondrial protein fraction in sweet potato mitochondria which becomes increasingly prevalent with ageing — and which is seemingly enriched with respect to CN insensitive electron transport. In this connection Arrigoni et al. (8) have observed that in potato slices the alkaloid, lycorine, a natural product found in certain members of the Amaryllidaceae, inhibits ascorbic acid biosynthesis and thereby totally represses the development of the induced respiration and the concomitant advent of CN resistance. Arrigoni et al. (9) subsequently reported that a hydroxyproline rich protein appears in mitochondria from aged slices, and ascorbate is of course involved in hydroxyproline synthesis. While Nakamura and Asahi attribute bile salt insolubility to a relative depletion of phospholipid, with the result that the remaining protein is more intensely hydrophobically bonded (10, 81), it remains to be seen whether the bile salt insoluble protein of Nakamura and Asahi bears a relationship to the hydroxyproline rich protein of Arrigoni et al.

Regulation of the Intensity and Quality of the Induced Respiration by External Factors

The evidence is incontrovertible that the development of the induced respiration depends at the very least upon RNA and protein synthesis. It is less obvious that the induced respiration may vary in its nature as a function of the conditions of ageing. The most evident external parameters in control of respiratory induction are temperature and oxygen availability. The induced respiration fails to develop at temperatures close to 0° and in the absence of oxygen. Oxygen tension has no bearing on the immediate rise in respiration which attends slicing, since tuber respiration is unaffected by elevated oxygen tensions, and the pO_2 in the center of a potato tuber at room temperature may be 15% or higher (25). Nor does oxygen tension influence the onset and early course of the induced respiration rise, especially in slices no thicker than 1 mm (61). As the induced respiration develops, however, oxygen becomes limiting, especially in thicker slices, and both further respiratory development and respiration per se are enhanced by raising the oxygen level above that in air (73, 74).

Some unexpected means are at hand to prevent the development of the induced respiration, e.g. the addition of low levels of lithium ion (60) or of acetaldehyde (or chloral) or a variety of other aldehydes (62, 64) to the slice incubation medium. The inhibitory effect of lithium ion may be selective. That is, when slices aged in lithium are removed therefrom, the subsequent time-dependent respiratory rise may be very much more rapid than the normal rise in untreated slices (60). Furthermore by controlling the lithium concentration short of concentrations which totally suppress the respiratory rise, one can control the malonate sensitivity of the induced respiration. In short, lithium exerts its effect on more than one metabolic system. At certain concentrations lithium fails to interfere with preliminary changes which lead to a malonate-resistant increment, while repressing other transformations which may take place as soon as lithium is withdrawn. Much the same situation obtains for control by CO_2/HCO_3^- and by diminished oxygen tensions (60).

Control by pCO_2, hydration and tuber age. As early as 1932 Steward et al. (99) recognized that at least the time-course of ageing in potato slices varied with respect to whether ageing was carried out in solution or in moist air. Much later Lee and Chasson (70) showed that the malonate resistance of potato slices aged in moist air is much greater than that of slices aged in solution. In the extremes, wound-induced respiration follows one of two courses — particularly in the first 24 to 48 hours. 1) When slices are aged in solution (i.e. submerged, albeit fully aerated) the respiration rises 3 to 4 fold in the space of 24 to 48 hours and the increment is largely malonate-sensitive, and by all indications TCAC mediated (63, 90). 2) When slices are aged in moist air in 10% CO_2 the respiration rate approximately doubles in something over 8 hours and is presumed to comprise the pentose phosphate, or phosphogluconate, pathway (57).

The criterion in the latter instance is the malonate resistance of the respiration (57) and to some extent the enhancement of C-1 over C-6 release from specifically labelled glucose (51). The alternatives are not always clear-cut. As noted above, slices aged in moist air alone (i.e. without enriched CO_2) may favor the development of malonate resistance (70) whereas slices which are physically abused or inadequately rinsed may favor the TCAC route even in CO_2-enriched moist air. Further, submerged slices in a mixture of bicarbonate-CO_2 develop a malonate-resistant respiration (67). According to Lange et al. (58) the one ineluctable correlation which they find is that slices destined to form periderm (and become suberized) are typified by a TCAC type increment, whereas slices destined to form callus (i.e. proliferating cells without suberization) develop a pentose phosphate type metabolism. The foregoing generalization is rough at best, since potato slices aged in solution assuredly increase their pentose phosphate path activity as well as the TCAC (7, 90).

As mentioned earlier, there is no reason to presume that malonate resistance per se denotes an explicit metabolic path. The extensive malonate resistance of fresh potato slices is observed in tissue with a vanishingly low capacity for the metabolism of exogenous glucose as measured by the evolution of $^{14}CO_2$ even from C-1 labelled glucose (90). On the other hand, in normally aged slices the rise in TCAC metabolism is underlain by a malonate-resistant rise as well, and the latter is associated with preferential C-1 release from C-1 labelled glucose — compared with the release of C-6 from C-6 labelled glucose (7, 90). Furthermore, slices aged in bicarbonate-CO_2 oxidize glucose handily while almost wholly malonate resistant (67). Thus, fresh slice malonate resistance may relate to a basal respiration yet to be elucidated — whereas aged slice malonate resistance may in fact reflect pentose phosphate pathway (PPP) metabolism as well.

The development of cyanide resistance. Whichever mode of wound-induced respiratory development takes place in potato slices (TCAC/periderm or PPP/callus) there is a development of CN resistance. Since CN resistance attends the TCAC route as well as the pentose phosphate path, different carbon paths are seemingly associated with the same electron transport path — at least insofar as the terminal oxidase is concerned. It is necessary to mention here that the demonstration of cyanide resistance by the addition of CN offers no indication of whether, or to what extent, the CN resistant, or alternate path, operates in the absence of CN. The foregoing information must be separately determined (see below). In aged potato slices the ultimate respiration rate is usually higher in the TCAC/periderm CN-resistant slices than in the PPP/callus type CN-resistant slices (57) — suggesting that carbon metabolism is ultimately rate limiting. In any event CN-insensitivity characterizes the respiration of intact tubers and aged slices. It is missing only in fresh slices of mature tubers.

Fresh slice behavior depends to some extent on the age of the parent tuber (3, 57, 103). Slices from young non-dormant tubers behave differently from slices from mature dormant tubers. Specifically the respiration of slices from non-dormant tubers is quite malonate-sensitive at the outset compared with that of dormant tubers (3). The wound respiration of slices of young tubers is largely CN-insensitive, and the induced respiration may follow a different time course from that of mature tubers (57). Slices from non-dormant tubers behave in moist air much as do slices from mature tubers kept in moist 10% CO_2. Thus it is noteworthy that when slices from non-dormant tubers are treated with NAA intense cell proliferation — which begins in 5 days — is accompanied by a sharp increase in the malonate sensitivity of respiration. In the first 5 days the malonate sensitivity drops to zero (3). Thus, in the latter instance slices preparing to proliferate show a diminishing sensitivity — whereas slices with actually proliferating cells show enhanced sensitivity (cf. 57, 58).

The Induced Respiration in Relation to Suberin Formation

The development of periderm is normally associated with suberin formation. What can be the basis of the apparent correlation between suberization and the development of periderm in connection with the wound-induced elicitation of both PPP and TCAC activity on the one hand, and a propensity for non-suberized callus in connection with predominant development of pentose phosphate pathway activity on the other? Suberization would seem to be a consequence rather than an effect of the determination of the quality of the induced respiration. Thus the distinction between periderm-committed and callus-committed slices is quite evident in 12 hours and fully evident in 24 hours (57, 58). Yet there is very little suberin formed in the first 3 days, intense suberization taking place in the fourth and fifth days (54). That is not to say that suberin is undetectable earlier (21), but its effectiveness in restricting water loss from slices is correlated with the time of its intense formation (54). Further, the effect of actinomycin D in suppressing the appearance of a major enzyme in suberin biosynthesis, ω-hydroxyacid dehydrogenase, is most marked in the third day after slicing (2). Ironically, suberin biosynthesis occurs more readily in slices kept in moist air than in solution (2). In his earliest work Steward (99) showed potato slices formed periderm only when aged in moist air (as compared with slices in solution), an observation which superficially is at odds with that of Lange et al. (58). It would seem that CO_2 (and HCO_3^-), rather than the state of hydration, is the causative agent in the determination of proliferative patterns (cf. 67). In this connection Borchert (private communication) has made the provocative discovery that a brief wash in KCN elicits a more rapid CN resistant respiratory rise than normal in potato slices, and that in the presence of elevated CO_2 a preliminary cyanide wash suppresses suberization and induces proliferation of surface cells. Parenthetically, the induction of CN-resistant respiration by low levels of CN has been reported in rice and wheat roots (50). CO_2 and CN seemingly act together on the determination of the respiratory path, which in turn is related to proliferation behavior.

Even though there seems to be a correlation between ultimate suberization and periderm formation, suberization and cell division coincide neither in time nor in space. Earliest cell division in connection with periderm formation is perceptible in 22 to 24 hours and occurs primarily 2 to 4 cell layers from the surface (22). By contrast, suberization occurs mainly in the surface layer (21) and becomes significant only after 4 days (54). DNA synthesis is perceived primarily in the 3 rows of cells below the suberized layer, and reaches a maximum 14 to 18 hours after cutting (22). On an appropriate nutrient substrate containing benzyl adenine and naphthylacetic acid, periderm-type cell divisions occur beneath the suberized layer yielding cells in ranks, in as many as 14 to 16 rows (Borchert, personal communication). Thus it is not simply a matter of whether cell division is favored under one or another condition of slice ageing — but of the <u>type</u> of cell division which is evoked as well.

The Role of Phenols in Slice Ageing

Suberin has been shown to contain a low level of covalently attached ferulic acid [0.16% in potato (52, 89)] which undoubtedly owes its origin to the shikimic acid pathway. The shikimic acid pathway, in turn, originates with the combination of phosphoenolpyruvate from glycolysis, and erythrose-4 phosphate from the PPP. Where suberization takes place it is to be expected that shikimic acid pathway activity will be evident as well. The augmentation of the PPP occurs with ageing both under conventional conditions (7, 90) and in the presence of high CO_2 (57). The control of the biosynthesis of simple phenols by way of the shikimic acid pathway must depend on more than the PPP.

Under ordinary conditions of ageing slices tend to turn pink. The most prevalent simple phenol in aged potato slices is chlorogenic acid — a depside of caffeic and quinic acids. Chlorogenic acid rises sharply when slices are aged in light (108) and it is almost surely the quinones and polymerized quinones from chlorogenic acid which cause the pink to brown color in aged potato slices. Agents which repress the development of the conventional induced respiration [lithium, acetaldehyde, actinomycin, etc. (60, 62, 105)] as well as agents which evoke a malonate-resistant induced respiration (CO_2/HCO_3; 57, 67) cause slices to remain white. The enzyme phenyl ammonia lyase (PAL), which deaminates phenylalanine to trans-cinnamic acid, is central to the biosynthesis of ferulic and caffeic acids. PAL increases sharply in the first 24 hours of slice ageing, and though dropping to perhaps half its maximal level in 2 days, remains plentiful. The spate of PAL biosynthesis is light dependent (109). Since the requisite light levels are not very high, however (500 fc is maximally effective), and since slice ageing is seldom carried out in the dark, the biosynthesis of simple phenols is probably a usual concomitant of conventional ageing. Is pinking due merely to the increase of oxidizable phenols? Perhaps not. Whenever anything is done to minimize the development or operation of the normal metabolic pathway (taken in this context to be the TCAC) pinking is precluded or minimized (for example, malonate given to lightly-pinked aged potato disks causes them to become white; Laties, unpublished). Thus there is at least the possibility of a basal electron transport system in which a hydroquinone/quinone couple intermediates. The development or release of an electron transport path that siphons electrons from the hypothetical quinone path would leave quinones unreduced, and pinking would ensue. In this connection there is both an NADH/chlorogenic acid oxido-reductase in potato (76) and a chlorogenic acid oxidase as well (4).

There appears to be an inconsistency in the observation that high CO_2 levels that putatively favor PPP activity lead to white disks rather than pink, when it might be expected that enhanced PPP activity would favor the shikimic acid pathway. It remains to be seen whether high levels of CO_2 directly affect the shikimic acid pathway or whether perhaps in the presence of high CO_2 a complete PPP recycles erythrose phosphate leaving none for the shikimic acid pathway. One would also wish to know whether high CO_2 influences the normally marked rise in PAL. In any event, both suberin formation and chlorogenic acid formation would seem to be consequences rather than causes of slice behavior during ageing – in the first instance because suberin formation follows so late behind the respiratory rise (54) and in the second instance because the induced respiration develops undiminished in the dark (37) under conditions where chlorogenic acid synthesis is suppressed (108). Whereas the course of the induced respiration seems to be indifferent to light when slices are aged normally, light, in particular red and far red light, has a profound effect on the development of the induced respiration when slices are aged in malonate or cyanide (37).

CYANIDE RESISTANT RESPIRATION IN SLICES: ESTIMATION AND SIGNIFICANCE

Whereas it has long been considered that the most salient developmental change in ageing storage organ slices is the transition from a cyanide-sensitive to a cyanide-resistant respiration, and while we have tacitly presumed that fresh slice respiration is akin to that of the parent organ, we have recently come to know that fresh slice respiration – particularly, though not solely, in potato – is an artefactual anomaly. Thus, cyanide-resistant respiration is a common characteristic of intact bulky storage organs, and of some fresh slices as well (Table I). In the latter instance there is still reason to believe fresh slice metabolism is anomalous (see above) albeit CN-resistant. In short, slicing is traumatic, and the consequences may range from the obvious loss of CN resistance to the less obvious loss of the ability to oxidize glucose and acids of the TCAC. Phospholipid – and presumably membrane – biosynthesis is involved in recuperation from what may in truth be called wounding effects, and in an object of extensive attention, the potato slice, regeneration of CN resistance accompanies the development of the induced respiration. There is every reason to believe that CN-resistant electron transport is mitochondrial in origin (94, 97). We have lately learned that even where the respiration is CN-sensitive, as in fresh white potato slices, access to the alternate path remains inviolate, albeit the full alternate path is not functional. Thus, fresh potato slices are indifferent to low levels of antimycin (0.2 μM) and inhibited, as would be expected, by the normally used higher levels of antimycin (2-10 μM). Yet when m-CLAM, an inhibitor of the alternate path, is added together with 0.2 μM antimycin, respiration is inhibited (Theologis and Laties, unpublished). In the presence of a low level of antimycin we envisage electron transport through a loop of the alternate path which, in the absence of a full alternate path, returns electrons to the cytochrome path. High antimycin levels block the return.

A question which has yet to be addressed pertains to the determination of the contribution of the CN-resistant electron transport path in the absence of cyanide. A distinction has been made between a resistant path and a compensatory path (67). In the former, the metabolic path in untreated tissue seemingly does not involve a step sensitive to the inhibitor in question. A case in point is the malonate-resistant respiration of potato slices aged in CO_2/HCO_3 (67). In the latter case, the inhibitor in question is effective, but inhibition of the sensitive pathway elicits an alternate or compensatory pathway. Cyanide resistance frequently exemplifies the latter situation.

Bahr (14, 15) has devised an effective means to estimate the contribution of the CN-resistant, or alternate, path in the absence of cyanide. He has used the method to analyze the contribution of the alternate path in the absence of CN in a variety of mitochondria of varying CN sensitivity. We have used the same technique on tissue slices. If the alternate path operates in part or in full in parallel with (i.e. in addition to) a fully operative cytochrome path, respiration will be inhibited to some extent by an inhibitor of the alternate path such as CLAM or SHAM (cf 100). An analysis involves first the estimation of the respiratory rate of a slice in the presence of CN as a function of CLAM or SHAM concentration. With the residual respiration rate (resistant to CN plus hydroxamate) subtracted, the rate of respiration at each concentration of inhibitor under these circumstances represents the maximal contribution of the alternate path at the given inhibitor concentration, and is designated g(i). The respiratory titration with CLAM or SHAM is now repeated in the absence of CN, and the total respiration rate (with V_{res} subtracted) is plotted against a linear series of g(i)s (Fig. 10). The graphic relationship describes the function: $V_T = V_{cyt} + \rho \, V_{alt}$, where V_T represents the total respiration, V_{cyt} the constant contribution of the cytochrome path (the intercept), V_{alt} the maximal contribution of the alternate path at a given CLAM or SHAM concentration, i.e. g(i), and ρ, the slope of the line, the fraction of the full alternate path which is in use. Thus, in the extremes, if $\rho = 0$ there is no contribution by the alternate path, and if $\rho = 1$ the alternate path is fully operative in addition to the cytochrome path. When $\rho = 0$ in CN-resistant tissue it is deduced that electron transport normally proceeds through the cytochrome path and is fully diverted through the alternate path in the presence of cyanide.

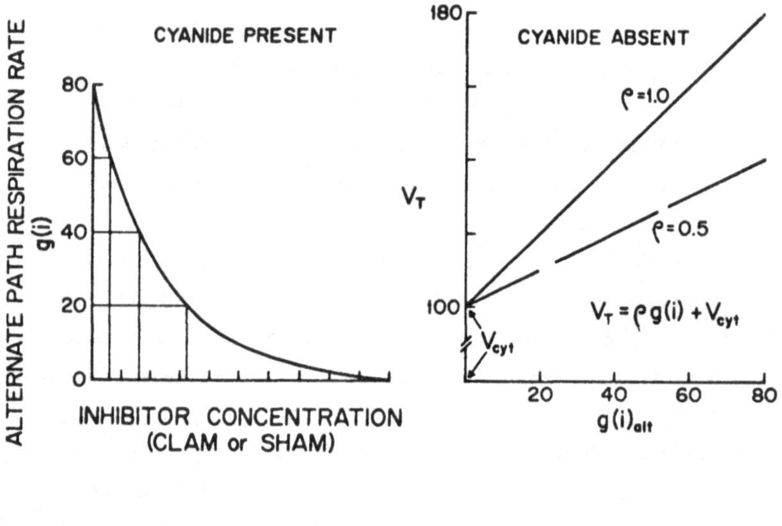

Fig. 10 Estimation of the Respiratory Contribution of the Alternate Path in the Absence of Cyanide

Occasionally ρ greater than 1 has been noted (27), an observation which on the face of it seems meaningless. The explanation lies in what appears to be a limited inhibitory effect of cyanide on the alternate path per se, or on access to the alternate path. In such a case g(i) is underestimated, and the subsequent plot of V_T against g(i) in the absence of cyanide yields an erroneous ρ greater than 1.

Inhibition by hydroxymates in the absence of CN indicates at once that the alternate path is operating. Such is occasionally the case in uncoupled sweet potato slices, and in sweet potato mitochondria, although the full extent of alternate path participation can only be determined by an analysis such as the aforementioned. In CN-resistant slices of aged white potato $\rho = 0$. By limiting the cytochrome path ever so slightly with a low level of CN (0.01 mM), ρ becomes 0.55, while 0.1 mM CN yields $\rho = 1$. Alternatively ρ is transformed from zero to 1 in the absence of CN by the addition of uncoupler. That is, the additional electron traffic released by uncoupler is more than the cytochrome system can handle, and the diverted electrons fully engage the alternate path.

It is a source of some consternation that $\rho = 0$ in ripe 2 cm thick banana slices, since it has been suggested that the operation of the alternate path is a critical element in the climacteric respiratory rise of ripening fruit (94, 97, 98). In fresh slices from preclimacteric avocado the respiration rate is almost 5 to 6 times that of the intact fruit yet $\rho = 0$. On uncoupler addition the rate increases another 60 percent and ρ becomes 0.85. Slices taken from ripe avocado respire at a rate no greater than that of fresh slices, and ρ is once again 0. When a slice is allowed to ripen, however, the respiration rate of the ripe slice is about 60% greater than the fresh slice rate, and $\rho = 1$ (Theologis and Laties, unpublished). Thus it is not so much the physiological condition of the tissue but rather the magnitude of respiration which determines the involvement of the alternate path.

In this regard it must be recalled that CN may cause respiratory stimulation — considerable in fruits and tubers (95, 96, 97) and significant, albeit occasional, in slices [Table I; (53)]. Stimulation in fruits and tubers, whether caused by cyanide or ethylene, is associated with an enhancement of glycolysis (95, 96), which in turn bespeaks substrate mobilization. Respiratory stimulation by CN (and uncouplers) can be explained in terms of a Pasteur effect, so long as the cytochrome and/or alternate path can cope with the additional electron transport which arises from enhanced glycolysis and presumably the attendant augmentation of TCAC activity. The manner of action of ethylene remains to be elucidated. Where respiration is stimulated in response to cyanide it follows that electron transport is mediated by the alternate path. Under ordinary conditions of ripening such need not be the case — as evidenced by the value of $\rho = 0$ in ripe banana slices and in slices from ripe avocado fruit. Since the respiration rate in fresh slices is at once several times that of the intact organ it would seem that a restraint on substrate flux is lifted by slicing whether the respiration is normal or anomalous. Thus, whereas respiration can still be enhanced by uncouplers and — in CN-resistant tissue — by cyanide, the stimulation is proportionately less than that witnessed in fruits or tubers. Where uncoupler stimulates more than does CN, as is usually the case in slices, the reason is presumably that the alternate path alone, in the presence of CN, cannot carry the full traffic.

Parenthetically we see in slices two disparate ways in which uncouplers may serve to stimulate respiration. In fresh white potato slices where there is reason to believe free fatty acids inhibit the mitochondrial ADP/ATP translocator, uncoupler releases respiration the way it would if electron transport were inhibited by oligomycin. Where fatty acid inhibition is less severe — e.g. in sweet potato slices — or not at issue — e.g. in intact fruits and tubers — uncoupler may serve to mobilize substrate. It is an open question whether in the examples which occupy our attention uncoupler ever acts to regulate electron transport by the effect of energy-charge on electron transport per se (56, cf 12).

Finally, what is the function of the alternate path? The evolutionary meaning of an unregulated, non-phosphorylative by-pass around a delicately controlled, phosphorylative electron transport path offers something to ponder. The effectiveness of the CN-resistant path in heat evolution in flowering Aroids is readily comprehensible (79). Its role in other higher plant tissues as well as in lower organisms (42) is far from clear. The alternate path branches from the conventional path well beyond the point where energy-charge exerts an effect on respiration rate in vivo, i.e. by the mobilization of substrate through glycolysis (1, 95, 96). Thus the alternate path does not seem to be a device to maintain minimal carbon traffic under conditions of high energy-charge. In fact, if it were, it would defeat the purpose of controlled oxidative phosphorylation linked to an effective Pasteur mechanism.

If we look to bulky organs, and fruits in particular, it is noteworthy that tissues which respond to ethylene with a sharp increase in respiration invariably respond similarly to CN. Tissues which fail to respond to ethylene are inhibited by CN (97, 98). One would like to think that the alternate path is a requirement for the ultimate manifestation of ethylene evoked phenomena, ripening in particular, but we are plagued with the observation that $\rho = 0$ in slices of ripe banana and avocado — the implication being that the alternate path is not operating therein. Is there a chance ρ is greater than zero in intact ripening fruits? Or is ρ so close to zero in the ripe slices in which we have measured it that we cannot rule out the participation of the alternate path, albeit at a low level? Parenthetically, what are we to say of the function of the alternate path in yeast and Neurospora, where a role for the alternate path vis-a-vis maturation and senescence, which we may adduce for fruit, is not at issue? In the fungi one knows only that impairment of the cytochrome path leads to the enrichment of the alternate path (42).

Fruit ripening has recently been shown to be related to the build-up of hydrogen peroxide (24). The ethylene induced respiratory rise in potato tubers is also attended by hydrogen peroxide formation, particularly at elevated oxygen tensions (28). The oxygen affinity of the alternate oxidase is sufficiently lower than that of cytochrome oxidase (94) that a response to elevated oxygen tension at high tuber respiration rates may be expected, albeit a second low-affinity oxidase has been described in potato tubers under ordinary conditions (75). Ripening-related ethylene formation, in turn, has been shown to be prevented by antioxidants, where inhibitors of ethylene formation from methionine are without effect (17). Peroxidative degradations and free-radical mediated degradations in general may take numerous forms, encompassing membrane breakdown with consequent decompartmentalization, ethylene formation from unsaturated fatty acids, and the degradation of pectins among other polymeric compounds (55). Thus a tentative case can be made for the alternate path as a possible source of superoxide (88, cf 23) and peroxide in connection with maturation and senescence.

The similar susceptibility of the CN-resistant respiration to hydroxamate inhibitors in certain tissue slices and mitochondria therefrom has led to the view that the mitochondrial alternate path is the basis of CN resistance (97). While this hypothesis commands precedence, it is well to note the stimulation by CN of the respiration of human polymorphonuclear leukocytes which attends phagocytosis, and which is characterized by the production of both hydrogen peroxide and superoxide (32). In phagocytosing leukocytes it is the oxidation of NADPH linked to the hexosemonophosphate shunt which is stimulated by CN. The oxidase is seemingly the source of superoxide and hydrogen peroxide, which in turn further peroxidize the substrate (see 23). CN stimulation takes place both in intact cells and in granules therefrom. Remarkably, CN stimulates the peroxidation of NADPH at the same time that it predictably inhibits the peroxidation of a conventional diagnostic peroxidase substrate such as o-dianisidine.

Finally, to return to plant tissue slices, certain peroxidase isoenzymes rise sharply in aged slices, particularly in the surface cell layers (21). Peroxide-mediated hydroxylation of coumaric acid has been reported (39), a prototypical event pertinent to the biosynthesis of simple phenols and to lignification.

The overall picture of coexisting respiratory pathways is not neatly simplified. We must face alternative carbon paths as well as alternative electron transport paths — and in all probability must include in our ultimate appraisal paths which have yet to be elucidated. The most provocative challenge is to fit the heterogeneous respiratory metabolism to the physiology of cells, tissues and organs.

Acknowledgement

Much of the work herein attributed to the author was generously supported by grants from the U. S. Public Health Service and the United States Energy Research and Development Administration.

REFERENCES

1. Adams, P. B.: Effect of adenine nucleotides on the respiration of carrot root slices. Plant Physiol. 45, 495-499 (1970).
2. Agrawal, V. P., Kolattukudy, P. E.: Biochemistry of suberization. ω-hydroxyacid oxidation in enzyme preparations from suberizing potato tuber disks. Plant Physiol. 59, 667-672 (1977).
3. Akemine, T., Kikuta, Y., Tagawa, T.: Respiratory changes during callus formation in potato tuber cultured in vitro. J. Faculty Agr. Hokkaido Univ. 56, 323-336 (1970).
4. Alberghina, F.A.M.: Chlorogenic acid oxidase from potato tuber slices: partial purification and properties. Phytochemistry 3, 65-72 (1964)
5. Annau, E., Banga, I., Blazsó, A., Bruckner, V., Laki, K., Straub, F. B. and Szent-Gyorgyi, A.: Über die Bedeutung der Fumarsäure für die tierische Gewebsatmung. Z. f. Physiol. Chem. 244, 105 (1936).
6. apRees, T.: Evidence for the widespread occurrence of induced respiration in slices of plant tissues. Aust. J. Biol. Sci. 19, 981-990 (1966).
7. apRees, T., Beevers, H.: Pentose phosphate pathway as a major component of induced respiration of carrot and potato slices. Plant Physiol. 35, 839-847 (1960).
8. Arrigoni, O., Arrigoni-Liso, R., Calabrese, G.: Ascorbic acid as a factor controlling the development of cyanide-insensitive respiration. Science 194, 332-333 (1976).

9. Arrigoni, O., De Santis, A., Arrigoni-Liso, R., Calabrese, G.: The increase of hydroxyproline-containing proteins in Jerusalem artichoke mitochondria during the development of cyanide-insensitive respiration. Biochem. Biophys. Res. Comm. 74, 1637-41 (1977).

10. Asahi, T.: Biogenesis of cell organelles in wounded plant storage tissue cells. This volume.

11. Asahi, T., Honda, Y., Uritani, I.: Increase of mitochondrial fraction in sweet potato root tissue after wounding or infection with Ceratocystis fimbriata. Plant Physiol. 41, 1179-84 (1966).

12. Atkinson, D. E.: Cellular Energy Metabolism and Its Regulation. Academic Press, 1977.

13. Atkinson, D.E.: The energy charge of the adenylate pool as a regulatory parameter. Interaction with feedback modifiers. Biochemistry 7: 4030-34 (1968).

14. Bahr, J.: Control of cyanide insensitive respiration. Ph.D. Dissertation, Univ. of Pennsylvania (1971).

15. Bahr, J.T., Bonner, W.D.: Cyanide insensitive respiration. I. The steady states of skunk cabbage spadix and bean hypocotyl mitochondria. J. Biol. Chem. 248, 3441-3445 (1973).

16. Bahr, J.T., Bonner, W.D.: Cyanide insensitive respiration. II. Control of the alternate pathway. J. Biol. Chem. 248, 3446-3450 (1973).

17. Baker, J.E., Lieberman, M., Kunishi, A.: Inhibition of ethylene production in tomato and avocado fruit slices by a rhizobitoxine analogue and free radical scavengers. Plant Physiol. 57 Suppl., p. 97 (1976).

18. Barker, J.: Studies in the respiratory and carbohydrate metabolism of plant tissues. XIII. The influence of oxygen at high pressures in increasing and decreasing the respiration of potatoes at 15°C. Proc. Roy. Soc. B 158, 143-155 (1963).

19. Benveniste, I., Salaun, J.P., Durst, F.: Wounding-induced cinnamic acid hydroxylase in Jerusalem artichoke tuber. Phytochemistry 16, 69-73 (1977).

20. Bieleski, R.L., Laties, G.G.: Turnover rates of phosphate esters in fresh and aged slices of potato tuber tissue. Plant Physiol. 38, 586-594 (1963).

21. Borchert, R.: Isoperoxidases as markers of the wound-induced differentiation pattern in potato tuber. Developmental Biol. 36, 391-399 (1974).

22. Borchert, R., McChesney, J.D.: Time course and localization of DNA synthesis during wound healing of potato tuber tissue. Developmental Biol. 35, 293-301 (1973).

23. Bors, W., Saran, M., Lengfelder, E., Spöttl, R., Michel, C.: The relevance of the superoxide anion radical in biological systems. Current Topics in Radiation Res. Quarterly 9, 247-309 (1974).

24. Brennan, T., Frenkel, C.: Involvement of hydrogen peroxide in the regulation of senescence in pear. Plant Physiol. 59, 411-416 (1977).

25. Burton, W.G.: Studies on the dormancy and sprouting of potatoes. I. The oxygen content of the potato tuber. New Phytol. 49, 121-134 (1950).

26. Chapman, A.G., Fall, L., Atkinson, D.E.: Adenylate energy charge in *Escherichia coli* during growth and starvation. J. Bact. 108, 1072-86 (1971).

27. Chauveau, M.: La chaîne respiratoire des mitochondries d'*Arum* II. Inhibition de la voie d'oxydation insensible au cyanure. Physiol. Vég. 14, 325-337 (1976).

28. Chin, C.K., Frenkel, C.: Induction of upsurge in respiration and peroxide levels in potato tubers as influenced by ethylene, propylene and cyanide. Plant Physiol. 59, 515-518 (1977)

29. Click, R. E., Hackett, D.P.: The role of protein and nucleic acid synthesis in the development of respiration in potato tuber slices. Proc. Natl. Acad. Sci. U.S. 50, 243-250 (1963).

30. Cotte-Martinon, M.G., Yahiel, V., Ducet, G.: Induction d'un cytochrome du type P_{450} et de peroxydase durant la survie du tubercule de pomme de terre. Phytochemistry 13, 2085-90 (1974).

31. Dalgarno, L., Birt, L.M.: Preparation and properties of a mitochondrial fraction from carrot tissue. Biochem. J. 83, 195-202 (1962).

32. De Chatelet, L.R., McPhail, L.C., Shirley, P.S.: Effect of cyanide on NADH oxidation by granules from human polymorphonuclear leukocytes. Blood 49, 445-454 (1977).

33. Dizengremel, P., Lance, E.: Control of changes in mitochondrial activities during aging of potato slices. Plant Physiol. 58, 147-151 (1976).

34. Duda, C.T., Cherry, J.H.: Chromatin -- and nuclei — directed ribonucleic acid synthesis in sugar beet root. Plant Physiol. 47, 262-68 (1971).

35. Galliard, T.: Lipolytic and lipoxygenase enzymes in plants and their action in wound tissues. This volume.

36. Hackett, D.P., Haas, D.W., Griffiths, S.K., Niederpruem, D.J.: Studies on development of cyanide-resistant respiration in potato tuber slices. Plant Physiol. 35, 8-19 (1960).

37. Hanebuth, W.F., Chasson, R.M.: The effect of light upon development in potato tissue slices. Plant Physiol. 49, 857-859 (1972).

38. Hanes, C.S., Barker, J.: The effects of cyanide on the respiration and sugar content of the potato at 15 C. Proc. Roy. Soc. (Lond.) B 108, 95-118 (1931).

39. Halliwell, B., Ahluwalia, S.: Hydroxylation of p-coumaric acid by horseradish peroxidase. The role of superoxide and hydroxyl radicals. Biochem. J. 153, 513-518 (1976).

40. Hasson, E.P., Laties, G.G.: Separation and characterization of potato lipid acylhydrolases. Plant Physiol. 57, 142-147 (1976).

41. Hasson, E.P., Laties, G.G.: Purification and characterization of an A type phospholipase from potato and its effect on potato mitochondria. Plant Physiol. 57, 148-152 (1976).

42. Henry, M.F., Nyns, E.J.: Cyanide-insensitive respiration. An alternative mitochondrial pathway. Sub-Cell Biochem. 4, 1-65 (1975).

43. Jackman, M.E., Van Steveninck, R.F.M.: Changes in the endoplasmic reticulum of beetroot slices during aging. Aust. J. biol. Sci. 20, 1063-68 (1967).

44. Jacobson, B.S., Smith, B., Epstein, S., Laties, G.G.: The prevalence of carbon-13 in respiratory carbon dioxide as an indicator of the type of endogenous substrate. J. Gen. Physiol. 55, 1-17 (1970).

45. Jacobson, B.S., Laties, G.G., Smith, B.N., Epstein, S., Laties, B.: Cyanide-induced transition from endogenous carbohydrate to lipid oxidation as indicated by the carbon-13 content of respiratory CO_2. Biochim. Biophys. Acta 216, 295-304 (1970).

46. Kader, J.C.: Proteins and the intracellular exchange of lipids. I. Stimulation of phospholipid exchange between mitochondria and microsomal fractions by proteins isolated from potato tuber. Biochim. Biophys. Acta 380, 31-44 (1975).

47. Kahl, G.: Metabolism in plant storage tissue slices. Botanical Rev. 40, 263-314 (1974).

48. Kahl, G.: Induction and degradation of enzymes in aging storage tissues. This volume.

49. Kahl, G., Wielgat, B.: Regulation of transcriptional activity in wounded potato tuber tissues. Physiol. Vég. 14, 725-738 (1976).

50. Kano, H., Kumazawa, K.: Studies on CN^--insensitive respiration in plant roots. III. Induction of CN^--insensitive respiration with low concentrations of respiratory inhibitors. Plant and Cell Physiol. 13, 237-244 (1972).

51. Kikuta, Y., Akemine, T., Tagawa, T.: Role of the pentose phosphate pathway during callus development in explants from potato tuber. Plant and Cell Physiol. 12, 73-79 (1971).

52. Kolattukudy, P.E.: Structure and biosynthesis of suberin. This volume.

53. Kolattukudy, P.E.; Reed, D.J.: Metabolism of red beet slices. II. Effects of aging. Plant Physiol. 41, 661-669 (1966).

54. Kolattukudy, P.E., Dean, B.B.: Structure, gas chromatographic measurement and function of suberin synthesized by potato tuber tissue slices. Plant Physiol. 54, 116-121 (1974).

55. Kon, S., Schwimmer, S.: Degradation of plant and other polysaccharides catalyzed by xanthine oxidase and its inhibition by catalase and superoxide dismutase. Fed. Proc. Abstracts 35(7), #1013 (1976).

56. Küster, U., Bohnensack, R., Kunz, W.: Control of oxidative phosphorylation by the extramitochondrial ATP/ADP ratio. Biochim. Biophys. Acta 440, 391-402 (1976).

57. Lange, H.: Atmungswege bei vernarbenden und proliferierenden Gebebefragmenten der Kartoffeknolle. Planta (Berl.) 90, 119-132 (1970).

58. Lange, H., Rosenstock, G., Kahl, G.: Induktionsbedingungen der Suberinsynthese und Zellproliferation bei Parenchymfragmenten der Kartoffeknolle. Planta (Berl.) 90, 109-118 (1970).

59. Laties, G.G.: The nature of the respiratory rise in slices of chicory root. Arch. Biochem. Biophys. 79, 364-377 (1959).

60. Laties, G.G.: The development and control of coexisting respiratory systems in slices of chicory root. Arch. Biochem. Biophys. 79, 378-391 (1959).

61. Laties, G.G.: The controlling influence of thickness on the development and type of respiratory activity in potato slices. Plant Physiol. 37, 679-690 (1962).

62. Laties, G.G.: Control of respiratory quality and magnitude during development. In Control Mechanisms in Respiration and Fermentation, B. Wright, ed. pp. 129-155. Ronald Press (1963).

63. Laties, G.G.: The onset of tricarboxylic acid cycle activity with aging in potato slices. Plant Physiol. 39, 654-663 (1964).

64. Laties, G.G.: Inhibition of RNA and protein synthesis by chloral in potato slices. Plant Physiol. 40, 1237-41 (1965).

65. Laties, G.G.: The inhibition of citrate, isocitrate and α-ketoglutarate oxidation in aged potato slices by γ hydroxy α-ketoglutarate. Phytochemistry 6, 181-185 (1967).

66. Laties, G.G.: Solute transport in relation to metabolism and membrane permeability in plant tissues. In Historical and Current Aspects of Plant Physiology. Publ. N. Y. State College of Agriculture and Life Sciences, pp. 98-151 (1975).

67. Laties, G.G., Hoelle, C.: Malonate and cyanide insensitivity in relation to respiratory compensation in potato slices. Plant Physiol. 40, 757-764 (1965).

68. Laties, G.G., Hoelle, C.: The α-oxidation of long-chain fatty acids as a possible component of the basal respiration of potato slices. Phytochem. 6, 49-57 (1967).

69. Laties, G.G., Hoelle, C., Jacobson, B.S.: α-oxidation of endogenous fatty acids in fresh potato slices. Phytochem. 11, 3403-3411 (1972).

70. Lee, S.G., Chasson, R.M.: Aging and the development of enhanced respiration in potato tuber tissue. Physiol. Plant. 19, 194-198 (1966).

71. Lee, S.G., Chasson, R.M.: Aging and mitochondrial development in potato tuber tissue. Physiol. Plant. 19, 199-206 (1966).

72. Loughman, B.C.: Uptake and utilization of phosphate associated with respiratory changes in potato slices. Plant Physiol. 35, 418-424 (1960).

73. MacDonald, I.R.: Oxygen tension a determining factor in the respiration of potato disks of varying thickness. Plant Physiol. 42, 227-232 (1967).

74. MacDonald, I.R.: Further evidence of oxygen diffusion as the determining factor in the relation between disk thickness and respiration of potato tissue. Plant Physiol. 43, 274-280 (1968).

75. Mapson, L.W., Burton, W.G.: The terminal oxidases of the potato tuber. Biochem. J. 82, 19-25 (1962).

76. Marré, E., Alberghina, F., Servettaz, G.: Variazione di attivata della clorogenico ossidasi e di altri enzimi interessati all' ossidoriduzione del TPN durante l'attivazione di fittine di tubero di potata. Giorn. Bot. Ital. 69, 250-266 (1962).

77. Martin, R.O., Stumpf, P.K.: Fat metabolism in higher plants. XII. α-oxidation of long chain fatty acids. J. Biol. Chem. 234, 2548-54 (1959).

78. Mazliak, P., Douady, D., Demandre, C., Kader, J.C.: Exchange processes between organelles involved in membrane lipid biosynthesis. In Recent Advances in the Chemistry and Biochemistry of Plant Lipids. T. Galliard and E.I. Mercer, eds. pp. 301-318. Academic Press (1975).

79. Meeuse, B.J.D.: Thermogenic respiration in Aroids. Ann. Rev. Plant Physiol. 26, 117-26 (1975).

80. Meigh, D.F., Filmer, A.A.E., Self, R.: Growth-inhibitory volatile aromatic compounds produced by Solanum tuberosum tubers. Phytochemistry 12, 987-993 (1973).

81. Nakamura, K., Asahi, T.: Changes in properties of the inner mitochondrial membrane during mitochondrial biogenesis in wounded

sweet potato root tissue and their relation to the development of cyanide-insensitive respiration. Arch. Biochem. Biophys. 174, 393-401 (1976).

82. Nakano, M., Asahi, T.: Biochemical studies on mitochondria formed during aging of sliced potato tuber tissue. Plant and Cell Physiol. 11, 499-502 (1970).

83. Ōmura, S.: The antibiotic cerulenin, a novel tool for biochemistry as an inhibitor of fatty acid synthesis. Bacteriological Rev. 40, 681-697 (1976).

84. Payes, B.: The ontogeny of the tricarboxylic acid cycle in potato slices. Ph.D. Dissertation, UCLA (1966).

85. Payes, B., Laties, G.G.: The inhibition of several tricarboxylic acid cycle enzymes by γ hydroxy α ketoglutarate. Biochem. Biophys. Res. Comm. 10, 460-466 (1963).

86. Pickard, B.G.: Action potentials in higher plants. Bot. Rev. 39, 172-201 (1973).

87. Radamoss, C.S., Uyeda, K., Johnston, J.M.: Studies on the fatty acid inactivation of phosphofructokinase. J. Biol. Chem. 251, 98-107 (1976).

88. Rich, P.R., Boveris, A., Bonner, W.D., Moore, A.L.: Hydrogen peroxide generation by the alternate oxidase of higher plants. Biochem. Biophys. Res. Comm. 3, 695-703 (1976).

89. Riley, R.G., Kolattukudy, P.E.: Evidence for covalently attached p-coumaric acid and ferulic acid in cutins and suberins. Plant Physiol. 56, 650-654 (1975).

90. Romberger, J.A., Norton, G.: Changing respiratory pathways in potato tuber slices. Plant Physiol. 36, 20-29 (1961).

91. Sampson, M.J., Laties, G.G.: Ribosomal RNA synthesis in newly sliced disks of potato tuber. Plant Physiol. 43, 1011-16 (1968).

92. Schonbaum, G.R., Bonner, W.D. Jr., Storey, B.T., Bahr, J.T.: Specific inhibition of the cyanide-insensitive respiratory pathway in plant mitochondria by hydroxamic acids. Plant Physiol. 47, 124-128 (1971).

93. Smith, B.N., Jacobson, B.S.: $^{2}H/^{1}H$ and $^{13}C/^{12}C$ ratios for classes of compounds isolated from potato tuber. Plant and Cell Physiol. 17, 1089-92 (1976).

94. Solomos, T.: Cyanide-resistant respiration in higher plants. Ann. Rev. Plant Physiol. 28, 279-97 (1977).

95. Solomos, T., Laties, G.G.: Similarities between the actions of ethylene and cyanide in initiating the climacteric and ripening of avocados. Plant Physiol. 54, 506-511 (1974).

96. Solomos, T., Laties, G.G.: The mechanism of ethylene and cyanide action in triggering the rise in respiration in potato tubers. Plant Physiol. 55, 73-78 (1975).

97. Solomos, T., Laties, G.G.: Induction by ethylene of cyanide-resistant respiration. Biochem. Biophys. Res. Comm. 70, 663-671 (1976).

98. Solomos, T., Laties, G.G.: Effects of cyanide and ethylene on the respiration of cyanide-sensitive and cyanide-resistant plant tissues. Plant Physiol. 58, 47-50 (1976).

99. Steward, F.C., Wright, R., Berry, W.E.: The absorption and accumulation of solutes by living plant cells. III. The respiration of cut discs of potato tuber in air and immersed in water, with observations upon surface:volume effects and salt accumulation. Protoplasma 26, 576-611 (1932).

100. Storey, B.T.: The respiratory chain of plant mitochondria. XVIII. Point of interaction of the alternate oxidase with the respiratory chain. Plant Physiol. 58, 521-525 (1976).

101. Theologis, A., Laties, G.G.: Membrane lipid integrity as a prerequisite element of cyanide-resistant respiration in potato slices. Plant Physiol. 57(5), Suppl. p. 93 (1976).

102. Tuan, D.Y.H., Bonner, J.: Dormancy associated with repression of genetic activity. Plant Physiol. 39, 768-772 (1964).

103. van der Plas, L.H.W., Wagner, M.J., Verleur, J.D.: Changes in respiratory pathways of potato tuber (Solanum tuberosum L.) after various times of storage of the tuber, comparison of wounded and non-wounded tissue. Z.f. Pflanzenphysiol. 79, 218-236 (1976).

104. Van Steveninck, R.F.M., Jackman, M.E.: Respiratory activity and morphology of mitochondria isolated from whole and sliced storage tissue. Aust. J. biol. Sci. 20, 749-760 (1967).

105. Waring, A.J., Laties, G.G.: The dependence of the wound-induced respiration in potato slices on the time-restricted actinomycin sensitive biosynthesis of phospholipid. Plant Physiol. 59(6), (1977).

106. Waring, A.J., Laties, G.G.: Inhibition of the development of the induced respiration and cyanide insensitive respiration in potato tuber slices by cerulenin and dimethylaminoethanol. Plant Physiol. 59(6), (1977).

107. Wojtczak, L., Bogucka, K., Sarzala, M.G., Zaluska, H.: Effect of fatty acids on energy metabolism and the transport of adenine nucleotides in mitochondria and other cellular structures. FEBS Symp. 17, 79-92. L. Ernster and Z. Drahota, eds. Academic Press (1969).

108. Zucker, M.: Induction of phenylalanine deaminase by light and its relation to chlorogenic acid synthesis in potato tuber tissue. Plant Physiol. 40, 779-784 (1965).

109. Zucker, M.: The influence of light on synthesis of protein and of chlorogenic acid in potato tuber tissue. Plant Physiol. 38, 575-580 (1963).

Slicing-Induced Alterations in Electron-Transport Systems During Aging of Storage Tissues

Claude Lance and Pierre Dizengremel

INTRODUCTION

It is now well established that bulky storage organs from many plant species, when sliced and exposed to an aerobic environment, develop a rise in their respiration severalfold above the initial respiratory rate of the resting tissue. This rise is generally refered to as "wound respiration" or "induced respiration" (39), but should be more accurately envisionned as an aspect of the process of "adaptative aging"(75), whereby derepression of enzyme activities brings about an overall change in the metabolic status of the tissue, resulting in a much higher respiratory activity (1, 39).

The mechanisms underlying this phenomenon have been described at length or summarized in a number of excellent reviews (5, 16, 27, 28, 38, 39, 45, 58, 75). Briefly stated, upon slicing of the tissues and subsequent washing in an aerated medium, carbohydrates are actively mobilized, presumably through the derepression or activation of various key enzymes. Then, enhanced glycolysis builds up large amounts of pyruvate which are not directed to alcoholic fermentation but, instead, are geared to their aerobic degradative pathway, the tricarboxylic acid cycle, due to increased oxygen availability. The final result is an impressive rise in respiratory activity (39).

Many aspects of cell biochemical mechanisms contribute to this overall and macroscopic phenomenon: derepression of enzyme activities, activation or switching-on of metabolic pathways, biosynthesis of new enzymes or new membrane systems, increase in the number of mitochondria or in the size of the endoplasmic reticulum, etc ... All these aspects of cell metabolism, which contribute to the increment in respiratory activity, should be con-

Biochemistry of Wounded Plant Tissues

sidered in order to accurately explain the quantitative aspects of respiration in aging slices of storage tissues. These various aspects have been the subjects of several reviews (38, 39, 45, 75) and are the topics of a number of articles in this monograph. To avoid unnecessary redundancy and overlapping, they will be considered here only incidentally, if not at all.

The scope of this chapter will be more restricted. One of the major features of respiration in aging tissue slices is that it is qualitatively different from the one preexisting in resting tissues. The difference bears on the nature of metabolic pathways sustaining the respiratory process (tricarboxylic acid cycle, hexose monophosphate pathway) and the nature of oxidized substrates (polysaccharides, lipids) but also on the mechanisms whereby oxygen uptake by the tissues occurs (5, 16, 27, 28, 33, 34, 36, 38, 39, 45, 52, 58, 75).

Early studies with specific inhibitors have clearly demonstrated that the process of oxygen uptake is qualitatively different in fresh and in aged storage tissues. With the progress of knowledge in cell biochemistry, this observation has been deemed to indicate that modifications occur in the composition or activity of the electron transport system of the respiratory chain on the inner mitochondrial membrane. This, *sensu stricto*, will be the topic of this article. In other words, we shall be mostly concerned here with the fate of electrons from succinate (or other Krebs cycle substrates and NADH to oxygen. In addition, as all electron transport systems are not exclusively located on the mitochondrial membranes, but are also found on some other cellular membrane systems, such as the endoplasmic reticulum, these systems will also be eventually considered.

Finally no attempt will be made to exhaustively cover the litterature on the subject. Instead, the purpose will be to present well established facts together with more recent findings, with the aim of stimulating discussion. Most of the observations reported here will refer to the two most widely studied storage organs, the white and the sweet potato tubers, with occasional references to other less studied plant materials (carrot, sugar or red beet, Jerusalem artichoke, chicory, swede, etc ...).

I. QUALITATIVE RESPIRATORY CHANGES IN SLICES OF STORAGE TISSUES.

The first indication of a change in the oxidases mediating O_2 uptake in aging tissue slices is probably to be found in early studies by Robertson et al. on carrot and beet tissues (59, 60). It should be recalled that the nature of the oxidases involved in the uptake of molecular oxygen by plant tissues was far from being well established at that time (36, 37). Many experiments showing that phenolic compounds could increase the rate of O_2 uptake by slices or homogenates of plant tissues, had emphasized the importance of various copper enzymes (phenolases, ascorbic acid oxidase, etc...). Their potential role in mediating O_2 uptake in plant tissues was then on an equal level with that of cytochrome oxidase, whose role in animal cell respiration was by that time rather firmly established (36, 37).

The first demonstration that cytochrome oxidase was the major enzyme responsible for O_2 uptake in potato slices came from experiments by Levy and Schade (49, 67), using photoreversible inhibition by carbon monoxide.These authors were able to demonstrate that the participation of cytochrome oxidase was reduced, and that of a pO_2 sensitive system increased, when the washing time was increased. However, these experiments were criticized on the grounds that the CO/O_2 gaz mixtures used could have altered the true pattern of respiration due to an excessive decrease in O_2 pressure.

In the early fifties, in relation with the demonstration that cytochrome oxidase was the terminal enzyme supporting growth in oat coleoptiles or pea stem sections (30) as well as water uptake in potato slices (31), Thimann and his group clearly established that a change was taking place in the terminal oxidation system of aging potato slices (70). Table I shows that CO in the dark inhibits respiration in fresh potato slices. The inhibition is fully reversible in the presence of light. By contrast, aged slices show a large increase of their respiratory rate. CO exerts no inhibition but, instead, slightly stimulates O_2 uptake by the tissues. From these results it was concluded that terminal oxidation was switching from cytochrome oxidase to a new system thought to be either a new cytochrome insensitive to CO and cyanide or a flavoprotein of high O_2 affinity.

Chicory root tissues, however, do not show any change of their sensitivity

Table I. CO inhibition of respiration in fresh or aged potato tissue slices.

Tissue	CO/O_2	QO_2 [a]		% of rate
		N_2/O_2	CO/O_2	in CO/N_2
Fresh	19/1	43	14	32
	9/1	48.3	29	59
Aged	19/1	80.5	85.5	107
	9/1	143	154.3	108

[a] µl O_2/h/g fresh weight. Inhibition by CO was measured in the dark at two different O_2 pressures. The control conditions were N_2/O_2 mixtures with partial O_2 pressures equal to those of CO/O_2 mixtures. Adapted from Thimann _et al_.(70).

to cyanide upon slicing (43). Instead, the respiration of aged slices becomes highly malonate-sensitive (_cf_. Fig. 2), indicating that significant changes are taking place at the level of intermediary metabolism (43, 44). These results were later confirmed (45, 61) and completed with the use of more refined techniques, using the discriminatory effect of some metabolic pathways towards $^{13}C/^{12}C$ (35). This aspect is dealt with in another section of this monograph.

The differences in tissue behaviour as to their responses to malonate and cyanide can be interpreted as suggesting that malonate-resistance reflects true sensitivity to the inhibitor whereas cyanide-resistance indicates the development of compensatory respiratory mechanisms (46).

Changes in terminal oxidase patterns or metabolic pathways are not specific to aging tissue slices, but are also known to occur under a large set of circumstances (see 51). The resistance to CO or cyanide can vary as aging proceeds (Fig. 1). As a rule, immediately upon slicing, the increase of respiration which then takes place (wound respiration _sensu stricto_) is highly cyanide-sensitive. The subsequent increase, which encompasses a period of several days, becomes strongly cyanide-resistant, but upon extended aging the respiration can recover its sensitivity to cyanide (40, 51). From all these experiments and many additional ones (see 39), it is

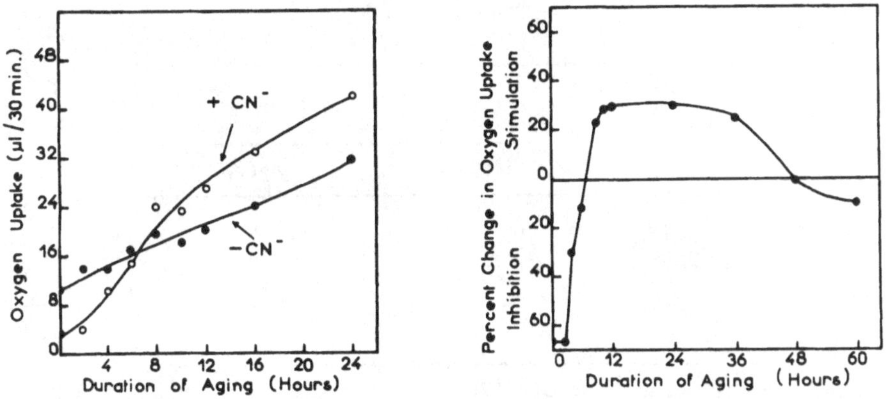

Fig. 1. Variation in cyanide sensitivity during aging of red beet tissue slices. From Kolattukudy and Reed (40).

clear that changes are occurring at the level of terminal oxidation in most tissue slices. Two kinds of approaches were then used, involving the use of inhibitors of electron transport or of phosphorylative reactions.

The first hint that changes could also affect parts of the respiratory chain other than the terminal oxidation step was provided by Hackett et al. (29) using antimycin as an inhibitor of electron transport acting between cyt b and c. Although the inhibition of electron transport is far from being complete in fresh tissues, probably because of the succession of permeability barriers the inhibitor has to go across before reaching its site of action, aged tissues appear to be much more resistant to antimycin (Table II). These results closely parallel those obtained in the presence of cyanide or CO (29, 39, 49, 67) and therefore strongly suggest that not only the terminal oxidation step but also intermediary parts of the respiratory chain can be altered during aging.

The situation is more confused when considering the phosphorylative activity of the respiratory chain. Either in beetroot (56), chicory root (44,45) or potato tuber slices (29, 45), dinitrophenol (DNP) strongly enhances the rate of oxygen uptake by fresh tissues (Fig. 2). On the contrary, it has no action on the respiration of aged tissue slices which display respiratory rates high above those of fresh tissues. The rise in respiratory activity has been shown to be coupled to a large increase in the rate of

Table II. Effect of antimycin A on respiratory rates of freshly cut and 24-h slices.

Antimycin A (μg/ml)	% Inhibition	
	Fresh	Aged
0.1	41	- 5
0.3	42	5
1.0	42	-17
3.0	56	18
10.0	80	57

From Hackett et al. (29).

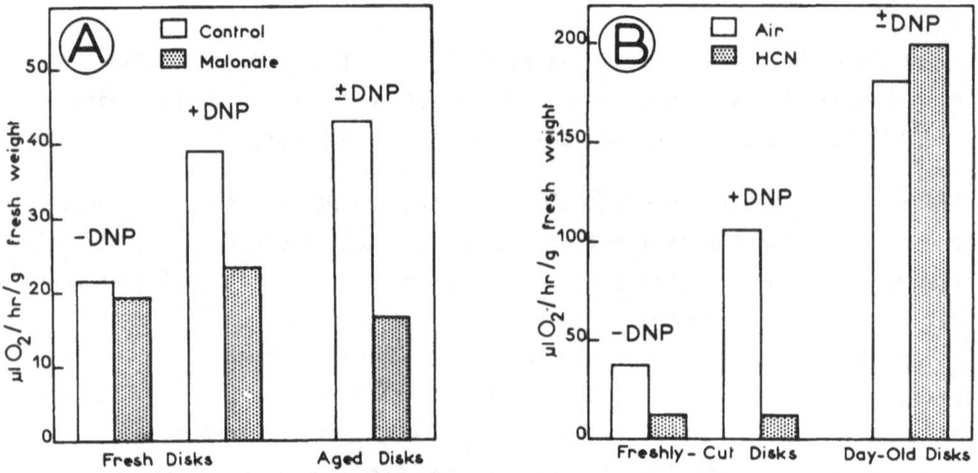

Fig. 2. Effect of dinitrophenol (DNP) on oxygen uptake by chicory root (A) or potato tuber (B) tissue slices in the presence of malonate or cyanide. A: from Laties (44). B: from Hackett et al. (29).

phosphate uptake and to its subsequent incorporation into high energy compounds (11, 29, 45, 50).

The positive action of DNP on fresh tissues can be interpreted as an indication that phosphorylative processes have a very low activity in resting tissues. On the other hand, in aging tissues, biosynthetic reactions consume large amounts of ATP, bringing about a high rate of regeneration of

ADP, which undoubtedly is the main controlling factor of respiratory intensity. However, the situation is complex. Thus, phosphate uptake is linked to the aerobic metabolism of the cell and, in aged tissues, is strongly inhibited in the presence of CO which, on the other hand, has no action on electron transport (29) (Table I). The interaction of DNP with other inhibitors such as malonate or cyanide (Fig. 2) shows that the action of DNP is mostly under the control of the general metabolic activity of the cell which regulates the turnover of the ATP/ADP couple. No direct conclusion can be easily drawn as to a potential effect of aging on the efficiency of the phosphorylative process at the level of the respiratory chain.

II. CHANGES IN ACTIVITY OR COMPOSITION OF THE RESPIRATORY CHAIN.

It is evident that any progress in our understanding of the changes affecting respiratory behaviour in aged tissues could be gained only by studies performed on isolated mitochondria. As a matter of fact, mitochondria isolated from CO- or cyanide-resistant aged tissues indeed show the same degrees (or even higher degrees) of resistance to electron transport inhibitors as the tissues from which they are isolated (29, 55). Fig. 3 shows that the increase in respiration and in cyanide resistance of aged potato tissues is paralleled by an equivalent increase in oxidative activity and resistance to cyanide in mitochondria isolated from these tissues.

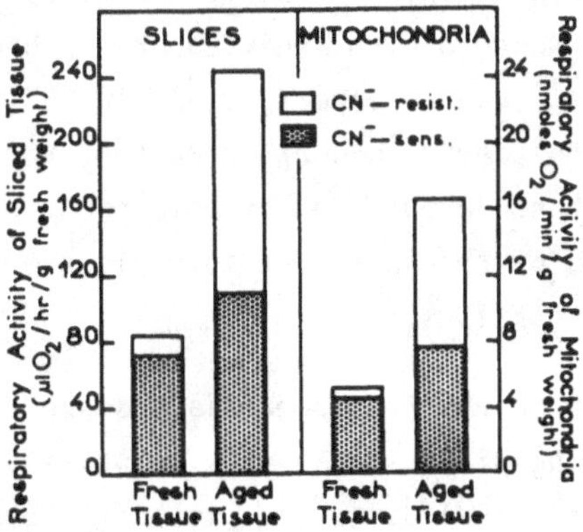

Fig. 3. Action of cyanide on oxygen uptake by fresh or aged potato tissue slices and by mitochondria isolated from these tissues. From Nakano and Asahi (55).

Table III. Oxidase and reductase activities in mitochondria from fresh or aged potato tissue slices.

Activity	Fresh	Aged
Cytochrome oxidase[a]	286	278
+ digitonin	1220	1321
NADH oxidase[a]	11	34
NADH-cyt *c* reductase[b]	250	1065
+ antimycin	68	607
Succinate-cyt *c* reductase[b]	234	186
+ antimycin	0	0
NADH-cyt *c* reductase[b] (microsomal)	177	860

[a] μmoles substrate oxidized/min/mitochondria from 1 g fresh weight.10^{-3}. [b] μmoles cyt *c* reduced/min/mitochondria from 1 g fresh weight.10^{-4}. From Hackett *et al.* (29).

The increase in mitochondrial activity can be interpreted either as an increase in the number of mitochondria per cell or as an increase in specific oxidative activities, or by a combination of both factors. The results are sometimes conflicting, depending on the techniques and criteria (fresh weight, protein nitrogen) used (29, 55). Similarly, it cannot be decided if the increase in cyanide resistance results from a change affecting the whole mitochondrial population or from the coexistence of two different mitochondrial populations within the cell. The question of the increase in the number of mitochondria with modified or new biochemical properties has been the subject of several papers (2, 3, 48, 55, 64, 76, 77, 78, 79) and is specifically considered in another part of this monograph. Here, we wish to deal only with new properties of mitochondria linked to the aging of tissues, regardless of the fact that this might occur in homogenous or dual mitochondrial populations.

Early studies on mitochondria from aged tissues soon showed that other changes were also taking place. In a very incentive paper, far ahead of his time, Hackett *et al.* (29) discovered most of the pecularities of mitochondria from aged tissues, which afterwards were the subject of many studies (Table III).

Although mitochondria from aged tissues are strongly resistant to cyanide or CO inhibition, they show no sensible change in their cytochrome oxidase activity either in the absence or in the presence of digitonin. On the other hand, the oxidation of NADH is strongly enhanced and so is the reductase activity. However, it can be noticed that this increased activity is much more resistant to antimycin inhibition than that of mitochondria from fresh tissues. This is not observed with succinate dehydrogenase activity. In addition, the microsomal system's capacity for NADH oxidation is also strongly enhanced, suggesting that NADH oxidation is probably a complex phenomenon in aged tissues. These aspects of mitochondrial physiology in aging tissues, together with some others dealing with the composition and phosphorylative capacity of the respiratory chain, have been rather extensively studied on isolated mitochondria and are reviewed below.

A. Changes in Mitochondrial Oxidative Activities.

When expressed on a fresh weight basis (*i.e.* mitochondria isolated from a given amount of tissue), the rates of oxidation of the major Krebs cycle intermediates are much higher in mitochondria from aged tissue slices than in those from fresh ones. This has been amply documented in white potato mitochondria (7, 21, 29, 48, 55, 72, 73, 77, 78, 79). However, as said above, this can result from an increase either in mitochondrial population or in mitochondrial oxidative activities.

It seems that the first hypothesis is probably closer to reality than the second. First, many workers have mentioned an increase in the amounts of mitochondrial proteins or in the number of mitochondria per cell in aged tissue slices (2, 3, 7, 21, 29, 48, 55, 64, 74, 77, 78, 79). Then, when expressed on a protein basis, mitochondrial activities seem to be rather similar in aged and fresh tissues. From Table IV, in spite of discrepancies due to the various techniques and preparative procedures used, it is clear that oxidative activities are rather similar in both types of mitochondria, with the exception of NADH oxidation (21).

If one considers some partial activities of the respiratory chain, such as

Table IV. Oxidase activities, on a protein basis, of mitochondria isolated from fresh or aged potato tissue slices.

Ref.	Mitochondria	Substrate	Fresh	Aged
(29)	washed	Succinate	77	74
		Malate	42	30
		α-ketoglutarate	43	33
		Citrate	36	32
(48)	washed	Succinate	37	35
(7)	washed	Succinate	30	35
	purified	Succinate	170	200
(14)	washed	Malate	90	90
(21)	purified	Succinate	244	239
		Malate	89	86
		NADH	225	454
(72)	washed	Succinate	93	60
		NADH	82	84

All data were converted into nmoles O_2/min/mg protein

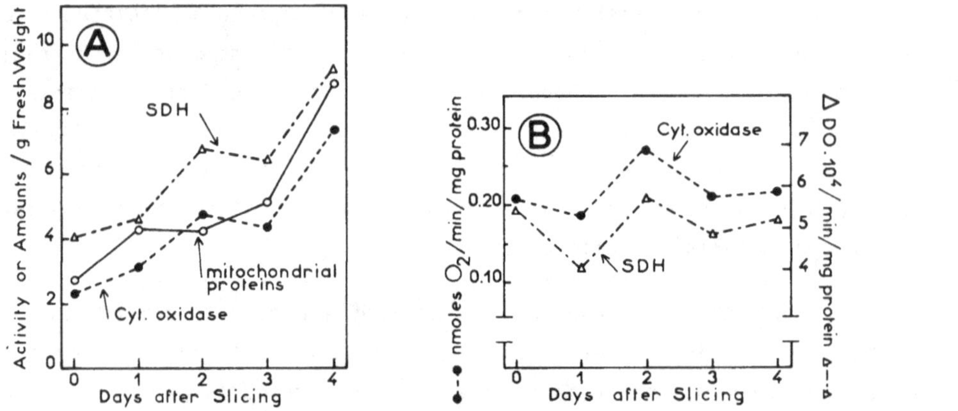

Fig. 4. Variation of enzyme activities during aging of sweet potato tissue slices. A. Activities expressed per g of fresh weight. One unit on the ordinate is equivalent to 62,5 µg mitochondrial protein or 15 nmoles O_2/min for cytochrome oxidase or 2.5 10^{-3} Δ DO at 400 nm for succinate dehydrogenase (SDH). B. Activities expressed per mg mitochondrial protein. Adapted from Asahi et al. (2).

succinate dehydrogenase, malate dehydrogenase or cyt c oxidase, the same pattern as with the overall oxidase activity is observed (2, 3, 64, 65,

77, 78, 79). One finds a large increase on a fresh weight basis, which is nearly nullified when activities are expressed on a protein basis. Parallel to the increase in enzyme activities is a concomitant increase in mitochondrial proteins. The result is that specific enzyme activities display but small variations (Fig. 4).

On the whole, it can be said that the increase in respiratory rate in aged slices is mostly the result of an increase in mitochondrial population rather than in mitochondrial specific activities. This conclusion agrees well with many other studies showing that protein and membrane synthesis is the main metabolic event during tissue aging (7, 18, 38, 39, 65, 66,75).

B. Changes in Composition of the Respiratory Chain.

Direct spectrophotometric examination of the various components of the respiratory chain may also be used to find differences in amounts of components and correlate them with variations in activity of the respiratory chain. As early as 1960, Hackett et al. (29) had indicated that the cyt b / cyt a ratio was much higher in aged tissue mitochondria than in mitochondria from fresh tissues. To a lesser extent this was also true of the cyt c / cyt a ratio. The absolute amounts of cytochromes in mitochondria did not show wide variations. This finding, concerning cyt a + a_3 in particular, was sufficient to rule out the hypothesis of an excess of cytochrome oxidase to account for the cyanide resistance of aged tissues and of their mitochondria.

The differential increase in cyt b, found in white potato (19, 21, 29) and sweet potato mitochondria (65), is a significant feature of tissue aging. Through purification on sucrose density gradient it was also observed that the increase in cyt b occured not only in the mitochondrial fraction, but also in the microsomal fraction (19). Moreover, if expressed on a protein basis, the mitochondrial contents in cyt b were not found to be higher in mitochondria from aged tissues when compared to those from fresh tissues (19, 21).

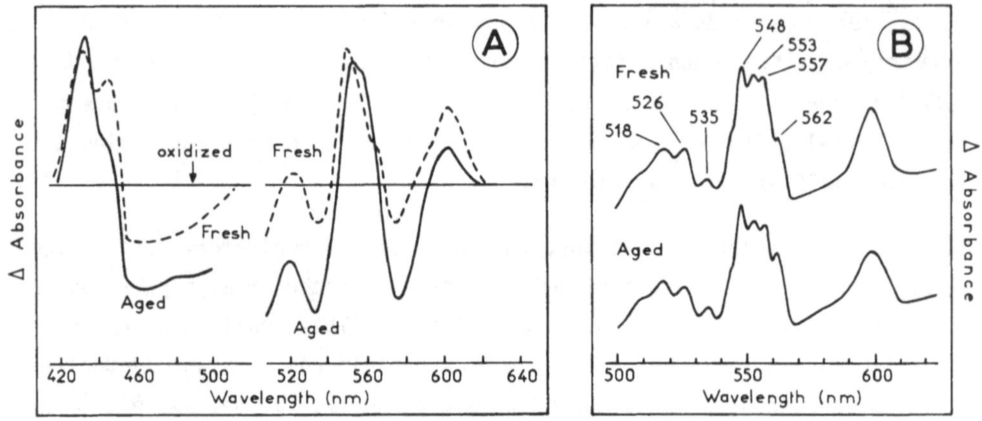

Fig. 5. Difference spectra of mitochondria from fresh or aged potato tissue slices. A. Room temperature difference spectra. Reduction by NADH.The increase in cyt b shows up as a higher peak at 430 nm and a higher shoulder around 560 nm. From Hackett et al. (29). B. Low temperature difference spectra. Reduction by hydrosulfite. 518 and 548 nm: β and α bands of cyt c; 526 and 553 and 557 nm: β and α bands of cyt b-556 and b-560, 535 and 562 nm: β and α bands of cyt b-565; 598 nm; α band of cyt a + a_3. For the nomenclature of cytochromes at room and low temperatures see (69). From Dizengremel (21).

Refinements in spectrophotometric techniques, particularly the possibility of producing difference spectra at liquid nitrogen temperatures (-196° C), led to the conclusion that no qualitative changes were taking place in cytochrome composition during aging (21, 65), the difference being only quantitative (Fig. 5). Mitochondria from aged tissue contain the usual set of cytochromes generally found in all plant mitochondria (12, 34, 42, 57).The results are identical in white or sweet potato mitochondria (21, 29, 65).

Quantitative determinations of the various cytochrome components can be made from spectra such as those of Fig. 5. Table V shows that cytochrome concentrations are lower in mitochondria from aged tissue, as already noticed by Cotte-Martinon et al. (19), except for cyt b-565. The general decrease in cytochrome concentrations in mitochondria from aged tissues indicates that these mitochondria have been the site of a synthesis of new proteins that are not cytochrome components. On the contrary, taking the

Table V. Concentrations of cytochromes in mitochondria from fresh or aged potato tissue slices.

Tissue	Cytochrome concentration[a]				
	$\underline{a}+\underline{a}_3$	$\underline{b}$-556	$\underline{b}$-560	$\underline{b}$-565	$\underline{c}$
Fresh	0.42	0.57	0.54	0.21	0.71
	(1.00)	(1.36)	(1.29)	(0.50)	(1.69)
Aged	0.31	0.50	0.48	0.25	0.58
	(1.00)	(1.61)	(1.55)	(0.81)	(1.87)

[a]nmoles/mg protein. Between brackets: concentrations relative to cyt $\underline{a}+\underline{a}_3$. From Dizengremel (21).

concentration of cyt $\underline{a}$ + $\underline{a}_3$ as a reference, one observes a general increase of the $\underline{b}$ and $\underline{c}$ cytochrome components relative to cyt $\underline{a}$ + $\underline{a}_3$, in good agreement with the results of Hackett et al. (29). This is particularly true of cyt $\underline{b}$-565 but, so far, this has not been related to a specific role of this cytochrome in the respiratory chain or to the oxidation of a particular substrate.

On the whole, these results show that no tremendous changes occur in the cytochrome composition of aging tissues. This is particularly true of cytochrome oxidase, whose variations could eventually be related to CO and cyanide resistance. Similarly, none of the cyt $\underline{b}$, which markedly increase during aging, could be equated with cyt $\underline{b}_7$, whose role was considered for a time as central in cyanide-resistant plant respiration (5, 9, 27, 39).

C. Phosphorylation in Mitochondria from Aged Tissues.

It has been shown above (Fig. 2) that DNP has no stimulatory effect on the increased rate of respiration of aged tissues, whereas it strongly enhances that of fresh tissues. This could equally mean that the high rate of respiration in aged tissues is due to an increase in the number or activity of the mitochondria, as seen above, or that respiration is endogenously uncoupled. If so, the study of phosphorylative activities in

Table VI. Phosphorylation in mitochondria from fresh or aged potato tissue slices.

Ref.	Mitochondria	Substrate	Fresh		Aged	
			P/O	R.C.	P/O	R.C.
(29)	washed	Succinate	1.32	-	1.57	-
		α-ketoglut.	2.30	-	2.52	-
(78)	washed	Succinate	1.70	3.00	1.20	2.50
(7)	washed	Succinate	1.70	4.00	1.30	3.00
	purified	Succinate	1.27	3.00	1.14	2.60
(21)	purified	Succinate	1.51	3.05	1.14	2.04
		Malate	2.08	3.17	1.76	2.09
		NADH	1.35	2.74	1.42	2.90

R.C.: respiratory control ratio. P/O: P/O ratio or ADP/O ratio (17).

mitochondria from aged and fresh tissues could answer the question.

Table VI shows that mitochondria from aged potato tissues are indeed perfectly coupled, thus ruining the above hypothesis. However, except for the results of Hackett et al. (29), obtained in early days (1960) with the manometric technique, modern measurements of oxidative phosphorylation with the Clark oxygen electrode, have shown that P/O ratios as well as respiratory control ratios (R.C.) (17) for all Krebs cycle substrates are consistently and significantly lower in mitochondria from aged tissues. One remarkable exception is NADH whose P/O and RC ratios are nearly as good as in mitochondria from fresh tissues (21, 24).

Further experiments have shown that the decrease of P/O and RC ratios in aged tissue mitochondria is not due to a lowered efficiency of phosphorylative reactions. As early as 1965, Verleur and Uritani (78) had noticed that the state 4 rate of oxidation (17) in these mitochondria had a relatively high value compared to that found in mitochondria from fresh tissues. As a matter of fact, the decrease in the RC value, which is the ratio of state 3 / state 4 rates (17), is mostly due to a relatively higher state 4 rate rather than to a relatively lower state 3 rate (21, 24, 78). We shall see that two electron transport pathways take part in the overall electron transport in mitochondria from cyanide-resistant tissues. One is the phosphorylative cyanide-sensitive cytochrome pathway; the other is a non-phosphorylative cyanide-resistant alternative pathway (12, 31, 33, 34, 57). It will be demonstrated that this alternative pathway plays a greater role in state 4 than in state 3 and thus accounts for the relatively high state 4 rate of oxidation in mitochondria from aged tissues.

As a conclusion, it should be emphasized that the decrease in phosphorylative efficiency in mitochondria from aged tissues is not due to a lesser efficiency of the phosphorylative process *per se*, but rather is the result of the occurrence of two electron transport systems, one being non-phosphorylating.

III. CYANIDE RESISTANCE.

It is not intended here to deal with the whole problem of cyanide resistance in plant tissues or mitochondria, amply documented in numerous reviews (5, 12, 16, 27, 28, 33, 34, 39, 52, 57), but rather to concentrate on its occurrence in aged tissue slices. At the present time, it can be taken for granted that a branched electron transport system operates in most plant mitochondria. The major system is the well known cytochrome pathway, sensitive to cyanide, CO and antimycin, with three phosphorylation sites. The other one, the alternative pathway, branches out from the cytochrome pathway at the level of the flavoprotein-ubiquinone region and contains no phosphorylation site. It is thought to involve one or several iron-sulfur proteins as its main electron transport components sensitive to hydroxamic acids or other chelating agents (9, 33, 34, 57, 68).

The oxidation of succinate or of other Krebs cycle intermediates by mitochondria isolated from fresh tissue slices shows a high degree of cyanide sensitivity, as does respiration in these tissues (*cf.* § I). This occurs in mitochondria isolated from white potato (21, 24, 29, 55, 72, 73) or Jerusalem artichoke (23) but, in fresh sweet potato tissue, mitochondria usually display some degree of resistance to cyanide (15, 71). In fresh tissues, NADH oxidation is also very sensitive to cyanide (15, 21, 24, 71,

Table VII. Inhibition (%) of electron transport in mitochondria from fresh or aged potato tissue slices.

Ref.	Inhibitor	Succinate		NADH	
		fresh	aged	fresh	aged
(29)	Cyanide	86	70	86	94
	Antimycin	85	-	87	89
(21)	Cyanide	98	65	98	89
	Antimycin[a]	95	56	95	90
(72)	Cyanide	90	85	90	91
	Antimycin	98	91	96	93
(73)	Cyanide	94	77	-	89

[a] unpublished results.

72). Antimycin gives the same results as cyanide (Table VII).

By contrast, in mitochondria from aged potato slices, the oxidation of succinate is rather resistant to cyanide and antimycin inhibition (Table VII). The same holds true for malate oxidation (21). Compared to succinate oxidation, NADH oxidation displays a strikingly different behaviour since mitochondria from aged tissues show the same sensitivity to cyanide or antimycin inhibition as mitochondria from fresh tissues. This point of discrepancy between the two substrates will be considered in the next paragraph.

Mitochondria from fresh sweet potato tissues behave more or less as those of aged white potato tissues. Succinate oxidation shows some degree of resistance to cyanide, whereas NADH oxidation is highly sensitive to this inhibitor (15, 71). HOQNO, which acts in a similar manner to that of antimycin, inhibits NADH oxidation but has less action on succinate or malate oxidation (71).

The fraction of electron transport that is resistant to cyanide can be inhibited by chelating agents such as 8-hydroxyquinoline (4, 9, 73). However, one of the best and most widely used inhibitors is salicyl hydroxamic acid (SHAM) which is much more specific with respect to the alternative pathway (4, 21, 68). Provided cyanide or antimycin are added first to inhibit the cytochrome pathway, SHAM strongly inhibits the cyanide-resistant

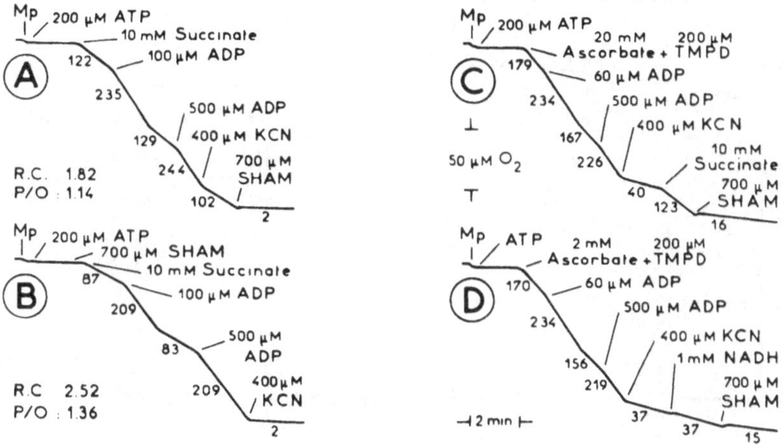

Fig. 6. Action of KCN and SHAM on the oxidation of succinate and ascorbate by mitochondria from aged potato slices. M_p: Mitochondria purified by density gradient centrifugation. Figures along the traces are nmoles O_2/min/mg protein. TMPD: tetramethyl-p-phenylenediamine. (Dizengremel, unpublished results).

electron transport in mitochondria from aged potato slices (21, 24).

Direct evidence for the participation of an alternative non-phosphorylative pathway in the respiration of aged potato tissue slices can be found in the experiments reported in Fig. 6. Cyanide partially inhibits the state 3 rate of succinate oxidation but the addition of SHAM almost completely inhibits the residual electron transfer (Fig. 6A). If SHAM is added first, the state 3 rate is only slightly reduced and cyanide then brings the total inhibition to the same extent as previously. The suppression of the alternative non-phosphorylative pathway in the presence of SHAM significantly increases the ADP/O and RC values (Fig. 6 B). The cyanide resistance of succinate oxidation is not due to a change in the sensitivity of cytochrome oxidase towards cyanide, since the oxidation of ascorbate-TMPD is strongly inhibited in the presence of cyanide (Fig. 6 C), but not totally because of reverse electron transport in the presence of ATP (41). Addition of succinate increases the electron flux, but SHAM then almost completely inhibits the oxidation of succinate since cyanide is present. Such an experiment indicates that the branching point of the alternative pathway is located between the sites of entry of the electrons from

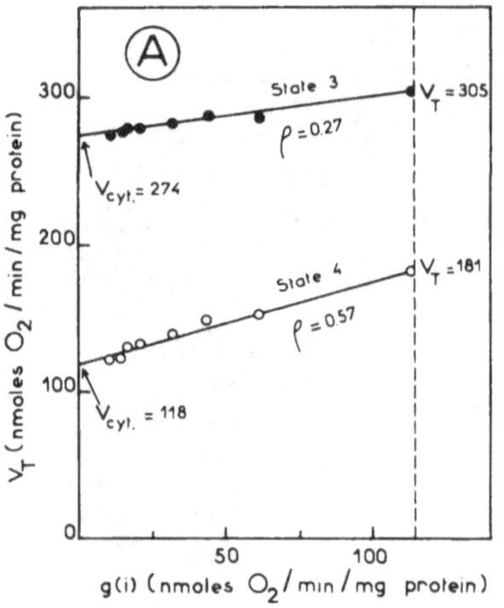

B

	Rates of Oxidation [a]		RC
	State 3	State 4	
V_T	305	181	1.68
$V_{cyt.}$	274	118	2.32
Δ	31	63	
ρ	0.27	0.57	
$V_{alt.}$	114	110	

[a] nmoles O_2 / min / mg protein

Fig. 7. Determination of the participation of the alternative pathway in the oxidation of succinate by mitochondria from aged potato tissue slices. A: for each concentration of SHAM, the rate in the absence of cyanide V_T is plotted versus the rate in the presence of cyanide g(i). For the rationale of this type of experiment, see Bahr and Bonner (4). B: from fig. A, one can determine V_T: maximal rate in the absence of the inhibitor, V_{cyt}: maximal rate of the cytochrome pathway, ρ : fraction of the alternative pathway taking part in the overall electron transport. V_{alt} (maximal rate of the alternative pathway) can be calculated, ρ and Δ being known (Dizengremel, unpublished results).

succinate and from ascorbate, i.e., between flavoproteins and cyt c. Antimycin action further indicates that the branching point must be located between flavoproteins and cyt b (9) (Table VII). Under the same conditions, NADH is not able to increase the rate of electron transport in cyanide-inhibited mitochondria and SHAM has no further inhibitory effect (Fig. 6D).

The contribution of the alternative pathway to the overall electron transport in mitochondria can be evaluated through the method of Bahr and Bonner (4). Applied to the conditions prevailing in mitochondria from aged potato tissues (Fig. 7), this procedure leads to the conclusion that the contribution of the alternative pathway is twice as large in state 4 as it is in state 3. This explain why state 4 rates in mitochondria from aged tissues are relatively high and contribute significantly to the decrease in the RC values observed in these mitochondria. However, if RC's are calculated on the basis of the true contribution of the sole cytochrome pathway, which carries out all oxidative phosphorylation, they are then much higher (Fig. 7B). As a conclusion, these results demonstrate that the existence of a non-phosphorylative electron transport pathway is the main explanatory cause of the reduced efficiency of oxidative phosphorylation observed in mitochondria from aged tissues.

IV. NADH OXIDATION

The oxidation of NADH is thus markedly different from that of succinate in mitochondria from aged tissues. During aging (Table III), the increase of activity is such that it cannot be accounted for by the increase in the amounts of mitochondrial membranes (21, 29) and it therefore implies an increase in specific mitochondrial activities (21, 24). This is paralleled by a concomitant increase of the levels of pyridine nucleotides (13, 32) and of the NADH dehydrogenase (NADH-cyt c reductase) activity (22, 29, 72). Contrary to succinate oxidation, NADH oxidation remains strongly linked to the cytochrome pathway. This is shown by a high degree of cyanide sensitivity (21, 22, 24, 29, 72, 73) (Table VII) and by unusually high ADP/O and RC ratios (21, 24) (Table VI), although only two phosphorylation sites are used with this substrate, as is the case in all plant mitochondria (16, 20, 25, 34, 57, 80). However, it should be noted that the dual behaviour of NADH and succinate is also observed in mitochondria from fresh sweet potato tissue (71).

During aging, the capacity to oxidize NADH also increases in microsomal fractions (6, 7, 29, 62, 63). When dealing with NADH dehydrogenase activity (usually measured as NADH-cyt c reductase activity) care should therefore be taken to work with density gradient purified preparations to clearly discriminate between the mitochondrial and microsomal activities and avoid discrepancy between NADH dehydrogenase (mitochondrial and microsomal) and NADH oxidase (only mitochondrial) activities (22, 29, 72).

As a rule, in aged tissues, the action of antimycin is closely parallel to that of cyanide (Table VII). However, in the case of oxidation of NADH by

Table VIII. Action of cytochrome _c_ on electron transport in mitochondria from fresh or aged potato tissue slices.

Inhibitor	Fresh				Aged			
	Succinate		NADH		Succinate		NADH	
	-[a]	+	-	+	-	+	-	+
None	12.5[b]	15.5	11.3	13.6	15.5	19.9	14.8	27.9
KCN	1.3	1.3	1.1	1.6	2.3	2.8	1.2	2.6
Antimycin	0.3	0.6	0.4	1.7	1.4	2.2	1.0	6.6

[a](±) Presence or absence of exogenous cyt _c_. [b] nmoles O_2/min/mitochondria from 1 g fresh weight. Adapted from Van der Plas _et al_. (72).

aged tissue mitochondria, it was observed that addition of cyt _c_ to the reaction medium could introduce significant differences in the inhibitor actions (22, 72). In carefully prepared mitochondria, cyt _c_ should have no (or only a slight) action on the rates of substrate oxidation, since no loss of cyt _c_ should occur during the preparation of mitochondria (12, 25). Van der Plas _et al_. could show that succinate oxidation in fresh and aged tissues and that of NADH in aged tissues are only slightly stimulated in the presence of exogenous cyt _c_ (Table VIII). By contrast, in aged tissues, NADH oxidation is strongly enhanced. Stimulation by cyt _c_ can be explained by a slightly damaged outer mitochondrial membrane becoming permeable to cyt _c_, but the fact remains that the effect is much more sensible in the case of NADH oxidation by aged tissues mitochondria. In addition, NADH oxidation appears to be rather sensitive to cyanide (91% inhibition) while being rather resistant to antimycin (76% inhibition). In all other instances succinate or NADH oxidations are strongly inhibited by cyanide and antimycin to approximately the same extent. From this, it can be concluded that the pattern of NADH oxidation in aged tissues is different from that in fresh tissues and also from that of succinate in both types of tissues.

NADH oxidation in plant mitochondria appears to be rather complex since

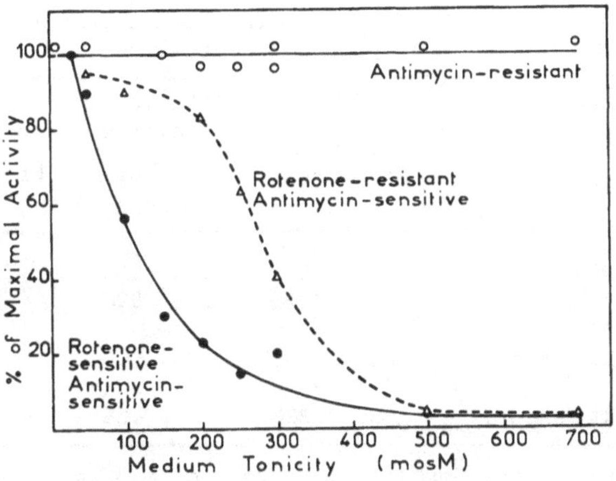

Fig. 8. Action of osmolarity on various NADH-dehydrogenase activities, measured as NADH-cyt _c_ reductase activities, in mitochondria from aged potato tissue slices. Adapted from Van der Plas _et_ al. (72).

three different dehydrogenases can take part in the oxidation of this substrate (20, 25, 57). These dehydrogenases can be discriminated on the basis of their sensitivity to various inhibitors, their mitochondrial location and their accessibility to exogenous cyt _c_. The outer mitochondrial membrane is deemed to be impermeable to cyt _c_ (20, 25), as is the inner mitochondrial membrane (20, 25, 57). Membrane rupture, under osmotic treatments, is necessary to make exogenous cyt _c_ more or less accessible to the dehydrogenases,

Fig. 8 shows that an antimycin-resistant NADH dehydrogenase activity is independent of medium tonicity, which points to its location on the outer mitochondrial membrane. The rotenone-resistant antimycin-sensitive activity reaches its maximum value at a tonicity of about 200 mosM. This is compatible with a location of this dehydrogenase on the outer side of the inner mitochondrial membrane. Rupture of the outer membrane makes it accessible to exogenous cyt _c_. Finally, the antimycin-sensitive rotenone-sensitive dehydrogenase shows its maximum activity at a very low tonicity ($\simeq$ 20 mosM). This requires, in addition, the rupture of the inner mitochondrial membrane and is consistent with a location of this dehydrogenase on the inner side of this membrane. From these results, it can be concluded that

Table IX. Oxidation of NADH by mitochondria and mitochondrial subfractions from aged potato tissue slices.

Fraction	Activity	Total		Antimycin-sensitive		Antimycin-resistant	
		Fresh	Aged	Fresh	Aged	Fresh	Aged
Mitochondria	NADH oxid.[a]	190	283	179	233	11	50
	NADH dehydr.[b]	610	1025	575	820	35	205
Inner membr.	NADH dehydr.	605	840	590	800	15	40
Outer membr.	NADH dehydr.	55	320	5	35	50	285

[a]nmoles O_2/min/mg protein. [b]nmoles cyt c reduced/min/mg protein. NADH oxidase and total NADH dehydrogenase activities were measured on the same mitochondrial preparations. Inner and outer membrane activities were measured on the same (but different from the above) preparations. Adapted from Dizengremel (22).

the site of the antimycin-resistant NADH oxidation in mitochondria from aged potato slices is probably located on the outer mitochondrial membrane.

Measurements of the antimycin-sensitive or resistant activities can be performed directly on isolated outer and inner mitochondrial membranes (Table IX). On a protein basis, NADH oxidase activity increases in mitochondria from aged tissues as does NADH dehydrogenase activity. Although the increase of activity concerns both the antimycin-sensitive and resistant activities, the increase is relatively much more sensible for the antimycin-resistant activity. This is confirmed by the distribution of these activities between the inner and outer membrane fractions: during aging the antimycin-sensitive activity of inner membranes is multiplied by 1.3 only whereas the antimycin-resistant activity of outer membranes is multiplied by 5.7.

The fact that NADH oxidation in aged potato mitochondria remains strongly cyanide-sensitive while being rather antimycin-resistant (22, 72) (Table VIII) has been correlated with the possibility of the occurrence of an

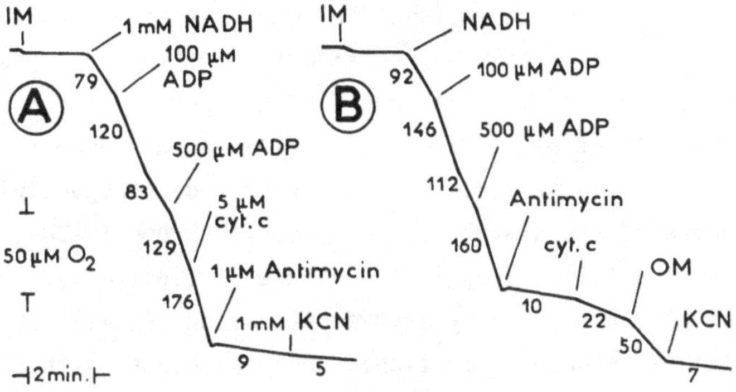

Fig. 9. In vitro electron transport between outer and inner mitochondrial membranes from aged potato tissue in the presence of cyt c. IM, OM: inner, outer mitochondrial membranes. From Dizengremel (22).

intermembrane electron transport. Such a possibility exists if a soluble electron carrier can provide a link between the antimycin-resistant NADH dehydrogenase on the outer membrane and the cyanide-sensitive cytochrome oxidase system on the inner membrane (20, 54). That cyt c might be such a link is demonstrated in Fig. 9. Inner membranes isolated in good structural conditions actively oxidize NADH and even show respiratory control in the presence of ADP. Cyt c increases the rate of oxidation, inner membranes being normally partly depleted of this easily soluble component during the time-consuming preparative procedure. The oxidation is very sensitive to antimycin and cyanide (Fig. 9A). If addition of cyt c follows that of antimycin, a slow antimycin-resistant electron transport occurs (Fig. 9B). It is further increased by the addition of outer membranes and is fully cyanide-sensitive. Identical results are observed if outer membranes are added prior to cyt c. An intermembrane electron transport between NADH and oxygen, sensitive to cyanide and resistant to antimycin, can then take place in the presence of cyt c under in vitro conditions. The possibility that such an electron transport pathway could be operational in mitochondria under in vivo conditions is currently considered (54). From the above results, its participation in mitochondria from aged tissue could amount to about 20% of total NADH oxidation (Fig. 9B).

The oxidation of NADH is still more complex in aged potato tissues because of the fact that, parallel to an increase in pyridine nucleotides (13,32), one also finds an increase of activity in some extramitochondrial systems (6, 7, 8, 29, 62, 63). Not only do the amounts of microsomal proteins increase during aging (6, 7, 19, 29), but also do specific dehydrogenase activities as Hackett et al. (29) were the first to discover (Table III). These systems, routinely measured as NADH-cyt c reductase activities, consist of flavoproteins associated to a b-type cytochrome (NADH-cyt b_5 reductase). The main uncertainty lies in the nature of electron acceptors. Some possible functions could be fatty acid desaturation (6, 7,8) or sterol hydroxylation (10). It should be mentioned that, associated with this type of reactions, another electron transport component, cytochrome P-450, has been found to increase, together with cyt b_5, during the aging of potato slices (26).

In some other plant species (turnip, swede, beet root), Rungie and Wiskich (63) have shown that microsomal NADH-cyt c reductase activities display dramatic decreases immediately upon slicing. Ten to 24 hours of aging are necessary for the tissues to recover their initial activities. As microsomal proteins do not show such spectacular variations, but rather increase slightly as aging proceeds, it is possible that the decrease of activity could be explained by an easy solubilization of the NADH dehydrogenase activity from microsomal membranes. As a matter of fact, these authors were able to demonstrate the presence of no less than three different NADH dehydrogenase activities in the soluble fraction of the cell: NADH-cyt c reductase, NADH-FeCN reductase and NAD(P)H-FeCN reductase activities (62). The first two activities are cytoplasmic, the last one being vacuolar. The question then arises of whether these activities are truly soluble or are the result of a solubilization from microsomal membranes.

V. CONTROL OF CHANGES DURING AGING.

It has been clearly shown by the early studies of Click and Hackett (18) that the development of wound respiration in storage tissues is dependent on the synthesis of new RNA and proteins. This observation applies not only to the overall rise of respiration, but also to individual enzymes and particular membrane systems and is amply documented at the present time

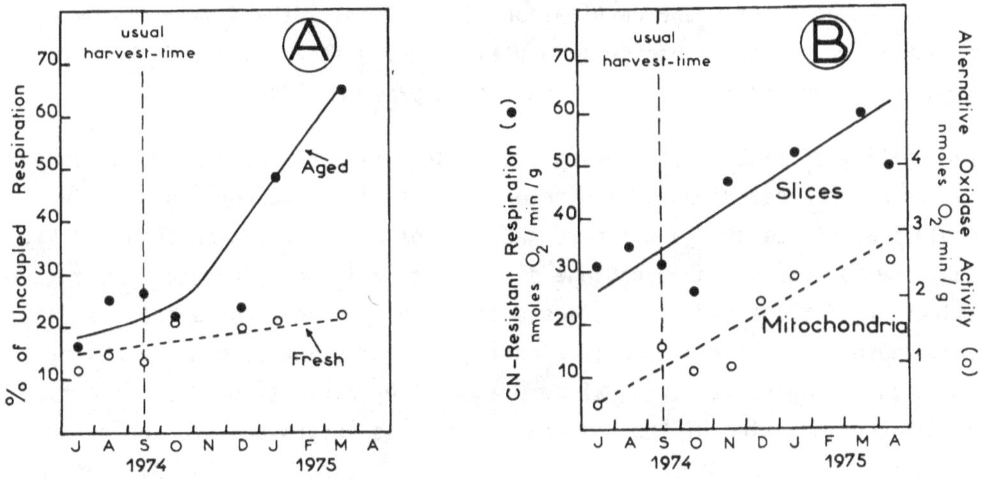

Fig. 10. Development of cyanide resistance and alternative oxidase activity during aging of potato slices from tubers collected at various times before or after the usual harvest period. A. Respiration resistant to cyanide as per cent of the maximal rate (uncoupled in the presence of DNP) for fresh or aged slices (cf. Fig. 2). B. Cyanide-resistant respiration in aged slices and alternative oxidase activity in mitochondria from aged tissues. Adapted from Van der Plas and Verleur (73).

(38, 39). As this point is considered in another part of this monograph, we shall examine here only various controlling factors interfering with the development of new respiratory patterns or enzyme activities in mitochondria from aged tissues.

One of these factors, rather ill-defined but well-known to experimentators on tissue slices, consists in the variations of responses to aging as a function of storage conditions. This point was recently examined in detail by Van der Plas *et al*. for the potato tuber (72, 73, 74).

During storage, many enzyme activities show quantitative variations. One remarkable fact is that all activities do not change in the same way (74). For instance, during storage, malate dehydrogenase remains constant but cytochrome oxidase decreases markedly. Succinate dehydrogenase shows only a slow decrease. These variations can be correlated with similar slow decreases in the respiration of the tubers and in their ability to develop a high intensity of wound respiration. Surprisingly, there seems to be a slight increase in mitochondrial proteins during storage, but part of it might be accounted for by contaminating peroxysomes (74).

The ability of tissues to develop a cyanide-resistant wound respiration during aging (regardless of its intensity) is also dependent on the physiological status of the tuber at the time of experimentation (Fig. 10 A). The same is true of the magnitude of cyanide resistance in aged tissue slices or of the alternative oxidase activity in mitochondria isolated from these tissues (Fig. 10 B). Clearly, the induction of cyanide resistance during aging is not equal before or after harvest. Before harvest, hardly any induction of the alternative oxidase occurs and cyanide resistance is low. After harvest, both phenomena increase markedly. By contrast, the activity of the antimycin-resistant NADH-cyt c reductase of the outer membrane shows opposite variations. Before harvest, it can be greatly increased by the aging of tissue slices; this increase gradually disappears during storage of the tubers (74). The factors controlling these systematic variations in relation to tuber development or storage duration are not known at the present time.

Other control mechanisms, more accessible to experimentation, are also involved in the development of cyanide resistance during aging.

Since the cyanide-resistant alternative pathway is specifically sensitive to SHAM or other iron-chelating compounds, this fact has been the basis for the hypothesis of the participation of iron-sulfur proteins (non-heme iron proteins) in the alternative pathway. Contrary to expectation, the levels of non-heme iron (or of labile sulfide) are not higher in mitochondria from aged tissues than in those from fresh tissues (21). Such is the general situation in cyanide-resistant mitochondria when compared to cyanide-sensitive ones (23). This observation leads to the conclusion that the induction of an alternative pathway containing iron-sulfur proteins is rather the result of the switching-on of a preexisting system than that of the de novo synthesis of new electron transport components (21, 23). However, it can be demonstrated that the switching-on of such a pathway is nevertheless dependent, directly or not, on the production of a protein fac-

Table X. Effect of chloramphenicol and cycloheximide on succinate and NADH oxidation by mitochondria from aged potato slices.

Substrate	Inhibitor	Fresh	Aged		
			Control	Chloram-phenicol	Cyclohex-imide
Succinate	None	349[a]	321	327	217
	KCN	2	135	126	11
	KCN+SHAM	2	2	4	3
NADH	None	194	397	356	142
	KCN	4	15	20	2
	KCN+SHAM	4	10	12	2

[a]nmoles O_2/min/mg protein. Adapted from Dizengremel and Lance (24).

tor synthesized in the cytoplasm.

Table X first shows well known facts occuring during aging of potato slices: the presence of an alternative pathway is evident in succinate oxidation and a large rise, highly sensitive to cyanide, occurs in NADH oxidation. If aging takes place in the presence of chloramphenicol,hardly any significant differences can be observed with mitochondria from tissues aged in the absence of inhibitor. The conclusion is that the mitochondrial protein-synthesizing system plays but an insignificant role in the development or expression of cyanide resistance in tissues. On the contrary, with cycloheximide, the picture is quite different; no cyanide resistance is observed in the oxidation of succinate and no increase in the rate of NADH oxidation takes place. In fact, the rates of succinate and NADH oxidation are both significantly lowered to 2/3 of their initial value in the presence of cycloheximide.

From this, it appears that both the development of the alternative pathway and the increase in NADH oxidase activity are dependent upon one or several protein factors synthesized in the cytoplasm. Whether these factors are true components of mitochondrial membranes or are necessary for the synthesis of such membrane components is not known at the present time.

CONCLUSIONS

If one takes the potato tuber tissue as a typical example for the description of respiration in aging storage tissues, the present state of knowledge on electron transport in such tissues can be summarized as in Fig. 11. This scheme is by no means complete, as far as true mechanisms of electron transport are concerned. In particular, it has been oversimplified as to the various flavoproteins, iron-sulfur proteins and b cytochromes intervening in the respiratory chain (57). Nevertheless it emphasizes the main typical aspects of electron transport in aged tissue slices.

This scheme, based on widely accepted concepts (16, 27, 33, 34, 57) and on results described in this review, assumes that the respiratory chain consists of a flavoprotein part and of a cytochrome part. The link between the two is ubiquinone, whose role has been recently emphasized to the point that it is now considered as the pivotal component controlling electron distribution between the cytochrome and alternative pathways (53, 69). Some components are known to be located on the inner or outer side of the inner mitochondrial membrane. The normal respiratory chain has three phosphorylation sites with specific inhibitors acting at their levels. The flavoprotein part also contains various iron-sulfur proteins (not represented on Fig. 11). Cyt c can be easily detached from the inner membrane and, since the outer membrane is impermeable to it, it stays in the intermembrane space, behaving as a shuttle carrier. Exogenous cyt c can act as an electron acceptor for the dehydrogenases located on either side of the inner membrane only after disruption of the outer or/and inner membranes. The alternative pathway, when it exists, branches out from the cytochrome pathway at the ubiquinone level and would contain a specific flavoprotein (69). It is insensitive to antimycin and cyanide and its inhibition by SHAM suggests that iron-sulfur proteins are implied in this pathway. The terminal oxidase is still unknown.

This being stated, the main characteristics of electron transport in mitochondria from aged (potato) tissues oxidizing typical substrates can be summarized as follows (Fig. 11).

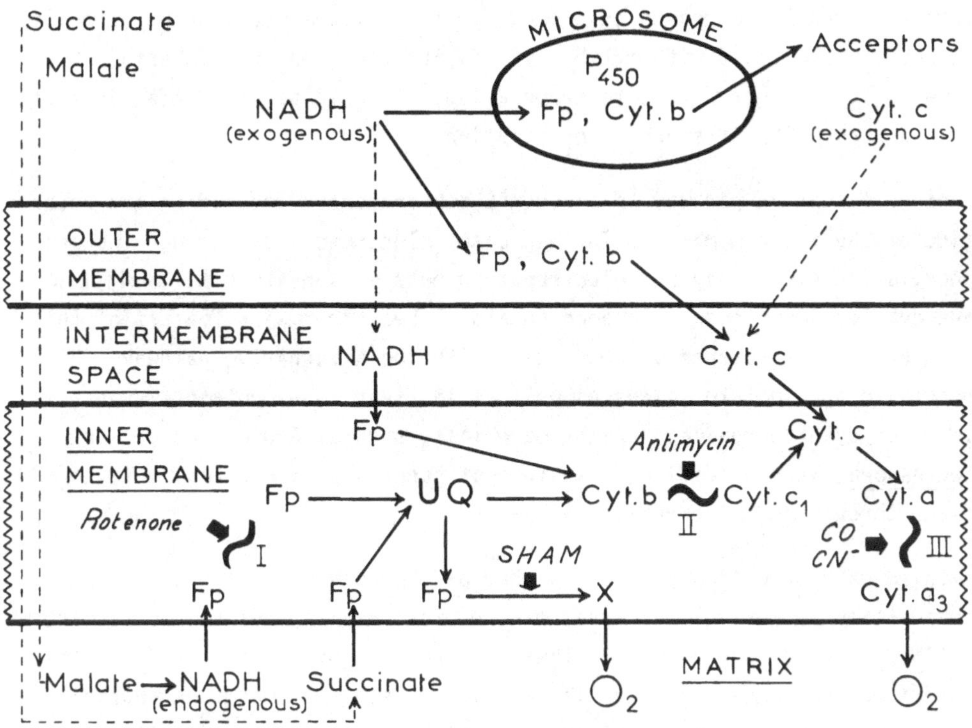

Fig. 11. Respiratory chain and substrate oxidation in mitochondria from aged potato slices. UQ: ubiquinone. Fp: various flavoprotein dehydrogenases (described in the text). X: terminal oxidase of the alternative pathway. I, II, III: phosphorylation sites.

Oxidation of malate in the matrix generates endogenous NADH which is oxidized by a rotenone-sensitive NADH dehydrogenase located on the inner side of the inner membrane. Three phosphorylation sites are involved in the oxidation of endogenous NADH. Phosphorylation site I is supposed to be located between two flavoproteins.

Succinate is oxidized via succinate dehydrogenase located on the inner side of the inner membrane. Electrons from succinate bypass site I, but can flow through the alternative pathway, when present.

The oxidation of exogenous NADH is much more complex, since no less than three pathways can contribute to it. One is the microsomal system, whose activity greatly increases during aging. This system, consisting of a flavoprotein associated with cyt $\underline{b}_5$, yields its electrons to hitherto unknown acceptors. Together with cytochrome P-450, it could be implicated in fatty acid desaturation or sterol hydroxylation.

Another oxidation pathway is through a dehydrogenase located on the outer side of the inner membrane. In this case, electrons flow through two phosphorylation sites only and electron transport is sensitive to antimycin and cyanide inhibitions. In aged potato slices, the major feature of this pathway is its absence of connection with the alternative pathway. This occurs in aged potato slices as well as in fresh sweet potato tissues, but in other cyanide-resistant plant materials, such as Aroid spadices or Neurospora, NADH oxidation, like that of succinate, can normally take place through the alternative pathway.

A third NADH oxidation pathway is that of the outer membrane system whose composition is analogous to that of the microsomal system (flavoprotein, cyt b) (54). Its operation requires the presence of soluble cyt c in the intermembrane space as a shuttle carrier between the two mitochondrial membranes. This pathway is antimycin-resistant and cyanide-sensitive and includes only one phosphorylation site (site III). It seems to normally exist and operate in standard plant mitochondria (20, 54, 80), but it intensifies to a remarkable extent during aging.

Although the picture of electron transport appears to be rather well known, at least as far as electron pathways are concerned, some problems are still unsolved and should provide matter for further research. Among them are the nature of the terminal oxidase of the alternative pathway, the significance of the general increase in NADH dehydrogenase activities, especially at the level of the outer mitochondrial membrane, and the nature of the control mechanisms that bring the alternative pathway into activity in aged tissues.

REFERENCES

1. Ap Rees T.: Evidence for the widespread occurence of induced respiration in slices of plant tissues. Aust. J. Biol. Sci. 19, 981-990 (1966).

2. Asahi T., Honda Y. and Uritani I.: Increase of mitochondrial content in sweet potato root tissue after wounding. Arch. Biochem. Biophys. 113, 498-499 (1966).

3. Asahi T., Honda Y. and Uritani I.: Increase of mitochondrial fraction in sweet potato root tissue after wounding or infection with Ceratocystis fimbriata. Plant Physiol. 41, 1179-1184 (1966).

4. Bahr J.T. and Bonner W.D.: Cyanide-insensitive respiration. I. The steady states of skunk cabbage spadix and bean hypocotyl mitochondria. J. Biol. Chem. 248, 3441-3445 (1973).

5. Beevers H.: Respiratory metabolism in plants. Row, Peterson and Company, Evanston, 232 p (1961).

6. Ben Abdelkader A.: Influence de la "survie" (ageing) sur la biosynthèse des phospholipides dans les "microsomes" de tubercule de pomme de terre (Solanum tuberosum L.) C. R. Acad. Sc.,Sér. D, 268, 2406-2409 (1969).

7. Ben Abdelkader A.: Biogénèse des lipides membranaires pendant la "survie" de tranches de tubercules de pomme de terre (Solanum tuberosum L.). Thèse Doctorat d'Etat, Paris. 184 p (1972).

8. Ben Abdelkader A., Cherif A., Demandre C. and Mazliak P.: The oleylcoenzyme-A desaturase of potato tubers. Enzymatic properties, intracellular localization and induction during "aging" of tuber slices. Eur. J. Biochem. 32, 155-165 (1973).

9. Bendall D.S. and Bonner W.D.: Cyanide-insensitive respiration in plant mitochondria. Plant Physiol. 47, 236-245 (1971).

10. Benveniste I. and Durst F.: Mise en évidence dans les tissus de tubercule de topinambour (Helianthus tuberosus L., variété blanc commun) d'une enzyme à cytochrome P-450, l'acide trans-cinnamique 4-hydroxylase (=CAH). C. R. Acad. Sci., Sér. D, 278, 1487-1490 (1974).

11. Bieleski R.L. and Laties G.G.: Turnover rates of phosphate esters in fresh and aged slices of potato tuber tissue. Plant Physiol. 38, 586-594 (1963).

12. Bonner W.D.: Mitochondria and electron transport. Plant Biochemistry, Academic Press, N.Y., 89-123 (1965).

13. Brinkman F.G., Van der Plas L.H.W. and Verleur J.D.: Pyridine nucleotide levels in potato tuber tissue and its mitochondrial fraction after wounding. Z. Pflanzenphysiol. 68, 364-372 (1973).

14. Cammack R. and Palmer J.M.: E.P.R. studies of iron-sulphur proteins of plant mitochondria. N.Y. Acad. Sci. 222, 816-823 (1973).

15. Carmeli C. and Biale J.B.: The nature of the oxidation states of sweet potato mitochondria. Plant Cell Physiol. 11, 65-81 (1970).

16. Chance B., Bonner W.D. and Storey B.T.: Electron transport in respiration . Ann. Rev. Plant Physiol. 19, 295-320 (1968).

17. Chance B. and Williams G.R.: The respiratory chain and oxidative phosphorylation. Advances Enzymol. XVII, 65-134 (1956).

18. Click R.E. and Hackett D.P.: The role of protein and nucleic acid synthesis in the development of respiration in potato tuber slices. Proc. Nat. Acad. Sci. 50, 243-250 (1963).

19. Cotte-Martinon M., Diano M., Meunier D. and Ducet G.: Etude de diverses fractions cellulaires dans le tubercule de pomme de terre au cours de la survie. Première partie: oxydations respiratoires et composition en cytochromes. Bull. Soc. Fr. Physiol. Vég. 15, 279-295 (1969).

20. Day D.A. and Wiskich J.T.: The oxidation of malate and exogenous reduced nicotinamide adenine dinucleotide by isolated plant mitochondria. Plant Physiol. 53, 104-109 (1974).

21. Dizengremel P.: La voie d'oxydation insensible au cyanure dans les mitochondries de tranches de pomme de terre (Solanum tuberosum L.) maintenues en survie. Physiol. Vég. 13, 39-54 (1975).

22. Dizengremel P.: Increased participation of the outer mitochondrial membrane in the oxidation of exogenous NADH during aging of potato slices. Plant Sci. Letters 8, 283-289 (1977).

23. Dizengremel P., Chauveau M. and Lance C.: Résistance au cyanure et teneur en ferrosulfoprotéines des mitochondries végétales. C. R. Acad. Sci., Sér. D, 277, 239-242 (1973).

24. Dizengremel P. and Lance C.: Control of changes in mitochondrial activities during aging of potato slices. Plant Physiol. 58, 147-151 (1976).

25. Douce R., Mannella C.A. and Bonner W.D.: The external NADH dehydrogenases of intact plant mitochondria. Biochim. Biophys. Acta 292, 105-116 (1973).

26. Gilewicz Cotte-Martinon M., Yahiel V. and Ducet G.: Induction d'un cytochrome du type P450 et de peroxydase durant la survie du tubercule de pomme de terre. Phytochem. 13, 2085-2090 (1974).

27. Hackett D.P.: Respiratory mechanisms in higher plants. Ann. Rev. Plant Physiol. 10, 113-146 (1959).

28. Hackett D.P.: Respiratory mechanisms and control in higher plant tissues. Control Mechanisms in Respiration and Fermentation, The Ronald Press Company, N.Y., 105-127 (1963).

29. Hackett D.P., Haas D.W., Griffiths S.K. and Niederpruem D.J.: Studies on development of cyanide-resistant respiration in potato tuber slices. Plant Physiol. 35, 8-19 (1960).

30. Hackett D.P. and Schneiderman H.A.: Terminal oxidases and growth in plant tissues. I. The terminal oxidase mediating growth of Avena coleoptile and Pisum stem sections. Arch. Biochem. Biophys. 47, 190-204 (1953).

31. Hackett D.P., Schneiderman H.A. and Thimann K.V.: Terminal oxidases and growth in plant tissues. II. The terminal oxidase mediating water uptake by potato tissue. Arch. Biochem. Biophys. 47, 205-214 (1953).

32. Hawker J.S. and Laties G.G.: Nicotinamide adenine dinucleotide in

potato tuber slices in relation to respiratory changes with age. Plant Physiol. 38, 498-500 (1963).

33. Henry M.F. and Nyns E.J.: Cyanide-insensitive respiration. An alternative mitochondrial pathway. Sub-Cell. Biochem. 4, 1-65 (1975).

34. Ikuma H.: Electron transport in plant respiration. Ann. Rev. Plant Physiol. 23, 419-436 (1972).

35. Jacobson B.S., Laties G.G., Smith B.N., Epstein S. and Laties B.: Cyanide-induced transition from endogenous carbohydrate to lipid oxidation as indicated by the carbon-13 content of respiratory CO_2. Biochim. Biophys. Acta 216, 295-304 (1970).

36. James W.O.: The terminal oxidases of plant respiration. Biol. Rev. 28, 245-260 (1953).

37. James W.O.: The use of respiratory inhibitors. Ann Rev. Plant Physiol. 4, 59-90 (1953).

38. Kahl G.: Genetic and metabolic regulation in differentiating plant storage tissue cells. Bot. Rev. 39, 274-299 (1973).

39. Kahl G.: Metabolism in plant storage tissue slices. Bot.Rev. 40, 263-314 (1974).

40. Kolattukudy P.E. and Reed D.J.: Metabolism of red beet slices. II. Effects of aging. Plant Physiol. 41, 661-669 (1966).

41. Lambowitz A.M., Smith E.W. and Slayman C.W.: Electron transport in Neurospora mitochondria. Studies on wild type and poky. J. Biol. Chem. 247, 4850-4858 (1972).

42. Lance C. and Bonner W.D.: The respiratory chain components of higher plant mitochondria. Plant Physiol. 43, 756-766 (1968).

43. Laties G.G.: The nature of the respiratory rise in slices of chicory root. Arch. Biochem. Biophys. 79, 364-377 (1959).

44. Laties G.G.: The development and control of coexisting respiratory systems in slices of chicory root. Arch. Biochem. Biophys. 79, 378-391 (1959).

45. Laties G.G.: Control of respiratory quality and magnitude during development. Control Mechanisms in Respiration and Fermentation, The Ronald Press Company, N.Y., 129-155 (1963).

46. Laties G.G. and Hoelle C.: Malonate and cyanide insensitivity in relation to respiratory compensation in potato slices. Plant Physiol. 40, 757-764 (1965).

47. Lee S.G. and Chasson R.M.: Aging and the development of enhanced respiration in potato tuber tissue. Physiol. Plantarum 19, 194-198 (1966).

48. Lee S.G. and Chasson R.M.: Aging and mitochondrial development in potato tuber tissue. Physiol. Plantarum 19, 199-206 (1966).

49. Levy H., Schade A.L., Bergmann L. and Harris S.: Studies in the respiration of the white potato. II. Terminal oxidase system of potato tuber respiration. Arch. Biochem. Biophys. 19, 273-286 (1948).

50. Loughman B.C.: Uptake and utilization of phosphate associated with

respiratory changes in potato tuber slices. Plant Physiol. 35, 418-424 (1960).

51. MacDonald I.R.: Changes in sensitivity to inhibitors of disks of storage tissue. Ann. Botany 23, 241-256 (1959).

52. Meeuse B.J.D.: Thermogenic respiration in Aroids. Ann. Rev. Plant Physiol. 26, 117-126 (1975).

53. Moore A.L., Rich P.R., Bonner W.D. and Ingledew W.J.: A complex EPR signal in mung bean mitochondria and its possible relation to the alternate pathway. Biochem. Biophys. Res. Commun. 72,1099-1107 (1976).

54. Moreau F.: Electron transfer between outer and inner membranes in plant mitochondria. Plant Sci. Letters 6, 215-221 (1976).

55. Nakano M. and Asahi T.: Biochemical studies on mitochondria formed during aging of sliced potato tuber tissue. Plant Cell Physiol. 11, 499-502 (1970).

56. Palmer J.M.: The influence of growth regulating substances on the development of enhanced metabolic rates in thin slices of beetroot storage tissue. Plant Physiol. 41, 1173-1178 (1966).

57. Palmer J.M.: The organization and regulation of electron transport in plant mitochondria. Ann. Rev. Plant Physiol. 27, 133-157 (1976).

58. Price C.A.: Respiration and development of vegetative plant organs and tissues. Encyclopedia of Plant Physiology XII/2, Springer Verlag, Berlin, 493-520 (1960).

59. Robertson R.N. and Turner J.S.: Studies in the metabolism of plant cells. III. The effects of cyanide on the accumulation of potassium chloride and on respiration; the nature of the salt respiration. Aust. J. Exp. Biol. Med. Sci. 23, 63-73 (1945).

60. Robertson R.N., Turner J.S. and Wilkins M.J.: Studies in the metabolism of plant cells. V. Salt respiration and accumulation in red beet tissue. Aust. J. Exp. Biol. Med. Sci. 25, 1-8 (1947).

61. Romberger J.A. and Norton G.: Changing respiratory pathways in potato tuber slices. Plant Physiol. 36, 20-29 (1961).

62. Rungie J.M. and Wiskich J.T.: Soluble electron transport activities in fresh and aged turnip tissue. Planta 102, 190-205 (1972).

63. Rungie J.M. and Wiskich J.T.: Changes in microsomal electron transport of plant storage tissues induced by slicing and aging. Aust. J. Biol. Sci. 25, 103-113 (1972).

64. Sakano K., Asahi T. and Uritani I.: Heterogeneity of mitochondrial particles in fresh and wounded tissues of sweet potato roots. Plant Cell Physiol. 9, 49-60 (1968).

65. Sakano K. and Asahi T.: Biochemical studies on biogenesis of mitochondria in wounded sweet potato root tissue. I. Time course analysis of increase in mitochondrial enzyme. Plant Cell Physiol. 12,417-426 (1971).

66. Sakano K. and Asahi T.: Biochemical studies on biogenesis of mitochondria in wounded sweet potato root tissue. II. Active synthesis of membrane-bound protein of mitochondria. Plant Cell Physiol. 12,427-436 (1971).

67. Schade A.L., Levy H., Bergmann L. and Harris S.: Studies on the respiration of the white potato. III. Changes in the terminal oxidase pattern of potato tissue associated with time of suspension in water. Arch. Biochem. Biophys. 20, 211-219 (1949).

68. Schonbaum G.R., Bonner W.D., Storey B.T. and Bahr J.T.: Specific inhibition of the cyanide-insensitive respiratory pathway in plant mitochondria by hydroxamic acids. Plant Physiol. 47, 124-128 (1971).

69. Storey B.T.: Respiratory chain of plant mitochondria. XVIII. Point of interaction of the alternate oxidase with the respiratory chain. Plant Physiol. 58, 521-525 (1976).

70. Thimann K.V., Yocum C.S. and Hackett D.P.: Terminal oxidases and growth in plant tissues. III. Terminal oxidation in potato tuber tissue. Arch. Biochem. Biophys. 53, 239-257 (1954).

71. Tomlinson P.F. and Moreland D.E.: Cyanide-resistant respiration of sweet potato mitochondria. Plant Physiol. 55, 365-369 (1975).

72. Van der Plas L.H.W., Jobse P.A. and Verleur J.D.: Cytochrome c dependent antimycin-A resistant respiration in mitochondria from potato tuber (Solanum tuberosum L.). Influence of wounding and storage time on outer membrane NADH-cytochrome c reductase. Biochim. Biophys. Acta 430, 1-12 (1976).

73. Van der Plas L.H.W. and Verleur J.D.: CN-resistant respiration in potato (Solanum tuberosum L.). Changes after storage of the tuber. Plant Sci. Letters 7, 149-154 (1976).

74. Van der Plas L.H.W., Wagner M.J. and Verleur J.D.: Changes in respiratory pathways of potato tuber (Solanum tuberosum L.) after various times of storage of the tuber. Comparison of wounded and non-wounded tissue. Z. Pflanzenphysiol. 79, 218-236 (1976).

75. Van Steveninck R.F.M.: The "washing" or "aging" phenomenon in plant tissues. Ann. Rev. Plant Physiol. 26, 237-258 (1975).

76. Van Steveninck R.F.M. and Jackman M.E.: Respiratory activity and morphology of mitochondria isolated from whole and sliced storage tissue. Aust. J. Biol. Sci. 20, 749-760 (1967).

77. Verleur J.D.: Observations on the induction of mitochondrial particles in potato tuber tissue after wounding. Z. Pflanzenphysiol. 61, 299-309 (1969).

78. Verleur J.D. and Uritani I.: Respiratory activity of the mitochondrial fraction isolated from healthy potato tubers and from tuber tissue incubated after cutting or infection with Ceratocystis fimbriata. Plant Physiol. 40, 1008-1012 (1965).

79. Verleur J.D., Van der Velde H.H. and Sminia T.: The heterogeneity of the mitochondrial fraction in relation to the increase of mitochondrial content in wounded and diseased potato tuber tissue. Z.Pflanzenphysiol. 62, 352-361 (1970).

80. Wilson R.H. and Hanson J.B.: Effect of respiratory inhibitors on NADH, succinate and malate oxidation in corn mitochondria. Plant Physiol. 44, 1334-1341 (1969).

Control of Ion Transport in Plant Storage Tissue Slices

Robert F. M. van Steveninck

1. INTRODUCTION

One important aspect of the wide spectrum of changes induced by slicing storage tissues concerns the osmotic components of the large vacuolated cells. The so-called osmotic volume (Briggs, Hope and Robertson, 1961) comprises approximately 80% of the total volume of the tissue and only 2.5% of this consists of cytoplasm, the rest being the large central vacuoles.

This Chapter will be mainly concerned with regulatory aspects controlling the contents of these vacuoles which are essential for the maintenance of turgor and the supply of materials for the sustenance of metabolism over prolonged periods following slicing. The osmotic components of vacuoles in fresh storage tissue consist largely of metabolic ions and organic compounds such as sugars, organic acids, etc. but often important shifts in contents occur as a result of slicing the tissue and during the subsequent adaptive aging of the slices (Van Steveninck, 1976a). The osmotic balance is maintained during these shifts and, largely because of the nature and consistency of the pattern of change which is intimately related to metabolism, membrane permeability and ultrastructure (Laties 1975, Van Steveninck, 1975), tissue slices provide excellent material for detailed studies of osmoregulation.

An excellent recent review on osmoregulation by Hellebust (1976) indicates that most of the information on this topic results from work on fungi and algae and that knowledge of mechanisms operating in higher plants is practically non-existent because of the greater complexity of higher plant systems. A further critical and highly relevant appraisal of negative feed-back regulation of ion fluxes is provided by Cram (1976), while more general aspects of ion transport in storage tissues have recently been reviewed by Cram (1975b), Laties (1975), Van Steveninck (1975, 1976a) and Poole (1976).

1.1. Osmoregulation Vs. Salt Saturation

Terms such as osmoregulation, turgor regulation and salt saturation have often been used indiscriminately and need to be clearly defined especially now that we are reaching a stage when mechanistic aspects can be described at a membrane molecular level (Coster *et al.*, 1976, Zimmermann *et al.*, 1977). Hellebust (1976) defines osmoregulation as a combination of processes and mechanisms by which plants regulate the osmotic pressure in their cells. It seems preferable in this context to use the term solute potential rather than osmotic pressure since it is easily confused with regulation of turgor pressure. The latter depends on the presence and elastic properties of cell walls and it appears that mechanisms underlying the control of turgor pressure (Coster *et al.*, 1976) may differ from those which involve osmotic adjustment or control of the solute potential. For instance, the control of turgor pressure seems to rely on a pressure transducing system, while the solute potential may depend on the actual concentration of a particular ion or solute in the vacuole (Cram, 1976), so that the nature of osmotic solutes as well as their compartmentation in the cell are important. In this connection the term salt saturation has often been used. It refers to a condition of flux equilibrium with respect to the ionic components of a particular salt which is present in the external solution. In other words when a tissue has been maintained in a constant external environment, ionic concentrations should remain at the same level in all cellular compartments because of equality of fluxes in opposite directions across any of the membrane barriers.

It is interesting that Poole (1976) states: "Although the ion content of tissue slices depends on the history of the plant, the cells are always far from salt saturation. Even when the tissue is at its most active stage it requires at least 4 days of continuous uptake to reach flux equilibrium with a constant external ion concentration." This is a somewhat simplistic view which does not fully consider the original state of the tissue and the subsequent dramatic changes with respect to osmoregulation and development of ion transport capacities which will result from slicing the tissue. Most dormant storage tissues could be considered to have reached a state of salt saturation. However, at the moment of slicing, drastic changes are induced which cause a rapid efflux of vacuolar contents from uncut cells for a period of at least 12 hours, quite apart from the immediate loss of contents from cut and damaged cells (Fig. 1). This initial period is generally followed by a condition of flux equilibrium at least for K^+, Na^+ and Cl^- for approximately 24 hours, and hence this latter period could be interpreted as a condition of salt saturation which applies until a further capacity for salt uptake is generated (end of lag phase, Van Steveninck, 1961a, b). Salt uptake may then resume until a new level of salt saturation is reached and hence it appears that at least 3 levels of salt saturation are pertinent (Table 1): one for the intact tissue, one for slices prior to the onset of net salt uptake capacity and one for the final equilibrium concentration which is reached *after* the completion of the lag phase.

TABLE 1.

Salt saturation levels of beetroot tissue and slices (Van Steveninck, unpublished data).

	Ion contents (μequiv. g^{-1})		
	K^+	Na^+	Cl^-
Intact beetroot	80.2	30.9	30.8
Tissue slices during stage II (see Fig.1).	63.2	24.7	22.0
Tissue slices aged in 20mM KCl or or NaCl until no further net uptake of K^+, Na^+ or Cl^- occurs (stage IV).	260	236	198

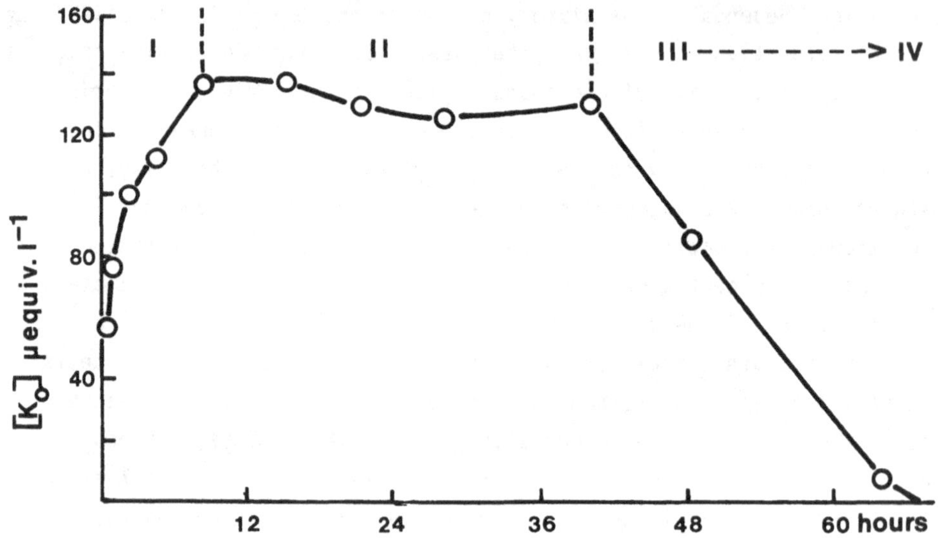

Fig. 1. K^+ contents of the external solution due to loss and re-absorption of K^+ by red beet slices

Stage I : Period of rapid loss of K^+ from cut and damaged cells
Net loss of K^+ from uncut cells occurs while $M_{co} > M_{oc}$

Stage II : Period of flux equilibrium or "salt saturation". That is, M_{co} and M_{oc} are approximately equal and presumably the same applies to M_{cv} and M_{vc}

Stage III : End of lag phase. Period of rapid net uptake of K^+. $M_{co} < M_{oc}$ and $M_{cv} > M_{vc}$. Provided sufficient K^+ is added to the external solution vacuolar concentrations of K^+ will increase dramatically over a period of several days until a new level of flux equilibrium or "salt saturation" is reached = Stage IV.

1.2 Compartmentation and Fluxes

It is highly desirable to acquire appropriate data regarding the ion contents of various membrane-enclosed compartments of tissue slice cells, especially to delineate the cytoplasmic phase as a separate entity from the vacuole. Hence it is important to distinguish the relative values of plasma membrane fluxes (M_{oc} and M_{co}) from the relative values of tonoplast fluxes (M_{cv} and M_{vc}), despite the fact that the cytoplasmic phase constitutes a relatively small volume with respect to the vacuole. For example, under conditions of very low K^+ supply in the presence of Na^+, the K^+ concentration in the cytoplasm could be relatively high and the Na^+ concentration relatively low in comparison with concentrations in the vacuole (Neirinckx and Bange, 1971, Hooymans, 1974). This is an important point when ion selectivity is discussed (section 4) and also when relative concentrations of organic acids in the cytoplasm and vacuole are considered with respect to the activity and location of K^+, Na^+ and H^+ pumps.

The first comprehensive flux analysis in storage tissues slices was carried out by Pitman (1963) and his data and subsequent work (Cram, 1968, Cram and Laties, 1971, Macklon, 1976) indicate that cytoplasmic exchange with the external solution is generally much more rapid ($t_{½} < 1$ h) than vacuolar exchange ($t_{½} >> 1$ h). It is therefore not unreasonable to assume that under conditions of quasi steady state flux equilibrium the cytoplasmic phase would reach salt saturation within a few hours [$(M_{oc} - M_{co}) = (M_{cv} - M_{vc})$] , but salt saturation with respect to the vacuole would be achieved only after some days ($M_{cv} = M_{vc}$). This is borne out by the comprehensive flux data of Macklon (1976) who carried out a thorough compartmental flux analysis in relation to the development of Na^+ and Cl^- absorption capacities in beetroot slices (Fig. 2). It can be seen that the magnitude of Na^+ flux (μeq g^{-1} h^{-1}) across the plasmalemma and tonoplast in fresh tissue is large with respect to the cytoplasmic concentration of Na^+. Hence, if the flux was largely in one direction only then the cytoplasmic concentration could be doubled in a matter of 1-2 hours. However, in fresh tissue this is not the case and therefore a condition of salt saturation exists because the fluxes in opposite directions are approximately equal both at the plasmalemma and at the tonoplast (Fig. 2). Aging leads to reduction of plasmalemma fluxes and a lowering of the cytoplasmic Na^+ concentration, but vacuolar accumulation of Na^+ sets in due to a dramatic decrease in efflux across the tonoplast. It is known that in aged beet slices Na^+ saturation levels may reach up to 250 μeq g^{-1}

(Table 1) and hence with a net inward Na^+ flux of 0.58 μeq $g^{-1} h^{-1}$, a state of salt saturation in the vacuole could be reached only after a period of unequal flux across the tonoplast extending over many hours. A similar case of contrasting time requirements for cytoplasmic and vacuolar salt saturation could be argued for Cl^-. Interestingly enough, the accumulation of Cl^- into the vacuole is not due to a lowering of efflux but to a large relative increase in influx across the tonoplast. Again the magnitude of the flux in comparison with vacuolar concentration at salt saturation is such (Table 1) that many hours would be required to reach this stage.

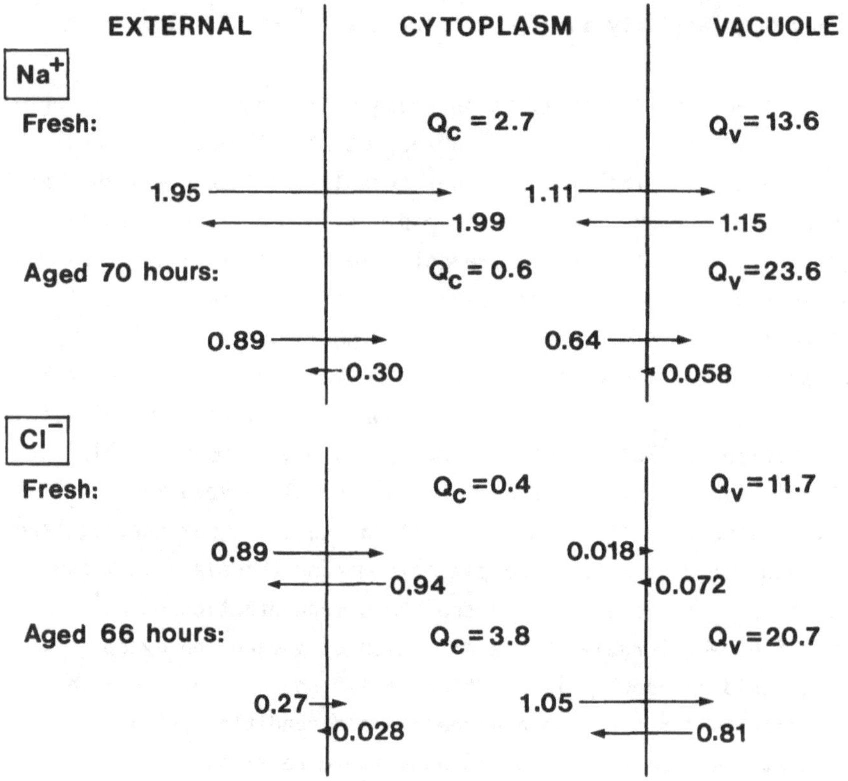

Fig. 2. Plasma membrane and tonoplast fluxes (μequiv. g^{-1} h^{-1}), Na^+ and Cl^- contents (μequiv. g^{-1}) of the cytoplasmic (Q_c) and vacuole (Q_v) compartments of fresh and aged red beetroot slices (data from Macklon, 1976).

Macklon's results confirmed earlier conclusions by Van Steveninck (1962, 1964) who showed that during adaptive aging of beet slices net K^+ uptake was initiated by a lowering of K^+ efflux from the tissue, and net Cl^- uptake by a relatively large increase of Cl^- influx. However, the great value of Macklon's (1976) contribution was to show that the changes in flux took place at the tonoplast rather than at the plasmalemma and to obtain accurate estimates of cytoplasmic and vacuolar concentrations for Na^+ and Cl^- (Fig. 2).

1.3. Metabolism and Energy Requirements

There is little doubt that cell metabolism depends on the buffering capacity of the cytoplasm and on the finely tuned regulatory mechanism of pH control, which in turn involves the maintenance of proton gradients across organelle membranes within the cytoplasmic compartment and across membranes separating the cytoplasm from free space and vacuole. Hence, of all the ions involved in cellular processes, protons or hydrogen ions (H^+), and of course hydroxyl ions (OH^-), should perhaps be considered to be the most important. Furthermore, it should be realised that although differences in H^+ concentration across membranes may differ by several orders of magnitude (pH-units), regulation apparently can occur at much lower concentrations (e.g. 10^{-8} - 10^{-5}M) than that of other ions such as K^+, Na^+ and Cl^- (e.g. 10^{-3} - 10^{-1}M).

Unfortunately, it is beyond the scope of this Chapter to deal with the many intricate aspects of H^+ (or OH^-) transport and regulation of cell pH, but an excellent review on this particular topic has recently been published (Smith and Raven, 1976). The topic of ions other than H^+ and OH^- and their interaction with enzymes and intermediary metabolism has also been reviewed recently by Osmond (1976).

H^+ or OH^- exchange between plant cells and their surroundings occurs in a wide range of metabolic processes, but one particular example of a mechanism of internal pH regulation ("pH stat") was proposed by Davies (1973) and may be relevant to the situation in storage tissues.

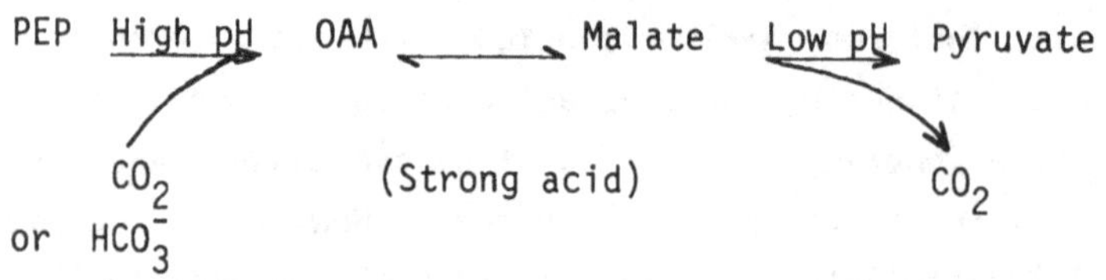

Because synthesis of malate would be expected to occur at a relatively high cytoplasmic pH and its breakdown to pyruvate at a relatively low cytoplasmic pH, a pH induced feedback mediated by changes of enzyme activity could provide a major role in maintaining cytoplasmic pH at a certain level. The scheme, of course, does not account for prolonged internal disposal of H^+ due to the production of H^+ during synthesis of malate. But it could account for continued organic acid synthesis while, for instance, a K^+/H^+ exchange mechanism operates at the plasmalemma. At the same time malic acid should be removed from the site of synthesis, for instance together with K^+ into the vacuole (see section 2.1).

Active proton efflux is becoming increasingly recognized as a basic function of plant cells, and it is now evident that it is also an important feature of storage tissue cells (see section 2.2). It has been suggested that transport of H^+ (and OH^-) evolved as a means of regulating intracellular pH and that this is its primary function (Raven and Smith, 1976). Evidence presented in the following sections will support this view and will show its importance throughout the developmental processes occurring during the adaptive aging of tissue slices. Nevertheless, the importance of H^+ and OH^- gradients in regard to energy relations should not be overlooked particularly in situations where a redox-driven H^+ pump exists alongside an ATP-driven H^+pump in the same membrane system.

Energy-coupling between redox reactions and ATP synthesis is a typical function of mitochondria and chloroplasts, and hence energy derived from respiratory metabolism or photosynthesis can result in either proton or electron gradients across membrane systems of these organelles or in the production of ATP (Robertson, 1968). Energy made available by photosynthesis and respiration can be utilized at other locations by means of further coupling between energy source and transport in particular membrane systems such as the endoplasmic reticulum (Rungie and Wiskich, 1972a, 1973), tonoplast and plasmalemma (Hodges, 1976). There has been some emphasis on the importance of H^+ pumps, especially those operating across the plasmalemma, in establishing ion gradients. This, of course, should not detract from the wealth of evidence existing in favour of ion-specific transport ATP-ases which mediate in energy transduction and ion transport across plant membranes (Hodges, 1976). The interchange between energy supply and maintenance of gradients of ions and solutes across membranes appears to be subjected to an intricate system of controls which may partly depend on the energy status of cells and on the levels of ions and solutes in a variety of cellular compartments.

In order to decide whether ions move passively along an existing electrochemical gradient or whether they are transported against a gradient with a concomitant expenditure of energy a number of parameters must be known and these can now be measured. The Nernst equation provides the simplest criterion to ascertain whether any particular ion is at equilibrium across a membrane:

$$E_{n_j} = \frac{RT}{z_j F} \ln \frac{a^o{}_j}{a^i{}_j} \quad \text{in which } E_{n_j}$$

denotes the potential difference across the membrane, z_j the valency of ion j and $a^o{}_j$, $a^i{}_j$ the chemical activities of ion j at the outside and inside of the membrane respectively. Reliable measurements of potential difference (PD) across plasmalemma and tonoplast respectively in storage tissue cells are still relatively scarce, but if it can be assumed that tonoplast PD is relatively small in most cases, with increasing knowledge of actual levels of ion concentrations in the cytoplasm and vacuoles respectively, predictions can be made whether energy must be expended to move a particular ion across the tonoplast or plasmamembrane.

An example of this approach may be found in the forceful arguments Poole (1974) used in his attempt to establish that an electrogenic proton pump operates across the plasmalemma of red beet storage tissue cells. On the basis of a measured hyperpolerization of -70mV induced by changing the pH of the external solution from pH 5.5 to 7.2 Poole argued that a cytoplasmic pH of less than 3.7 would be required to allow for a passive diffusion gradient of H^+ across the plasmalemma, which is obviously not the case.

Other arguments in favour of the electrogenic nature of the proton pump were based on the immediate sensitivity of the potential to temperature and its correlation with metabolic and transport activities during adaptive aging (Poole, 1974). In a later section of this Chapter (2.1.) however, it will become apparent that some caution may be necessary regarding these conclusions since pH changes were achieved using tris (tris-hydroxymethyl amino methane) buffer.

The energetic requirements for Cl^- transport provide an interesting aspect. Concentration and electrical measurements show that for initial net influx and to achieve the final steady state (salt saturation) active transport of Cl^- is required in beetroot, carrot and potato tissue slices (Cram, 1975b). Although these conclusions are based mainly on observed electrical potentials of the vacuole relative to the external solution and on vacuolar concentrations, certain assumptions, e.g. that the electrical potential difference is located largely at the plasmalemma, and estimates of cytoplasmic concentrations, were made to establish that Cl^- must be actively transported across the plasmalemma and possibly also at the tonoplast (Cram, 1975b). These estimates were based on cytoplasmic concentrations of Cl^- between 20-50 mM, which are at least an order of magnitude different from those determined by Macklon (1976) for beetroot slices (0.4 and 3.8 mM Cl^- in cytoplasm of fresh and aged slices respectively). With the Cl^- concentrations being generally over-estimated, especially with respect to the vacuolar concentration, it now appears that the accent may fall more strongly on active transport of Cl^- at the tonoplast. This conclusion, however, was already expected on the basis of experiments using inhibitors (oligomycin and anaerobiosis) as well as uncouplers (CCCP) of energy metabolism on tonoplast and plasmalemma Cl^- fluxes in carrot slices (Cram, 1969, 1975a). For instance, Cl^- influx at the plasmalemma was unaffected by CCCP and oligomycin, but reduced by N_2, while Cl^- influx at the tonoplast was reduced by all three agents. The direct effect of anaerobiosis on plasmalemma Cl^- influx indicated that a close and immediate link exists between this influx and respiratory redox reactions (Cram, 1969). It was also shown that in carrot slices Cl^- influx across the tonoplast increases approximately 3 days after slicing. This Cl^- influx in well-washed tissue was inhibited by CCCP while influx of Cl^-

across the plasmamembrane remained constant. It was therefore decided that in carrot slices the transport of Cl^- within the cell did not appear to be electrogenic (Cram, 1975a).

2. MECHANISMS OF ION TRANSPORT DURING ADAPTIVE AGING OF SLICES

It should be stressed that the term aging applies to changes in physiological competence which may occur during the passage of time, and that these changes find expression in the degree of membrane development (Jackman and Van Steveninck, 1967, Laties, 1975; Van Steveninck, 1975) or compartmentation of the component cells of the sliced storage tissue (Van Steveninck, 1976a). It has already been pointed out (Van Steveninck, 1976a) that processes taking place during aging may lead either to an ultimate, irreversible break-down of permeability barriers and hence loss of compartmentation *(senescent aging)* or to a further elaboration of physiological competence through a stimulation of protein sythesis involving fresh genetical information *(adaptive aging)*. Aging of storage tissue slices usually belongs to the adaptive type especially when storage tissues derive from roots, stolons or hypocotyl tissue (Poole, 1976). Thus the acquisition of new biochemical and physiological functions involves a further elaboration of membrane structures, compartmentation and the development of ion uptake capacities.

It is of particular interest that net ion uptake capacities develop sequentially for different ions. This property and the possibility of examining these sequential patterns under strictly controlled conditions would be most useful in the study of mechanisms of ion transport and of turgor or osmoregulation. A number of examples are now available and will therefore receive special attention in this section. For instance, it was shown (Van Steveninck and Van Steveninck, 1972) that in beetroot slices the *onset* of net Na^+ uptake usually precedes that of net K^+ uptake and that the latter may either coincide with or precede the net uptake of Cl^- (Table 2). Similar observations were made by Poole (1976) with regard to the times required for the development of maximal rates of transport of cations vs anions in red beet slices (e.g. 2-3 days for cations and 6-14 days for

TABLE 2

Times for the development of net Na^+, K^+ and Cl^- uptake by fresh beetroot slices in 10 separate experiments (Van Steveninck, unpublished results).

	Hours required for onset of net uptake		
	Na^+	K^+	Cl^-
Experiment 65	26	41	45
" 101	14	20	33
" 102	13	18	26
" 103	14	32	50
" 104	26	18	16
" 121	18	24	24
" 122	18	25	32
" 123	32	25	30
" 124	26	26	38
" 128	35	26	25

anions), although the tissues used by this author seemed to preferentially develop a capacity for net K^+ uptake well before net Na^+ uptake (Poole, 1976). It has since been shown that a particular sequence in the development of net K^+ and Na^+ uptake may be altered by the presence of abscisic acid (ABA) in the external medium (Van Steveninck, 1972a). It seems feasible that either genetical make-up or environmental conditions during the growth of the roots may determine the selective properties of the component parenchyma cells. For example, two types of root were selected:

a) the more common Na^+-selective roots with a K^+/Na^+ ratio in the tissue at the time of slicing of less than 2.0, and

b) K^+-selective roots with a K^+/Na^+ ratio of 3.0 or higher.

After completion of the lag phase the onset of net K^+ uptake in sliced tissue obtained from K^+-selective roots would normally precede that of Na^+, but this sequence could readily be reversed through the addition of 3.8 x 10^{-5} M ABA to the external medium. The rates of net K^+ and Na^+ uptake were also affected by the ABA treatment (Table 3). Generally, with both K^+ and Na^+ present in the external medium, net Na^+ uptake was stimulated by ABA treatment while net K^+ uptake was stimulated only in the absence of Na^+ or when the external Na^+ concentration fell to a very low level.

TABLE 3

Effect of 3.8 x 10^{-5} M ABA (added at 0 hours) on rates of net K^+ and Na^+ uptake by slices obtained from Na^+ selective ($[K_i^+]$ / $[Na_i^+]$ = 0.30) or K^+ selective ($[K_i^+]$ / $[Na_i^+]$ = 3.76) beetroots (Van Steveninck 1972a).

	(μequiv. g^{-1} h^{-1})			
Na^+ selective beetroot:	K^+		Na^+	
Time from slicing (h)	Control	ABA	Control	ABA
20 - 43	0.06	0.01	0.19	0.22
43 - 81	0.10	0.10	0.30	0.44
81 - 116	0.08	0.28	- *	- *
K^+ selective beetroot:				
20 - 27	0.14	0.05	0.02	0.15
27 - 50	0.15	-0.003	0.15	0.21
50 - 68	0.09	0.16	0.23	0.33
68 - 92	0.26	0.42	0.17	- *

* Na^+ becoming depleted in external solution.

It was also shown that benzyladenine (an analog of kinetin) prevented the development of net K^+ or Na^+ uptake during the aging of beetroot slices while no such effects were observed during the aging of slices of swede tissue *(Brassica napo-brassica* Mill.) (Van Steveninck, 1972b). An interesting aspect was that a closely similar effect was obtained by adding cycloheximide (CHM) or actinomycin D to the external medium at the start of the aging period (Van Steveninck and Van Steveninck, 1972). Actinomycin D, when added after the net uptake capacity had developed, was ineffective in causing further inhibition of net K^+, Na^+ or Cl^- uptake. However, the effects of later applications of benzyladenine or kinetin (Van Steveninck, 1972b) or cycloheximide (Van Steveninck and Van Steveninck, 1972) were similar in that either substance caused an inhibition of net K^+, Na^+ or Cl^-

uptake in aged tissue. But the period of time required for the inhibition to become apparent differed for each ion and became progressively longer when the application was made during a more advanced stage of adaptive aging (e.g. from 3 hours for Cl^- uptake up to a maximum of 25 hours for K^+ uptake). It was shown by means of ^{14}C-kinetin that these delays of apparent inhibition were not due to lack of penetration of the kinetin into the tissue (Van Steveninck, 1972b) and also it has been recorded that CHM applied to beetroot tissue slices will cause an inhibition of protein synthesis within half an hour after its application (Polya, 1968).

Another example of sequential changes in Na^+ and K^+ uptake has been observed during the adaptive aging of bean stem slices (Rains, 1969, Rains and Floyd, 1970). Fresh slices showed a net uptake of Na^+ over the range of 0.02 - 50mM while net K^+ uptake was zero at external concentration of less than 0.5mM. This situation was reversed in aged slices (20 hours in 0.5mM $CaSO_4$) where net K^+ uptake occurred over the entire range of 0.01-50mM, while net Na^+ uptake was zero at concentrations of less than 0.2mM. Thus, the capacity for net Na^+ uptake was lost and replaced by a capacity for net K^+ uptake under conditions of low external ion concentrations. It appeared that the presence of Ca^{++} in the aging medium was important in promoting the development of the K^+ absorption capacity (Rains and Floyd, 1970), while benzyladenine suppressed the development of the K^+ absorption capacity, but did not prevent loss of the Na^+ absorption capacity (Rains, 1969).

In one instance it was also recorded (Van Steveninck, 1976b) that gibberellic acid (GA3) caused a distinct shortening (22 $\rightarrow$ 8 hours) of the lag phase in the development of a net Na^+ uptake capacity in beetroot slices, while the duration of the lag for K^+ uptake was unaffected (38 hours).

Although the collection of information on this topic is still in its infancy it seems safe to conclude that plant growth substances exert an important regulatory effect on the onset of net uptake capacities and on the K^+/Na^+ selectivity of tissue slices (for further details see Van Steveninck, 1974, 1976b and section 3).

2.1. K^+/H^+ or Na^+/H^+ Antiport

In the light of growing evidence for the widespread occurrence of electrogenic proton pumps in plant cells it seems well justified to pay particular attention to what has been referred to as the "tris-effect" (tris-hydroxy-methylamino methane or 2 amino-2 hydroxymethyl-1:3 propanediol) in storage tissue slices. In the series of events involved in the development of ion uptake capacities in adapting storage slices, this effect allows the immediate rapid net uptake of K^+ or Na^+ in freshly cut discs (Van Steveninck, 1961, 1967a, b). Hence, the "tris-effect" bypasses any changes which might normally take place during the process of adaptive aging.

Although most of its features have been discussed in earlier reviews (Van Steveninck, 1973, 1975) it is pertinent to summarise those aspects which relate to the probable existence of electrogenic proton pumps in most tissue:

a) The "tris-effect" is specific for certain tissues. For instance, slices of beetroot, parsnip and carrot immediately respond to 10^{-2} M tris at pH 8 with a rapid and sustained rate of K^+ or Na^+ uptake, while uptake in swede, artichoke, potato and turnip discs is relatively unaffected by this treatment (Van Steveninck, 1961a, b). This is an important observation since it may provide a valuable indicator for identifying the actual mechanisms of ion transport in operation in a particular type of storage tissue, e.g. ion transport in beetroot, parsnips and carrots may primarily operate through an electrogenically driven K^+/H^+ antiport, perhaps supplemented to varying degrees by a K^+/Na^+ or Na^+/H^+ antiport, which in turn may determine ion selectivity (see section 3). Swedes, artichokes, potato and turnips on the other hand may depend on an entirely different system as a pacemaker, e.g. a Cl^-/OH^- or Cl^-/HCO_3^- antiport or a H^+/Cl^- symport.
b) The structure of tris and that of related molecules suggests that the effect may depend on certain structural requirements and specific interactions between tris and component molecules of membranes, possibly through a process of H-bonding. This notion is reinforced by the fact that any evidence of plasmalemma structure is consistently lacking in

storage tissue which has been exposed to tris buffer solutions and subsequently fixed in OsO_4 for electron microscopy (Van Steveninck *et al.*, 1973). A similar effect on OsO_4 fixation caused by the presence of urea has been reported by Littman and Barrnett (1972). It should therefore be interesting to check whether tissue specificity also applies to this type of fixation artifact induced by tris.

c) Experiments with tris at various pH and at various concentrations have shown that the non-protonated form (RNH_2) is effective in stimulating cation uptake in freshly sliced tissue (Van Steveninck, 1965, 1966a). Raising the pH of the external solution (e.g. by means of phosphate buffer) has a relatively insignificant effect on net K^+ uptake in sliced beetroot tissue (Van Steveninck, 1964), and it was also found that varying the pH of tris has little effect on the Cl^- accumulatory system during or after its development (Van Steveninck, 1964). Hence tris-stimulated net K^+ uptake appears to operate through a K^+/H^+ antiport system.

d) It has been asserted on previous occasions (Hurd, 1959) that HCO_3^- may play an important role in maintaining electrical neutrality during net K^+ uptake. This notion was dismissed by Poole (1974) on the basis of experiments in which HCO_3^- concentrations and pH were varied independently. Also HCO_3^- is not present as such in an alkaline tris solution because tris-base (RNH_2) rapidly combines with CO_2 to form tris carbamate ($RNHCOO^-$), in fact more rapidly than the hydration of CO_2 to form bicarbonate (Edsall and Wyman, 1958). Tris carbamate does not seem to play a role in accompanying K^+ uptake since tissues exposed to tris buffer show no increase in nitrogen content while significant amounts of K^+ are accumulated (Van Steveninck, 1966a).

e) Tris-induced excess cation uptake in fresh slices results in an increase in organic acid content (mainly of malate and citrate) of the tissue with the amounts involved showing a high degree of stoichiometry (Van Steveninck, 1966a). It appears that the monovalent cation content of the cytoplasm tends to decrease rather than increase during the process of adaptive aging (Macklon, 1976), hence in terms of the pH stat mechanism one might expect that a primary stimulation of the K^+/H^+ antiport system would result in the production of an appropriate amount of organic acid which, by using a K^+-organic acid symport system, is subsequently secreted into the vacuole. This system does not require participation of HCO_3^- transport because more than sufficient CO_2 is produced by cellular respiration to account for the formation of oxaloacetate and malate (Cram, 1974). It is therefore suggested that the primary action of tris involves the stimulation of an already existing K^+/H^+ antiport, which is supported by the cellular pH stat. This system would be expected to operate in intact mature storage tissues since most of these contain excess monovalent cations as K^+-organate. However, it should be stressed that because of relatively low internal Cl^- concentrations (Cram, 1975b) there may be an essential difference between this tris-stimulated operation of the cellular pH stat and the electrogenic pH pump which becomes operational after a period of adaptive aging (see section 2.2).

f) Accumulation ratios for K^+ observed in the presence of tris are of the order of 5×10^5 (Van Steveninck, 1961b), i.e. a flux equilibrium is reached with a K^+ concentration in the vacuole of approximately 100 mM and that of the external solution 2×10^{-4} mM. Macklon (1976) has determined the Na^+ contents for the cytoplasmic compartment of fresh and 70 hour aged beetroot tissue equilibrated in 1 mM NaCl at 2.7 and 0.6 μequiv. g^{-1} respectively, while Pitman (1963) determined a K^+ concentration of 4.3 μequiv. g^{-1} in 7 - 10 day old slices which were equilibrated in 5 mM KCl. With overall K^+/Na^+ ratios in beetroot tissue of the order of 1-5 it seems reasonable to assume cytoplasmic K^+ contents to be of the order of 10 μequiv g^{-1} in fresh slices. It thus appears that with an electrical potential difference of about -160mV (Poole, 1966) which is largely located at the plasmalemma (Cram, 1975b), tris-induced K^+ transport must be energized both at the plasmalemma and at the tonoplast. Unfortunately, this conclusion is based on a number of assumptions and requires verification of the actual cytoplasmic K^+ contents under conditions when tris buffer is used and external K^+ concentrations are allowed to drop to extremely low levels (2×10^{-4} mM). As long as active transport across the tonoplast remains operative it seems quite feasible for cytoplasmic K^+ contents to drop to below 1 μequiv. g^{-1}. Some early evidence that this form of cytoplasmic depletion can take place was produced by Laties (1959) for Cl^- when he observed the so-called "absorption shoulders" in his ion uptake experiments using aged potato tissue slices. This effect was later confirmed by Van Steveninck (1964) and ascribed to ion depletion of the cytoplasmic phase. Furthermore, Poole (1974) has measured hyperpolarizations of up to -50mV in beetroot slices by adding HCO_3^- and raising the pH of the external solution by means of tris buffer. Higinbotham *et al.* (1964) on the other hand observed a depolarization when exposing *Avena* coleoptile tissue to tris buffer, but in view of the tissue specificity of the tris effect (see section 2.1.a) this observation may have no relevance to the tris effect in fresh beetroot slices. Considering the above evidence it still seems uncertain whether tris-stimulation of K^+ uptake involves an active process across the plasmalemma or a passive process of facilitated K^+/H^+ exchange through the capacity of tris to act as a proton acceptor at the external side of the plasmalemma membrane. This is an important question since it will help in establishing whether the tris effect primarily occurs at the cell surface through a stimulation of K^+/H^+ antiport, or in the cytoplasmic phase through its effect on a cellular pH stat mechanism.

g) While there is considerable uncertainty over the question of whether the measured potential difference equates with the K^+ equilibrium potential (E_k) across the plasmalemma, this does not apply to proton gradients. Poole (1966) claimed that if the HCO_3^--induced hyperpolarization was to be attributed to passive diffusion of H^+ and HCO_3^-, the cytoplasmic pH would need to drop to less than 3.7, which under any circumstances would be most unlikely. However, Poole's conclusions are based on experiments which were conducted in the presence of tris buffer and also apply to tissue slices which have been aged more than one day and hence could have developed an electrogenic proton pump as part of the adaptive aging process. This, in fact, appears quite likely since measured potential differences increased from -170 mV to -200 mV, to -212 mV, and to -220 mV on days 1,2,3 and 4 respectively, while the estimated K^+ equilibrium potential (E_k) was only -140 mV. Although additional evidence will be required to establish the precise nature of the tris effect in fresh slices, it appears that at least in aged slices K^+ influx is balanced by apparent net efflux of H^+ and that it is primarily the flux of H^+ which is against an electrochemical diffusion gradient (see section 2.2).

h) The fact that labeled amines such as ^{14}C-methylamine are used to measure light-induced proton gradients across chloroplast membranes indicates the rapid penetration of the non-protonated amine (RNH_2) across membrane barriers. It is tempting to equate the effectiveness of tris-base (RNH_2) with its capacity for rapid penetration into the cytoplasmic phase, but its effectiveness may, of course, depend primarily on its capacity to act as a proton acceptor. From a structural point of view the presence of OH groups seems another essential requirement for its activity (Van Steveninck *et al.*, 1973), while the degree of substitution at the N-atom is relatively unimportant. This, together with the ultrastructural effects on the plasmalemma mentioned earlier (see 2.1.b), seems to indicate that tris primarily operates through molecular interactions with membrane components at the cell surface, e.g. those resulting from interactions between OH groups and H-bonds and other side chains containing $-COO^-$ and $\equiv N^+H$ groups. The tris effect would then be primarily on the K^+/H^+ antiport system at the plasmalemma, possibly combined with other effects on the permeability properties of the membrane towards a range of other ions. On the other hand, clear ultrastructural evidence is available (Mittelheuser and Van Steveninck, 1972) which shows that tris readily penetrates chloroplast and tonoplast membranes in leaf tissue. In these tissues significant proton gradients are expected to exist across chloroplast membranes in a direction favouring accumulation of protonated tris molecules on the inside of the chloroplast. It should be noted that such accumulation of tris does not appear to occur in mitochondria which would be expected to provide a proton gradient in the opposite direction to that of chloroplasts (Mittelheuser and Van Steveninck, 1972). Similarly, proton gradients could occur across the tonoplast in leaf cells which would normally be absent or in the opposite direction in cells of beetroot tissue. The fact that no ultrastructural evidence exists for accumulation of tris in beetroot storage slices (tris and the glutaraldehyde used for fixation are known to form a reaction product, cf. Hauptmann and Gabler, 1968) should not be taken as evidence that tris does not readily equilibrate across the plasmalemma. Furthermore, since the cytoplasm occupies only 2-5% of the tissue volume its equilibration in a 0.01 M tris solution should not present a measurable increase in the total N-content of the tissue (see section 2.1.d). Thus, the question whether the primary effect of tris is on the K^+/H^+ antiport system or primarily on a cytoplasmic pH stat mechanism can not be resolved on the basis of ultrastructural evidence alone.

2.2. The Onset of Electrogenic Proton Pumps

In this section an *a priori* assumption is made that enough evidence exists to justify the existence of electrogenic proton pumps in slices of storage tissue. It is required to establish whether these pumps exist in intact storage tissue or whether they arise during adaptive aging. A special feature of this process is that the tissue slices exercise a certain buffering capacity over the pH of the external solution. The effectiveness of this buffering capacity naturally depends on the ratio of weight of tissue to volume of solution in the same way as the concentration of a buffer. It was observed by Van Steveninck *et al.* (1973) that slices of certain tissues (carrot, parsnip, beetroot) when incubated caused a progressive rise in pH of the external solution precisely until the onset of net K^+ or Na^+ uptake. From then onwards the external pH drops to a much lower value (Fig. 3). Thus it seems that a close relation exists between K^+ or Na^+ uptake and H^+ extrusion. It is interesting to note that the tissues which show this kind of response are also the ones which are capable of showing an immediate response to tris buffer through a high rate of net K^+ or Na^+ uptake immediately after slicing. In this response to tris buffer a fairly close stoichiometric relationship is found to exist between excess cation (K^+ or Na^+) uptake and H^+ output (Table 4). On the other hand fresh slices from tissues which normally do not respond to tris treatment (e.g. potato, artichoke, swede) do not show the above pattern of pH change against time. In fact the pH in these cases tends to increase over the entire duration of the adaptive aging period (Fig. 3). It is suggested that this observation is based on an essential difference in the mechanism of development of ion uptake capacities, and it appears that, at least in carrot, parsnip and beetroot slices, the capacity for increased levels of K^+ and Na^+ accumulation is based on the operation of an electrogenic proton pump which becomes operative after a defined time of adaptive aging.

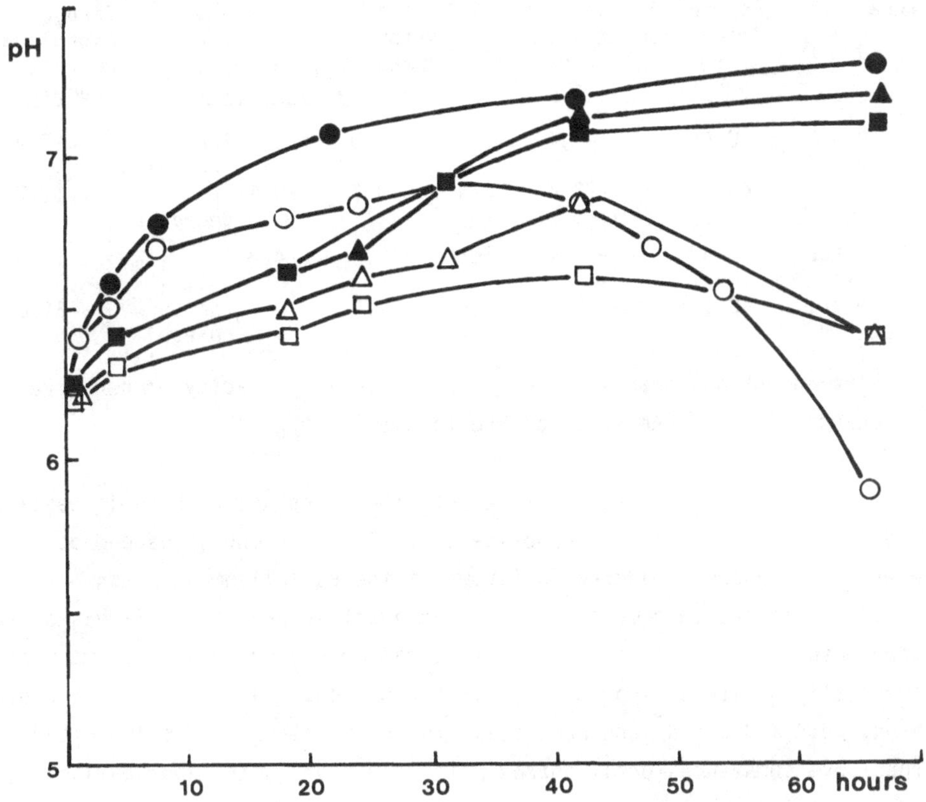

Fig. 3. pH changed external solution when slices of beetroot (○), carrot (△), parsnip (□), potato (●), artichoke (▲) and swedes (■) are placed in aerated deionized water.

It might be argued that the acidification of the external solution is the inevitable result of net cation uptake which is not accompanied by net Cl^- uptake. Yet, this does not indicate why it occurs only after a number of hours of adaptive aging and not earlier or immediately after slicing.

TABLE 4

Effect of tris on excess cation uptake, H^+ output and change of organic acid contents during adaptive aging of beetroot slices (Van Steveninck *et al.*, 1973 and unpublished data).

Treatment	Period of Incubation (h)	Net ion movement (μequiv.g^{-1}) K^+	Cl^-	Excess cation uptake (μequiv.g^{-1})	Estimated H^+ output (μequiv.g^{-1})		Change in organic acid content (μequiv.g^{-1})
10^{-3}M KCL	0-48	-7.4	-2.4	-5.0	-0.1	(pH increase)	-15.2
	48-117	+27.6	+24.4	+3.2	+0.1	(pH decrease)	-11.7
10^{-3}M KCL	0-48	+21.4	-5.8	+27.2	24.8	*	+17.0
+ 10^{-2}M tris pH 7.9	48-117	0	+21.6	-21.6	0	(no pH change)	-34.6

* Plus an unknown quantity due to the buffering capacity of negative charges in the free space of the tissue.

Also the likelihood that K^+ uptake into the cytoplasmic phase is passive and the H^+ movement in the opposite direction is bound to depend on energized transport, argues in favour of the establishment of an electrogenic proton pump as part of the adaptive process. It has already been pointed out that this development can be prevented by treatment of fresh slices with actinomycin D, kinetin or benzyl adenine. On the other hand, tris induced K^+ and Na^+ uptake in fresh slices is *not* inhibited by the above substances until approximately 10 hours after application. In fact, distinct promotions of tris-induced net K^+ and Na^+ uptake have been observed over the first few hours after treatment of fresh slices with benzyladenine (Van Steveninck, 1972b). It appears therefore that the tris effect in fresh slices, and the development of an electrogenic pump during adaptive aging, are two phenomena which are quite distinct from one another. Yet, they seem related in that they are both specific to certain tissues. Thus the tris effect may merely depend on the triggering of an existing cellular pH stat system resulting in excess production of protons which are subsequently exchanged for K^+ through an existing K^+/H^+ antiport system at the plasmalemma, whereas the electrogenic proton pump may depend primarily on the necessity of removing excess H^+ produced by a change in metabolism during adaptive aging. Although evidence of a negative type should be regarded with caution, the absence of a tris effect in potato, swede and artichoke slices, coupled with an absence of distinct proton output during adaptive aging seems to underline the fact that a close relation may exist between the cellular pH stat system and the development of an electrogenic proton pump. Further evidence for the apparent absence or lack of operation of these systems in swede, potato and artichoke tissues may be taken from the fact that kinin-like substances have no effect whatsoever on the development of ion uptake properties of swede slices during adaptive aging (Van Steveninck, 1972b).

2.3 Onset of a Cl^- Pump

In beetroot slices the onset of net Cl^- uptake generally follows the onset of the electrogenic proton pump usually after a delay of 3-12 hours. It is also easy to demonstrate that net cation uptake and net Cl^- uptake can operate quite independently, i.e. tris-induced net K^+ and Na^+ uptake and net loss of Cl^- may occur simultaneously in fresh slices, while in aged slices significant net Cl^- uptake can take place in the absence of net K^+ or Na^+ uptake (Table 4). These inequalities in cation and anion uptake closely relate to changes in organic acid content of the tissue slices. Often a close stoichiometry can be demonstrated between change in organic acid content and excess cation uptake or excess anion uptake respectively. It should be realized that as a rule total organic acid contents progressively diminish during the period of adaptive aging (Table 4).

This measure of independence between net K^+, Na^+ and Cl^- uptake indicates that the mechanism of Cl^- transport which evolves during adaptive aging does not necessarily rely on K^+ or Na^+ for the maintenance of electrical neutrality, and hence it may depend on co-transport of protons (Cl^-, H^+ symport) or on exchange with hydroxyl or bicarbonate ions (Cl^-/OH^- or Cl^-/HCO_3^- antiport). Existing data on Cl^- concentrations in the cytoplasm and vacuole of beetroot and carrot tissue slices (Cram, 1968, 1975b, Macklon, 1976) indicate that Cl^- is transported against an electrochemical gradient both at the plasmalemma and at the tonoplast (Fig. 4).

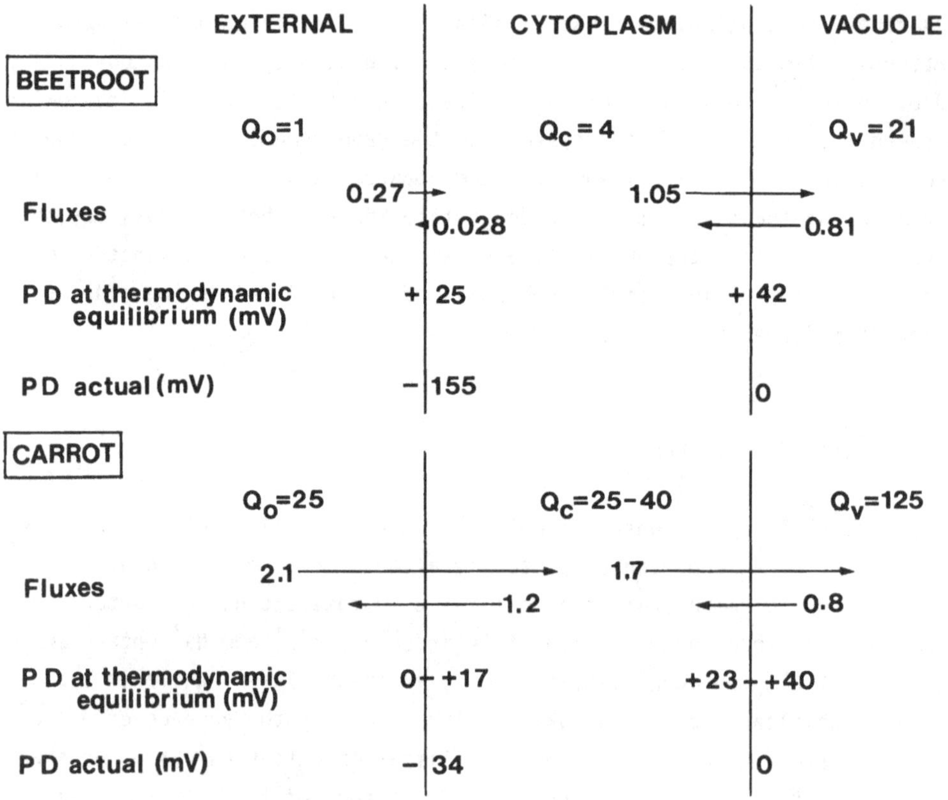

Fig. 4. Cl^- fluxes and electrochemical gradients at the plasmamembrane and tonoplast of beetroot and carrot slices. Fluxes at plasmalemma and tonoplast in μequiv.g^{-1}. h^{-1}. Cytoplasmic and vacuoler contents in μequiv.g^{-1}. Potential differences (PD) relative to the external solution in mV. Tonoplast potential differences are assumed to be zero (data from Cram 1975b and Macklon 1976).

Considering that OH^- and HCO_3^- movement generally appears to follow downhill gradients at cytoplasmic boundaries, the energized aspect of transport would appear to directly relate to the Cl^- ion. In this respect it is of considerable interest that the first example of induction of a Cl^- specific transport ATP-ase system has been detected in leaf discs of *Limonium* which were exposed to a high level of Cl^- ions (Hill and Hill, 1973). As in beetroot tissue, the induction of this Cl^- pump could be prevented by pretreatment of the leaf discs with actinomycin D or puromycin (Hill and Hill, 1973).

Van Steveninck (1964) showed that in contrast to the observed changes in monovalent cation fluxes (section 1.2), net Cl^- uptake during adaptive aging is triggered by a considerable increase in influx rather than a decrease in efflux. This conclusion was confirmed by a much more exhaustive analysis (Macklon, 1976) which showed that these changes mainly took place at the tonoplast (Fig.2). At the same time the cytoplasmic contents of Cl^- increased from 0.4 μequiv.g^{-1} during a 66 hour period of adaptive aging. Cl^- contents in the vacuole are also usually low in freshly sliced beet, carrot or potato tissue (about 10 μequiv. g^{-1}, see Cram, 1975) and hence the relatively high amounts of K^+ (60-120 μequiv. g^{-1}) are generally counter-balanced by relatively high concentrations of organic acid (about 100 μequiv. g^{-1}, Cram, 1975b). Hence, one of the most important osmotic components of fresh tissue slices consists of K^+ organate and sugars (20 μequiv. g^{-1} of the latter in potato, 100 μequiv. g^{-1} in beetroot and 200 μequiv. g^{-1} in carrot; for further details see Cram, 1975b). Slices of storage tissue generally exhibit a considerable increase in respiratory activity during adaptive aging (MacDonald and de Kock, 1958; Laties, 1963; ap Rees, 1966) and hence it may be expected that much of the organic material initially present in the vacuole is required to provide the appropriate substrate. Thus we can expect that sugar and organic acid contents will diminish during the process of aging.

From a purely teleological point of view it could be expected that the tissue or cells would require an alternative solute to sugars and organic acids in order to maintain the solute potential in the relatively large storage vacuoles when the organic contents become depleted by respiratory metabolism. Cram (1973), however, has shown that Cl^- influx is in no way determined by the organic acid and/or reducing sugar concentrations or by the turgor of the storage vacuoles in aged carrot slices, while on the other hand, the Cl^- influx did seem to be determined by the combined level of Cl^- and NO_3^- concentrations. The role of NO_3^- in controlling Cl^- influx is not understood. A presumably related but also unexplained observation is that influxes of NO_3^- and of malate are reduced in a similar way by high Cl^- concentrations in the vacuoles of carrot cells (Cram, 1974).

Another point worth considering is that from the moment tissue slices are cut nothing is added to the system in the way of metabolizable energy-containing substrate. Hence, the process of adaptive aging progresses on the basis of a finite energy supply. Smith and Raven (1976) make the point that increased cytoplasmic OH^- availability resulting from K^+/H^+ exchange could favour Cl^- transport either as a simple result of an increased pH gradient in a simple Cl^-/OH^- exchange system, or by the operation of a cytoplasmic pH controlled feed-back on a (Cl^-/OH^-) transport ATP-ase. As a means of countering internal alkalinity Cl^-/OH^- exchange or OH^- extrusion would be approximately 10 times less expensive energetically than malate synthesis (Smith and Raven, 1976). Thus, in a system where the stock of organic acids and sugars cannot be replenished, and energy may become in such short supply that utilization of these organic solutes becomes imperative, it would seem to make sense that Cl^-/OH^- exchange would become one of the major means of maintaining cytoplasmic pH and the solute potential of the vacuoles.

3. ION SELECTIVITY AND SATURATION LEVELS

Already brief reference has been made to the concept of salt saturation with respect to osmo-and turgor regulation (section 1.1). In this section it is proposed to explore the effect of salt saturation on ion selectivity, bearing in mind that both the activity of certain ion pumps and the saturation levels may change during the process of adaptive aging. Also, it should be remembered that various storage tissues may differ substantially in their ion selectivities (Table 5). For instance, the relative preferences of different storage tissues for K^+ rather than Na^+ may range from artichoke > potato > swede > carrot > red beet with artichoke at one end of the spectrum having a K^+/Na^+ ratio in fresh tissue of up to 30 and beetroot at the other end with ratios as low as 0.3.

TABLE 5

Ratios of K^+ to Na^+ contents in slices of various storage tissues (Van Steveninck, unpublished data).

	$[K_i^+]$	$[Na_i^+]$	$[K_i^+] / [Na_i^+]$
Artichoke	187	7.0	26.8
Beetroot (K^+ selective)	94	17	5.5
(Na^+ selective)	24	84	0.29
Carrot	86	22	3.9
Parsnip	48	7.2	6.7
Potato	93	6.7	13.9
Turnip	40	12.4	3.2

Large differences may also exist within one particular type of tissue e.g. K^+/Na^+ ratios in fresh beetroot have been found to range from 0.3 to 5.5 (Van Steveninck, 1972a and Table 5). In aged tissue slices the ratios may again be greatly different depending on whether tissue slices have been exposed exclusively to NaCl or KCl, or to a mixture of both. In spite of these variations it seems worthwhile to consider aspects of K^+/Na^+ selectivity since it may be important for ion regulation, especially in connection with effects of Ca^{++} on relative membrane permeabilities (P_{K^+} and P_{Na^+}), effects of ABA on ion pumps, and adaptive aging in general. Further knowledge of these aspects may improve our understanding of problems related to salinity tolerance and osmoregulation under stress conditions, of which relatively little is known for storage tissue slices. But observations by Pitman *et al.* (1968, 1971) on barley roots may be relevant. With respect to K^+/Na^+ selectivity, defined by Pitman (1967) as

$$S_{K,Na} = \frac{K^+ \text{ uptake}}{Na^+ \text{ uptake}} \times \frac{[Na_o]}{[K_o]}$$

it was shown that the overall salt status of the seedlings was of considerable importance. Although internal ($[K_i^+] / [Na_i^+]$) ratios might tend towards unity in "low salt" roots (seedlings grown in $CaSO_4$ only, Jackson and Stieff, 1965), Pitman (1967) found that at salt saturation $[K_i^+] / [Na_i^+]$ could reach 6 - 10 times $[K_o^+] / [Na_o^+]$ when seedlings were grown on a nutrient solution ("high salt" roots).

Rates of salt transport were thought to be influenced by the levels of reducing sugars present, some typical values for "low salt" roots being 15 μequiv. g^{-1} K^+ and 60 μmoles g^{-1} reducing sugars, and for "high salt" roots 95 μequiv. g^{-1} K^+ and 15 μmoles g^{-1} reducing sugars respectively (Pitman *et al.*, 1971). Progressive changes in K^+/ Na^+ selectivity became apparent as "low salt" roots approached salt saturation, i.e. during the first 5 hours there was little preference for K^+ over Na^+, but after that time until when salt saturation was reached, K^+ uptake exceeded Na^+ uptake due to a progressive reduction in Na^+ influx. Thus values for selectivities of K^+ over Na^+ ($S_{K,Na}$) of up to 18 were observed (Pitman 1967).

It was suggested that this change of selectivity is due to an increased $[K_c^+]$ / $[Na_c^+]$ in the cytoplasm which also leads to altered proportions of K^+ and Na^+ taken up into the vacuole.

At this point it should be stressed that a clear distinction must be made between changes in ion selectivity which relate to salt status and changes in selectivity which result from the development of ion pumps during the adaptive aging of tissue slices. The former is brought about by progressive changes in relative magnitudes of influx and efflux as salt saturation is approached, while the latter depends on developmental changes which can be prevented by exposure of the tissue to substances such as actinomycin D or cycloheximide, e.g. in aging bean stem slices (Rains and Floyd, 1970) and beetroot slices (Van Steveninck, 1972b).

As stated previously data in the former category are relatively scarce for storage tissue slices. Some selectivity ratios for slices have been calculated on the basis of existing data (Table 6).

It has been shown that abscisic acid (ABA) has very little or no effect on the duration of the lag phase (Van Steveninck, 1972a), but ABA has a remarkable effect on K^+/Na^+ selectivity once both the K^+ and Na^+ pumps are operative (Table 6). The experiments were carried out on two types of beetroot, i.e. "Na^+-selective" and "K^+-selective" beetroot with K^+/Na^+ ratios of 0.3 and 3.76 respectively at the time of slicing. K^+/Na^+ selectivity figures ($S_{K,Na}$) decreased from 0.45 to 0.06 as a result of ABA treatment in slices originating from a "Na^+-selective" root, while the decrease in selectivity was even more astonishing in the "K^+-selective" root (0.89 and 0.03 respectively for control and ABA treatments respectively; Table 6). These results stress the importance of controlling and clearly defining conditions used for growth of experimental material (for other examples see Pitman, 1967; Kähr and Kylin, 1974).

TABLE 6

The effect of ABA (3.8×10^{-5}M) on selectivity ratios ($S_{K,Na}$) of slices obtained from either "Na^+-selective" ($[K_i^+] / [Na_i^+] = 0.30$) or "$K^+$-selective" ($[K_i^+] / [Na_i^+] = 3.76$) beetroot.

	Uptake over a period of 20-43 hours incubation (μequiv. g^{-1} h^{-1})		Mean external ion concentration (mM)		$S_{K,Na}$
	K^+	Na^+	$[K_o^+]$	$[Na_o^+]$	
"Na-selective" beetroot					
Control	0.06	0.19	0.61	0.86	0.45
+ ABA	0.01	0.22	0.65	0.91	0.06
"K^+-selective" beetroot					
Control	0.17	0.14	0.61	0.44	0.89
+ ABA	0.01	0.23	0.72	0.41	0.03

Poole (1971a) has provided a precise analysis of the effect of Na^+ on cytoplasmic K^+ content and K^+ fluxes across the plasmalemma and tonoplast in 5-6 day aged beetroot slices. It was shown that 0.4 - 0.5 mM Na^+ strongly inhibits net K^+ uptake (1.19 μequiv. g^{-1} h^{-1} and 0.11 μequiv. g^{-1} h^{-1} in the absence and presence of Na^+ respectively with 0.3 mM $KHCO_3$; 1.98 μequiv. g^{-1} h^{-1} and 0.51 μequiv. g^{-1} h^{-1} in the absence and presence of Na^+ respectively with 1.0 mM $CaCl_2$). This inhibition affects mainly the K^+ influx across the plasmalemma, although presence or absence of Ca^{++} also had a significant effect in reducing K^+ efflux from the cytoplasm. It was quite evident that, although Poole (1971a) used a particularly "Na-selective" beetroot, the levels of K^+ in the cytoplasm remained high in spite of external Na^+ because K^+ transfer to the vacuole was inhibited under conditions of rapid Na^+ accumulation.

The advantages of the flux analysis carried out by Poole (1971a) are obvious but it should be realised that individual roots may give widely different results particularly when "K^+-selective" and "Na^+ selective" roots are used. Even so, it should prove rewarding to accumulate further data for other types of tissue slices (carrot, potato, swedes, etc.) which show a high degree of "K^+-selectivity", and also to further extend this work to include a study of the effects of ABA and kinin-like substances. This, together with an examination of the endogenous levels of these substances (as recently carried out by Conrad and Köhn (1975) on potato slices) may then further define the role of hormones in the regulation of ion selectivity and transport (for a relevant review see Van Steveninck, 1976b).

4. CONCLUSION: REGULATION OF ION TRANSPORT IN TISSUE SLICES

One intriguing aspect of the fascinating range of developmental processes taking place in sliced storage tissue concerns changes in the transport and compartmentation of inorganic ions. As is the case with metabolic changes, particular pathways, either existing or newly developed, become predominant. However, in ion transport flux rates are determined either by energized transport (ion pumps) or by a process of passive diffusion which is in turn controlled by membrane permeability and by the steepness of the electrochemical gradients. Evidently, the equilibrium concentrations depend on the relative magnitudes of the various fluxes and it has become a major challenge to try to understand the regulatory mechanism or feed-back components which control such ion movements.

It appears that feed-back may depend on some factors which provide gross adjustments and others which are responsible for fine adjustments during the period of adaptive aging. For instance, a typical example of a gross adjustment in slices of beetroot storage tissue may be the development of an electrogenic proton pump leading to a sustained rapid uptake of Na^+ and/or K^+, closely followed by the development of a Cl^- pump. One type of tissue (beetroot, carrot, parsnip) already possesses a K^+ or Na^+/H^+

antiport which can be stimulated by tris induced pH changes in the cytoplasm immediately after slicing (tris effect on cellular pH stat). This antiport system seems to be linked in some way with a propensity for the tissue slices to develop an electrogenic proton pump during adaptive aging. This induction can be prevented by the addition of kinin-like substances, actinomycin D and cycloheximide. Slices of another tissue type (swede, potato, artichoke) do not show a tris effect, neither do they have the capacity to develop an active electrogenic proton pump.

Gross adjustments may closely interact with other metabolic changes induced by slicing. For instance, respiratory rates are greatly stimulated by slicing and important shifts in the respiratory pathway are known to take place (Kahl, 1974; Laties, 1975). Further, the principal substrate utilized may alter drastically, e.g. initial respiratory stimulations may depend on α-oxidation of fatty acids (Laties *et al.*, 1972), while during later stages a close correlation seems to exist between levels of reducing sugars and rate of respiration (MacDonald and de Kock, 1958). Naturally, the relative proportions of readily utilizable energy substrate in the form of starch, sugars, organic acids etc. are determined by the type and condition of the storage tissue and for this reason alone experimental results for ion transport and its regulation may show substantial variations between species. It seems inevitable that changes in level of salt saturation and total solute-potential will be interrelated with the regulation of carbohydrate metabolism. For instance, kinins, ABA and GA_3 are shown to have marked effects on storage and hydrolysis of starch (Mittelheuser and Van Steveninck, 1971; Tasseron-de Jong and Veldstra, 1971; Hadačová *et al.*, 1973; Jones, 1973; Kull and Unger, 1974), and much evidence is available concerning the induction or activation of enzymes involved in the breakdown of starch and reducing sugars (Palmer, 1966; Vaughan and MacDonald, 1967a, b; Ricardo and Ap Rees, 1970; Rutherford and Bard, 1971). However, it should be noted that Cram (1973) has not been able to observe any effects of sugar concentration on the rate of Cl^- influx during adaptive aging in carrot tissue slices, neither did there seem to be any evidence that hydrostatic pressure could have a direct effect on the development of the Cl^- pump (Cram, 1972). This result seems to be in sharp contrast with some earlier observations on *Valonia* by Gutknecht (1968) who showed that an inhibition of K^+ influx could be induced by small increases in hydrostatic pressure. Of course, one should avoid making comparisons between exceedingly different cell types. But perhaps further exploration for the existence of mechanisms based on turgor regulation are justified when it is realized that Cram's conclusions are related to an aspect of gross regulation, i.e. the effect of turgor on the *development* of a Cl^- pump, rather than on feed-back or the fine adjustment of an already existing pump. However, Cram (1973) was able to demonstrate an example of fine adjustment based on the dependence of Cl^- and NO_3^- influx on the sum of Cl^- and NO_3^- concentrations in the vacuole of carrot slices. It was also shown that no relation existed between Cl^- influx and vacuolar concentrations of K^+, Na^+, malate, sugar and Cl^-, or NO_3^- separately.

Other examples relating to the regulation of already operating transport systems may be found in the effects of fusicoccin (FC) on electrogenic proton pumps (Marré *et al.*, 1974). This FC-induced stimulation of H^+ efflux seemed to be dependent on the presence of K^+ in the external medium, and it was suggested that it might involve a K^+ activated ATP-ase in the plasmalemma. Pitman *et al.* (1975a, b) found that FC was equally effective in stimulating H^+ efflux in the presence of 5 mM KCl or 5 mM NaCl when low salt roots were used, but in salt saturated roots the stimulation of H^+ efflux in 5 mM NaCl was only 20% of that in 5 mM KCl. Thus the differences in sensitivity seemed to be related to changes in K^+/Na^+ selectivity of the roots and the ratio of P_{Na^+} / P_{K^+} (low salt roots: $P_{Na^+} / P_{K^+} = 0.8 - 0.9$; high salt roots: $P_{Na^+} / P_{K^+} = 0.3 \rightarrow 0.03$). It seems an exciting proposition to extend these studies to slices of storage tissue. Some effects of ABA on storage tissue slices have already been mentioned. ABA seems to have profound effects on K^+/Na^+ selectivity in beetroot slices (Van Steveninck, 1972a), but in addition up to 10-fold stimulations of Cl^- influx by ABA have been recorded in aged beetroot slices which have acquired a capacity for net Cl^- uptake (Van Steveninck, 1974).

The effect of divalent ions in maintaining the physial properties of membranes should always be considered (Van Steveninck, 1965; Epstein, 1973). For example, Ca^{++} has profound effects on K^+/Na^+ selectivity in bean stem slices (Rains and Floyd, 1970), in this case, mainly through its stimulatory effect on the development of K^+ uptake capacity during the adaptive aging of the slices. Fisher and Albersheim (1973) have also shown a specific requirement for Ca^{++} in the operation of H^+ efflux pumps in the plasmalemma of cultured *Sycamore* and bean cells. These effects are likely to relate to stimulations of ion specific transport ATP-ases (Hodges, 1976).

Finally, in respect to ion regulation it remains important to monitor overall changes in structure and properties of membrane systems during the process of adaptive aging (Van Steveninck, 1975). These include cell membrane potentials such as wound-induced hyper-polarizations (Koopowitz *et al.*, 1975), the properties of the endoplasmic reticulum (Jackman and Van Steveninck, 1967; Rungie and Wiskich, 1972b) and the properties of mitochondria (Van Steveninck and Jackman, 1967b; Nakano and Asahi, 1970). However, these topics fall outside the scope of this Chapter and the reader is therefore invited to consult the relevant Chapters of this volume.

REFERENCES

Ap Rees, T. : Evidence for the wide spread occurrence of induced respiration in slices of plant tissue. Aust. J. Biol. Sci. 19, 981-990 (1966).

Briggs, G.E., Hope, A.B., Robertson, R.N. : Electrolytes and plant cells. Blackwell Scientific Publications, Oxford, (1961).

Conrad, K., Köhn, B. : Zunahme von Cytokinin und Auxin in verwundetem Speichergewebe von *Solanum tuberosum*. Phytochem. 14, 325-328 (1975).

Coster, H.G.L., Steudle, E., Zimmermann, U. : Turgor pressure sensing in plant cell membranes. Plant Physiol. 58, 636-643 (1976).

Cram, W.J. : Compartmentation and exchange of chloride in carrot root tissue. Biochim. Biophys. Acta. 163, 339-353 (1968).

Cram, W.J. : Respiration and energy-dependent movements at plasmalemma and tonoplast of carrot root cells. Biochim. Biophys. Acta. 173, 213-222 (1969).

Cram, W.J. : The initiation of developmental drifts in excised plant tissues. Aust. J. Biol. Sci. 25, 855-859 (1972).

Cram, W.J. : Internal factors regulating nitrate and chloride influx in plant cells. J. Exp. Bot. 24, 328-241 (1973).

Cram, W.J. : Effects of Cl^- on HCO_3^- and malate fluxes and CO_2 fixation in carrot and barley root cells. J. Exp. Bot. 25, 253-268 (1974).

Cram, W.J. : Relationships between chloride transport and electrical potential differences in carrot root cells. Aust. J. Plant Physiol. 2, 301-310 (1975a).

Cram, W.J. : Storage tissues. In : Ion transport in plant cells and tissues (D.A. Baker and J.L. Hall eds), pp 161-191, North-Holland Publishing Company, Amsterdam, (1975b).

Cram, W.J. : Negative feedback regulation of transport in cells. The maintenance of turgor, volume and nutrient supply, In : Encyclopedia of Plant Physiology, New Series (U. Luttge and M.G. Pitman, eds), II A, pp 284-316, Springer Verlag, Berlin, (1976).

Cram, W.J., Laties, G.G. : The use of short-term and quasi steady influx in estimating plasmalemma and tonoplast influx in barley root cells at various external and internal chloride concentrations. Aust. J. Biol. Sci. 24, 633-646 (1971).

Davies, D.D. : Control of and by pH. Symp. Soc. Exp. Biol. 27, 513-529 (1973).

Edsall, J.T., Wyman, J. : Biophysical Chemistry, Vol. 1, Academic Press, New York (1958).

Epstein, E. : Mechanisms of ion transport through plant cell membranes. Intern. Rev. Cytol. 34, 123-168 (1973).

Fisher, M.L., Albersheim, P. : A calcium dependent hydrogen pump in plant plasma membranes. Plant Physiol. 51 (Suppl), 2 (1973).

Gutknecht, J. : Salt transport in *Valonia* : Inhibition of potassium uptake by small hydrostatic pressures. Science, N.Y. 160, 68-70 (1968).

Hadačová, V., Luštinec, J., Kamiňek, M. : Kinetin and naphthalene acetic acid controlled starch formation in isolated roots of *Pisum sativum*. Biol. Plant Acad. Sci. Bohemoslov. 15, 427-429 (1973).

Hauptmann, S., Gabler, W. : Zur Reaktion von Tris (hydroxymethyl) Amino Methan (Tris) mit einigen Aldehyden. Z. Naturf. 23b, 111-112 (1968).

Hellebust, J.A. : Osmoregulation. Ann. Rev. Plant Physiol. 27, 485-505 (1976).

Higinbotham, N., Etherton, B., Foster, R.J. : Effect of external K, NH_4,

Na, Ca, Mg and H ions on the cell transmembrane electropotential of *Avena* coleoptile. Plant Physiol. 39, 196-203 (1964).

Hill, B.S., Hill, A.E. : ATP-driven chloride pumping and ATP-ase activity in the *Limonium* salt gland. J. Membrane Biol. 12, 145-158 (1973).

Hodges, T.K. : ATP ases associated with membranes of plant cells. In : Encyclopedia of Plant Physiology, New Series (U. Lüttge and M.G. Pitman eds), II A, pp 260-293, Springer Verlag, Berlin, (1976).

Hooymans, J.J.M. : Role of cell compartments in the distribution of K^+ and Na^+ ions absorbed by the roots of intact barley plants. Z. Pflanzenphysiol. 73, 234-242 (1974).

Hurd, R.G. : An effect of pH and bicarbonate on salt accumulation of disks of storage tissue. J. Exp. Bot. 10, 345-358 (1959).

Jackman, M.E., Van Steveninck, R.F.M. : Changes in the endoplasmic reticulum of beetroot slices during aging. Aust. J. Biol. Sci. 20, 1063-1068 (1967).

Jackson, P.G., Stief, K.J. : Equilibrium and ion exchange characteristics of potassium and sodium accumulation by barley roots. J. Gen. Physiol. 48, 601-616 (1965).

Jones, R.L. : Gibberellins : their physiological role. Ann. Rev. Plant Physiol. 24, 571-598 (1973).

Kahl, G. : Metabolism in plant storage tissue slices. Bot. Rev. 40, 263-314 (1974).

Kähr, M., Kylin, A. : Effects of divalent cations and oligomycin on membrane ATP-ases from roots of wheat and oat in relation to salt status and cultivation. In : Membrane transport in plants, (U. Zimmermann and J. Dainty eds), pp 321-325, Springer and Verlag, Berlin, (1974).

Koopowitz, H., Dhyse, R., Fosket, D.E. : Cell membrane potentials of higher plants : changes induced wounding. J. Exp. Bot. 26, 131-137 (1975).

Kull, U., Unger, M. : Wirkungen von Abscisinsäure auf den Kohlenhydrat-und Fettsäurehaushalt von *Coleus blumei*. Z. Pflanzenphysiol. 72, 135-140 (1974).

Laties, G.G. : The generation of latent-ion-transport capacity. Proc. N.A.S. 45, 163-172 (1959).

Laties, G.G. : Control of respiratory quality and magnitude during development. In : Control mechanisms in respiration and fermentation (B. Wright, ed), pp 129-155, Ronald Press, New York (1963).

Laties, G.G. : Solute transport in relation to metabolism and membrane permeability in plant tissues. In : Historical and current aspects of plant physiology : A symposium honoring Prof. F.C. Steward. Cornell University : New York State College of Agriculture and Life Science Publication, pp 98-151 (1975).

Laties, G.G., Hoelle, C., Jacobson, B.S. : α-Oxidation of endogenous fatty acids in fresh potato slices. Phytochemistry 11, 3403-3411 (1972).

Littman, R.B., Barnett, R.J. : The mechanism of the fixation of components by osmium tetroxide via hydrogen bonding. J. Ultrastr. Res. 38, 63-86 (1972).

Mac Donald, I.R., de Kock, P.C. : Temperature control and metabolic drifts in ageing disks of storage tissue. Ann. Bot. 22, 429-448 (1958).

Mac Donald, I.R., Laties, G.G. : Kinetic studies of anion absorption by potato slices at $0^{o}C$. Plant Physiol. 38. 38-44 (1963).

Macklon, A.E.S. : An examination, by compartmental flux analysis of the sodium and chloride absorption capacities in beetroot disks. J. Exp. Bot. 27, 651-657 (1976).

Marré, E., Lado, P., Rasi-Caldogno, F., Colombo, R., de Michelis, M.I. : Evidence for the coupling of proton extrusion to K^+ uptake in pea internode segments treated in fusicoccin or auxin. Plant Science Letters 3, 365-379 (1974).

Mittelheuser, C.J., Van Steveninck, R.F.M. : The ultrastructure of wheat leaves. II. The effects of kinetin and ABA on detached leaves incubated in the dark. Protoplasma 73, 253-262 (1971).

Mittelheuser, C.J., Van Steveninck, R.F.M. : Light-induced effects of tris on the ultrastructure of leaf tissue. Aust. J. Biol. Sci. 25, 517-530 (1972).

Nakano, M., Asahi, T. : Biochemical studies on mitochondria formed during ageing of sliced potato tuber tissue. Plant Cell Physiol. 11, 499-502 (1970).

Neirinckx, L.J.A., Bange, G.G.J. : Irreversible equilibration of barley roots with Na^+ ions at different external Na^+ concentrations. Acta. Bot. Neérl 20, 481-488 (1971)

Osmond, C.B. : Ion absorption and carbon metabolism in cells of higher plants. In : Encyclopedia of Plant Physiology, New Series (U. Lüttge and M.G. Pitman eds), II A,pp 347-372, Springer Verlag, Berlin, (1976).

Palmer, J.M. : The influence of growth regulating substances on the development of enhanced metabolic rates in thin slices of beetroot storage tissue. Plant Physiol 41, 1173-1178 (1966).

Pitman, M.G. : The determination of the salt relations of the cytoplasmic phase in cells of beetroot tissue. Aust. J. Biol. Sci. 16, 647-668 (1963).

Pitman, M.G. : Conflicting measurements of sodium and potassium uptake by barley roots. Nature 216, 1343-1344 (1967).

Pitman, M.G., Courtice, A.C., Lee, B. : Comparison of potassium and sodium uptake by barley roots at high and low salt status. Aust. J. Biol. Sci. 21, 871-881 (1968).

Pitman, M.G., Mowat, J., Nair, H. : Interactions of processes for accumulation of salt and sugar in barley plants. Aust. J. Biol. Sci. 24, 619-631 (1971).

Pitman, M.G., Schaefer, N., Wildes, R.A. : Effect of salt status of roots on stimulation of K^+ and Na^+ uptake by fusicoccin. Plant Physiol, 56 (Suppl), 82 (1975a).

Pitman, M.G., Schaefer, N., Wildes, R.A. : Relation between permeability to potassium and sodium ions and fusicoccin - stimulated hydrogen ion efflux in barley roots. Planta 126, 61-73 (1975b).

Polya, G.M. : Inhibition of protein synthesis and cation uptake in beetroot tissue by cycloheximide and cryptopleurine. Aust. J. Biol. Sci. 21, 1107-1118 (1968).

Poole, R.J. : The influence of the intracellular potential on potassium uptake by beetroot tissue. J. Gen. Physiol, 49, 551-563 (1966).

Poole, R.J. : Effect of sodium on potassium fluxes at the cell membrane and vacuole membrane of red beet. Plant Physiol, 47, 731-734 (1971a).

Poole, R.J. : Development and characteristics of sodium-selective transport in red beet. Plant Physiol 47, 735-739 (1971b).

Poole, R.J. : Ion transport and electrogenic pumps in storage tissue cells. Can. J. Bot. 52, 1023-1028 (1974).

Poole, R.J. : Transport in cells of storage tissues. In : Encyclopedia of Plant Physiology, New Series (U. Lüttge and M.G. Pitman eds), II A, pp 229-248, Springer Verlag, Berlin, (1976).

Rains, D.W. : Sodium and potassium absorption by bean stem tissue. Plant Physiol. 44, 547-554 (1969).

Rains, D.W., Floyd, R.A. : Influence of calcium on sodium and potassium absorption by fresh and aged bean stem slices. Plant Physiol. 46, 93-98 (1970).

Ricardo, C.P.P., Ap Rees, T. : Invertase activity during the development of carrot roots. Phytochemistry 9, 239-247 (1970).

Robertson, R.N. : Protons, electrons, phosphorylation and active transport. pp 1-96, Cambridge University Press, London (1968).

Rottenberg, H., Grunwald, T., Avron, M. : Determination of a pH in chloroplasts. I. distribution of ^{14}C methylamine. Eur. J. Biochem. 25, 54-63 (1972).

Rungie, J.M., Wiskich, J.T. : Soluble electron-transport activities in fresh and aged turnip tissue. Planta 102, 190-205 (1972a).

Rungie, J.M., Wiskich, J.T. : Changes in microsomal electron transport of plant storage tissues induced by slicing and aging. Aust. J. Biol. Sci. 25, 103-113 (1972b).

Rungie, J.M., Wiskich, J.T. : Salt-stimulated adenosine triphosphatase from smooth microsomes of turnip. Plant Physiol 51, 1064-1068 (1973).

Rutherford, P.P., Bard, D.R. : Water uptake and invertase and hydrolase activities induced in chicory root disks by treatment with various plant growth regulators. Phytochemistry, 10, 1635-1638 (1971).

Smith, F.A., Raven, J.A. : H^+ transport and regulation of cell pH. In : Encyclopedia of Plant Physiology, New Series (U. Lüttge and M.G. Pitman eds), IIA, pp 317-346, Springer Verlag, Berlin, (1976).

Sutcliffe, J.F. : The influence of internal ion concentration of potassium accumulation and salt respiration of red beet tissue. J. Exp. Bot. 3, 59-76 (1952).

Sutcliffe, J.F. : The relation uptake of alkali cations by red beet tissue. J. Exp. Bot. 8, 36-49 (1957).

Tasseron-de Jong, J.G., Veldstra, H. : Investigations on cytokinins I. Effect of 6-benzylamino purine on growth and starch content of *Lemna minor*. Physiol. Plant. 24, 235-238 (1971).

Van Steveninck, R.F.M. : The lag phase in ion uptake by plant tissue. Ph.D. Thesis, University of London (1961a).

Van Steveninck, R.F.M. : The "lag-phase" in salt uptake of storage tissue. Nature 190, 1072-1075 (1961b).

Van Steveninck, R.F.M. : Potassium fluxes in red beet tissue during its "lag phase". Physiol. Plant. 15, 211-215 (1962).

Van Steveninck, R.F.M. : A comparison of chloride and potassium fluxes in red beet tissue. Physiol. Plant. 17, 757-770 (1964).

Van Steveninck, R.F.M. : The effects of calcium and tris (hydroxymethyl) aminomethane on potassium uptake during and after the lag phase in red beet tissue. Aust. J. Biol. Sci. 18, 227-233 (1965).

Van Steveninck, R.F.M. : Some metabolic implications of the tris effect in beetroot tissue. Aust. J. Biol. Sci. 19, 271-281 (1966a).

Van Steveninck, R.F.M. : Further aspects of the tris effect in beetroot tissue during its lag phase. Aust. J. Biol. Sci. 19, 283-290 (1966b).

Van Steveninck, R.F.M. : Abscisic acid stimulation of ion transport and alteration in K^+/Na^+ selectivity. Z Pflansenphysiol. 67, 282-286 (1972a).

Van Steveninck, R.F.M. : Inhibition of the development of a cation accumulatory system and tris-induced uptake in storage tissue by N - benzyladenine and kinetin. Physiol. Plant. 27, 43-47 (1972b).

Van Steveninck, R.F.M. : Hormonal regulation of ion transport in parenchyma tissue. In : Membrane transport in plants, (U. Zimmermann and J. Dainty eds), pp 450-456, Springer Verlag, Berlin, (1974).

Van Steveninck, R.F.M. : The "washing" or "aging" phenomenon in plant tissues. Ann. Rev. Plant Physiol. 26, 237-258 (1975).

Van Steveninck, R.F.M. : Cell differentiation, aging and ion transport. In : Encyclopedia of Plant Physiology, New Series (U. Luttge and M.G. Pitman eds), II B pp 343-371, Springer Verlag, Berlin, (1976a).

Van Steveninck, R.F.M. : Effect of hormones and related substances on ion transport. In : Encyclopedia of Plant Physiology, New Series (U. Luttge and M.G. Pitman eds), II B pp 307-342, Springer Verlag, Berlin, (1976b).

Van Steveninck, R.F.M., Jackman, M.E. : Respiratory activity and morphology of mitochondria isolated from whole and sliced storage tissue. Aust. J. Biol. Sci. 20, 749-760 (1967).

Van Steveninck, R.F.M., Van Steveninck, M.E. : Effects of inhibitors of protein and nucleic acid synthesis on the development of ion uptake mechanisms in beetroot slices (*Beta vulgaris*). Physiol. Plant. 27, 407-411 (1972).

Van Steveninck, R.F.M., Mittelheuser, C.J., Van Steveninck, M.E. : Effects of Tris-buffer on ion uptake and cellular ultrastructure. In : Ion transport in plants (W.P. Anderson ed), pp 251-269, Academic Press, New York, (1973).

Vaughan, D., Mac Donald, I.R. : Invertase development in storage tissue disks of *Beta vulgaris* : its nature, extent and location. J. Exp. Bot. 18, 578-586 (1967a).
Vaughan, D., Mac Donald, I.R. : Development of soluble and insoluble invertase activity in washed storage tissue. Plant Physiol. 42, 456-458 (1967b).
Zimmermann, U., Beckers, F., Coster, H.G.L. : The effect of pressure on the electrical breakdown in the membranes of *Valonia utricularis*, Biochim. Biophys. Acta. 464, 399-416 (1977).

Studies on DNA and RNA Polymerase Activities and Ribosomal RNA in Plant Storage Tissue

Joe H. Cherry

INTRODUCTION

When intact storage tissue, such as roots or tubers, is sectioned and washed, a dramatic increase in metabolic activity ensues. This appears to be a general phenomenon in higher plants and has been regarded as a derepression phenomenon (10, 12, 13, 20, 25, 39). Potato discs incubated in aerated, distilled water undergo a time-dependent increase in respiration (10), which is dependent upon the synthesis of new RNA and protein. The synthesis of invertase and of other enzymes in washed discs of sugar beet, artichoke, carrot, and red beet is also dependent upon the synthesis of new RNA and protein (1, 7, 12, 21, 39). The synthesis of nucleic acid precedes the rise in enzyme activities. The development of invertase activity can be enhanced further by addition of exogenously applied GA and can be inhibited by added auxin (7, 21).

Recent investigations of DNA polymerase activity in eukaryotic tissues indicated the presence of multiple DNA polymerases in different cellular locations having distinct properties (5, 26, 27, 33). In addition, the cellular concentration of certain of these enzymes changes during specific stages of normal or induced growth (6, 9, 14). In an attempt to elucidate the controls of nucleic acid synthesis in higher plants, we have chosen sugar beet storage tissue as an experimental system. Following slicing of the tissue, washing for extended time periods in sterile phosphate buffer results in a dramatic increase in nucleic acid synthesis (11, 21). To investigate the possible presence and properties of multiple DNA polymerases in plants, the isolation of chromatin from unwashed and washed sugar beet tissue was selected as a starting point and a source of the enzyme(s).

Furthermore, both the type of RNA formed and the rate of RNA synthesis during development have been shown to vary with stages of differentiation

Biochemistry of Wounded Plant Tissues

(8, 38). Such fluctuations in RNA synthesis may result from changes in chromatin (4) such as alterations in the DNA template available for transcription. Based on evidence from several laboratories (30, 35), however, regulation of the synthesis and activity of RNA polymerase may also result in the observed changes in RNA synthesis. For example, multiple RNA polymerases have been shown to exist in several eukaryotic organisms (16, 28, 31, 32) and the presence of certain of these enzymes has been shown to vary during development (30). Therefore, an investigation of RNA synthesis in sugar beet tissue has included a study of chromatin-bound RNA polymerase activity.

DNA POLYMERASE

Chromatography on DEAE-Sephadex

DNA polymerase activity isolated from unwashed and washed tissue was dependent on Mg^{2+}, added DNA and all four deoxynucleoside triphoshates. Activity was linear through 30 minutes and proportional to added protein up to 50 µg. Extracts (from unwashed and washed tissue) were separately adsorbed to columns of DEAE-Sephadex A-25 (1 x 20 cm) previously equilibrated with TGMED buffer (31), pH 7.9. The columns were washed with 15 ml of 0.032 M $(NH_4)_2SO_4$ in TGMED and then eluted with an 80 ml linear gradient of $(NH_4)_2SO_4$ (0.032 to 0.08 M) in TGMED (Figure 1). Four major activity peaks were observed from unwashed and washed tissues. Peaks I and II of both tissues appeared to elute from the columns at the same salt concentration. Peak III from washed tissue, a major activity peak, appeared to be absent in unwashed tissue. Although similar amounts of protein were solubilized from both tissues, washing resulted in a tremendous increase in DNA polymerase activity (Figure 1). Activity peaks from unwashed tissue were less stable than those from washed tissue. Activity was reduced by 50% after one week in liquid nitrogen for peaks isolated from unwashed tissue. With the exception of peak I, the activity from washed tissue was reduced only 10% after two weeks with repeated freezing and thawing. Other attempts at purification using DEAE-Sephadex in batch columns with two salt steps followed by chromatography with linear gradients on the salt step fractions have resolved four activity peaks at the salt concentrations observed in Figure 1.

Properties of DNA Polymerase Activities

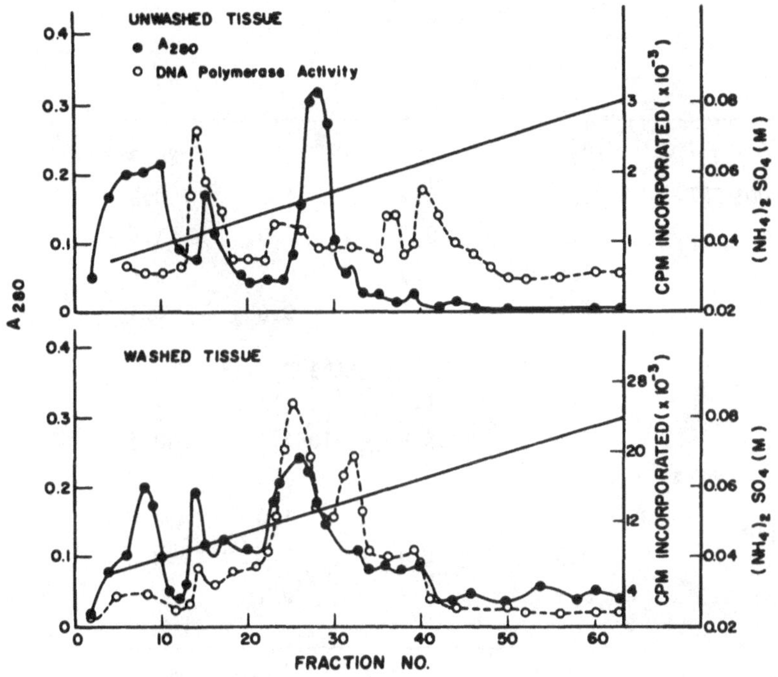

Figure 1. DEAE-Sephadex chromatography of DNA polymerase activity solubilized from chromatin of unwashed and washed sugar beet tissue. Unwashed tissue profile: Peak I = tubes 14-16, Peak II = tubes 23-25, Peak III = tubes 36, 37 and Peak IV = tubes 40-42. Washed tissue profile: Peak I = tubes 14-16, Peak II = tubes 24-26, Peak III = tubes 31-34 and Peak IV = tubes 38-40.

Table 1. Properties of DNA polymerase activity peaks following DEAE-Sephadex chromatography.

	Unwashed tissue			
	Peak			
	I	II	III	IV
Complete	15.5*	38.9	11.3	20.3
$-Mg^{2+}$ + 12 mM EDTA	0.5	0.9	1.0	0.2
-dNTP	1.1	0.8	0.0	1.5
-DNA	0.5	0.3	0.2	0.2
-enzyme	0.2	0.1	0.0	0.1
	Washed Tissue			
	I	II	III	IV
Complete	72.6	1936.5	1054.2	668.1
$-Mg^{2+}$ + 12 mM EDTA	0.0	0.0	0.0	6.0
-dNTP	3.0	4.1	1.0	0.6
-DNA	0.0	1.0	1.6	0.8
-enzyme	0.2	0.0	0.1	0.7

*Activity expressed as pmoles/mg of protein in 30 minutes employing saturating levels of template yielding the most activity for each peak.

Peak tubes of the four major activity peaks from both tissues were separately pooled and the DNA polymerase activity characterized (Table 1). Activity in each peak was dependent on all four deoxynucleoside triphosphates, Mg^{2+}, added DNA and enzyme. Activity was linear through 45 minutes and proportional with respect to enzyme concentration (up to 40 μg in assay). Further characterization of the activity of each peak was possible due to the high activity in peaks isolated from washed tissue. Activity in all four peaks was completely sensitive to actinomycin D (40 μg/ml). The products of the reaction employing each peak separately were completely sensitive to pancreatic DNase (20 μg/ml) and insensitive to RNase (20 μg/ml). Activity was approximately 80% of that observed with activated calf thymus DNA when poly (dA-T) was employed as template. No activity was observed when poly (rA-dT) was employed as the primer/template for any of the activity peaks.

RNA POLYMERASE

Solubilization of RNA Polymerase from Chromatin

To determine a procedure which would routinely solubilize the most activity from chromatin and still maintain an active RNA polymerase, three different procedures were compared (Table 2). As can be seen, all three procedures removed 80% of the total chromatin-bound RNA polymerase activity. The specific activity of the solubilized enzyme, however, was the highest when the stirring procedure was employed. In the sonication and homogenization experiments the solution frothed easily, and this probably resulted in enzyme denaturation. The remainder of the experiments reported here utilized soluble RNA polymerase obtained by the stirring procedure.

Characterization of RNA Polymerase

Sugar beet RNA polymerase using denatured calf thymus DNA as template gave a linear rate of activity 6-10 min then plateaued. If additional substrates (nucleoside triphosphates), DNA or enzyme were added to the assay after 10 min, little increase in activity was observed. The enzyme reaction was linear with enzyme concentration (up to 40 μg of protein) and depended on the presence of each nucleoside triphosphate, added DNA and divalent metal ions (Table 3). The activity was completely inhibited by actinomycin-D and insensitive to rifampin. As can be seen in Fig. 2, the enzyme exhibited a sharp Mn^{2+} optimum of 1.25 mM with higher concentrations

Table 2. Solubilization of RNA polymerase from chromatin.

Procedure	% of chromatin activity remaining on pellet	Specific activity of soluble enzyme*
Sonication	21	5
VirTis homogenizer	23	17
Magnetic stirrer	20	25

*pmol of ^{3}H-UMP incorporated/min/mg protein.

Two procedures for solubilization of enzyme from chromatin were compared to that described in Experimental (Magnetic stirrer). Chromatin from 600 g of 20 hr washed tissue was either sonicated for a total time of 1 min (in 10 sec intervals to reduce frothing) in the presence of 0.3 M $(NH_4)_2SO_4$ with a Branson S125 (microtip attachment, setting 3), or homogenized in the presence of 0.3 M $(NH_4)_2SO_4$ for 30 min in a 5 ml VirTis homogenizing flask using microblades.

Table 3. Components of RNA polymerase assay.

Assay components	% of control activity	Assay components	% of control activity	Assay components	% of control activity
Complete	100	-GTP	2	-Enzyme	0
-DNA	10	$-Mg^{2+}$	40	+Heated enzyme (60°, 10 min)	0
-ATP	3	$-Mn^{2+}$	80	+Actinomycin D (200 µg/ml)	0
-CTP	1	$-Mg^{2+}/Mn^{2+}$	5	+Rifampin (200 µg/ml)	100

Control activity was 392 pmol of UMP incorporated per mg protein in 10 min.

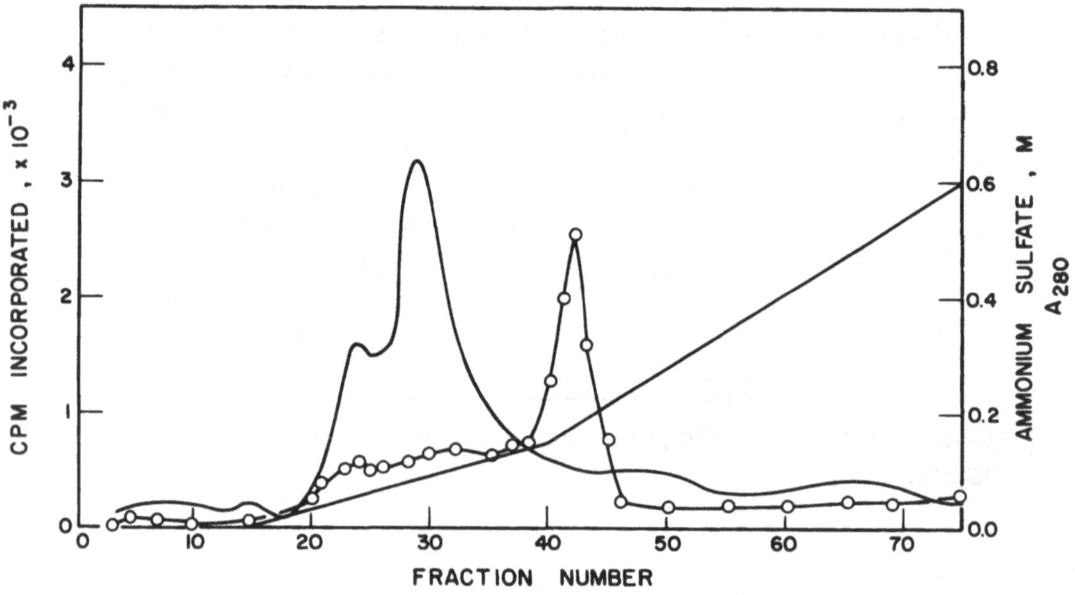

Figure 2. Enzyme (2-3 mg protein) was chromatographed on a DEAE-Sephadex column. Fractions were assayed employing denatured calf thymus DNA as template. RNA polymerase activity o—o and A_{280} values are plotted.

being extremely inhibitory. An approximate concentration of 10 mM Mg^{2+} resulted in maximum enzyme activity, and higher concentrations (up to 35 mM) of Mg^{2+} were not inhibitory. The Mn^{2+}/Mg^{2+} activity ratio (activity at optimum concentrations) was found to be approximately 2.0. The optimum $(NH_4)_2SO_4$ concentration for the enzyme (50 mM was quite low (2750 cpm incorporated/10 min). Higher concentrations were inhibitory, at 75 mM and 100 mM $(NH_4)_2SO_4$, incorporation was 2650 and 2520 cpm/10 min. Without $(NH_4)_2SO_4$ activity was 2580 and 2700 cpm respectively.

The enzyme had a slight preference for native calf thymus DNA as a template when compared to denatured DNA (Table 4). Native sugar beet DNA served as a template significantly better than denatured DNA. Activities were very similar on native calf thymus DNA and native sugar beet DNA.

α-Amanitin Studies

Several workers (24) have shown that α-amanitin, a toxin extracted from the basidiomycete, Amanita phalloides, specifically inhibits RNA polymerase II which is located in the nucleoplasm of eukaryotic cells. RNA polymerase I, the nucleolar enzyme, has been shown to be insensitive to this toxin. Employing the standard assay, maximum inhibition of the enzyme activity was observed at concentrations ranging from 0.01 to 0.05 μg/ml. At the maximum inhibitory concentration (25 mg/ml), the polymerase activity was inhibited by 23%. It has been shown, however, that the various RNA polymerases have different salt and metal ion concentration optima (3, 30, 31). Therefore, further experiments employing α-amanitin were performed in the presence of optimum salt and metal ion concentrations specific for the polymerases as reported in other eukaryotic organisms. The results of these experiments are presented in Table 5. If the values obtained in the presence of high salt concentrations, Mn^{2+} and α-amanitin are subtracted from the activity when high salt concentration and Mn^{2+} only are present, the value obtained is an indication of the activity due to polymerase II (56 pmol/mg/10 min). Polymerase I activity is then assumed to be that activity observed in the presence of low salt concentrations, Mg^{2+} and α-amanitin (288 pmol/mg/10 min). If these assumptions are valid for this crude preparation, RNA polymerase I represents the major portion of the total activity. The remaining activity may be attributable to RNA polymerase II and another nuclear RNA polymerase (30, 31) or perhaps a

Table 4. Template specificity of RNA polymerase activity.

Template	pmol incorporated/ mg/protein	Template	pmol incorporated/ mg/protein
Calf thymus (native)	350	Sugar beet (native)	368
Calf thymus (denatured)	325	Sugar beet (denatured)	246

Assays were as described in Experimental. Calf thymus DNA was purchased from Sigma Co., and sugar beet DNA was extracted by a procedure which included ribonuclease and pronase treatment followed by hydroxyatite chromatography. DNA was denatured at 100° for 5 min and cooled on ice.

Table 5. Effect of α-amanitin on RNA polymerase activity at various salt and metal ion concentrations.

Salt concentration	Metal ion	α-Amanitin (0.025 mg/ml)	Activity (pmol/mg/10 min)
-	Mg^{2+}, Mn^{2+}	-	393
Low	Mg^{2+}	+	288
High	Mn^{2+}	-	272
High	Mn^{2+}	+	216

$(NH_4)_2SO_4$ concentrations--Low = 50 mM; high = 300 mM, Metal ion concentrations--Mg^{2+} = 10 mM; Mn^{2+} = 1.25 mM.

polymerase from an organelle (28, 36). α-Amanitin was observed to have little effect on the activity solubilized from unwashed tissue. It should also be noted that in a 20½ hr wash period the specific activity of the solubilized enzyme(s) increased from 2.4 pmol/mg/min of protein in unwashed tissue to 39.2 pmol/mg/min for tissue washed for 20 hr.

DEAE-Sephadex Chromatography

To elucidate the possible presence of multiple polymerases indicated by the above inhibitor studies, the enzyme preparation was subjected to DEAE-Sephadex chromatography. The enzyme was eluted from the column in one major peak of activity at a salt concentration of 0.18 M (Fig. 2).

POLYRIBOSOME FORMATION

One of the most rapid and dramatic biochemical changes associated with cutting tuber tissue is the formation of polyribosomes (22). Translation of messenger RNA (mRNA) requires that ribosomes and mRNA bind to form a stable complex. Ribosomes when in the presence of charged transfer RNA (tRNA) and other appropiate factors, move along the mRNA chain to form peptide bonds. Freshly cut sugar beet root tissue contains a fairly high level of ribosomes which consists mainly of the monomer population (Table 6). After excision of the tissue, monoribosomes associate instantaneously into polyribosome structures, and this process continues up to the 6th hour. At this time, polyribosome formation seems to be complete, and the percentage distribution of mono- and polyribosomes is 34 and 66%, respectively; this remained nearly constant thereafter (Table 6). It should be noticed that this transition from free ribosomes to polyribosomal aggregates presumedly proceeded within the preexisting ribosome pool, since only a relatively small increase in the total RNA isolated from the polyribosomal complex was observed over the 0-18 hr period (Table 6). After breaking tissue dormancy, as early as within the first hour of aging, the newly synthesized RNA fraction entered the polyribosomal complex (Fig. 3a). This early synthesized RNA (after being released from polyribosomes) sedimented in 5-25% sucrose gradient as a rather broad zone located in the upper portion of the gradient, clearly apart from 23S and 28S rRNAs (Fig. 3b). On prolonged incubation, the polyribosomal complex becomes progressively labelled and after 18 hr its radioactivity increased several-fold as compared with the 1 hr stage of tissue aging (Fig. 4a).

Table 6. Rate of polyribosome formation over 0-18 hr period after breaking the dormancy of sugar beet root tissue.

Time of aging (hr)	Number of determinations	Polyribosomes (%)	Total ribosomal RNA
0	8	12	0.54
0.5	2	37	0.52
1	2	51	0.54
2	5	57	0.53
6	6	66	0.61
8	2	68	0.64
12	2	68	0.67
18	6	70	0.69

Percentage of polyribosomes was calculated taking as 100 the total u.v. absorbing material. RNA in the ribosomal fraction is expressed as milligrams of total ribosomal RNA (i.e. mono- + poly-ribosomes) isolated from 10 g of fresh tissue.

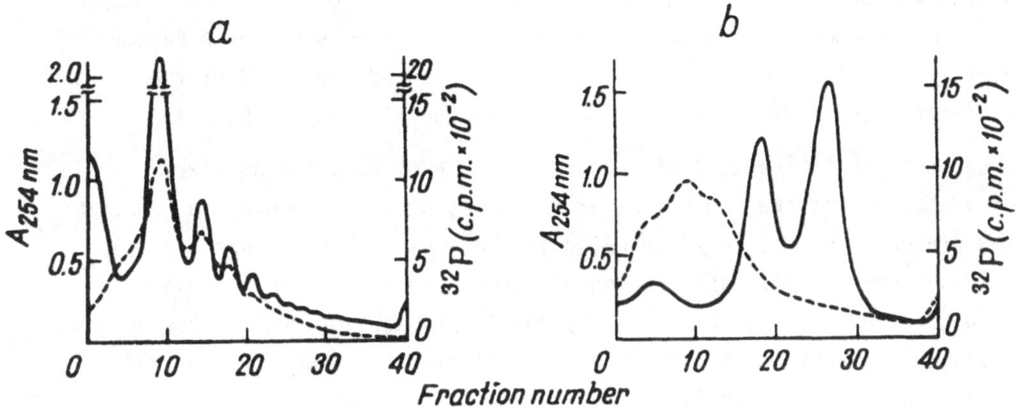

Figure 3. Sucrose density gradient profiles of polyribosomes (a) and RNA released from polyribosomes (b). The tissue sample (15 g) was aged for 1 hr in the presence of 1 mCi of (^{32}P) phosphate. ——, absorbance at 254 nm; ----, radioactivity.

The radioactivity distribution pattern of the RNA released from polyribosomes derived from the late stage (18 hr) of tissue aging (Fig. 4b) differed from that obtained at the very early stage (1 hr) (Fig. 3b). Under the same labelling conditions, in addition to rapidly labelled RNA, sedimenting as a heterodisperse fraction between 18S and 4S RNA, a significant labelling of lighter and heavier rRNAs was observed even after short exposure (1 hr) to the radioactive RNA precursor (Fig. 4b). When the time of incubation with isotope was prolonged to 6 hr, rRNA became heavily labelled and its radioactivity represented about 75% of the total radioactivity incorporated into the polyribosomal complex (Fig. 4b).

Incorporation of labelled precursors into RNA of polyribosomes was totally prevented during the first 6 hr of tissue aging if actinomycin D (20 μg/ml) was present from the very beginning in the medium (Fig. 5a and 5b), although some limited association of monoribosomes into polyribosome structures still occurred, probably due to preexisting mRNA in the dormant tissue. In addition to the inhibitory effect of actinomycin D on RNA synthesis, a significantly smaller amount of ribosomes could be extracted from the tissue treated with the antibiotic (Fig. 5a and 5b). On the other hand, 5-fluorouracil (10-3 M) only slightly reduced ^{3}H - uridine incorporation, not suppressing appreciably the assembly of ribosomes into polyribosome structures (Fig. 5c).

As can be seen from Fig. 6a and 6b, brief treatment with a very low concentration (0.1 μg/ml) of pancreatic RNase converted heavy polyribosomes almost quantitatively to monomers and dimers, most of the radioactivity being shifted to the top of the gradient. The easy dissociation of polyribosomes by RNase probably indicates that mRNA is involved in the polyribosomal complex rather than the polymer being aggregates. The radioactivity sedimenting with monomers and dimers probably represents only short fragments of mRNA still attached to ribosomes and, in part, the labelled rRNA which seems to be preserved under very mild conditions of RNase digestion. However, raising the RNase concentration to 2 μg/ml resulted in ribosome destruction followed by complete RNA degradation, both the radioactivity and the u.v. absorbing material being shifted to the upper portion of the centrifuge tube (Fig. 6c).

To ascertain whether polyribosomes formed during tissue aging inhibited

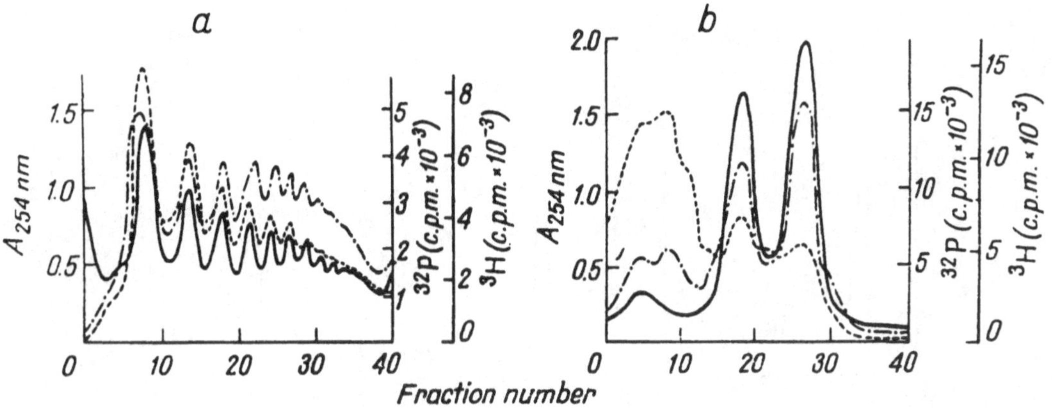

Figure 4. Sucrose density gradient profiles of polyribosomes (a), and RNA released from polyribosomes (b). The tissue sample (15 g) was aged for 18 hr under standard conditions; 100 µCi of (5-^{3}H)uridine was introduced 6 hr, and 1 mCi of (^{32}P)phosphate 1 hr before the end of each incubation. ——, absorbance at 254 nm; -·-·-, (^{3}H)uridine radioactivity; ----, (^{32}P)phosphate radioactivity.

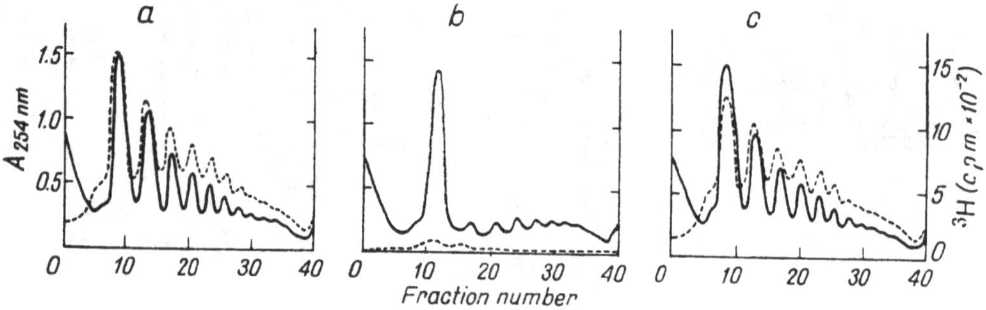

Figure 5. Effect of actinomycin D and 5-flourouracil treatment on polyribosome formation and (^3H)uridine incorporation into RNA of polyribosomes in sugar beet slices aged for 6 hr. The tissue samples (7.5 g) were aged under standard conditions (a), in the presence of 20 µg/ml of actinomycin (b), or 10^{-3} m-5-fluorouracil (c). One hour before the end of incubation 50 µCi of (5-^3H)uridine was introduced. ——, absorbance at 254 nm; ----, radioactivity.

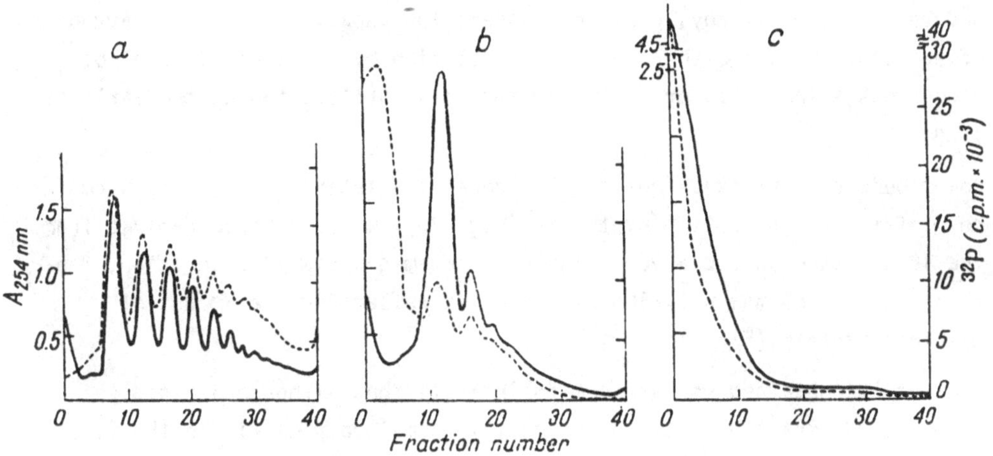

Figure 6. Effect of RNase treatment on sedimentation and radioactivity profiles of polyribosomes. The tissue sample (10 g) was aged for 18 hr, 1.2 mCi of (^{32}P) phosphate was introduced to the medium 1 hr before the end of incubation. Control (a); RNase-treated at concn of: 0.1 μg/ml (b), and 2 μg/ml (c). Polyribosomes were fractionated by sucrose density gradient centrifugation. ——, Absorbance at 254 nm; ----, radioactivity.

the protein synthesizing capacity, a series of experiments were performed in which the slices of sugar beet root at three various stages of aging were incubated for 2 hr with ^{14}C-labelled protein hydrolysate. As shown in Fig. 7a, polyribosomes from the early stage (2 hr) of tissue aging incorporated labelled amino acids very poorly. This might indicate that they had very little, if any, protein synthesizing capacity, or the observed effect could be due either to isotope dilution by non-labelled pool of endogenous amino acids or to its preferential utilization at the earliest stage.

Polyribosomes extracted from the tissue at the later stage (6 hr) were far more efficient in protein synthesis (Fig. 7b), whereas those derived from the 18 hr stage showed a very high rate of amino acid incorporation, the radioactivity of the polyribosomal complex reflecting the amount of nascent protein (Fig. 7c).

The results obtained so far seem to indicate that although a rapidly labelled RNA fraction entered the polyribosomal complex within the first hour of aging at least another hour was necessary to demonstrate protein synthesis in vivo. This indication, however, has not been confirmed by experiments in vitro.

In somewhat different experiments Watanabe and Imaseki (37) demonstrated that inhibition of protein synthesis by cycloheximide also greatly inhibited DNA synthesis when the agent was added during the first 6 hours after cutting. On the other hand, cycloheximide had little affect on DNA synthesis if added after 6 hours of cutting.

The pathway of ribosome formation in plant cells has recently been studied in some detail. Before 1967 there was no evidence for large rRNA-precursor molecules in plants (2). However, using radioactive-tracer techniques combined with the high-resolution fractionation afforded by polyacrylamide-gel electrophoresis, Leaver and Key (22) studied some of the events involved in the formation of rRNA in plants. Carrot-root discs were the experimental plant tissue of choice because previously their nucleic-acid metabolism had been studied in detail, and the cell fractionation techniques required to study ribosome synthesis established (21, 22). The tissue discs rapidly incorporate radioactivity into RNA after suitable periods of pre-incubation, and the separation of

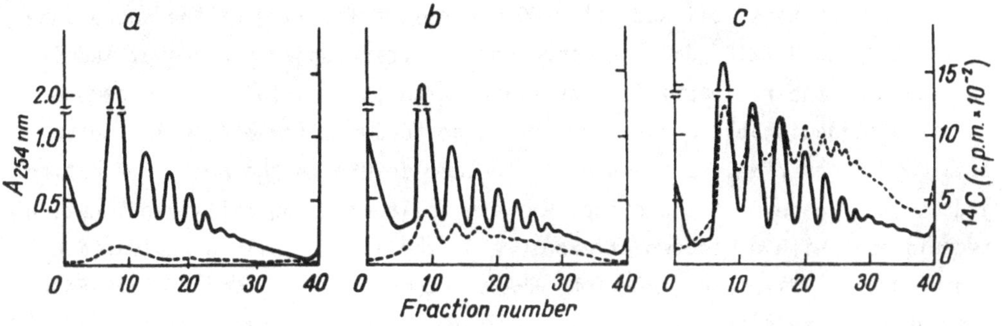

Figure 7. Effect of tissue aging on activation of protein synthesis *in vivo*. Three 7.5 g samples of sugar beet slices were aged for 2 hr (a), 6 hr (b), and 18 hr (c); 100 µCi of ^{14}C-reconstituted protein hydrolysate was added 2 hr before the end of incubation. ——, Absorbance at 254 nm; ----, radioactivity.

ribosomal and nuclear fractions was found to be comparatively easy with minimum degradation of nucleic acids. The so called nuclear fraction contained all of the nuclear DNA of the cell and also all of the detectable precursors of rRNA.

Observations of the kinetics of rRNA formation suggested that the initial event is the synthesis of RNA species with molecular weights of $2.8x10^6$ and $2.2x10^6$ daltons followed after 20 minutes of labeling by the appearance of $1.4x10^6$ and $0.7x10^6$ daltons components. After longer periods of labeling (60/min) the radioactivity associated with the $1.4x10^6$ daltons component is not distinguishable because of the large accumulation of ^{32}P in the $1.3x10^6$ daltons rRNA. Synthesis of ^{32}P-labeled RNA in the nuclear fraction and its emergence into the cytoplasm were followed using cell-fractionation techniques. Within minutes of exposure of the intact cell to radioactive precursor, polydisperse (AMP-rich) RNA's accumulate, and some of this RNA emerges into the cytoplasm and becomes associated with preexisting ribosomes to form polysomes (23). The rate of synthesis of the polydisperse AMP-rich RNA's appears to be about twice that of the rRNA precursors. The $2.8x10^6$ and $2.2x10^6$ daltons RNA's are distinguishable soon afterwards and are found only in the nuclear fraction. Labeled $1.4x10^6$ daltons nuclear RNA and the $0.7x10^6$ daltons rRNA appear after about 20 minutes, and the $0.7x10^6$ daltons rRNA rapidly appears in the ribosomal fraction, being released from the nucleus into the cytoplasm. The $1.4x10^6$ daltons RNA remains in the nucleus where it is apparently converted to the $1.3x10^6$ daltons rRNA before being transported to the cytoplasm. The base-composition analyses offer further evidence that the high molecular-weight nuclear RNA's are precursors to the $1.3x10^6$ and $0.7x10^6$ daltons rRNA's. The base compositions of the $2.8x10^6$, $2.2x10^6$ and $1.4x10^6$ daltons nuclear RNA's are similar to those of the cytoplasmic ribosomal RNA's and are characterized by a high GMP content, and GMP:AMP ratios between 1.15 and 1.51. The close similarity in base composition between the $1.4x10^6$ daltons nuclear RNA and $1.3x10^6$ daltons rRNA indicates that the former is almost certainly a precursor to the latter.

Although the data does not allow a definitive conclusion as to the pathway of precursor conversion to the $1.3x10^6$ and $0.7x10^6$ daltons rRNA's, the $2.8x10^6$, $2.2x10^6$ and $1.4x10^6$ daltons components probably serve as precursors to the mature rRNA's. In addition other high molecular-weight components, especially a 0.9×10^6 daltons component which does accumulate above normal levels in the presence of inhibitors may be involved in the synthesis of plant rRNA's. Since the 2.8×10^6 daltons components show essentially the same kinetics of accumulation, the possibility exists that they are independently associated with rRNA synthesis, with the 2.8×10^6 daltons RNA not being a precursor of the 2.2×10^6 daltons component.

DISCUSSION

Wounding sugar beet storage tissue results in increased DNA synthesis as evidenced by an increase in DNA polymerase activity. Based on the results of Watanabe and Ismaseki (37) it appears that protein synthesis is necessary for the observed increase in polymerase activity. If enzyme activation is involved, increased activity could result from alteration of existing polymerase molecules or by the simple removal of an inhibitor during the washing period. Based on the DEAE-Sephadex profiles presented in this paper and the various peak characteristics, the increased activity may have resulted from _de novo_ synthesis, enzyme activation or both. DNA polymerase peaks I and II from both tissues elute at the same salt concentrations and have similar properties and, therefore, may be the same proteins. Peak III of washed tissue, however, is not detectable in unwashed tissue and its activity could be a result of _de novo_ synthesis. This does not, however, preclude the possibility of an alteration of an existing protein.

The washing of sugar beet tissue results in increased chromatin-bound RNA polymerase activity. When solubilized from chromatin of washed tissue and compared with the enzyme solubilized from chromatin isolated from unwashed beets, the specific activity of the 'washed' RNA polymerase was much higher. It is difficult, based on these data, to determine whether this increase is due to increased synthesis of the enzyme or activation of the enzyme or both. Unpublished data, in which cycloheximide was placed in the washing buffer at concentrations as high as 10 mg/ml, indicated that protein synthesis may not be necessary for the increase in RNA polymerase activity. A quantitative measurement of the total activity per g fresh weight is difficult due to the isolation technique employed to obtain chromatin. Even though scraping of the pellets to remove the crude chromatin layer may vary, consistent results are obtained between experiments.

The rapid association of monoribosomes into polyribosome structures is one of the earliest metabolic events triggered by breaking the dormancy of sugar beet root. Polyribosomes hardly detectable (12%) in the freshly cut tissue begin to develop almost instantaneously after the sliced tissue has been exposed to the optimal temperature (30°C). The rate of polyribosome formation, very high during the first two hours, decelerates gradually and the maximum level of polyribosomes (66%) reached at the 6th hour of tissue aging remains nearly constant and aging prolonged for the next 12 hr. Our results are essentially consistent with the data obtained for carrot roots (23) and potato tubers (19). Polyribosomes isolated by us from the aged slices of sugar beet roots were much less degraded (in our preparations dodecamers were present, whereas in the preparations from the carrot and potato, at most hexa- and heptamers could be detected).

These data suggest that the synthesis of a new mRNA during the initial phase might not be strictly required and most likely messenger molecules already preexisting in dormant tissue could be utilized for the earliest polyribosome formation. Such a performed mRNA might play a similar role in plant storage organs as does the so-called "maternal" mRNA which was found to be stored awaiting activation in the unfertilized sea urchin eggs (15, 34). The stored message is released for translation in the early post-fertilization stages (18, 29), and is responsible for at least 85% of protein synthesizing capacity (17).

SUMMARY

DNA polymerase preparations solubilized from chromatin of unwashed and washed sugar beet storage tissue exhibit multiple activity peaks when separated by DEAE-Sephadex chromatography. Activity peaks isolated from unwashed and washed tissue required Mg^{2+}, all four deoxynucleoside triphosphates and added DNA. Activity from peaks isolated from washed tissue was completely sensitive to pancreatic DNase and insensitive to RNase. Enzyme activity was increased in the tissue during the washing period. In addition, there was a drastic change in template specificity from single-stranded primer in unwashed tissue to double-stranded primer in unwashed tissue to double-stranded DNA in washed tissue.

RNA polymerase exhibited a sharp Mn^{2+} optimum of 1.25 mM and a Mg^{2+} optimum at 10 mM. The Mn^{2+}/Mg^{2+} activity ratio (activity at optimum concentrations) was 2.0 with an optimum salt concentration of 50 mM. Based on data including inhibition with α-amanitin (0.025 μg/ml), the majority of the total activity appeared to be RNA polymerase I. Subsequent fractionation by DEAE-Sephadex column chromatography resulted in one peak of activity eluted with 0.18 M $(NH_4)_2SO_4$.

An instantaneous association of preexisting ribosomes into polyribosome structures occurs within one hr and increases to a maximum level by six hr. RNA synthesis is observed after 1 hr of aging but early polyribosome development does not entirely depend on RNA synthesis. It is suggested that preexisting mRNA is used for this early polyribosome formation. It appears that depression of genome activity proceeds with increasing formation of certain RNA species not entering a polyribosomal complex; these species might represent species of RNA confined to the nucleus.

Cytoplasmic ribosomal RNA's ($1.3x10^6$ and $0.7x10^6$) of carrot discs arise by the selective cleavage of larger precursor molecules. The main precursor RNA species tentatively were characterized by polyacrylamide-gel electrophoresis as having molecular weights of $2.3x10^6$, $2.2x10^6$, $1.4x10^6$ daltons, and base compositions similar to ribosomal RNA. Isotopic labelling studies combined with cell-fractionation techniques have shown that these molecules are localized in the nucleus and have kinetics of synthesis consistent with precursor ribosomal RNA.

REFERENCES

1. Bacon, J.S.D., MacDonald, I.R., Knight, A.H.: The development of invertase activity in slices of the root of Beta vulgaris L. washed under asceptic conditions. Biochem. J. 94, 175-182 (1965).
2. Birnstiel, M.: The nucleolus in cell metabolism. Ann. Rev. Plant Physiol. 18, 25-58 (1967).
3. Blatti, S.P., Incles, C.J., Lindall, T.J., Morris, P.W., Weaver, R.F., Weinberg, F., Rutter, W.J.: Structure and regulatory properties of eukaryotic RNA polymerase. Cold Spring Harbor Symp. Quant. Biol. 35, 649-657 (1970).
4. Bonner, J., Dahmus, M.E., Fambrough, D., Huang, R.C., Marushige, K., Tuan, D.Y.H.: The biology of isolated chromatin. Science 159, 47-56 (1968).
5. Chang, L.M.S., Bollum, F.J.: Low molecular weight deoxyribonucleic acid polymerase in mammalian cells. J. Biol. Chem. 246, 5835-5837 (1971).
6. Chang, L.M.S., Bollum, F.J.: Deoxynucleotide-polymerizing enzymes of calf thymus gland. J. Biol. Chem. 246, 909-916 (1971).
7. Cherry, J.H.: Regulation of invertase in washed sugar beet tissue. IN: Biochemistry and Physiology of Plant Growth Substances, F. Wightman and G.W. Setterfield, eds. The Runge Press, OTAWA. 417-431 (1968).
8. Chetsanga, C.J., Poccia, D.L., Hill, R.J., Doty, P.: Stage specific RNA transcription in developing sea urchins. Cold Spring Harbor Symp. Quant. Biol. 35, 629-634 (1970).
9. Chiu, J., Chiu, S.C.: Pattern of developmental changes in two DNA polymerases of rat brain. Biochim. Biophys. Acta 269, 364-369 (1972).
10. Click, R.E., Hackett, D.P.: The role of protein and nucleic acid synthesis in the development of respiration in potato tuber slices. Proc. Nat. Acad. Sci. U.S. 50, 243-250 (1963).
11. Duda, C.T., Cherry, J.H.: Chromatin- and nuclei- directed ribonucleic acid synthesis in sugar beet root. Plant Physiol. 47, 262-268 (1971).
12. Edelman, J., Hall, M.A.: Enzyme formation in higher plant tissue. Development of invertase and ascorbate oxidase activities in

mature storage tissue of Helianthus tuberosum L. Biochem. J. 95, 403-410 (1965).

13. Ellis, R.J., MacDonald, I.R.: Activation of protein synthesis by microsomes from aging beet discs. Plant Physiol. 42, 1297-1302 (1967).

14. Grippo, P. LoScavo, A.: DNA polymerase activity during maturation in Xenopus laevis oocytes. Biochem. Biophys. Res. Comm. 48, 280-285 (1972).

15. Gross, P.R., Malkin, L.I., Moyer, W.A.: Templates for the first proteins of embryonic development. Proc. Nat. Acad. Sci. U.S. 51, 404-414 (1964).

16. Horgen, P.A., Griffin, D.H.: Specific inhibitors of the three RNA polymerases from the aquatic fungus Blastocladiella emersonii. Proc. Nat. Acad. Sci. U.S. 68, 338-341 (1971).

17. Humphreys, T.: Measurement of messenger RNA entering polysomes upon fertilization of sea urchin eggs. Develop. Biol. 26, 201-208 (1971).

18. Infante, A.A., Nemer, M.: Accumulation of newly synthesized RNA templates in a unique class. Proc. Nat. Acad. Sci. U.S. 58, 681-688 (1967).

19. Kahl, G.: Metabolism in plant storage tissue slices. Bot. Review 40, 263-314 (1974).

20. Kamisaka, S., Masuda, Y.: Auxin-induced growth of tuber tissue from Jerusalem artichoke. IV. Significance of gibberellin biosynthesis and basic proteins in chromatin in aging process. Plant Cell Physiol. 9, 61-67 (1968).

21. Leaver, C. J., Edelman, J.: Nucleic acid synthesis in carrot tissue slices. Biochem. J. 97, 27P (1965).

22. Leaver, C.J., Key, J.L.: Ribosomal RNA synthesis in plants. J. Mol. Biol. 49, 671-680 (1970).

23. Leaver, C.J., Key, J.L.: Polyribosome formation and RNA synthesis during aging of carrot-root tissue. Proc. Nat. Acad. Sci. U.S. 57, 1338-1344 (1967).

24. Lindell, T.J., Weinberg, F., Morris, P.W., Roeder, R.G., Rutter, W.J.: Specific inhibition of nuclear RNA polymerase II by α-amanitin. Science 170, 447-449 (1970).

25. MacDonald, I.R., Bacon, J.S.D., Vaughan, D. Ellis, R.J.: The relation between ion absorption and protein synthesis in beet disks. J. Exp. Bot. 17, 822-837 (1966).

26. Matsukage, A., Bohn, E.W., Wilson, S.H.: Multiple forms of DNA polymerase in murine myeloma. Fed. Proc. 32, 451 (1973).

27. Momparler, R.L., Rossi, M., Labitan, A.: Partial purification and properties of two forms of deoxyribonucleic acid polymerase from calf thymus. J. Biol. Chem. 248, 285-293 (1973).

28. Polya, G.M., Jagen dorf, A.T.: Wheat leaf RNA polymerases. Arch. Biochem. Biophys. 146, 635-648 (1971).

29. Rinaldi, A.M., Monroy, A.: Ribosome formation and RNA synthesis in the early post-fertilization of the sea urchin egg. Develop. Biol. 19, 73-86 (1969).

30. Roeder, R.G., Rutter, W.J.: Multiple ribonucleic acid polymerases and ribonucleic acid synthesis during sea urchin development. Biochem. 9, 2543-2553 (1970).

31. Roeder, R.G., Rutter, W.J.: Specific nucleolar and nucleoplasmic RNA polymerases. Proc. Nat. Acad. Sci. U.S. 65, 675-682 (1970).

32. Roeder, R.G., Reeder, R.H., Brown, D.D.: Multiple forms of RNA polymerase in *Xenopus laevis*: their relationship to RNA synthesis *in vivo* and their vitality of transcription *in vitro*. Cold Spring Harbor Symp. Quant. Biol. 35, 727-735 (1970).

33. Spencer, D., Whitfeld, R.R.: The characteristics of spinach chloroplast DNA polymerase. Arch. Biochem. Biophys. 132, 477-488 (1969).

34. Spirin, A.S., Nemer, M.: Messenger RNA in early sea urchin embryos: cytoplasmic particles. Science 150, 214-217 (1965).

35. Travers, A.: Control of transcription in bacteria. Nature New Biol. 229, 69-74 (1971).

36. Tsai, M., Michaelis, G., Criddle, R.S.: DNA-dependent RNA polymerase from yeast mitochondria. Proc. Nat. Acad. Sci. U.S. 68, 473-477 (1971).

37. Watanabe, A., Ismaseki, H.: Induction of ribonucleic acid synthesis in potato tuber slices. Plant Physiol. 57, 568-571 (1976).

38. Whiteley, A.H., McCarthy, B.J., Whiteley, H.R.: Changing populations of messenger RNA during sea urchin development. Proc. Nat. Acad.

Sci. U.S. 55, 519-525 (1966).

39. Willemot, C., Stumpf, P.K.: Fat metabolism of higher plants. XXXIV. Development of fatty acid synthetase as a function of protein synthesis in aging potato tuber slices. Plant Physiol. 42, 391-397 (1967).

Ribosome Metabolism in Excised Slices of Jerusalem Artichoke Tuber

George Setterfield, Joachim Sparkuhl and Henry Byrne

INTRODUCTION

Early autoradiographic (1) and biochemical (2) studies using excised slices of Jerusalem artichoke and potato tubers, respectively, revealed that one of the main responses in such slices was rapid onset of RNA synthesis. The autoradiographic results in particular indicated that much of the RNA synthesis involved the nucleolus and resulted in stable new cytoplasmic RNA. This indicated that alterations in the ribosome complement of the cells were occuring. Since these beginnings considerable study of ribosome metabolism in excised slices of storage tissues has been undertaken.

Our studies have concentrated on Jerusalem artichoke tuber slices since these offer the experimental advantage that growth responses of the cells may be clearly separated from initial metabolic responses following slicing (1,3). Excised artichoke tuber slices "aged" in water show rapid increases in a number of aspects of metabolism, including protein and RNA synthesis, but do not undergo cell division, expansion, or increases in protoplasmic mass (1,3,4). If, after this initial activation is well established (i.e. following 24 h water aging), auxin (with or without cytokinin) is supplied to the tissue further metabolic stimulation results and cell growth and organelle replication ensue (1,3,4). It is also possible to stimulate the tissue to undergo extensive cell division (1,3). The commonest method for this is to supply auxin continuously after excision. We have not studied RNA metabolism in this situation to any extent but it

*Present address: Department of Microbiology and Parasitology, University of Toronto, Toronto, Ontario, Canada M5S 1A1.

has been thoroughly analysed by others (5,6,7). By manipulation of exogenous auxin and kinetin it is thus possible to study aging, growth and division responses separately.

This short paper will attempt to summarize studies on the ribosomal apparatus, i.e. the nucleolus, cytoplasmic ribosomes and polyribosomes, and rough endoplasmic reticulum, of Jerusalem artichoke tuber slices during both aging and hormonally-stimulated growth periods. Some reference to results of relevant studies with other tissues will be included but no exhaustive review is intended. The studies in artichoke have involved both cytological and biochemical approaches and the results are remarkably complementary. For sake of clear exposition these two approaches will be considered separately.

CYTOLOGICAL STUDIES

The cells of cold-stored intact Jerusalem artichoke tubers are physiologically quiescent and show only rudimentary cytological structure (4). A large vacuole appresses the nucleus and a very thin layer of cytoplasm to the wall. Within the cytoplasm the only organelles present with reasonable frequency are plastids, mitochondria and free ribosomes. Other organelles such as dictyosomes, ER, microtubules, etc., are extremely rare. On excision of tissue slices from the tubers and aging in water, the main morphological changes which occur involve organelles of the ribosomal apparatus, i.e. the nucleolus, ribosomes and ER.

Nucleolus

In living quiescent cells freshly excised from the tuber and viewed by phase or Nomarski interference optics the nucleoli are relatively small, solid, homogeneous structures (8). Electron microscopy shows them to be mainly composed of compact ribonucleoprotein fibrils. In some cases no granular zone is present while in others a thin peripheral layer of granules is seen (8,9,10). Distinct masses of specialized nucleolar-associated chromatin embedded in the nucleolar periphery are clearly visible by both light and electron microscopes (8,10). In the EM these zones appear as lightly staining regions composed of fibrils 10 nm in diameter (10).

This basic form of nucleolus has been observed in a number of plant cells which are relatively inactive in ribosome synthesis: e.g. freshly excised carrot-root cells (11); cells of ungerminated and fully grown wheat coleoptiles (12); cells of the quiescent centre and fully expanded parenchyma cells in root tips (13,14); meiocytes of the bluebell (15).

Excision and water-aging of artichoke tuber cells rapidly induces changes in the nucleolus. By three hours post-excision nucleolar volume is measurably increased, and the increase continues to a peak at about 24 h, followed by a decline if the water-aging treatment is continued (4,8,16). Gibberellic acid treatment of aging slices causes an almost two-fold enhancement of the volume increase (16). Treatment of slices aged 24 h in water with 2,4-D and kinetin causes a further sharp increase in nucleolar volume by 25 h post-hormones followed by a decline to near control values at 44 h (8).

The nucleolar volume changes during aging are paralleled by very distinct changes in structural organization of the organelle. In the main ribonucleoprotein area of the nucleolus a distinct granular zone develops. This zone is initially peripheral but by 24 h becomes extensively intermingled with fibrillar regions (4,8,10). Starting at about 8 h aging nucleolar vacuoles appear and progressively increase in mean size up to 24 h. The vacuoles contract and expand over periods of 1-2 h. They occur mainly in the granular zone of the nucleolus and contain a mixture of granules and fine filaments. Giant vacuoles are produced by treatment with 10 µg/ml 5-fluorouracil; these vacuoles contain RNA, probably as ribonucleoprotein (8). We presume that the normal vacuoles are involved in movement of ribosomal precursor particles from the nucleolus. Since they only appear as the nucleoli become large they probably represent a means of increasing surface area for transport of nucleolar granules. Connections of the vacuoles to the exterior of the nucleolus are commonly seen (8). In post-age hormone-treated tissue the nucleoli show essentially the same fine structure as the nucleoli of 24 h aged tissues (4,8).

At the same time as the ribonucleoprotein regions of the nucleoli are undergoing the above changes, the nucleolar-associated chromatin, termed the L zone by Jordan and Chapman (10), also changes markedly in form. By 3 h post-excision this distinct lightly staining chromatin moves from the periphery into the interior of the nucleolus and then it disperses and by 24 h is seen mainly in small electron-transparent lacunae spread through the fibrillar zone (10). In many nucleoli of fully activated tissue the lacunae are quite small and difficult to define, indicating extreme dispersion of the chromatin. Jordan and Luck (15) have discussed the problems in interpreting the role of this chromatin. The simplest hypothesis is that at least part of the lightly-staining nucleolar chromatin contains ribosomal cistrons, which are extensively withdrawn from the main nucleolar region in the quiescent condition. As nucleoli are activated these cistrons disperse and spread through the nucleolus and commence transcription of pre-rRNA. The transition of the nucleolar chromatin during activation of artichoke tuber cells constitutes some of the best evidence for this model.

The morphological changes involving nucleoli seen in artichoke tuber cells during activation are essentially identical to those seen during activation of cells of carrot-root discs (11) and embryonic wheat coleoptiles (12). In addition, similar changes in reverse have been described for cells ceasing active RNA synthesis: e.g. older wheat coleoptiles (12); maturing root cells (13,14); bluebell meiocytes (15).

One other cytological approach to nucleolar activity in artichoke tuber cells has been the use of autoradiography following feeding of tritiated nucleosides. Rose et al (8) showed that the rate of incorporation of uridine into all cell compartments i.e. nucleolus, chromatin and cytoplasm, was detectable by 2.5 h post-excision, rose steadily to a peak between 7.5 and 12 h and then declined. If hormones are given after 24 h aging further stimulation of nucleolar incorporation occurs (1). Recently we have done a more detailed autoradiographic analysis of nuclear labelling in early times of aging (17). At 20-40 min post-excision most nuclear label is over chromatin, not nucleoli. By 60 min clear nucleolar labelling is seen and by 100 min the relative labelling of chromatin and nucleoli is similar to that in cells aged 12 h. Clearly, the onset of rRNA synthesis in excised artichoke slices is quite rapid, although there is some lag relative to the synthesis of other types of RNA (see Biochemical section below).

Cytoplasmic Ribosomes and Endoplasmic Reticulum (ER)

The thin layer of cytoplasm in cells fixed directly from the tuber is reasonably rich in ribosomes. These ribosomes, however, are scattered through the cytoplasm with little or no association with ER or with each other (4,9). This situation changes markedly in cells aged 24 h. In such cells a large proportion of the ribosomes are associated with ER and of the remaining free ribosomes many are in polysomal aggregates (4,9). On the basis of EM observations we originally suggested that the total population of ribosomes increases during aging (4). Biochemical measurements of ribosomal RNA (18) (see below) do not support this view, indicating that after an initial drop the total number of ribosomes remains constant through prolonged water-aging. With post-aging hormone treatment the total number of ribosomes, polysomes and ribosomes associated with ER remains more or less constant on a per area basis in EM sections (4). However, since the total volume of cytoplasm increases in response to hormones (4) the total number of all of these classes of ribosomes must increase. Numerous large spiral polysomes were observed free in the cytoplasm of hormone-treated cells (9).

The origin of the ER during aging is of some interest. In both of our published studies of cytoplasmic structure (4,9) we failed to observe significant amounts of endomembranes in the quiescent cells and assumed most of the ER appearing during aging arose *de novo*. However, recently we have encountered stacks of smooth ER-like membranes (not dictyosomes) in initial cells allowing the possibility that at least part of the RER formed during aging results from redistribution of pre-existing membranes. This situation needs further evaluation.

Increases in RER have been noted in the cells of a number of other tissues in relation to developmental situations: e.g. aged carrot-root slices (11); germinating lettuce embryos (19); growing wheat coleoptiles (12); rapidly expanding cells of pea roots (14). In most of these cases appearance of substantial amounts of RER is correlated with cell growth. However, in the case of artichoke tuber slices no growth is immediately involved, although the RER formed may be a prerequisite for subsequent stimulation of growth by auxin (9).

Inhibitor Studies

The main inhibitor which has been used in cytological studies of artichoke activation and growth is 5-fluorouracil (5-FU), a compound which is presumed to interfere preferentially with ribosome metabolism (7, 20). If given during aging this compound inhibits subsequent auxin-induced growth at concentrations of 10 μg/ml and above. On the other hand if supplied to the tissue after aging, with hormones, it has relatively little effect on cell growth even at 100 μg/ml (9).

Supplied either during aging or growth 5-FU causes marked increases in nucleolar volume over water and hormone-treated controls, respectively (8, 21). At concentrations between 5 - 25 μg/ml giant nucleolar vacuoles appear which qualitatively resemble the smaller vacuoles of untreated cells. At concentrations of 5-FU above 25 μg/ml, vacuoles are reduced or absent and the nucleoli appear much as in the initial quiescent cells, i.e. solidly fibrillar without a granular zone. The nucleolar chromatin does appear to ramify partially in such nucleoli, however, forming internal lacunae (8).

In the cytoplasm of cells aged 24 h in 5-FU (100 μg/ml) the total number of ribosomes and RER is noticeably reduced. If the 5-FU treatment is continued beyond aging into the hormone period cells almost devoid of ribosomes and RER result. Such cells are incapable of growth. In cells given 5-FU with hormones only after aging considerable RER remains although the free ribosomes appear reduced in number. Such cells grow readily (9).

Autoradiographic analysis (9) indicates that 5-FU present during aging reduces incorporation of labelled nucleosides and amino acids into cytoplasmic RNA and protein, respectively. A similar effect is obtained in growing tissue if 5-FU treatment is continued after aging. On the other hand, 5-FU given only after aging reduces incorporation into cytoplasmic RNA but not protein.

We interpret these results with 5-FU to indicate that the inhibitor reduces synthesis of stable new ribosomes which normally replace the preexisting ribosomes of the quiescent tuber cells during the aging and early growth periods. Failure of proper replacement leads to cells denuded of ribosomes and RER which are incapable of growth in response to hormones. On the other hand, if normal synthesis of new ribosomes and formation of RER during aging is allowed, subsequent interference with ribosome synthesis by 5-FU does not substantially influence protein synthesis or growth. This situation is clarified somewhat further by biochemical studies reported in the next section (7, 22). The precise action of 5-FU on ribosome production is as yet unclear. Rose and Setterfield (9) originally suggested that the main result of 5-FU treatment was instability of ribosomes incorporating 5-FU. The inhibitor does become incorporated into nucleoli (1) and RNA (7). It has been argued, however, (21) that the data do not rule out the possibility of reduced synthesis and transport of rRNA at the nucleolar level. We have been unable to settle this question, even using biochemical methods, and conclude that either or both models of inhibition may be operative.

The effects of two other inhibitors, actinomycin D and cycloheximide, have been examined cytologically in artichoke tuber slices. Both completely prevent the increase in nucleolar volume which occurs normally during aging (21). We have confirmed these results (23). In fact as

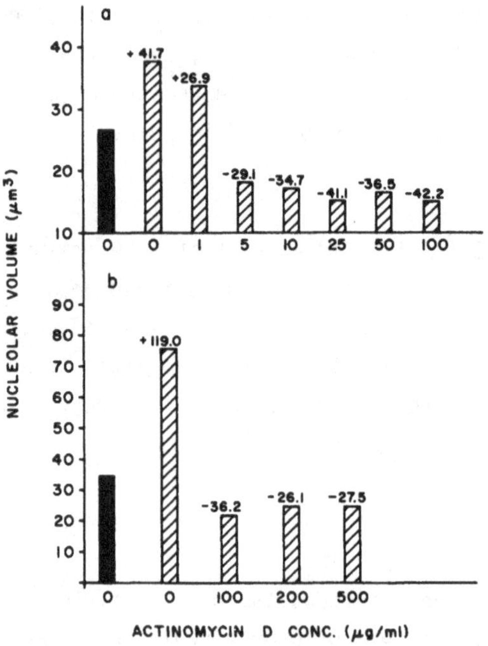

Figure 1. The effect of various concentrations of actinomycin D on nucleolar volume in artichoke slices. Two separate experiments are shown (a and b). Solid bars represent initial, untreated tissue; hatched bars represent tissue incubated for 24 h post-excision. Numbers give percent increase (+) or decrease (-) in volume.

Fig. 1 shows, actinomycin D at a wide range of concentrations actually can lead to reduction in nucleolar volume. When these nucleoli were examined in the EM (23) it was found that the nucleoli of the initial quiescent cells had a substantial peripheral zone of granules surrounding the dense central fibrillar region. In cells treated with actinomycin D for 24 h this granular zone was absent, leaving only the fibrillar zone. Presumably then, actinomycin D treatment blocked RNA transcription but did not prevent transport of highly processed RNA in nucleolar granules. Interestingly, the nucleolar-associated chromatin in these nucleoli partially dispersed and spread out over the nucleolar surface but did not invade the interior (Fig. 2a).

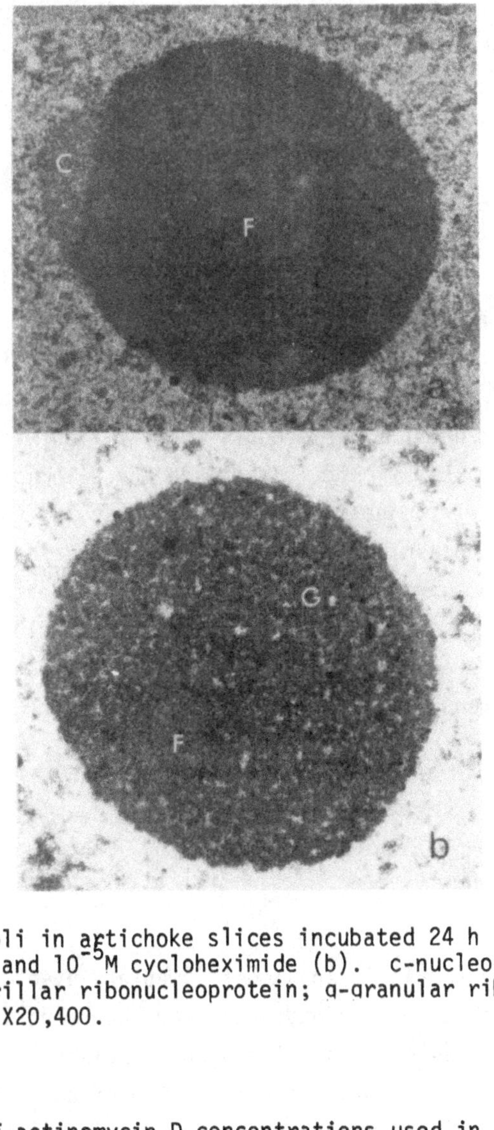

Figure 2. Nucleoli in artichoke slices incubated 24 h in 100 μg/ml 5-fluorouracil (a) and 10^{-5}M cycloheximide (b). c-nucleolar associated chromatin; f-fibrillar ribonucleoprotein; g-granular ribonucleoprotein. a - X22,400; b - X20,400.

The wide range of actinomycin D concentrations used in the experiment of Fig. 1 were designed to test the hypothesis that cessation of rRNA synthesis in intact tubers might depend on shortage of a particular nucleolar (or ribosomal) protein necessary for processing pre-rRNA. In this way one simple genetic locus might regulate transcription of the complex highly repetitive rDNA locus. We reasoned that at low concentrations actinomycin D might restrict transcription of rRNA but allow production of mRNA (24, 25). In this case the limiting protein might be produced and lead to processing and transport of most of the pre-rRNA in the nucleolus. In the absence of further transcription of pre-rRNA the nucleolus should drop sharply in volume. As Fig. 1 shows this condition was not attained and the hypothesis is not supported. Probably regulation of nucleolar activity involves RNA transcription directly.

At the EM level treatment with cycloheximide (10^{-5}M) during 24 h aging gave strikingly different results than actinomycin D (23). While the nucleoli remained small and compact they became highly granular rather than remaining fibrillar (Fig. 2b). In some sections a large granular zone was distinct from an adjacent fibrillar zone while in other cases the entire nucleolar profile was homogeneously filled with granules. It appears that in this situation cycloheximide may allow processing of preexisting fibrillar ribonucleoproteins to a granular form but prevent transport to the cytoplasm. Presumably, continued transcription of pre-rRNA is also blocked. In these nuclei the nucleolar-associated chromatin remained as a compact body at the periphery of the nucleolus. These preliminary observations are interesting and suggestive but require further confirmation and extension.

BIOCHEMICAL STUDIES

Early RNA Synthesis

Initiation of RNA synthesis is an almost immediate event on excision of cells from artichoke tubers; we have detected incorporation of ^{3}H-adenosine into ATP and RNA as early as 10 min post-excision. When polysomes labelled with ^{3}H-adenosine during the first 2 h of aging are dissociated with EDTA into mRNA and ribosomal subunits and fractionated on 10 - 30% EDTA-sucrose gradients the bulk of the incorporated radioactivity is not found with ribosomal subunits, but with a polydisperse fraction which is presumably mRNA (17). Tissue aged for longer periods prior to 2 h pulse labelling and polysome dissociation shows a progressive increase in the proportion of radioactivity incorporated into ribosomal subunits. These biochemical data suggest that mRNA is synthesized preferentially early in aging. This is in agreement with the autoradiographic data described previously (17) which show that up to 1 h post-excision most adenosine incorporation is in the chromatin region of the nucleus rather than in the nucleolus. Leaver and Key (26) using the inhibitors actinomycin D and 5-FU in excised carrot discs, came to the conclusion that one of the earliest changes with aging is formation of polyribosomes using pre-existing ribosomes and newly-synthesized mRNA.

Fractionation of total RNA prepared from artichoke slices exposed to a 2 h pulse of ^{3}H-adenosine at different times up to 14 h post-excision shows that low-molecular weight RNA accounts for about 50% of the incorporated label at all times (17). Fraser (7) demonstrated that considerable 4S RNA, i.e. presumed tRNA, is synthesized during the first 2.5 h of aging of artichoke discs. Of the remaining RNA, poly(A)+RNA declines from 30% of total during 0-2 h to 13% at 12-14 h while poly(A)-RNA rises from 20% during the first 2 h to over 30% during 12-14 h (17).

Analysis of labelled adenosine incorporated into polysomes shows that newly synthesized poly(A)+RNA (i.e. mRNA) and poly(A)-RNA are present at all times from 2 to 14 h (17). The presence of labelled poly(A)-RNA at later times of aging is not surprising since rRNA synthesis is occuring after 2 h. However, its occurrence in polysomes early in the aging period, (0-2 h), before substantial rRNA synthesis has begun, is of interest since it indicates that much newly synthesized mRNA in this tissue may not be polyadenylated. In support of this proposal is the previous finding that in the presence of 5-FU, which inhibits rRNA synthesis, RNA synthesized during the first 2.5 h of aging contains a polydisperse poly(A)-fraction presumed to be mRNA (7).

We have also examined the rate of total RNA synthesis during the first 12 h of aging (17). As in the experiments described above, tissue was exposed to ^{3}H-adenosine for 2 h periods at various times following excision. At the end of each pulse the specific activities of total RNA (cpm/μg) and ATP (cpm/pmol) were determined and the average specific activity of ATP calculated. The specific activities of both RNA and ATP increase during aging but during the first 6 h the number of pmoles of AMP incorporated per μg RNA remains constant, indicating a constant rate of RNA synthesis in this period. By 12-14 h a modest 50% increase in rate of total RNA synthesis is evident.

Ribosomal RNA Synthesis

Synthesis of rRNA in artichoke proceeds via an initial high molecular weight precursor of 2.3×10^6 daltons which is cleaved into molecules of 1.4×10^6 daltons and 0.9×10^6 daltons. These intermediate molecules are subsequently processed to rRNA molecules of 1.3×10^6 daltons and 0.7×10^6 daltons, respectively, the RNA molecules of the large and small ribosomal subunits (27). The same RNA processing system occurs in carrot root slices (28). Artichoke is a hexaploid with about 260 copies of ribosomal genes per haploid genome (29).

Synthesis of rRNA can be detected 2-4 h after excision. Aging of excised slices for 2 h in water followed by 2 h in ^{3}H-uridine results in significant labelling of cytoplasmic ribosomes. When such ribosomes are dissociated on 10-30% sucrose density gradients containing 0.8M KCl the small ribosomal subunits show about 4-times the specific activity of the large subunits. This is consistent with the generally accepted scheme of rRNA processing in which 18S rRNA enters the cytoplasm from the nucleolus more rapidly than 28S rRNA (27, 28).

The rate of incorporation of ^{3}H-uridine into ribosomes increases during 24 h of aging in water. As in the case of total RNA however, uptake of soluble precursor into the tissue parallels increased incorporation into rRNA, the ratio of uptake to incorporation remaining essentially constant. This indicates that the rate of net rRNA synthesis remains stable from 2 to 24 h activation (18). Fraser and Loening (6) came to the same conclusion using ^{32}P-orthophosphate as a precursor.

Cells of dormant artichoke tubers contain an abundance of ribosomes as determined by cell fractionation (18). Ribosomal RNA per g fresh weight of tissue drops by 30% in the first 2 h after excision probably as a result of degradation of ribosomes prior to the onset of new rRNA synthesis (17). As ribosome production commences (2-4 h) a balance between synthesis and turnover is established leading to a constant level of ribosomes throughout subsequent water aging to 48 h (18). Even in tissue supplied auxin immediately after excision, the ribosome population remains essentially constant during the first 20 h (6). These results indicate that a basic event in the first 20-24 h post-excision, regardless of the presence or absence of auxin, is a rapid turnover of rRNA, presumably through degradation of old pre-existing ribosomes and replacement by newly synthesized ones. The biochemical results then, support the EM observations on tissue aged in 5-FU (9). The purpose and control of such turnover is unclear but should warrant further study.

Addition of hormones to 24 h aged slices stimulates ribosome synthesis leading to about a 3-fold increase in total ribosomes by 24 h after hormone application (22). Similar increases in ribosomes are seen between 20-40 h in tissue supplied auxin immediately on excision and which is entering cell division (6). In the latter case it was established that the rate of rRNA synthesis rose sharply during the period of rapid ribosome accumulation. These increases in rRNA synthesis are accomplished with existing ribosomal genes since no amplification of rDNA takes place in response to hormone treatment (29).

Activity of Ribosomes

Using fractionation techniques two classes of ribosomes may be isolated from artichoke tuber cells; free ribosomes and ribosomes bound to membranes (mb) of the ER (18). As predicted from EM observations only about 5% of ribosomes in dormant tubers are mb. After excision this proportion climbs to a maximum of 25% after 24 h aging and then declines partially (18, 22). The value of 25% is comparable to figures reported for non-secretory animal tissues. If hormones are added to 24 h aged tissue the percentage of mb ribosomes is stabilized at the 25% level (22). Since the total ribosome population increases 3-fold in response to hormones the mb ribosome population is increased by this factor also.

Membrane-bound ribosomes appear to be derived from the pool of free ribosomes. If tissue aged in water for 12 or 22 h is incubated for 2 h in ^{3}H-uridine and then free and mb ribosomes are extracted and analyzed for incorporation, free ribosomes have twice the specific activity of the mb fraction. On the other hand, if labelling is followed by a 10 h chase in water, the specific activity of the two ribosome fractions is about equal (18). The simplest model to explain these observations has newly-synthesized ribosomes enter the free ribosome pool directly, with some subsequently becoming mb through exchange between the two populations. Since equilibration is not instantaneous most labelled ribosomes are found in the free fraction directly after a 2 h pulse. The chase period allows time for equilibration, resulting by 10 h in equal specific activities for free and mb ribosomes. Ribosome exchange of this sort has been suggested in other tissues on the basis of turnover kinetics.

Aging of tuber slices results in enhancement in protein synthetic activity (for reviews of this vast subject see references 30-32). The ribosomes of dormant tubers are virtually all in monosome form (18) but polysomes can be isolated from tissue by 1 h after excision. In water, both free and mb polysomes increase during the first 24 h of aging. Polysomes isolated from aged tissue are functional (33), synthesizing protein <u>in vitro</u> when supplied with wheat germ supernate and the necessary co-factors prepared according to Marcus (34, 35). ^{3}H-leucine is incorporated into a complex array of proteins ranging in MW from 18,000 to 65,000 as determined by SDS polyacrylamide gel electrophoresis (36).

The degree of ribosome involvement in polysomes is usually estimated from profiles obtained in sucrose-density gradients. This method becomes unreliable when endogenous RNase levels increase, as occurs in artichoke slices with aging. An alternative method of estimating synthetically active and inactive populations introduced by Martin (37) is relatively insensitive to RNase levels. The method depends on stability of ribosomes in sucrose gradients containing high salt concentrations such as 0.8M KCl. In this salt concentration ribosomes not active in protein synthesis dissociate into 40S and 60S subunits while active ribosomes are stabilized by nascent polypeptides and sediment as 80S monomers or dimers. Application of this technique to aging artichoke slices (18) indicates that in dormant cells only 12-13% of ribosomes are synthetically active. This figure rises dramatically to 30% at 1 h, 46% at 4 h and peaks at 65-70% at 12 - 24 h aging. These values are remarkably similar to those reported a decade ago for freshly excised and aged carrot root slices (26). Only slight decline occurs during continued water aging to 48 h. The percentage of active ribosomes is the same in both free and mb populations. The dramatic rise in polysomes at 0 - 4 h, when synthesis of new ribosomes is minimal, clearly shows that ribosomes pre-existing in the dormant storage cells are capable of participating in extensive protein synthesis at least for a short period, as proposed for carrot slices by Leaver and Key (26). Increases in polysomes in excised storage tissues have also been reported for beet root (38) and potato (39).

Application of 2,4-D and kinetin to 24 h aged artichoke slices maintains the level of active ribosomes at 70-75% in both free and mb populations for 24 h (when water controls decline) and then a decline to 50-58% occurs during the next 24 h (22). Again, since the total ribosome population increases 3-fold in the first 24 h of hormone treatment the total number of active ribosomes in both free and mb populations rises markedly. In other words, total protein synthetic capacity within hormone-treated tissue rises by about 3-fold. This is manifested by a doubling of total protein per g fresh weight of tissue within 48 h (22). This protein increase is apparently necessary for continued growth since protein synthesis inhibitors block cell expansion (40). Cytological observations indicate that much of the net protein increase represents synthesis of new structural cytoplasm and organelles. Increases in these components do not occur in water-treated tissue where no net increase in protein occurs (4). Basically similar results on protein accumulation were obtained by Yeoman and Mitchell (5). Tissues aged in water or treated immediately on excision with 2,4-D showed little increase in protein over 24 h. On continued incubation in water only a small increase in protein occurred while after 24 h the auxin-treated tissue showed a steady increase in total protein. Treatment of 5-day water-aged tissue with 2,4-D again led to a rapid rise in protein. These results seem somewhat at variance with those of Fraser and Loening (6), who found almost the same rise in protein content after 67 h incubation with and without 2,4-D. As emphasized by the latter authors, and by Adamson (3) in earlier physiological studies, storage history of tubers can influence responses of excised slices; this factor might account for these differences. In autoradiographic studies substantial variation in protein synthesis was also encountered in different experiments (9). Increased protein synthesis in excised slices of storage tissues has been reported for a wide variety of species (30, 31). Possibly most of these diverse tissues respond more or less similarly during short aging periods up to 24 h, but beyond this differences in responses are to be expected. Apparently not all storage tissues lack endogenous auxin as does artichoke tuber; many tissues such as potato and chickory roots will undergo substantial growth without added hormones. Such tissues are more comparable to artichoke slices treated continuously with auxin than those aged in water. Furthermore, as emphasized by Kahl (30), the growth response of different tissues varies. Some undergo both cell division and expansion while in others one or other of the responses predominates. Other storage tissues such as turnip and Dahlia root show little or no growth during short term treatment with auxin and kinetin. Presumably, ribosome metabolism and protein synthesis in tissues with differing hormone responses will show variations, which need further study.

As shown by cytological analysis of 5-FU treated artichoke tissue (9) mb ribosomes seem to be essential for normal growth response. In addition, development of substantial RER seems to be a common phenomenon in growing plant cells (12, 14). It is therefore of some interest to question the functional role of mb ribosomes in artichoke slices. In specialized secretory cells mb ribosomes are responsible for synthesizing most proteins destined for secretion (41). In such cases the newly-synthesized secretory proteins are made on the mb ribosomes and immediately enter either the ER membrane or cisternal space by a process termed vectorial transport (42). Vectorial transport thus results in segregation of specific classes of proteins and constitutes the first step in their secondary processing leading to secretion. In non-secretory tissues, however, the occurence and possible roles of vectorial transport are problematic (42).

Recently we have examined the extent to which mb ribosomes of artichoke slices are involved in vectorial transport (33). Using a rapid flotation procedure a fraction enriched in mb ribosomes and NADH cytochrome C reductase activity was prepared from water-aged and subsequently hormone-treated tissues. This fraction supported completion of synthesis of pre-initiated polypeptides in vitro, and using the procedure of puromycin and detergent release of incorporated radioactive leucine (42), vectorial transport of labelled polypeptides was measured. The results indicate that mb ribosomes of 24 h water-aged tissue release the bulk of their polypeptide products directly into the medium; only 12% of the incorporated label was released by detergent. On the other hand, with tissue treated with hormones 42% of the newly synthesized proteins required detergent for release from the membranes. The assumed intra-membrane location of these proteins was confirmed by measurements of resistance to protease digestion (33).

These preliminary studies indicate that despite the fact that a substantial population of mb ribosomes develops during aging of artichoke slices, little vectorial transport takes place in this sytem. Most proteins are released directly from the mb ribosomes into the cytoplasm. On stimulation of growth by hormones, however, the RER system changes its function and almost half of the proteins synthesized by mb ribosomes enter the ER membrane system. The particular functions of these vectorially-transported proteins are unknown as yet but assuming they are involved in the growth responses of the tissue several possibilities exist. Cell-wall modification leading to expansion may require enzymes such as cellulase (43) while considerable increases in amounts of plasma membrane, tonoplast and membranous organelles (4) would presumably require constituent proteins derived, after secondary modification, via vectorial transport (44). Interestingly, dictyosomes are extremely rare in water-aged artichoke cells but increase markedly after hormone treatment (4) at the time when vectorial transport becomes significant. The functional role of the substantial population of mb ribosomes in non-growing cells remains unexplained. Similar populations of mb ribosomes not apparently involved in vectorial transport occur in tissues such as rat muscle and brain (42). It appears that studies to relate synthesis of specific proteins to particular ribosome populations are now in order.

Effects of 5-Fluorouracil

As described in connection with cytological studies, 5-FU has been used extensively as an inhibitor in the Jerusalem artichoke slice system, giving some unexplained results. Recently, using biochemical techniques we have examined the effects of this compound on RNA and protein synthesis, when applied with hormones to slices which were aged for 24 h in water (22). As described above, in this situation 5-FU shows little or no inhibition of growth. At 100 μg/ml 5-FU reduces incorporation of ^{3}H-uridine into ribosomes of hormonally-stimulated cells by 70%. In addition, analysis by dissociation in 0.8M KCl indicates that most of the ribosomes made in the presence of 5-FU accumulate in the inactive fraction, as expected if the inhibitor is incorporated into rRNA, leading to functionally inert ribosomes. Despite these strong inhibitory effects at a time when the ribosome population is normally increasing very markedly, the total ribosome population is only reduced by about one third and the proportion of ribosomes active in protein synthesis is unaffected. Furthermore, as expected from the lack of effect by 5-FU on growth, protein accumulation over 48 h is almost normal. The only explanation we are able to offer for these somewhat anomalous results is that in the presence of stimulating hormones 5-FU directly or indirectly reduces turnover of functionally competent ribosomes leading to retention of near normal protein synthesizing capacity in the face of a sharply reduced production of new functional ribosomes (22).

Fraser (7) has carried out a detailed analysis of effects of 5-FU in artichoke slices treated with 2,4-D immediately from excision, i.e. to tissue destined normally to enter cell division. Ribosomal RNA synthesis measured at 22-23.5 h post-excision was strongly inhibited by pre-exposure to 5-FU for 5-22 h post-excision. Synthesis of mRNA was much less sensitive to the inhibitor. In such 5-FU treated tissue subsequent DNA synthesis and cell division are inhibited also and the inhibition is not reversed by thymidine. On the other hand, inhibition of DNA synthesis by 5-FU given at 47 h is reversed by thymidine. Since selective inhibition of rRNA synthesis by low concentrations of actinomycin D did not lead to inhibition of DNA synthesis it is suggested that the non-reversible inhibitory effect of 5-FU given early after aging is not through its effect on ribosome synthesis but rather due to its incorporation into mRNA. This presumably leads to **aberrant** forms of essential proteins made during the early post-excision period. A similar explanation might also apply to inhibition of cell expansion through application of 5-FU during water aging. However, in this case ribosome depletion does occur through breakdown without replacement (9) leading to reduced capacity for protein synthesis. 5-FU given with auxin immediately after excision appears to prevent rRNA turnover (7). This situation warrants some further biochemical analysis.

SUMMARY

The combination of cytological, cytochemical and biochemical approaches described above has given us a reasonably coherent picture of the events involving ribosomes which occur in excised artichoke tuber slices. The following is a summary of these events.

Almost immediately, i.e. within minutes, after excision from the intact tuber the quiescent cells form new polyribosomes and commence protein synthesis. The polysomes are largely free in the cytoplasm and contain pre-existing ribosomes. Some pre-existing mRNA may be used in the polysomes but in addition new mRNA, part of which is not polyadenylated, is synthesized very shortly after excision. New low-molecular weight RNA is also made at this time. In this early period breakdown of pre-existing ribosomes starts so that total RNA content of the tissue declines. Following these very early events, synthesis of pre-rRNA in the nucleolus commences at about 1 h post-excision and increases rapidly. The first transcription probably depends on rDNA cistrons which are extended through the fibrillar nucleolus of the quiescent cell. By about 3 h the nucleolus enlarges noticeably, and the nucleolar chromatin bodies move into the nucleolus and begin to disperse and form lacunae. Presumably this represents activation of additional ribosomal cistrons leading to increased synthesis of pre-rRNA. By 4 h a distinct granular zone is appearing at the periphery of the nucleolus. During this period the rate of ribosome production achieves balance with ribosome breakdown, giving a stable level of RNA in this tissue. For about the next 20 h (i.e. to 24 h post-excision) these processes continue actively. The nucleolus continues to enlarge, nucleolar chromatin becomes more finely dispersed, an extensive intermingled granular zone develops, and pulsating nucleolar vacuoles appear. The latter probably aid in transport of ribosome precursors to the cytoplasm. The number of polyribosomes in the cytoplasm increases rapidly so that by 12-24 h 65-70% of total ribosomes are in polysomes. At the same time the proportion of ribosomes bound to membranes increases from less than 5% at excision to about 25% at 24 h. The proportion of polysomal ribosomes is the same in free and mb populations. Newly formed ribosomes enter the pool of free ribosomes initially and then gradually equilibrate with the mb fraction. The mb ribosomes show <u>in vitro</u> protein-synthesizing activity but only about 12% of the resulting polypeptides enter the ER membranes or cisternae, i.e. show vectorial transport. The above activation of the slice reaches a peak at about 24 h post-excision. This is the classical "aging" period and the processes probably occur quite independently of the presence or absence of auxin and kinetin; gibberellic acid, however, may be involved. No growth or cell division occurs in this period, even in the presence of auxin. If the 24 h aged tissue is maintained for further periods a slow decline in nucleolar activity, number of polysomes and number of mb ribosomes takes place.

If the tissue is exposed to auxin following water aging, rapid cell expansion ensues after a short lag period. At the same time the total number of ribosomes increases about 3-fold in 24 h. The proportion of ribosomes in free and mb polysomes remains constant so these populations increase by 3-fold also. Total protein in the tissue increases steadily in this period, doubling after 24 h. Much of this represents newly synthesized structural protoplasm. Of the protein made by mb ribosomes 42% is now vectorially transported, probably representing proteins which modify cell walls and/or enter growing membranes. Rapid growth continues for about 72 h after auxin treatment but even by 24 h some decline in ribosome metabolism is evident. Nucleolar activity drops and total number of ribosomes and polysomes decrease between 24 and 48 h.

In tissue supplied auxin continuously after excision the main response is cell division between 20-60 h. In this case the same activation of ribosome and protein synthesis after the initial 24 h aging events are seen. However as the cells enter division more complex changes in metabolism occur and continued synthesis of all types of RNA and protein are observed. Eventually a continuously growing tissue culture may arise.

The events involving ribosomes in excised artichoke tuber slices show many similarities to events in intact growing parts of plants, e.g. coleoptiles and roots. It is hoped that the data and theories emerging from studies of slices will help interpret activities and responses occurring in intact organs. Further, since in the tissue slice many processes are starting or stopping, often under external physiological control, the excised tissue slice still provides one of the better systems for understanding regulation of basic processes such as rRNA synthesis, ribosomal protein synthesis, ribosome assembly, ribosome compartmentalization, vectorial transport, etc. It therefore warrants continued interest by plant biochemists and cell biologists.

REFERENCES

1. Setterfield, G.: Growth regulation in excised slices of Jerusalem artichoke tuber tissue. Symp. Soc. Exp. Biol. 17, 98-126 (1963).

2. Click, R.E., Hackett, D.P.: The role of protein and nucleic acid synthesis in the development of respiration in potato tuber slices. Proc. Natl. Acad. Sci. U.S. 50, 243-250 (1963).

3. Adamson, D.: Expansion and division in auxin-treated plant cells. Can. J. Bot. 40, 719-744 (1962).

4. Fowke, L., Setterfield, G.: Cytological responses of Jerusalem artichoke tuber slices during aging and subsequent auxin treatment. In Biochemistry and Physiology of Plant Growth Substances, p. 581-602, Eds. F. Wightman and G. Setterfield. Runge Press, Ottawa (1968).

5. Yeoman, M.M., Mitchell, J.P.: Changes accompanying the addition of 2, 4-D to excised Jerusalem artichoke tuber tissue. Ann. Bot. 34, 799-810 (1970).

6. Fraser, R.S.S., Loening, U.E.: RNA synthesis during synchronous cell division in cultured explants of Jerusalem artichoke tuber. J. Exp. Bot. 25, 847-859 (1974).

7. Fraser, R.S.S.: Synchronous cell division in cultured explants of Jerusalem artichoke tubers: the effects of 5-fluorouracil on messenger RNA synthesis and the induction of cell division. J. Exp. Bot. 26, 555-568 (1975).

8. Rose, R.J., Setterfield, G., Fowke, L.C.: Activation of nucleoli in tuber slices and the function of nucleolar vacuoles. Exp. Cell Res. 71, 1-16 (1972).

9. Rose, R.J., Setterfield, G.: Cytological studies on the inhibition by 5-fluoro-uracil of ribosome synthesis and growth in Jerusalem artichoke tuber slices. Planta (Berl.) 101, 210-230 (1971).

10. Jordan, E.G., Chapman, J.M.: Ultrastructural changes in the nucleoli of Jerusalem artichoke (Helianthus tuberosus) tuber discs. J. Exp. Bot. 22, 627-634 (1971).

11. Jordan, E.G., Chapman, J.M.: Nucleolar and nuclear envelope ultrastructure in relation to cell activity in discs of carrot root (Daucus carota L.). J. Exp. Bot. 24, 197-209 (1973).

12. Rose, R.J.: Changes in nucleolar activity during the growth and development of the wheat coleoptile. Protoplasma 79, 127-143 (1974).

13. Hyde, B.B.: Changes in nucleolar ultrastructure associated with differentiation in the root tip. J. Ultrastruct. Res. 18, 25-54 (1967).

14. Chaly, N.M., Setterfield, G.: Organization of the nucleus, nucleolus and protein-synthesizing apparatus in relation to cell development in roots of Pisum sativum. Can. J. Bot. 53, 200-218 (1975).

15. Jordan, E.G., Luck, B.T.: The nucleolus organizer and the synaptonemal complex in Endymion non-scriptus (L.). J. Cell Sci. 22, 75-86 (1976).

16. Chapman, J.M., Jordan, E.G.: Influence of gibberellic acid on nucleolar size changes in storage-tissue discs. J. Exp. Bot. 22, 620-626

(1971).

17. Byrne, H., Setterfield, G.: Activation of ribosomal and messenger RNA synthesis in excised Jerusalem artichoke tuber slices. Planta (Berl.) 136, 203-210 (1977).

18. Sparkuhl, J., Gare, R.L., Setterfield, G.: Metabolism of free and membrane-bound ribosomes during aging of Jerusalem artichoke tuber slices. Planta (Berl.) 129, 97-104 (1976).

19. Srivastara, L.M., Paulson, R.E.: The fine structure of the embryo of Lactuca sativa. II. Changes during germination. Can. J. Bot. 46, 1447-1453 (1968).

20. Key, J.L.: Effect of purine and pyrimidine analogues on growth and RNA metabolism in the soybean hypocotyl -- the selective action of 5-fluorouracil. Plant Physiol. 41, 1257-1264 (1966).

21. King, B., Chapman, J.M.: The effect of inhibitors of protein and nucleic acid synthesis on nucleolar size and enzyme induction in Jerusalem artichoke tuber slices. Planta (Berl.) 104, 306-315 (1972).

22. Sparkuhl, J., Setterfield, G.: Ribosome metabolism in hormone-treated Jerusalem artichoke tuber slices in the absence and presence of 5-fluorouracil. Planta (Berl.) 135, 267-273 (1977).

23. Philip, N., Setterfield, G.: Unpublished results.

24. Fraser, R.S.S.: Studies on messenger and ribosomal RNA synthesis in plant tissue cultures induced to undergo synchronous cell divisions. Eur. J. Biochem. 50, 529-537 (1975).

25. Craig, N., Perry, R.P.: Persistent cytoplasmic synthesis of ribosomal proteins during the selective inhibition of ribosomal RNA synthesis. Nature New Biol. 229, 75-80 (1971).

26. Leaver, C.J., Key, J.L.: Polyribosome formation and RNA synthesis during aging of carrot-root tissue. Proc. Natl. Acad. Sci. U.S. 57, 1338-1344 (1967).

27. Rogers, M.E., Loening, U.E., Fraser, R.S.S.: Ribosomal RNA precursors in plants. J. Mol. Biol. 49, 681-692 (1970).

28. Leaver, C.J., Key, J.L.: Ribosomal RNA synthesis in plants. J. Mol. Biol. 49, 671-680 (1970).

29. Ingle, J., Sinclair, J.: Ribosomal RNA genes and plant development. Nature 235, 30-32 (1972).

30. Kahl, G.: Genetic and metabolic regulation in differentiating plant storage tissue cells. Bot. Rev. 39, 274-299 (1973).

31. Kahl, G.: Metabolism in plant storage tissue slices. Bot. Rev. 40, 263-314 (1974).

32. Van Steveninck, R.F.M.: The washing or aging phenomenon in plant tissues. Ann. Rev. Plant Physiol. 26, 237-258 (1975).

33. Sparkuhl, J., Setterfield, G.: Vectorial transport of proteins by membrane-bound ribosomes of non-growing and growing Jerusalem artichoke tuber cells. Can. J. Biochem. In Press (1978).

34. Marcus, A., Seal, S.N., Weeks, D.P.: Protein chain initiation in wheat embryo. Meth. Enzymol. 30, 94-101 (1974).

35. Marcus, A., Efron, D., Weeks, D.P.: The wheat embryo cell-free system. Meth. Enzymol. 30, 749-754 (1974).

36. Sparkuhl, J.: Metabolism of free and membrane-bound ribosomes during activation and auxin-induced growth of Jerusalem artichoke tuber slices. Ph.D. Thesis, Carleton University, Ottawa, Canada (1977).

37. Martin, T.E.: A simple general method to determine the proportion of active ribosomes in eucaryotic cells. Exp. Cell Res. 80, 496-498 (1973).

38. Cherry, J.H.: Regulation of invertase in washed sugar beet tissue. In Biochemistry and Physiology of Plant Growth Substances, p. 417-431. Eds. F. Wightman and G. Setterfield. Runge Press, Ottawa (1968).

39. Kahl, G.: Activation of protein synthesis in aging potato tuber slices. Z. Naturforsch. 26b, 1064-1067 (1971).

40. Nooden, L.D.: Studies on the role of RNA synthesis in auxin induction of cell enlargement. Plant Physiol. 43, 140-150 (1968).

41. Rolleston, F.S.: Membrane-bound and free ribosomes. Sub-Cell Biochem. 3, 91-117 (1974).

42. Andrews, T.M., Tata, J.R.: Protein synthesis by membrane-bound and free ribosomes of secretory and non-secretory tissue. Biochem. J. 121, 683-694 (1971).

43. Verma, D.P.S., Maclachlan, G.A., Byrne, H., Ewings, D.: Regulation and in vitro translation of messenger ribonucleic acid for cellulase from auxin-treated pea epicotyls. J. Biol. Chem. 250, 1019-1026 (1975).

44. Shore, G. C., Tata, J.R.: Two fractions of rough endoplasmic reticulum from rat liver. II. Cytoplasmic messenger RNA's which code for albumin and mitochondrial proteins are distributed differently between the two fractions. J. Cell Biol. 72, 726-743 (1977).

The Physiology of Ethylene in Wounded Plant Tissues

Shang Fa Yang and Harlan K. Pratt

I. INTRODUCTION

It is now generally agreed that ethylene is an endogenous plant hormone involved in many phenomena of growth regulation and senescence (2, 75); its role in fruit ripening appears to be particularly important. Not only is endogenous ethylene important in growth and development, but the effects of ethylene treatments are physiologically and biochemically dramatic (2).

Nearly all plant tissues appear to be capable of producing some ethylene. In some tissues, particularly fruits, a massive ethylene production at the appropriate time is part of the normal life cycle (74). In others, it appears to be a reserve capability which enables the plant to cope successfully with such crises as the shedding of infected parts (2) or growing in a difficult situation (28, 49). Ethylene produced by plants under such conditions is usually referred to as "stress ethylene," and Abeles (2) has devoted a chapter to the subject. The list of stresses which can induce increased ethylene production by plant tissues suggests that any condition deviating from the normal environment of the intact healthy plant may lead to such an increase.

Metabolic changes in plant tissues are regularly studied with tissue slice techniques. Because the preparation of such tissue samples inevitably involves injury, tissue slice techniques may be used to study metabolic changes associated with injury as well as normal metabolism. Among the consequences of tissue injury can be an increased synthesis of ethylene. If the metabolism under investigation may be influenced by ethylene, it is necessary to consider that manipulation of the material may of itself be introducing an unwanted variable into the system.

Perhaps the most striking established example of the role of stress ethylene in

Biochemistry of Wounded Plant Tissues

complicating experimental interpretations is in the use of abscission zone explants to study abscission of leaves or other organs. Burg (13) reviewed the role of ethylene in abscission, as understood to that time, and pointed out that freshly prepared explants produce wound ethylene at a rate high enough to affect the tissue and mask any response to applied ethylene. Jackson and Osborne (47) sought to avoid this complication by preventing accumulation of ethylene around their explants. Jackson and Campbell (45) emphasized the wound effect when trying to use petiole segments to estimate ethylene in the intact plant. Because ethylene production was greatest between 1 and 2 hr after excission, measurements should be made within 20 to 30 minutes after cutting; the early phase of low ethylene production presumably represents the time required before the wound stimulus takes effect.

In this chapter we shall focus our attention on induction of ethylene production caused by the stress inflicted on the plant tissue and the physiological consequences of the ethylene so produced.

II. STRESS ETHYLENE

In "unstressed" plants ethylene is usually produced in those parts which have the highest auxin content, presumably because the production of the gas is stimulated or regulated by auxin. Trauma-induced ethylene production, however, whether resulting from physical, chemical, or biological stress, does not appear to be directly associated with the normal auxin-mediated ethylene production exhibited by young developing plant parts.

Physical Stress

Among the physical stresses that have been shown to induce increased ethylene production are mechanical events (cutting, bruising, pressure), temperature deviations (including freezing, chilling, and high temperature injury), stress due to either water deficit or excess, radiation injury, light, and alteration of gravity. Simply handling intact plants, as in setting up an experiment, can induce ethylene formation which may complicate the results (44); for example, gentle stroking of the leaf blades caused lily plants to produce more ethylene, resulting in reduced growth (34). Of greatest concern is the mechanical injury implicit in excising experimental tissue. It is interesting to note that the first comprehensive qualitative survey of ethylene production by plant materials (24) probably depended on injury inflicted on the tested materials. Denny used as a bioassay the epinasty induced in leaves of potato seedlings when various kinds of plant tissues were chopped and positioned around the test plants. Most of the plant materials tested, including chopped potato seedlings, did induce epinasty, yet the uninjured seedling, similarly confined in a sealed chamber, did not produce enough ethylene for self induction. Although the test seedlings were cuttings severed from the plant, this amount of injury did not compare with shredding leaves or chopping up massive tissues such as roots and tubers.

Cutting. Williamson (97) was apparently the first to record direct evidence of increased ethylene production by injured plant material, but serious study of the phenomenon of wound ethylene had to wait for the adequate analytical capability which has been provided by flame ionization gas chromatography. Using this technique, Meigh et al. (69) determined evolution of ethylene by various tissue components (Table I); no respiration data were presented, but it is clear that any degree of tissue damage may be significant. McGlasson and Pratt (67) found that muskmelon fruit slices have higher rates of respiration and ethylene production than comparable intact organs, and Burg (12) referred to unpublished data of the same sort from his laboratory. McGlasson and Pratt (67) suggested that the stimulated production of ethylene may account for at least part of the observed physiological differences between tissue slices and whole fruits, and that the higher rates of ethylene production in slices were apparently great enough to cause the onset of a respiratory climacteric in the slices. They further pointed out that wounding apparently overcame whatever factor limits ethylene production in the intact (younger) fruits. McGlasson (65) studied ethylene production by slices of green banana fruit which produces a barely detectable amount of ethylene (less than 50 nl/kg/hr) while intact. The initial rates, measured about 2 hr after cutting, were several times higher than the rate in intact fruit. A second rise in the rate of ethylene production was detected 4 hr after cutting, and the maximum rate was reached within 6 to 8 hr. The rate of ethylene production was substantially greater with 2-mm than 4- or 6-mm slices. Ethylene production by disks of tomato pericarp was about 20-fold higher than by whole fruit within 15 to 20 min after cutting (55). During the next 4 hr, ethylene production increased a further 5-fold and then declined to a stable level 20 to 50 times that of whole fruit. Respiration rate showed a similar pattern, the rate of carbon dioxide production being 7 times higher in disks from

Table I. Ethylene production by components of fruits[a]

	nl ethylene/g/hr at 30°	
	Apple	Tomato (ripe)
Whole fruit	29.6	4.4
Segments[b]	35.4	58.6
Tissue slices[c]	20.3	102.0
Homogenates	6.8	2.0
Mitochondria	*d*	*d*
Microsomes	*d*	*d*
Supernatant	*e*	*e*

a Data from Meigh et al. (69).

b A whole fruit was cut into 16 equal segments.

c Disks 1 cm in diameter and 2 mm thick for apples; 4 mm cubes of pericarp wall for tomatoes.

d Below limit for detection: <0.003.

e Trace: <0.03.

immature fruit than in intact fruits within 15 to 20 min after cutting. The more immature fruits might show a second rise beginning 1 hr after cutting. Although the nonripening mutant tomatoes, *rin* and *nor*, produce little ethylene during natural aging of intact fruits, wounding stress induced ethylene production in fruits of these mutants (33, 66).

In work reported by Burg and Thimann (16), plugs of postclimacteric apple tissue produced ethylene at the same rate as whole fruits but 1-mm tissue slices produced much less. Disks of the peel of preclimacteric apples did not produce ethylene but could be induced to do so by periods of incubation ("aging") in buffer solution (27). Ethylene production began about 5 hr after the incubation was started and reached the maximum rate in 9 hr. In contrast, when a plug (1 cm in diameter and 2 cm long) of the cortex of preclimacteric apple was incubated up to 8 hr, D. O. Adams in this laboratory (unpublished observation) was not able to detect enhanced ethylene production.

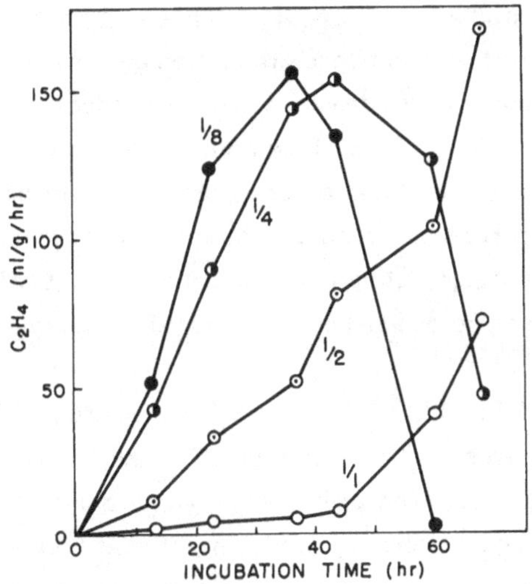

Fig. 1. Ethylene production by excised segments of young 'Satsuma' mandarin fruit during incubation at 26°. Whole fruit (1/1, ○) were cut into halves (1/2, ⊙), quarters (1/4, ◑) and eighths (1/8, ●). (From Hyodo, 37).

It is known that intact citrus fruits produce ethylene at a very low rate unless subjected to some form of stress (94). Ben-Yehoshua and Eaks (8) observed that the rate of ethylene production by orange fruits was approximately doubled when incisions were made in the skin. Riov et al. (79) showed rapid increases in the rate of ethylene production from disks following excission from grapefruit peel. Hyodo (37) found that excised mandarin peel tissue produced a large amount of ethylene even if the fruit from which it was taken was incapable of producing the gas. The remarkable increase in ethylene production by young mandarin orange fruits after excission into segments is illustrated in Figure 1.

Cooper (21) observed that mechanical wounding of woody stem tissues of citrus also caused a large increase in ethylene production. When the bark was peeled from the wood of 'Robinson' tangerine terminal shoots, 5 to 10 times as much ethylene was produced in the peeled bark and wood as in the intact stems. Slitting the stems into two pieces, or macerating the stems increased the rate of ethylene production more than 10 times.

Imaseki et al. (42, 43) studied the possible inducers of metabolic change in sweet potato root slices and found that ethylene production increased in response to cutting. Increasing the cut surface area increased ethylene production, and the amount was proportional to the logarithm of the surface area. They suggested that this increased ethylene may account for some of the observed enhancement of metabolic activities. Other fleshy tissues reported to produce stress-ethylene when induced by incision include potato (65), cucumber seedling (26), and carrot (48). Leaf and stem tissue also produced stress ethylene when excised or shredded, although in smaller amounts.

Bruising. Lougheed and Franklin (57) showed that bruising substantially increased ethylene production in freshly harvested preclimacteric apples, although earlier workers found no such effect. The discrepancy was attributed to the use of storage apples which were presumably already producing ethylene at a high rate. Dropping grapefruit from heights up to 2 m increased both respiration and ethylene production in fruits that normally produce little ethylene (94). Ethylene production, respiration, and rate of ripening were all increased by bruising mature green tomato fruits (60); the response was proportional to the number of impacts.

Temperature Stress. There appears to be little known about the effect of temperature on ethylene production by tissue slices. One should note, however, that intact fruits are often unable to produce ethylene at temperatures above about 35 C (12), and so do not ripen normally. Chilling injury, however, has induced ethylene production in tissues not normally producing significant amounts, notably in the citrus fruits (94).

Elstner and Konze (25) froze various areas of sugar beet leaf disks and found that ethylene production increased with increase in the area of frozen tissue until 50% of the leaf area was damaged. Further increase in damage reduced ethylene production. It was suggested that ethylene production was related to the number of perturbed cells adjacent to those killed by freezing.

Radiation. The effects of radiation on plant tissues were thoroughly reviewed by Maxie and Abdel-Kader (64). Injurious doses of gamma radiation usually caused an immediate small peak of ethylene production in apple and pear fruits, but subsequently the ethylene production rate dropped to normal or was reduced below normal by the injury. Sensitivity to applied ethylene was also reduced. On the other hand, radiation increased ethylene production and hastened ripening changes in peaches and in the nonclimacteric citrus fruits. These general observations on the citrus fruits have been confirmed by Riov et al. (80).

Moisture Stress

A plant tissue system may be subjected to stress by flooding of tissue slices, by maintaining them in an environment conducive to wilting, or by vacuum infiltration of solutions. Even intact plants may show increased ethylene production in response to water stress, and use of detached organs in systems with low humidity may lead to increased ethylene production, as has been shown in immature cotton bolls and bracts (30), mature but preclimacteric banana and avocado fruit (5), and in orange leaves (7).

Kawase (48) tested various plant organs and pieces thereof and found that submergence in water increased the ethylene level beyond that caused by the injury involved in preparation of the samples.

Wright (98) found that dehydration of detached leaves stimulated ethylene production; chilling increased ethylene production in leaves that had been partially dehydrated but had little or no effect on fully hydrated leaves.

Jackson and Campbell (46) found that low 0_2 in waterlogged soil (up to 3%) caused production of ethylene by plants, even though 0_2 is required for ethylene production. The ethylene produced by roots when under low 0_2 conditions causes epinasty in the tops of tomato plants.

Chemical Stress

Chemical stresses which may affect ethylene production by plant tissues may be construed to include not only toxic chemicals but also auxins and other plant regulators applied to a tissue system. Hormonal and herbicidal regulation of ethylene has been adequately reviewed by Abeles (2). Webb et al. (96) have proposed ethylene production by leaf disks after various chemical treatments as a means of evaluating the defoliant potentials of the tested chemicals.

Increased ethylene production may occur when plants are exposed to air pollutants. Craker (22) demonstrated that increased ethylene production could be detected within an hour after treatment of tomato plants with ozone, and an ozone-susceptible cultivar produced more than a resistant type. Similarly Bressan et al. (10) found that ethylene production by cucurbits was related to SO_2 injury. Ethylene production increased greatly in slightly injured leaves but declined in severely injured leaves. The ethylene production response was correlated with sensitivity to SO_2, so ethylene production may be a useful indicator of injury by SO_2.

Various forms of mercury are phytotoxic, and it might be supposed that mercury vapor would induce ethylene formation as a general injury response. Goren and Siegel (29), however, believe that mercury vapor induces ethylene production in a more specific way because it causes no obvious disturbance of the respiratory gas exchange rates; concentrations of mercury vapor as low as 1 ppm were effective.

The other inorganic salts which are known to be effective promoters of ethylene production are Cu^{2+}, Fe^{3+}, and KI. Some organic compounds, such as monoiodoacetamide, trichloroacetic acid, sodium ethylmercurithiosalicylate, ascorbic acid, and iodoacetic acid are also known to increase ethylene production (2).

An interaction of metal ions with plant hormones in promoting ethylene production is known. Although only a slight increase in ethylene resulted from the application of either kinetin or Ca^{2+} to mungbean hypocotyl segments, a remarkable synergistic increase in ethylene production was observed when they were applied together (53). Among the divalent cations tested, only Sr^{2+} was able to replace Ca^{2+}, while all cytokinins tested were as effective or nearly as effective as kinetin (52). Although several metal ions, most notably copper and iron, promoted ethylene production, some of them, such as Hg^{2+}, Co^{2+} and Ni^{2+}, inhibited production (54).

Of the chemicals which induce a stress situation, cycloheximide is perhaps the most interesting. Cycloheximide can block the ethylene production induced by auxin, slicing, or Cu^{2+} (2). This effect is interpreted as an indication that stress ethylene production in plants requires new enzyme synthesis, which is blocked by cycloheximide. When citrus trees were sprayed with a solution containing cycloheximide, however, small amounts of the deposited chemical caused injury of the cells of the rind, and as a result wound ethylene was produced. This in turn caused fruit drop or loosening (21), and the practice has now been used commercially.

Biological Stress

The stresses referred to in this section are those induced by other organisms, including insects and microbes, and especially those bacteria and fungi associated with plant disease. It is becoming apparent that ethylene may be involved in many plant diseases, perhaps as a product of the pathogen, but more commonly as a product of the ailing tissue. Many plant disease symptoms, such as yellowing, epinasty, and abscission, appear to be the result of this induced production of ethylene from infected plants. This subject has been well reviewed by Hislop et al. (35).

Shain and Wheeler (86) showed that resistant strains of oats produced much less ethylene after treatment with a pathotoxin than did susceptible strains. Marco et al. (62) found that cucumber mosaic virus caused enhanced production of ethylene in infected hypocotyls, and this in turn retarded hypocotyl elongation. Similar results have been obtained with virus infection of cowpea hypocotyls (56). Abbattista Gentile and Matta (1) confirmed that *Fusarium* wilt greatly increased the ethylene production rate by susceptible tomato plants, while infected resistant plants and plants infected by a nonvirulent strain did not show such an increase. They believe that the ethylene production was a consequence of the water stress rather than a direct cause of disease symptoms.

A number of investigators have shown that plants suffering microbial infection produced large quantities of ethylene. Usually the host tissue produces the ethylene, although in some cases a fungus can be the source. It is known that several fungi can produce ethylene in vitro (39). Among the microbes which produce ethylene in culture, *Penicillium digitatum*, which infects citrus fruits, appears to be most prolific. Another example of an ethylene-producing pathogen is *Fusarium oxysporum* (91).

Accelerated senescence and shedding of plant organs following insect infestation has been observed in a variety of plants. Shain and Hillis (85) found that attack by *Sirex* wasps caused pine sapwood to produce more ethylene than did an equivalent purely mechanical injury, suggesting a reaction to the mucus secretion of the insect. This reaction was related to the resistance of the tree to attack by fungi introduced by the insect.

III. BIOCHEMISTRY OF ETHYLENE FORMATION

Most of the work on the mechanism of ethylene biosynthesis in higher plants has been carried out with ripening fruit tissues which produce large amounts of ethylene. Although cut tissues were used in these studies, there is no indication that the biosynthetic pathway might differ from that of intact fruit, since wound-induced ethylene production was of minor significance when compared to the massive ethylene output of the ripening fruit. It is now fairly well established that methionine is the precursor of ethylene in higher plants. Recent work by Murr and Yang (70) and Adams and Yang (4) have indicated that S-adenosylmethionine is an intermediate in the conversion of methionine to ethylene. A postulated mechanism for the biological production of ethylene from methionine is shown in Figure 2.

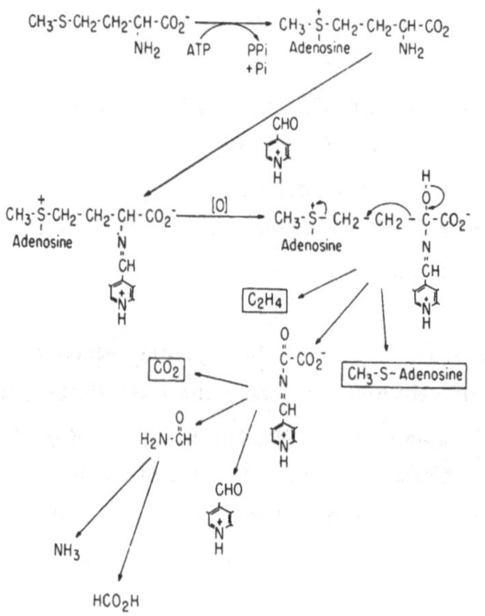

Fig. 2. A postulated mechanism for the biological production of ethylene from methionine mediated by pyridoxal phosphate (70).

Cell-free enzymatic systems capable of ethylene production from methionine or its analogues have been sought actively in several laboratories but have not yet been established. Although cutting into segments or slices of many organs or tissues often resulted in increased ethylene release or production, as illustrated in Table I, the capacity for ethylene synthesis is completely destroyed upon homogenization. The slight ethylene output shown for the homogenates in Table I is considered to be due to the release of residual ethylene which was synthesized prior to cell disruption.

In addition to higher plants, many microorganisms are known to produce ethylene. Ethylene is thought to be a common metabolic product of fungi (39), and it may play an important role in some host-pathogen interactions. Most of the work on the biosynthesis of ethylene by microorganisms has been carried out with *Penicillium digitatum*, which produces ethylene at high rates. It is now clear that the biosynthetic pathway in *P. digitatum* is entirely different from that of higher plants. *P. digitatum* utilizes C-3,4 of glutamic acid (or α-ketoglutarate) but not methionine as its precursor, whereas higher plants utilize C-3,4 of methionine but not glutamic acid (20). These results indicate that the TCA cycle is closely related to the biosynthesis of ethylene by *P. digitatum*. The sequence of the biosynthetic pathway is as follows:

$$\text{Acetate} \rightarrow \text{citrate} \rightarrow \text{isocitrate} \rightarrow \alpha\text{-ketoglutarate} \rightarrow \rightarrow \text{ethylene}$$

Methionine has been claimed to be the precursor of ethylene in *M. hiemalis* (58) and other microbes (9). It is worth noting, however, that this ethylene production occurred extracellularly, required very high concentrations of methionine, and was stimulated by light, FMN, Fe^{2+}, and ascorbate. These characteristics are typical of non-enzymatic ethylene production from methionine or its analogs. For this reason, it must be considered unlikely that *M. hiemalis* produced ethylene biologically and there exists no valid evidence to indicate that methionine served as the biological precursor.

For detailed accounts dealing with the mechanism and regulation of ethylene biosynthesis in general, the reviews of Abeles (2) and Yang (100) may be consulted.

The biochemistry of ethylene formation in wounded tissues has been studied in a number of plant materials, and all of them confirmed the validity of the methionine pathway. Abeles and Abeles (3) have shown that ethylene production from bean and tobacco leaves increased rapidly following the application of toxic compounds such as $CuSO_4$, Endothal, or ozone. The compounds which increased ethylene evolution also increased the conversion of methionine into ethylene. In these experiments, however, the conversion efficiency (specific radioactivity of ^{14}C-ethylene produced/ specific radioactivity of ^{14}C-methionine administered) was below 1% for unstressed plants, and further decreased (50%) after chemical stress. An extensive dilution of labeled methionine by endogenous methionine must have occurred. Since data on such a dilution were not presented, it is not possible to evaluate whether an alternative pathway may have operated during the production of stress ethylene. If all of the ethylene evolved was derived from methionine, more unlabeled endogenous methionine must have become available as the rate of ethylene production increased following the chemical injury. It is interesting to note that cycloheximide effectively inhibits the production of chemical stress-induced ethylene, suggesting the possibility that protein synthesis may be required for its production.

Recently Hyodo (36) reported that albedo disks prepared from mandarin fruit produced ethylene at an increasing rate until a maximum (about 60 nl/g-hr) was reached after incubation for 30 hr. The ability of the albedo discs to convert labeled methionine to ethylene paralleled the rate of endogenous ethylene production. No measurable conversion of labeled methionine into ethylene was found in freshly prepared disks, but aged disks actively converted methionine. This conversion was markedly inhibited by cycloheximide, but not by actinomycin D. This was thought to suggest that m-RNA, once formed, is stable in this tissue while the ethylene-producing enzyme system is not.

It is known that injury stimulates ethylene production by some unripe fruit tissues such as apple, cantaloupe, banana, and tomato, although intact fruits at the same stage of development produce no significant amount of ethylene. Cantaloupe harvested 30 days after anthesis produced very little ethylene (about 0.1 nl/g-hr), and the low rate was maintained for the period of about 5 days preceding the climacteric stage. When plugs (1 cm in diameter and 2 cm in length), were cut from the freshly harvested cantaloupe, however, a rapid increase in ethylene production began. M. L. Bliss (unpublished results from this laboratory) found that within 6 hr the ethylene production rate of the plug increased by a factor of more than 10 and reached a maximum (about 110 nl/g-hr) in 18 hours. This exceeded the rate of the intact melon at the ripened stage. As in the citrus albedo tissue reported by Hyodo (36), the freshly cut cantaloupe plug, which produces little ethylene (0.8 nl/g-hr), did not convert measurable amounts of labeled methionine into ethylene, while the aged plug, which produced ethylene at a high rate, successfully converted methionine. When 10 uM cycloheximide was administered at the beginning of a 21 hr incubation of the cut plug, it inhibited ethylene production by 88% and the conversion of labeled methionine into ethylene by 90%. The application of 10 uM L-2-amino-4-(aminoethoxy)-trans-butenoic acid, an ethoxy analog of rhizobitoxin, inhibited ethylene production by 98% and the conversion of labeled methionine into ethylene by 92%. These data support the view that methionine is also the precursor of wound ethylene in the unripe fruit tissue of cantaloupe.

Although intact fruits of *rin and nor* mutant tomatoes show very low rates of ethylene production and lack a respiratory climacteric, pericarp disks cut from these strains produced ethylene and were able to convert radioactive methionine into ethylene at rates comparable to disks from a normal cultivar (66).

Linolenate was at one time suggested as a precursor of ethylene in plant tissue. Such a view has not been supported by experimental data, however, and in several cases has been refuted (for review see ref. 100). Galliard et al. (27) observed that disks of peel of unripe apples, which did not themselves produce ethylene, can be induced to produce ethylene by periods of aerobic incubation ("aging"). Based on their observation that this stress ethylene evolution was stimulated by peroxidation products from linolenic acid but not by methionine, they proposed that linolenate may serve as a precursor in this system. The system was characterized further by Mapson et al. (61), who compared the ability of labeled methionine or linolenic acid to serve as precursors of ethylene in aged apple peel and concluded that methionine, but not linolenate, served as a precursor of ethylene, although linolenic acid was capable of stimulating the conversion of methionine into ethylene.

Biosynthesis of wound ethylene in rib segments excised from morning glory flower buds was studied by Hanson and Kende (32). These immature segments evolved only a small quantity of ethylene during incubation, but when they were wounded mechanically by cutting or crushing them gently, the rate of ethylene production rose more than 10-fold within 1 hour and returned to a low rate after 3 hours. Production of ethylene by both control (untreated) and wounded rib segments was inhibited more than 95% by overnight pretreatment with the ethoxy analog of rhizobitoxine at 0.03 or 0.1 mM. After overnight exposure of segments to 9 uM L-[U-^{14}C]methionine, the specific radioactivity of ethylene evolved by wounded tissue was determined and compared to the specific radioactivity of methionine extracted from the segments. Their results are interpreted to indicate that wounding causes a temporary breakdown in compartmentation between two methionine pools: one serves as the precursor of ethylene while the other does not. This temporary increase in methionine concentration at the site of ethylene generation would cause the transient rise in the rate of ethylene production. However, it was not determined whether synthesis of new protein is required in this system.

From the evidence presented above, we may conclude that the pathway for synthesis of wound ethylene is similar, if not identical, to the pathway for synthesis of ethylene in ripening fruit tissue. In both cases, methionine serves as the precursor. It is not known how the synthesis of wound ethylene is induced. The data from kinetic studies (lag period required for the onset of ethylene production) and inhibitor studies (cycloheximide) have been interpreted to indicate that synthesis of specific enzymes is required for the onset of wound ethylene formation. Since exogenous labeled methionine is only slightly converted to ethylene in unwounded tissues which produce little ethylene, but active conversion is observed in wounded tissue, it is suggested that the control of ethylene production occurs at a step subsequent to the synthesis of methionine. Cycloheximide is known to exert several effects in addition to inhibition of protein synthesis (68), however, and the data from inhibitor studies do not constitute adequate proof that *de novo* enzyme synthesis is required for production of wound ethylene. Of particular concern is the effect of cycloheximide as a phosphorylation uncoupler (59). It should be noted that a phosphorylation uncoupler would also inhibit the production of ethylene from methionine (14, 70, see also Fig. 2.). It is quite possible that a part of the inhibition of wound-ethylene production by cycloheximide may be attributed to interference with the process of phosphorylation. Alternatively, wounding may cause cellular disorganization, such as membrane modification and destruction of compartmentalization. This could in turn lead to an enhanced metabolism including protein synthesis, which cycloheximide may influence, and is essential for the synthesis of wound ethylene. There is, however, no solid experimental evidence supporting such a view.

Since intact citrus fruits do not produce ethylene, and *P. digitatum*, which is isolated from infected citrus fruits, produces large amounts of ethylene, it is assumed that ethylene is produced by the fungus in the infected citrus fruits. However, wounded citrus fruits can produce ethylene from the peels; it is likely that the ethylene of infected citrus fruits comes from both the fungus and the host. Furthermore, citrus fruits can be induced to produce large amounts of ethylene by inoculation with *P. italicum*, which is another important citrus fruit pathogen but has been shown not to produce ethylene in pure culture. In this case, also, it is not known whether the ethylene is produced by the host alone or by both host and fungus. A biochemical method which would provide an answer to this question is readily available. As previously described, the fungus *P. digitatum* and its host tissue differ in biosynthetic pathway and ethylene precursor. The relative labeling of ethylene produced by the infected fruit following administration of labeled glutamic acid or methionine would therefore reveal the respective contributions of the two distinct pathways.

The biosynthetic pathway of ethylene in freshly cut and black rot-infected tissues of sweet potato was investigated by Sakai et al. (81). They found that [1-^{14}C]acetate and [2-^{14}C]acetate were poorly but equally incorporated into ethylene by fresh tissue, but [2-^{14}C]acetate was preferentially incorporated into ethylene from diseased tissue. They therefore concluded that the pathways of ethylene production in healthy and in fungus-infected tissues are different. The TCA cycle is closely related to the biosynthetic pathway of diseased tissue but not of healthy tissue. Considering the finding that black rot fungus (*Ceratocystis fimbriata*) produces ethylene (18), it is very likely that this difference was caused by ethylene production from the fungus in the infected tissues. It has been pointed out that, in the conversion of acetate into ethylene via the methionine pathway, both carbons of acetate should be incorporated into ethylene with equal facility (99). If the mechanism of ethylene biosynthesis in black rot fungus is identical to that in *P. digitatum*, it is understandable that the pattern of ethylene biosynthesis changes in sweet potato root tissue after it is infected. The data of Sakai et al. (81) are consistent with the interpretation that, in the healthy sweet potato slices, ethylene is produced from methionine, while in the infected tissue most of the ethylene is synthesized via the glutamic acid pathway characteristic of the fungus.

IV. PHYSIOLOGICAL SIGNIFICANCE

Excepting a few cases , the function of the enhanced ethylene production which results from trauma or injury is unknown. Certain physiological consequences of the activity of stress ethylene, such as the accelerated senescence or abscission of damaged organs, are readily apparent. The response to stress ethylene may be the activation of a system enabling the plant to cope with the initial stress. An example of such a device was reported by Goeschl et al. (28). Germinating seedlings were shown to respond to a mechanical load (stress), which was applied in a manner designed to oppose growth, with an increase in ethylene production. The response of the seedling to the incremental ethylene was a decreased elongation and an increased radial expansion, resulting in a sturdier axis capable of exerting and withstanding the forces required for emergence from a compacted or crusted soil.

In other circumstances stress ethylene may act in a manner enabling the plant to eliminate the stress rather than to cope with it directly. The healing of wounds is an example. Ethylene is known to produce intumescences in apple twigs (95) and callusing and root formation in stem cuttings of tomato (101). It is in this respect that ethylene may be considered to fulfill the requirements of a wound hormone: It is a product of damaged cells, diffuses to neighboring cells, and induces cell division leading to healing of the wound.

Internal blackspot of potato tubers is a physiological disorder resulting from bruising during harvest or postharvest handling. The severity of blackspot may be reduced by conditioning the tubers by exposure to 1 ppm of exogenous ethylene. Recently Ilker et al. (40) presented cytological evidence showing that bruised potato tubers which had been preconditioned with exogenous ethylene developed wound-healing responses different from those of the control tubers. It is not clear, however, to what extent the stress ethylene produced by wounded potato tubers might contribute to the wound-healing processes.

Stress ethylene may also influence respiration and metabolic activities, as well as the effectiveness of disease-resistance mechanisms. Induced respiration and induced metabolic activities may, in turn, participate indirectly in the functioning of wound-healing and disease-resistance mechanisms.

Ethylene and Induced Respiration

The ability of ethylene to cause increased respiration in fruits, vegetative tissues, storage organs, and flowers is known (2, 75), and the mode of ethylene action to increase respiration in several fruits and in potato tubers has been studied (82, 87, 88, 89). It was found that phosphofructokinase is the key glycolytic enzyme which imposes regulatory restraint. Since there exists a close relationship between the ability of ethylene to induce the respiratory climacteric and the activity of the cyanide-resistant alternate pathway, it was proposed that the alternate path is necessary for realization of the stimulatory effect of ethylene on plant respiration. Ethylene might induce the respiratory climacteric by activating the alternate path and decontrolling glycolysis.

Since wounding is often accompanied by increases in respiration and ethylene production, it is pertinent to ask whether ethylene might play a causal role in the onset of wound-induced respiration. To show that ethylene plays an essential role in induction of the respiratory climacteric, as has been demonstrated in ripening fruits, it is necessary to establish that exogenous ethylene is able to induce the respiratory increase, that removal of endogenous ethylene suppresses the increase and, finally, that the increase in ethylene within the tissue reaches a threshold concentration prior to the increase in respiration. In fruit slices, the induced rise in respiration is always accompanied by an induced rise in ethylene production (65, 67). There is little doubt that wound ethylene plays a hormonal role in the induction of increased respiration in the tissue slices (55), as ethylene does normally during the ripening of intact fruit. In citrus fruits, also, the increased ethylene production resulting from fungus infection hastens the onset of a respiratory climacteric (83).

In underground storage tissue, slicing caused the onset of induced respiration (51); however, ethylene does not appear to play an essential role. In slices of potato tuber (65) or sweet potato root (42) stress ethylene production was observed during the initial few hours, but it decreased to a low level without the induced rise in ethylene production as was observed in fruit tissue. A part of the high ethylene production observed immediately after cutting might not be attributed to the increased synthesis but simply to the facilitated diffusion and release of ethylene present within the tissue. A time course of ethylene production by slices of potato tuber is shown in Figure 3. Ethylene is known to stimulate respiration of intact potato tubers. Reid

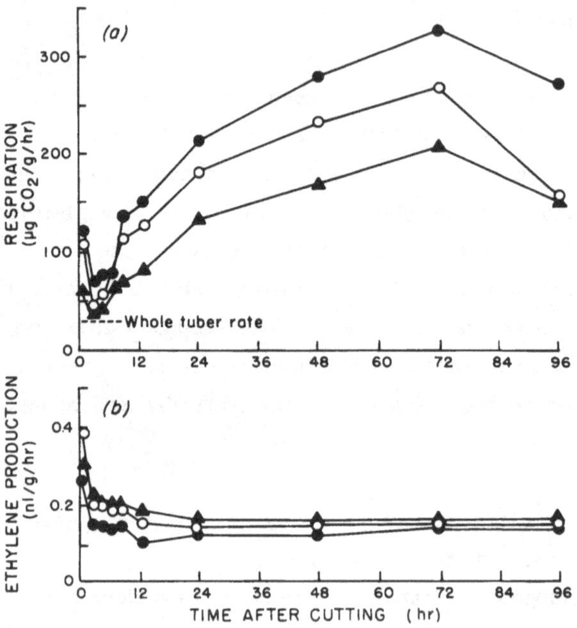

Fig. 3. Respiration (a) and ethylene production (b) by composite samples of potato slices. Each sample consisted of eight 2-mm (●), six 4-mm (○) or four 6-mm (▲) slices. (From McGlasson, 65).

and Pratt (76) have shown that 0.15 ppm of ethylene affords partial stimulation, and any concentration above 2 ppm affords full stimulation. It is not known, however, whether the low ethylene production rate (less than 0.2 nl/g/hr) of potato slices would provide an internal ethylene concentration sufficient to cause onset of the respiratory increase (15). In the absence of such information, it cannot be assumed that stress ethylene plays a specific role in triggering the onset of induced respiration in these tissues.

Ethylene and Metabolic Activation

The effect of exogenous ethylene on enzyme induction in a variety of plant tissueshas been reviewed (2), and the present discussion will be confined to changes in metabolism induced by stress ethylene.

In many plant tissues, wounding is followed by a lag period preceding a rise in various metabolic activities. Elevation of the activity of peroxidase, phenylalanine ammonia-lyase (PAL), polyphenol oxidase, and mitochondria, rise in respiration rates, and increased chlorogenic acid have been reported to occur in sweet potato roots after cutting (42). Seeking the factor(s) which induced such metabolic activation in cut sweet potato root tissue, Imaseki et al. (42) found that ethylene played an important role. This conclusion was based on the observation that low levels of exogenous ethylene effectively stimulated increases in oxygen uptake, peroxidase and PAL activities, and the level of chlorogenic acid; maximum stimulation was obtained with 1 to 10 ppm of ethylene. Removal of endogenous ethylene by mercuric perchlorate solution during the incubation resulted in less activation than in the controls in which endogenously-produced ethylene was not removed. Imaseki (41) further showed that when the slices were incubated under reduced pressure, the increase in peroxidase activity was slightly less than in the control. Imaseki et al. (42) claimed that the level of ethylene necessary for the induction of metabolic changes was so small that the mere effect of wounding would explain most of the changes observed. However, it should be noted that the ethylene production pattern in the disks of sweet potato root was similar to that in potato tuber. The rate of ethylene production by disks of sweet potato root increased to 0.5 nl/g/hr during the first 6 hr after slicing, but later declined to less than 0.07 nl/g/hr during a 20-to 46-hr incubation period. During this period, peroxidase activity and the level of chlorogenic acid continued to increase. They observed that the application of exogenous ethylene as low as 0.01 ppm to sweet potato disks was enough to elicit the increase in peroxidase activity, indicating that the endogenous ethylene concentration must not have been much higher than 0.01 ppm; this is apparently too low to activate other metabolic changes, such as the development of PAL activity and the synthesis of chlorogenic acid. Although stress ethylene may contribute to the induction of metabolic changes in these underground storage tissues, there are other factors which regulate the processes. An example of such a limited regulatory role of ethylene may be seen in the cut tissue of potato tuber in which PAL activity is known to increase after cutting (102). Hyodo and Yang (38) observed that the tuber disks produced very little ethylene, that the development of PAL activity was not retarded by the removal of evolved ethylene from the system, and that it was not increased by the addition of exogenous ethylene. PAL activity obviously was not dependent upon stress ethylene in the cut potato tuber.

Carrot root disks, in contrast, develop very slight PAL activity, but the enzyme activity increased greatly when exogenous ethylene was applied (17). It would appear from this evidence that although the enzyme was freely responsive to exogenous ethylene, the low level of stress ethylene produced was not physiologically effective in this cut tissue (18). Stress ethylene has been shown to perform a regulatory function in the induction of PAL activity in citrus (79). Very little PAL activity was found in the flavedo of intact grapefruit peel, but excision of flavedo disks was followed by a vigorous surge of ethylene production and a parallel increase in PAL activity by the disks, as illustrated in Figure 4. Exposure of intact fruits to exogenous ethylene resulted in an increase in PAL activity, and subsequent withdrawal of exogenous ethylene was followed by a decrease in the activity. It was concluded that, in this tissue, cutting results in the formation of stress ethylene which is fully capable of causing an increase in PAL activity. It is interesting to note that ethylene plays different roles in the development of PAL activities in the disks of potato tuber, carrot root, and citrus peel.

Rhodes et al. (78) found that excision of disks of preclimacteric apple peel initiated the development of the malate decarboxylation system after a 6-hr lag. The lag period was abolished if ethylene was added to the system. Since excision of the disk also caused a burst of ethylene production, they postulated that the stress ethylene was the triggering substance.

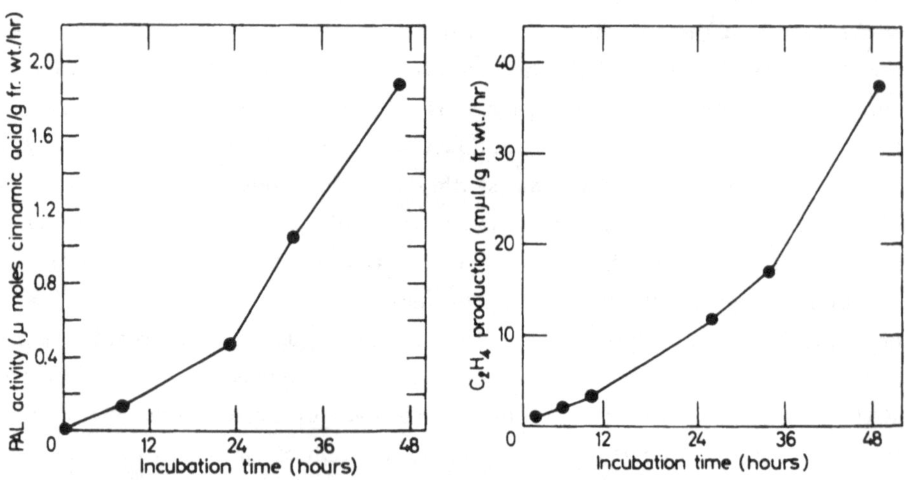

Fig. 4. Time course of development of PAL activity and increase in ethylene production rate by grapefruit flavedo disks. (From Riov et al., 79).

Ethylene and Disease Resistance

The significance of stress ethylene in relation to disease resistance and to phytoalexins has been reviewed (6, 35, 84). Although there is little question that increased ethylene production is a characteristic feature of numerous host-parasite interactions, the relationship of this phenomenon to metabolic alterations which may result in containment of the parasite is not at all clear. Wounded underground storage tissues characteristically produce little ethylene. If these tissues were infected, however, a marked ethylene production was observed in carrot root (18), sweet potato root (43), and potato tuber (23).

Stahmann et al. (90) first reported that ethylene plays a role in the defense mechanism of host plants and that the ability of plant tissues to produce ethylene is correlated with disease resistance. It was suggested that the protective mechanism activated during infection by a pathogen was dependent upon an increase in ethylene production and upon a secondary increase in peroxidase activity induced by the ethylene. Chalutz and DeVay (18), however, were not able to confirm Stahmann's conclusion that ethylene treatment increased resistance of sweet potato tissues to *C. fimbriata*. Although induction of increased peroxidase activity by ethylene is recognized in a wide variety of plant tissues, it is not a general phenomenon, and it is not yet sufficiently clear that peroxidase is involved in the induced resistance of plant tissue. The correlation between increased peroxidase activity and disease resistance has been discussed by Kosuge (50). Matsuno and Uritani (63) found increased peroxidase activity in ethylene-treated, cut-injured, and infected sweet potato root tissue. They demonstrated an electrophoretic form of peroxidase in cell layers near affected sites which was associated with formation of lignin-like material.

The increased activity of PAL following ethylene treatment has been mentioned, and the key role of this enzyme in the biosynthesis of aromatic compounds is well known. Correlations between increased activity of this enzyme and of phytoalexin production in pea tissues has been claimed (31), but the correlation appears to be inconsistent (72). Although Chalutz and Stahmann (19) reported that ethylene treatment stimulated pisatin formation in pea pods, the role of stress ethylene in the production of this phytoalexin is uncertain. It is quite clear, however, that ethylene was responsible for increased formation of 3-methyl-6-methoxy-8-hydroxy-3,4-dihydroisocoumarin (MMHD) in carrot root tissue infected with *C. fimbriata*. Uninfected cut tissue produced little ethylene and little MMHD, but application of exogenous ethylene elicited the accumulation of a very large quantity of the phytoalexin. Inoculation with the fungus caused a massive formation of ethylene, and the concentration of MMHD produced was closely related to the level of ethylene produced which, in turn, varied with the extent of tissue invasion by the fungus. Thus the infection-induced ethylene production appears to be responsible for the massive accumulation of MMHD. Although MMHD has been considered a phytoalexin, Chalutz and DeVay (18) questioned its role in the resistance of carrot roots to*C. fimbriata*, because the amount of MMHD produced was proportional to the extent of tissue penetration by the fungus. It should be noted that phytoalexin production can be elicited by a wide variety of chemical agents and, in many cases, ethylene is not particularly effective. Sequeira (84) commented that there is little reason to single out ethylene as the specific inducer of phytoalexin production.

Ethylene treatments are known to have evoked increased disease resistance in some systems. Brown and Barmore (11) demonstrated that ethylene increased the resistance of green mandarin oranges to anthracnose. The causal organism, *Colletotrichum gloeosporioides*, apparently required ethylene before infection hyphae were formed. Green fruit was susceptible to infection if the fungus received ethylene stimulation. Orange (de-greened) fruit was resistant to infection. Ethylene treatment, in the absence of *Colletotrichum*, increased resistance by de-greening the fruit. Resistance appears due to decreased germination of fungal spores and decreased penetration of infection hyphae because of lignification in the flavedo layer. Ethylene treatment increased the resistance of tomatoes to vascular wilt caused by *Verticillium* (73) and *Fusarium* (71). The increased resistance may be due to increased xylem vessels, to increased activity of β-1, 3-glucanase in lysing fungal cell walls (73), and/or to deposition of lignin-like material around penetrating fungal hyphae (92). Reduced growth of *Sclerotinia fructigena in vivo* on apples pretreated with ethylene or *in vitro* on apple tissue in ethylene atmospheres has been reported (35).

PAL, tyrosine ammonia-lyase (TAL), and peroxidase were associated with lignification and resistance to *Helminthosporium avenae* in reed canarygrass (93). Infection of leaf discs induced increased levels of PAL, TAL, peroxidase, and lignin. Although the role of ethylene in stimulating lignification has been reported in some plant tissues

(77), the relationship of infection-induced ethylene production and the disease resistance associated with lignification is not clear.

V. CONCLUSION

Many plant tissues appear to be potentially capable of synthesizing ethylene, but this synthesis is suppressed by some as yet unknown control mechanism in unstressed tissues. Wounding and other forms of stress are able to relieve such restraint, so ethylene production increases; this in turn may cause a variety of metabolic activations. Stress ethylene is of metabolic origin and its production may be great or small, depending upon the tissue and upon the nature and intensity of the stress. The biosynthetic pathway of stress ethylene does not appear to differ from normal ethylene; methionine is the precursor common to both. Fungi, in contrast, utilize a different pathway.

The production of stress ethylene by cut tissue follows one of two general patterns. In tissue excised from underground storage organs, production quickly increases to a maximum soon after cutting and then declines to a steady slow rate. In such tissues, therefore, stress ethylene activates only those metabolic systems which are capable of responding to low concentrations of the gas. In other tissues an induced rise in ethylene production develops after a lag period, and the concentration of stress ethylene in these tissues is so great that it can activate a number of metabolic systems.

The physiological consequences of stress are known in a few cases, such as accelerated senescence and abscission. It has been suggested that stress ethylene may be involved in wound-healing, production of phytoalexins, and increase in disease resistance; however, these responses are not general phenomena.

Some of the physiological differences known to exist between excised tissue and intact organs are undoubtedly attributable to the stress ethylene induced by cutting. Plant physiologists using tissue slices as experimental material should alert themselves to this complication.

VI. ACKNOWLEDGEMENTS

We wish to thank Gordon J. VonAbrams who has critically read this manuscript and Hiroshi Hyodo and Larry L. Strand for helpful discussion. Part of our work, cited herein, has been supported by research grants from the National Science Foundation.

VII. LITERATURE CITED

1. Abbattista Gentile, I., Matta, A.: Production of and some effects of ethylene in relation to *Fusarium* wilt of tomato. Physiol. Plant Path. 5:27-35. (1975)
2. Abeles, F. B.: Ethylene in Plant Biology. Academic Press, New York. 302 p. (1973)
3. Abeles, A. L., Abeles, F. B.: Biochemical pathway of stress-induced ethylene. Plant Physiol. 50:496-498. (1972)
4. Adams, D. O., Yang, S. F.: S-Adenosylmethionine as the direct precursor of etyhlene. Plant Physiol. 57(5, sup.):51. (1976)
5. Adato, I., Gazit, S.: Water-deficit stress, ethylene production, and ripening in avocado fruits. Plant Physiol. 53:45-46. (1974)
6. Archer, S. A., Hislop, E. C. : Ethylene in host-pathogen relationships. Ann. Appl. Biol. 81:121-126. (1975)
7. Ben-Yehoshua, S., Aloni, B.: Effect of water stress on ethylene production by detached leaves of Valencia orange (*Citrus sinensis* Osbeck). Plant Physiol. 53:863-865. (1974)
8. Ben-Yehoshua, S., Eaks, I. L.: Mode of action of ascorbic acid in inducing selective abscission of citrus fruit, *Citrus sinensis*, Osbeck. Jour. Amer. Soc. Hort. Sci. 94:292-298. (1969)
9. Bird, C. W., Lynch, J. M.: Formation of hydrocarbons by microorganism. Chemical Society Rev. 3:309-328. (1974)
10. Bressan, R., Wilson, L., Filner, P.: Mechanism of sulfur dioxide toxicity in plants. Report of the Michigan State University/Energy Research and Development Administration Plant Research Laboratory. p. 128-129. (1965)
11. Brown, G. E., Barmore, C. R.: The effect of ethylene on susceptibility of Robinson tangerines to anthracnose. Phytopathology 67:120-123. (1977)
12. Burg, S. P.: The physiology of ethylene formation. Ann. Rev. Plant Physiol. 13:265-302. (1962)
13. Burg, S. P.: Ethylene, plant senescence and abscission. Plant Physiol. 43:1503-1511. (1968).
14. Burg, S. P.: Ethylene in plant growth. Proc. Nat. Acad. Sci. 70:591-597. (1973)
15. Burg, S. P., Burg, E. A.: Gas exchange in fruits. Physiol. Plant. 18:870-884. (1965)
16. Burg, S. P., Thimann, K. V.: Studies on the ethylene production of apple tissue. Plant Physiol. 35:24-35. (1960)
17. Chalutz, E.: Ethylene-induced phenylalanine ammonia-lyase activity in carrot roots. Plant Physiol. 51:1033-2036. (1973)
18. Chalutz, E., DeVay, J. E.: Production of ethylene in vitro and in vivo by *Ceratocystis fimbriata* in relation to disease development. Phytopathology 59:750-755. (1969)

19. Chalutz, E., Stahmann, M. A.: Induction of pisatin by ethylene. Phytopathology 59:1972-1973 (1969)
20. Chou, T. W., Yang, S. F.: The biogenesis of ethylene in *Penicillium digitatum*. Arch. Biochem. Biophys. 157:73-82. (1973)
21. Cooper, W. C.: Trauma-induced ethylene production by citrus flowers, fruit and wood. In: Plant Growth Substances. (Ed. D. J. Carr). Springer-Verlag, Berlin. p. 543-548. (1972)
22. Craker, L. E.: Ethylene production from ozone injured plants. Environ. Pollut. 1:299-304. (1971)
23. Creech, D. L., Workman, M. O., Harrison, M. D.: The influence of storage factors on endogenous ethylene production by potato tubers. Amer. Potato J. 50:145-150. (1973)
24. Denny, F. E.: Testing plant tissue for emanations causing leaf epinasty. Boyce Thompson Inst., Contrib. 7:341-347. (1935)
25. Elstner, E. F., Konze, J. R.: Effect of point freezing on ethylene and ethane production by sugar beet leaf disks. Nature 263:351-352. (1976)
26. Engelsma, G., van Bruggen, J. M. H.: Ethylene production and enzyme induction in excised plant tissues. Plant Physiol. 48:94-96. (1971)
27. Galliard, T., Rhodes, M. J. C., Wooltorton, L. S. C., Hulme, A. C.: The development of ethylene biosynthesis during the aging of disks of apple peel. Phytochemistry 7:1465-1470. (1968)
28. Goeschl, J. D., Rappaport, L., Pratt, H. K.: Ethylene as a factor regulating the growth of pea epicotyls subjected to physical stress. Plant Physiol. 41:877-884. (1966)
29. Goren, R., Siegel, S. M.: Mercury-induced ethylene formation and abscission in *Citrus* and *Coleus* explants. Plant Physiol. 57:628-631. (1976)
30. Guinn, G.: Water deficit and ethylene evolution by young cotton bolls. Plant Physiol. 57:403-405. (1976)
31. Hadwiger, L. A., Hess, S. L., von Broembsen, S.: Stimulation of phenylalanine ammonia-lyase activity and phytoalexin production. Phytopathology 60:332-336. (1970)
32. Hanson, A. D., Kende, H.: Biosynthesis of wound ethylene in morning-glory flower tissue. Plant Physiol. 57:538-541. (1976)
33. Herner, R. C., Sink, K. C., Jr.: Ethylene production and respiratory behavior of the *rin* tomato mutant. Plant Physiol. 52:38-42. (1973)
34. Hiraki, Y., Ota, Y.: The relationship between growth inhibition and ethylene production by mechanical stimulation in *Lilium longiflorum*. Plant Cell Physiol. 16:185-189. (1975)
35. Hislop, E. C., Hoad, G. V., Archer, S. A. The involvement of ethylene in plant diseases. In: Fungal pathogenecity and the plant's response. (Ed. R. J. W. Byrde and C. V. Cutting). Academic Press, New York. p. 87-117. (1973)
36. Hyodo, H.: Ethylene production by albedo tissue of Satsuma mandarin (*Citrus unshiu* Marc.) fruit. Plant Physiol. 59:111-113. (1977)
37. Hyodo, H.: Ethylene production and respiration of Satsuma mandarin (*Citrus unshiu* Marc.) fruit harvested at different stages of development. Jour. Japanese Soc. Hort. Sci. 45:427-432. (1977)
38. Hyodo, H., Yang, S. F.: The effect of ethylene on the development of phenylalanine ammonia-lyase in potato tuber disks. Z. Pflanzenphysiol. 71:76-79. (1974)
39. Ilag, L., Curtis, R. W.: Production of ethylene by fungi. Science 159:1357-1358. (1968)
40. Ilker, R., Spurr, A. R., Timm, H.: Ethylene pretreatment and blackspot of potato tubers: histochemistry and histology of wound healing. Pflanzen-

physiologie (in press). (1977)
41. Imaseki, H.: Induction of peroxidase activity by ethylene in sweet potato. Plant Physiol. 46:172-174. (1970)
42. Imaseki, H., Asahi, T., Uritani, I.: Investigations on the possible inducers of metabolic changes in injured plant tissues. In: Biochemical Regulation in Diseased Plants or Injury. (Ed. Tokuzo Hirai). Phytopathological Society of Japan, Tokyo. p. 189-201. (1968)
43. Imaseki, H., Uritani, I., Stahmann, M. A.: Production of ethylene by injured sweet potato root tissue. Plant Cell Physiol. 9:757-768. (1968)
44. Irvine, R. F., Osborne, D. J.: The effect of ethylene on (1-^{14}C)glycerol incorporation into phospholipids of etiolated pea stems. Biochem. J. 136:1133-1135. (1973)
45. Jackson, , M. B., Campbell, D. J.: Production of ethylene by excised segments of pant tissue prior to the effect of wounding. Planta 129:273-274. (1976)
46. Jackson, M. B., Campbell, D. J.: Waterlogging and petiole epinasty in tomato: the role of ethylene and low oxygen. New Phytol. 76:21-29. (1976)
47. Jackson, M. B.., Osborne, D. J.: Ethylene, the natural regulator of leaf abscission. Nature 225:1019-1022. (1970)
48. Kawase, M. : Effect of flooding on ethylene concentration in horticultural plants. Jour. Amer. Soc. Hort. Sci. 97:584-588. (1972)
49. Kays, S. J., Nicklow, C. W., Simons, D. H.: Ethylene in relation to the response of roots to physical impedance. Plant and Soil 40:565-571. (1974)
50. Kosuge, T.: The role of phenolics in host response to infection. Ann. Rev. Phytopathology 7:197-222. (1969)
51. Laties, G. C.: Control of respiratory quality and magnitude during development. In: Control Mechanism in Respiration and Fermentation (Ed. B. Wright). Ronald Press, p. 129-155 (1963)
52. Lau, O. L., John, W. W., Yang, S. F.: Effect of different cytokinins on ethylene production by mung bean hypocotyls in the presence of indole-3-acetic acid or calcium ion. Physiol. Plant. 39:1-3. (1977)
53. Lau, O. L., Yang, S. F.: Synergistic effect of calcium and kinetin on ethylene production by the mung bean hypocotyl. Planta 118:1-6. (1974)
54. Lau, O. L., Yang, S. F.: Inhibition of ethylene production by cobaltous ion. Plant Physiol. 58:114-117. (1976)
55. Lee, T. H., McGlasson, W. B., Edwards. R. A.: Physiology of disks of irradiated tomato fruit. I. Influence of cutting and infiltration on respiration, ethylene production and ripening. Rad. Bot. 10:521-529. (1970)
56. Lockhart, B. E., Semancik, J. S.: Growth inhibition, peroxidase and 3-indoleacetic acid oxidase activity and ethylene production in cowpea mosaic virus-infected cowpea seedlings. Phytopath. 60:553-554. (1970)
57. Lougheed, E. C., Franklin, E. W.: Ethylene production increased by bruising of apples. HortScience 9:192-193. (1974)
58. Lynch, J. M.: Mode of ethylene formation by *Mucor hiemalis*. J. Gen. Microbial. 83:407-411. (1974)
59. MacDonald, I. R., Ellis, R. J.: Does cycloheximide inhibit protein synthesis specifically in plant tissues? Nature 222:791-792. (1969)
60. MacLeod, R. F., Kader, A. A., Morris, L. L.: Stimulation of ethylene and CO_2 production of mature-green tomatoes by impact bruising. HortScience 11:604-606. (1976)
61. Mapson, L. W., March, J. F., Rhodes, M. J. C., Wooltorton, L. S. C.: A comparative study of the ability of methionine or linolenic acid to act as precursors of ethylene in plant tissues. Biochem. J. 117:473-479. (1970)
62. Marco, S., Levy, D., Aharoni, N.: Involvement of ethylene in the suppression of hypocotyl elongation in CMV-infected cucumbers. Physiol. Plant Path.

8:1-7. (1976)

63. Matsuno, H., Uritani, I.: Physiological behavior of peroxidase isozymes in sweet potato root tissue injured by cutting or with black rot. Plant Cell Physiol. 13:1091-1101. (1972)
64. Maxie, E. C., Abdel-Kader, A.: Food irradiation--physiology of fruits as related to feasibility of the technology. Adv. Food Res. 15:105-145. (1966)
65. McGlasson, W. B.: Ethylene production by slices of green banana fruit and potato tuber tissue during the development of induced respiration. Australian Jour. Biol. Sci. 22:489-491. (1969)
66. McGlasson, W. B., Poovaiah, B. W., Dostal, H. C.: Ethylene production and respiration in aging leaf segments and in disks of fruit tissue of normal and mutant tomatoes. Plant Physiol. 56:547-549. (1975)
67. McGlasson, W. B., Pratt, H. K.: Effects of wounding on respiration and ethylene production by cantaloupe fruit tissue. Plant Physiol. 39:128-132. (1964)
68. McMahon, D.: Cycloheximide is not a specific inhibitor of protein synthesis in vivo. Plant Physiol. 55:815-821. (1975)
69. Meigh, D. F., Norris, K. H., Craft, C. C., Lieberman, M.: Ethylene production by tomato and apple fruits. Nature 186:902-903. (1960)
70. Murr, D. P., Yang, S. F.: Inhibition of in vivo conversion of methionine to ethylene by L-canaline and 2,4-dinitrophenol. Plant Physiol. 55:79-82. (1975)
71. Orion, D., Hoestra, H.: The effect of root-knot nematodes and Ethrel on *Fusarium* wilt of tomato. Neth. J. Plant Pathol. 80:28-36. (1974)
72. Partridge, J. E., Keen, N. T.: Soybean phytoalexins: rates of synthesis are not regulated by activation of initial enzymes in flavonoid biosynthesis. Phytopathology 67:50-55. (1977)
73. Pegg, G. F.: The response of ethylene-treated tomato plants to infection by *Verticillium albo-atrum* . Physiol. Plant Pathol. 9:215-226. (1976)
74. Pratt, H. K.: The role of ethylene in fruit ripening. In: Facteurs et Regulation de la Maturation des Fruits. (Ed. R. Ulrich). Colloq. Internat. C. N. R. S. No. 238, Paris. p. 153-160. (1975)
75. Pratt, H. K., Goeschl, J. D.: Physiological roles of ethylene in plants. Ann. Rev. Plant Physiol. 20:541-584. (1969)
76. Reid, M. S., Pratt, H. K.: Effects of ethylene on potato tuber respiration. Plant Physiol. 49:252-255. (1970)
77. Rhodes, M. J. C., Wooltorton, L. S. C.: Stimulation of phenolic acid and lignin biosynthesis in swede root tissue by ethylene. Phytochemistry 12:107-118. (1973)
78. Rhodes, M. J. C., Wooltorton, L. S. C., Hulme, A. C.: Some enzyme systems involved in ripening of apples. Qual. Plant. Mater. Veg. 19:167-183. (1969)
79. Riov, J., Monselise, S. P., Kahan, R. S.: Ethylene-controlled induction of phenylalanine ammonia-lyase in citrus fruit peel. Plant Physiol. 44:631-635. (1969)
80. Riov, J., Monselise, S. P., Kahan, R. S.: Radiation damage to grapefruit in relation to ethylene production and phenylalanine ammonia-lyase activity. Rad. Bot. 10:281-286. (1970)
81. Sakai, S., Imaseki, H., Uritani, I.: Biosynthesis of ethylene in sweet potato root tissue. Plant & Cell Physiol. 11:737-745. (1970)
82. Salminen, S. O., Young, R. E.: The control properties of phosphofructokinase in relation to the respiratory climacteric in banana fruit. Plant Physiol. 55:45-50. (1975)

83. Schiffmann-Nadel, M.: Relation between fungal attack and postharvest fruit maturation. In: Facteurs et Regulation de la Maturation des Fruits. (Ed. R. Ulrich). Colloq. Internat. C.N.R.S. No. 238, Paris. p. 139-145. (1974)
84. Sequeira, L.: Hormone metabolism in diseased plants. Ann. Rev. Plant Physiol. 24:353-380. (1973)
85. Shain, L., Hillis, W. E.: Ethylene production in *Pinus radiata* in response to *Sirex amylostereum* attack. Phytopathology 62:1407-1409. (1972)
86. Shain, L., Wheeler, H.: Production of ethylene by oats resistant and susceptible to victorin. Phytopathology 65:88-89. (1975)
87. Solomos, T., Laties, G. G.: Similarities between the action of ethylene and cyanide in initiating the climacteric and ripening of avocado. Plant Physiol. 54:506-511. (1974)
88. Solomos, T., Laties, G. G.: The mechanism of ethylene and cyanide action in triggering the rise in respiration in potato tubers. Plant Physiol. 55:73-78. (1975)
89. Solomos, T., Laties, G. G.: Induction by ethylene of cyanide-resistant respiration. Biochem. Biophys. Res. Comm. 70:663-671. (1976)
90. Stahmann, M. A., Clare, B. G., Woodbury, W.: Increased disease resistance and enzyme activity induced by ethylene and ethylene production by black rot-infected sweet potato tissue. Plant Physiol. 1505-1512. (1966)
91. Swart, A., Kamerbeek, G. A.: Ethylene production and mycelium growth of the tulip strain of *Fusarium oxysporum* as influenced by shaking of and oxygen supply to the culture medium. Physiol. Plant 39:38-44. (1977)
92. Talboys, P. W.: Resistance to vascular wilt fungi. Proc. Roy. Soc. Lond. B 181:319-332. (1972)
93. Vance, C. P., Sherwood, R. T.: Regulation of lignin formation in reed canarygrass in relation to disease resistance. Plant Physiol. 57:915-919. (1976)
94. Vines, H. M., Grierson, W., Edwards, G. J.: Respiration, internal atmosphere, and ethylene evolution of citrus fruit. Proc. Amer. Soc. Hort. Sci. 92:227-234. (1968)
95. Wallace, R. H.: The production of intumescence upon apple twigs by ethylene gas. Bull. Torrey Bot. Club 53:385-401. (1926)
96. Webb, D. T., Darrow, R. A., Leather, G. R.: Appraisal of ethylene production as a test for defoliants. Jour. Agr. Food Chem. 23:1113-1115. (1975)
97. Williamson, C. E.: Ethylene, a metabolic product of diseased or injured plants. Phytopath. 40:205-208. (1950)
98. Wright, M.: The effect of chilling on ethylene production, membrane permeability and water loss of leaves of *Phaseolus vulgaris*. Planta 120:63-69. (1974)
99. Yang,S. F.: Biosynthesis of ethylene. In: Biochemistry and Physiology of Plant Growth Substances. (Eds F. Wightman, G. Setterfield). Runge Press, Ottawa. p. 1217-1228. (1968)
100. Yang,S. F.: The biochemistry of ethylene: biogenesis and metabolism. In: Recent Advances in Phytochemistry. (Ed. V. C. Runeckles) 7:131-164. Academic Press, New York. (1974)
101. Zimmerman, P. W., Wilcoxon, F.: Several chemical growth substances which cause initiation of roots and other responses in plants. Boyce Thompson Inst., Contrib. 7:209-229. (1935)
102. Zucker, M.: Induction of phenylalanine deaminase by light and its relation to chlorogenic acid synthesis in potato tuber tissue. Plant Physiol. 40:779-785. (1965)

Phytohormones and the Regulation of Cellular Processes in Aging Storage Tissues

Günter Rosenstock and Günter Kahl

INTRODUCTION

Numerous papers witness the suitability of tissue slices from tuberous plant organs such as potato tubers, carrot roots or sugar beets and others for various problems in nearly all areas of plant physiology. However, only few have really taken into consideration, that slicing such an organ represents an injury, which in turn gives rise to various transient wound reactions, ultimately leading to the closure of the wound and hence to woundhealing quite comparable to similar processes in human beings. Slicing in fact is a severe perturbation of the normal structure-function relationship of the cell and would almost certainly result in its permanent damage, if the organ or organism concerned would not react appropriately. As a rule, the injured tissues react upon the trauma with specific physiological, biochemical as well as histological processes. These all work together to restore the status quo ante (7, 8, 48, 49, 75).

HISTOLOGICAL AND HISTOCHEMICAL REACTIONS AND CAPABILITIES OF WOUNDED POTATO TUBER PARENCHYMA.

The cytology, histology and histochemistry of wound reactions were most thoroughly studied in small tissue fragments of the white potato tuber (Solanum tuberosum L.). Principally, three types of histological reactions are distinguishable from one another (50, 67, 71, 97).

Biochemistry of Wounded Plant Tissues

Most frequently the formation of a wound periderm ensues with the induction of mitotic activity and concomitant suberization processes near the wound surface. This is true for tissue fragments, which are kept under normal atmospheric conditions (Fig. 1A). If the slices are exposed to increased water stress (i.e. partial desiccation), then mitotic activity is slowed down in favor of more intense suberization.

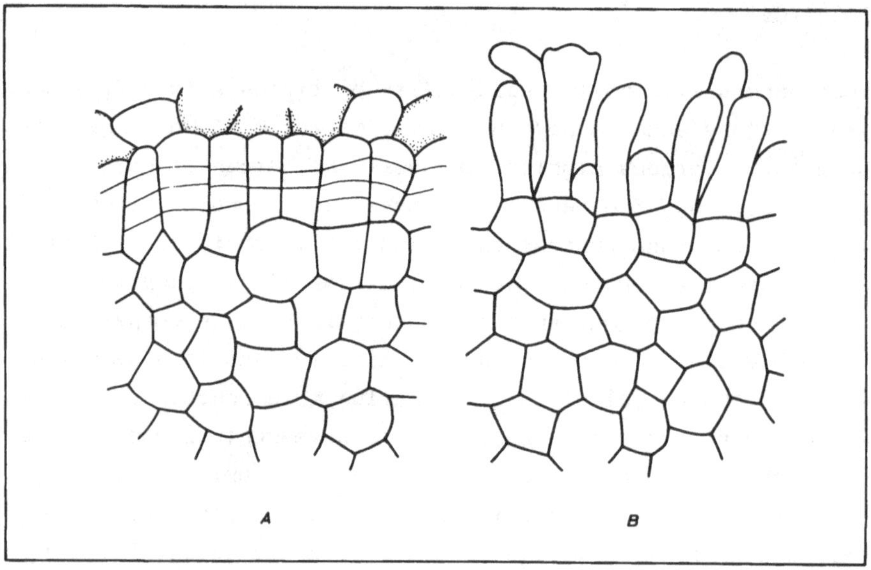

Figure 1. Histological reactions below the wound-surface of potato tuber tissue.
A. Typical wound-periderm with suberized cell walls in a tissue slice of 1 mm thickness after 7 days of woundhealing.
B. Cell proliferations in a tissue slice of 1 mm thickness after 32 hours of wound-healing (modified after Lange et al. 1970).

However, should the external atmosphere to which the slices are exposed mimic the internal atmosphere of the intact tuber (i.e. high water tension and a relatively high concentration of CO_2), then no suberization occurs. The cells at the wound surfaces of such tissue slices proliferate and occasionally divide mitotically and form a callus-like tissue. This type of reaction represents the more regenerative one (Fig. 1B).

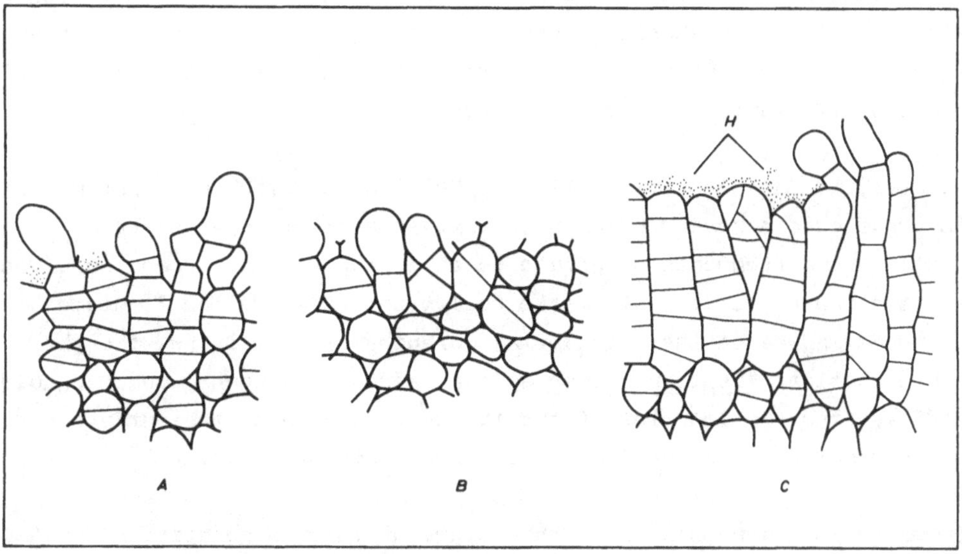

Figure 2. Mitotic activity below the wound-surface of tissue fragments from the Kohlrabi tuber
A. without any washing
B. after extensive washing
C. after application of a Kohlrabi tissue homogenate (H).
Modified after Haberlandt, 1921.

The cells of the storage parenchyma of potato tubers then are capable of reacting differentially upon wounding, the differences being dictated by the quality of the environment. Now, the results presented in the first part of this paper should shed some light on the causative relationship between traumatically induced mitoses on one hand and cell proliferations on the other.

THE HISTORY OF THE WOUND-HORMONE CONCEPT: THE "HABERLANDT PHENOMENON".

Mitoses. FRANK (29) firstly discussed the principles, which he thought to be responsible for the wound-induced transitions of differentiated, mitotically inactive cells to mitotically active cells.

According to his theory a highly hypothetical "Wundreiz" should induce mitoses in otherwise resting cells. WIESNER (133) tried to substantiate this hypothesis: resorption of degraded material from the wounded cells by the neighbouring cells are thought to trigger mitosis and a dedifferentiation process there.

However, it was not until 1921 that HABERLANDT performed experiments to prove or disprove WIESNER's theory. In earlier experiments with small tissue fragments of potato tuber parenchyma he already observed a promoting effect of phloematic cell types (then called "leptome") on wound-induced mitoses in normal parenchymatic cells. He thought a hypothetical "lepto-hormone" as the responsible agent, which should diffuse from the interior of the tissue fragments to the surface and in combination with the "Wundreiz" induce mitoses (34).

In order to learn more about the nature of this "Wundreiz", HABERLANDT carried out a series of really "classical" experiments (35, 36). If the degradation products from wounded or killed cells in fact should represent hormones inducing mitosis in other cells ("wound-hormones"), then extensive washing of the wound surfaces of tissue fragments should remove these cytoplasmic residues and consequently reduce the number of mitoses below the wound surface. Actually this was the case. In leaves of succulent plants, tuber tissues of white potato plants (Solanum tuberosum L.) and tissues of the Kohlrabi (Brassica oleracea cv. gongylodes), the wound surfaces of which were thoroughly washed, mitotic frequency was considerably reduced. Contrastly, mitotic frequency was accelerated if tissue homogenate was applied to washed wound-surfaces (Figure 2). All these observations lend credence to the hypothesis that the inducing principle "Wundreiz" was substantial and originated from the degradation products of wounded or killed cells: "wound-hormone" or "nekro-hormone".

Later on HABERLANDT (37) tried to consolidate his conception with experimental and theoretical evidence and thought the "wound-hormone" to be responsible for the induction of mitosis during the periderm formation of woody plants, natural and artificial parthenogenesis, nucellar embryogenesis as well as the development of fertilized eggs. The induction of mitosis in all these cases was postulated to proceed as follows (Table 1):

Cause	Sequence			Reactions	
	Generation of stimulus	Transmission of stimulus	Perception of stimulus	Sequence of reactions	Last step of the sequence
Wounding	Wound-induced liberation of degradation products of the cytoplasm ("wound-hormones")	Diffusion of wound-hormones from the wound-surface into the reacting cells	unknown (plasma-bound)	unknown	cell division

Table 1: Theory of wound-induced mitotic induction (HABERLANDT)

Although still highly hypothetical and incomplete - the interaction of the "lepto-hormone" could not be explained and the sequence of reactions following the perception of the wound-hormone was totally unknown - the proposed conception was accepted by plant, animal and human physiologists as well (30; 130).

TRAUMATIC ACID, A WOUND-HORMONE IN PLANT ORGANISMS? CRITICISM OF THE WOUND-HORMONE THEORY.

Many experiments were dedicated to the chemical nature of the proposed "wound-hormone". The HAAGEN-SMITH group isolated an active principle from bean pericarp tissues, which was characterized as a 1-decene-1,10-dicarboxylic acid ("traumatin", "traumatic acid") and capable of inducing mitoses in the pericarp itself (9, 25, 33). Since this substance or its precursors were thought to be of ubiquitous occurence and responsible for the induction of traumatic processes, it was termed "traumatin". However, synthetic traumatin was only weakly inductive in the bean pericarp system ("WEHNELT-test"), but highly active if supplied together with the otherwise inactive glutamic acid ("cofactor"). Reproductions of these results were impossible. On the contrary, it could definitely be shown that glutamic acid is more active in the bean pericarp tissue than traumatic acid itself (56). Frequently traumatin was totally inactive (19). In view of the critical comments of WEHNELT himself (131), who could induce mitoses in the pericarp system with a wide variety of unrelated substances, i.e. ovalbumin, horse serum or sugar solutions, the test system itself certainly is unreliable and therefore not qualified for probing "wound-hormones" (13). Similarly mitoses can be induced by a series of different substances in pith tissues of various other plants (56, 134).

All these experiments then raised doubt in the validity of the "wound-hormone" concept of HABERLANDT. Opposing ideas were published (7, 8, 63, 76, 77, 78, 79, 80, 92, 104, 112, 113, 130) but could not seriously invalidate the "wound-hormone" theory, since they were in themselves inconclusive, mostly hypothetical and frequently merely theoretical. All this opposition was hampered by the fact, that the existence of "wound-hormones" and their stimulative effects on mitosis had apparently been demonstrated experimentally. So it becomes understandable that the theory of "wound-hormones" predominated in literature (7, 8, 11, 12, 54, 63, 106, 109, 125, 130).

A NEW CONCEPT: VARIATIONS OF THE OSMOTIC POTENTIAL AS PART OF THE "WUNDREIZ".

A set of experiments by ROSENSTOCK and coworkers imitating the "classical" investigations of HABERLANDT lead to a new understanding of the problem: Wound-induced mitoses may very well be the consequence of variations of the osmotical state of the cell rather than the action of obscure wound-hormones (68, 97). Hypothetically, the stimulating effect of tissue homogenates on mitotic frequency could equally well be explained by their osmotic efficiency. If so, then the inductive capacity of these tissue breis should be exerted by osmotically active chemicals as well. In fact, this is exactly the case.

Since potato tuber tissue responded the same way to synthetic osmoticals as do leaves of Echeveria derenbergii and Sedum spectabile, only experiments with both succulents will be reported here. If leaf fragments are incubated in an atmosphere of about 100% relative humidity, then mitoses are induced below the wound-surfaces. This is especially true for regions, which are in close contact with residues from the wounded cells. Usually 1-4 mitoses are performed successively (Fig. 3A). However, if the leaves are not injured mechanically, but carefully torn into pieces, which leaves the cells unabated and does not generate any free cell sap constituents, then the cells adherent to this surface proliferate extensively. The length of these proliferations frequently exceeded the length of a normal cell by a factor of 6 (Fig. 3B). Occasionally the proliferated cells form club-like swellings with small evaginations, only exceptionally mitoses are performed. However, if these very surfaces are covered with tissue homogenate immediately after preparation, then the enormous cell proliferation is inhibited and instead a coherent wound periderm is formed (Fig. 3A).

So far these experiments certainly confirm the report of HABERLANDT, who regarded these results as impressive prove for his theory (64). However, if the leaf fragments are incubated together with the synthetic Polyethyleneoxide Lutrol, which is osmotically active, but never comes into direct contact with the tissue itself, then a differential wound-reaction can be evoked according to the molarity of the drug. So a 0.5 M solution of Lutrol greatly retards the proliferations (Fig. 4A) so characteristic for the control tissue (Fig. 3B). A molar solution of Lutrol finally prevents any proliferations, but allows cell divisions in the same array as in samples treated with tissue homogenate (Fig. 4B). It can be concluded from all these experiments, that one of the primary factors inducing mitosis must be a change in the osmotic balance of the respective cell. Although the existence of "wound-hormones" as cytoplasmic degradation products cannot be excluded with certainty, their proof in experiments resembling those of Haberlandt is not possible. As can be proved, however, effects ascribed to a hypothetical "wound-hormone" can be readily demonstrated to be a consequence of variations in the cellular osmotic balance.

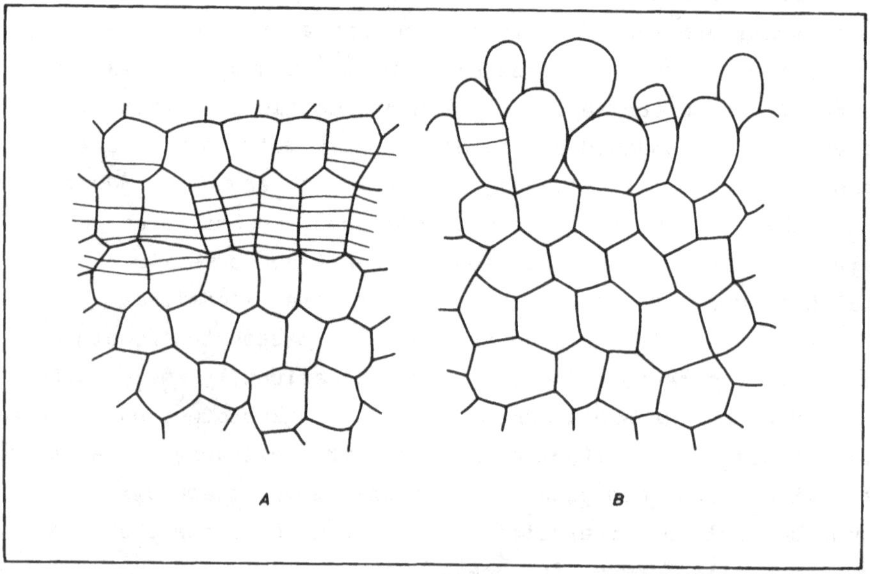

Figure 3. Histological wound-reactions in the mesophyll of a leaf from Echeveria derenbergii after incubation in humid atmosphere.
A. Formation of a wound-periderm below a wound-surface produced by cutting. Microscopic examination after a 9-days-incubation.
B. Cell proliferations at a wound-surface produced by careful tearing of the leaf mesophyll. Microscopic examination after a 9-days-incubation (modified after Lange and Rosenstock, 1963).

SUBSTANTIATION OF THE "LEPTO-HORMONE" THEORY OF HABERLANDT.

After all it seems important to prove or disprove the existence of HABERLANDTs "lepto-hormones", which after diffusion to the wound-surface and in combination with the "Wundreiz" should induce mitoses in the cells adherent to the wound. This question was most thoroughly investigated by ROSENSTOCK and coworkers with wounded potato tuber tissue, the reaction of which is restricted to 1-5 cell layers immediately below the wound surface and finally leads to the formation of a periderm and consequently to the closure of the wound. Now the higher the number of cells present beneath this reactive zone, the more pronounced the reaction should be, since more inductive substances should be provided and vice versa (98, 99).

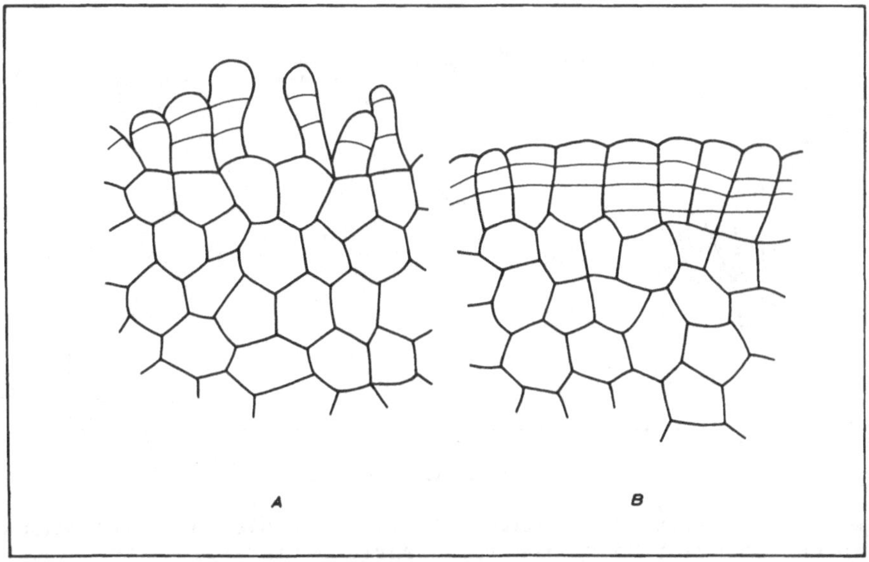

Figure 4. Histological wound-reactions in the mesophyll of a leaf from Echeveria derenbergii under different osmotic conditions.
A. Cellular reactions at a wound-surface produced by careful tearing of the leaf mesophyll. Microscopic examination after a 9-days-incubation together with 0.5 M Lutrol.
B. Cellular reactions at a wound-surface - which in contrast to Fig. 3A is essentially free of tissue residues - after a 9-days-incubation together with 1 M Lutrol (modified after Lange and Rosenstock, 1963).

This is indeed the case. Potato tissue slices of about 1 mm thickness (5-7 cell layers) form a wound-periderm at both wound surfaces, which in the mean consists of 2,4 new cell walls per initial cell. So nearly all the slice volume is engaged in the performance of mitoses. No non-reactive cells exist in the interior of the fragment and consequently no source for an additional supply with inductive substances. However, if the tissue slice comprises some 30 cell layers, then the mitotic capacity doubles. This is true for tissue slices taken from the tuber pith. If excised from the cortex of the tuber, a highly active tissue, only about 10-12 cell layers (2 mm) are sufficient to form an optimal periderm (Fig. 5A,B).

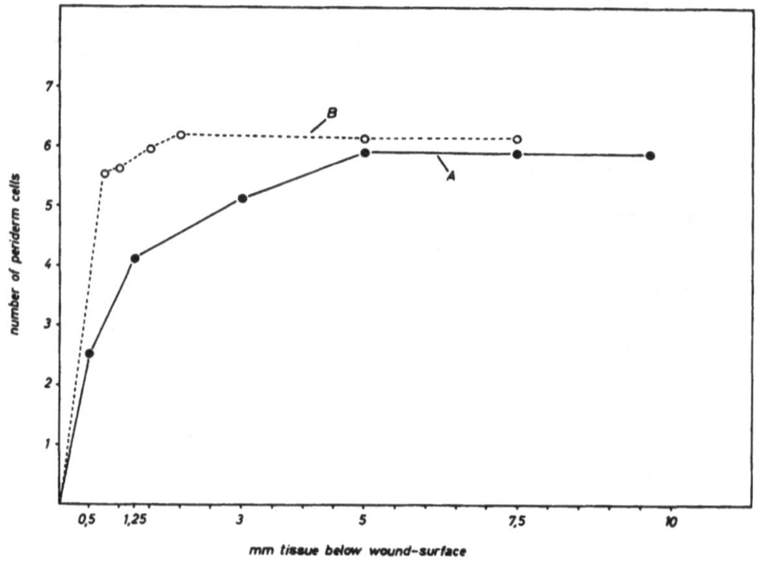

Figure 5. Intensity of traumatogenic cell division as a function of the tissue thickness below the wound-surface (modified after Rosenstock et al. 1972).
A. Pith tissue
B. Cortex tissue

It is almost certain, that the interior of such a tissue slice does not function as a source for nutrients. Even in extremely thin slices many starch grains remain undegraded, although the wound-reaction is gone to completion (126). However, it becomes more and more likely, that phytohormones play a crucial role in these processes. So the nucleoli of potato tuber cells at the wound surface form giant "vacuoles" (4), which are characteristic for other hormon-treated tissues as well (18, 45, 60, 61, 96). Consequently the application of phytohormones should give an answer to the question, whether such hormones could modify the wound-reaction or even induce mitoses where they are otherwise absent. Actually such phyto-hormones

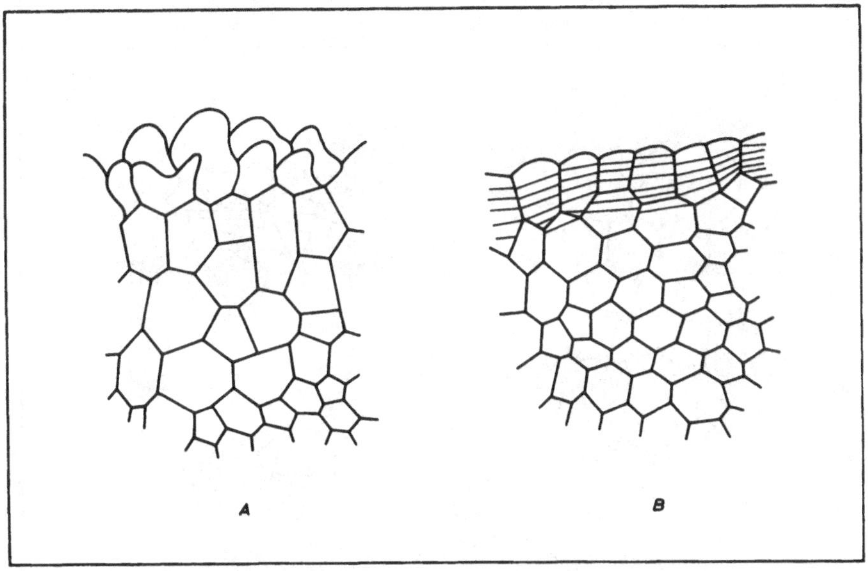

Figure 6. The effect of IAA on histological reactions at the wound-surface produced by careful tearing of leaves from the 5. node of Sedum spectabile (modified after Rosenstock and Sondermann 1974).
A. Control
B. After application of 10^{-7} M IAA

could be identical to the postulated "lepto-hormones", especially since hormones are reported to be synthesized and/or transported preferentially in the phloem (42, 107, 108) and capable of inducing cell divisions(2, 3, 17, 26, 38, 42, 87, 107).

Actually the extent of wound-induced cell divisions can be strengthened by the application of naphthalene acetic acid (NAA) under certain conditions. NAA was used since it represents one of the most active auxins for the induction of callus in potato tuber tissue (3, 26). One of the parameters determining the ultimate response of the cells to the auxin is the duration of the incubation time respectively the auxin concentration itself (Tables 2 and 3). Taken altogether these experiments show, that auxins may well overcome the reduced capability of thin tissue slices to perform cell divisions.

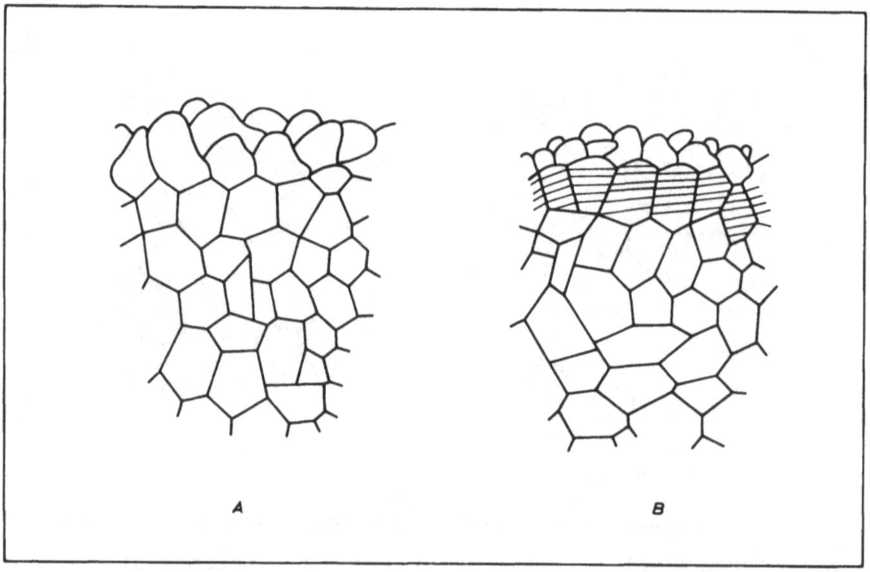

Figure 7. The effect of GA_3 on histological reactions at the wound-surface produced by careful tearing of leaves from the 5. node of Sedum spectabile (modified after Rosenstock and Sondermann 1974).
A. Control
B. After application of 10^{-7} M GA_3

Tissue slices, which are exposed to the auxin for longer periods of time tend to form a callus, i.e. exhibit cell proliferations and cell divisions side by side. Possibly a second hormone is effective in such long-time experiments as has been shown with fragments of succulent leaves (100). In these systems IAA and/or Gibberellic acid

(GA_3) accellerate mitoses (Fig. 6,7), whereas kinetin promotes proliferative growth (Fig. 8). By combination of the various hormones it could be confirmed that the hormonal balance is indeed the determining factor: increasing the concentration of GA_3 or IAA above a basal level intensifies mitotic activity of the cells, increasing the kinetin concentration consequently leads to enhanced proliferation.

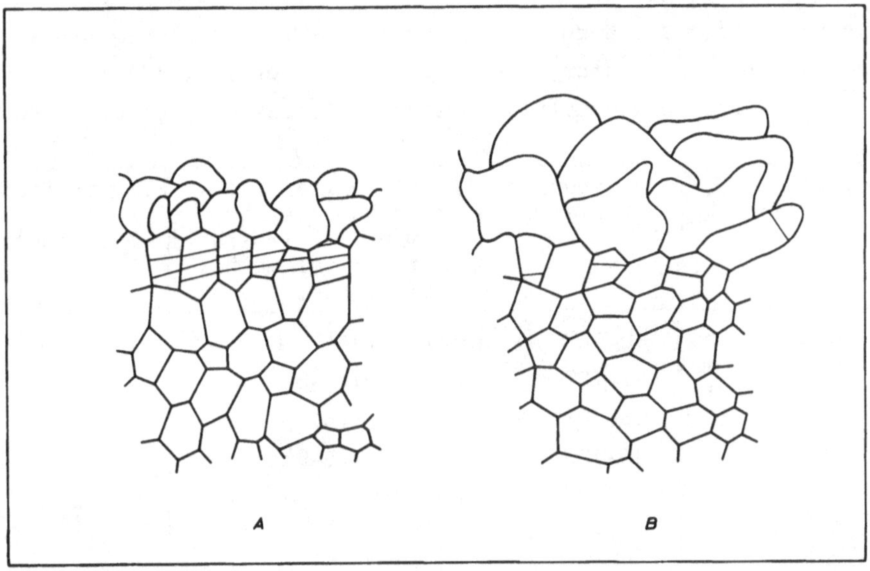

Figure 8. The effect of kinetin on the histological reactions at the wound-surface produced by careful tearing of leaves from the 3. node of Sedum spectabile (modified after Rosenstock and Sondermann 1974).
A. Control
B. After application of 10^{-7} M kinetin.

So the apparently unreactive tissue beneath the reactive cells in wounded tissues may well serve as potent source for phytohormones. Evidently this source is increasingly depleted with increasing age of the tissue. One indication for this process of senescence is the reduced capability of older tissues to induce mitoses at the wound-surface. However, this phenomenon has been treated contradictory in the literature. Whereas De VRIES (127) and SCHNEIDER-ORELLI (105) failed to observe any wound periderm formation in older tissues, no major differences were found by SCHICK and coworkers (103) between young growing tissues and quiescent as well as germinating, older tubers. A careful reinvestigation of this problem by LANGE and ROSENSTOCK (69, 70) lead to the conclusion, that "the potency of a tissue to react upon wounding with induced mitoses is subject to an aging process, which already begins at an early stage in the development of an organ". Deficiency of nutrients has been rendered unlikely (70, 126) yet shortage of phytohormones as inducers of cell divisions may be the limiting factor, since application of i.e. NAA restores the reduced mitotic capability of senescent tissue after wounding (Table 4).

Duration of incubation (hours)	Degradation of starch	Histological reaction and average number of mitoses per initial mother cell. (s.d. ± 0.5)	Mitotic frequency (% of control)
control	60-70%	periderm(4)	100
0.5	60-70%	periderm(4-5)	110
1	60-70%	periderm(5)	125
2	60-70%	periderm(8-9)	150
4	60-70%	periderm(5) and irregular cell divisions	175
6	80%	periderm(5) and irregular cell divisions	200
13	95%	periderm(5) and irregular cell divisions	250
24	95%	periderm(4-5) and irregular cell divisions	350
48	95%	callus	450
96	95%	callus	450

Table 2: Duration of NAA-treatment and its effect on the histological reactions of wounded potato tuber parenchyma.
Tubers of the variety "Saskia" were cut into 0.5 mm thick slices immediately after harvest and incubated in 2.5×10^{-5} M NAA for the time indicated. Subsequently they were cultured on NAA-free agar for 8 days at 20 ± 1°C in the dark. Control tissue was treated with water only.

NAA concentration (M)	Incubation time (hours)	Rate of mitosis (average number of mitosis per initial mother cell; s.d. $\pm$0.5)
control		4
5.0×10^{-6}	3	7
2.5×10^{-5}	1	7-8
5.0×10^{-5}	0.5	6-7

Table 3: Optimal periderm formation as influenced by the time of incubation in and the concentration of NAA. Conditions are identical to those described in the legend of table 2, except that tubers of the variety "Hansa" were used.

RELATIONSHIP BETWEEN AUXINS AND DNA CONTENT OF CELLS IN WOUNDED POTATO TUBER PARENCHYMA.

Formation of new cell walls is the ultimate result of a long sequence of biochemical events after wounding. One of the more central event doubtless concerns the nucleic acids more generally and the metabolism of DNA in particular.

The cells of potato tuber fragments consisting of few cell layers only as well as those of tissue slices from senescent tubers react upon wounding with the performance of only few mitoses and an increase in DNA content. However, the increase in DNA content is greater than the increase in mitotic activity (Table 5). This fact can be interpreted in terms of wound-induced endomitoses (10, 65, 114, 126, 129). If two factors responsible for endomitosis come together, their effect is additive: in thin slices of senescent tissue the endomitotic processes are more pronounced (126).

Storage time (days)	Treatment	Starch degradation	Rate of mitosis (average number of mitoses per initial mother cell; s.d. ±0.5)	Rate of mitosis (% of control)
20	control	50%	periderm (3)	100
	NAA	90%	periderm as well as irregular cell divisions	500
62	control	80%	periderm (2-3)	100
	NAA	95%	periderm as well as irregular cell divisions	600
200	control	95%	periderm (0.5)	100
	NAA	95%	numerous irregular cell divisions	800-1000

Table 4: The modulating influence of NAA on the mitotic activity of wounded parenchyma from potato tubers of different age. Tubers of the variety "Hansa" were adapted to 20°C, sliced into 0.5 mm thick slices, which in turn were incubated for 15 days on NAA-free (control) and NAA-containing agar (2.5×10^{-5}M) at 20 ± 1°C.

However, if very thin potato tuber slices or tissues from senescent tubers, both of which show reduced mitotic frequency upon wounding, are exposed to auxins, then the DNA content does not change dramatically, albeit formation of new cell walls is strongly stimulated (126; Table 5). Clearly the auxin somehow prevents any endomitoses in these tissues (27, 28, 74).

Tissue slice thickness (mm)	Duration of storage (days)	DNA-content/rate of mitosis control	NAA
1.0	20	3.2	1.2
0.5	20	3.1	1.2
0.25	20	95	3.9
0.5	20	3.2	1.2
0.5	82	3.6	1.1
0.5	220	4.3	1.1

Table 5: The influence of NAA on the relationship between increase in DNA-content and mitotic activity in white potato tuber slices of different thickness and age. Experimental conditions are identical to those of Table 4.

The reduced mitotic activity in these thin tissue slices or fragments of senescent tubers then is not the consequence of a reduced potency to synthesize DNA, but is mainly due to the lack of hormones, which in turn prevents the mitotic cycle to run to an end. Of course, a direct effect on the genome cannot be excluded, especially since the synthesis of a special kind of RNA seems to be necessary for the induction of mitoses, but not endomitoses (86).

The importance of endomitotic activity for the cell in general is largely obscure and the role of this process in wound-healing particularly remains to be elucidated.

All these experiments demonstrate, that the wound-reactions may be modulated in one way or the other by the interplay of two or more distinct phytohormones, two of which are auxin and kinetin respectively. There is no need whatsoever to postulate specific "wound-hormones". This is congruent with the ideas of Kefford and Goldacre (55) as well as Lipetz (75). Last not least the existence and increased synthesis of some of these phytohormones in wounded tissues has been demonstrated. HEMBERG (40, 41) reports on an increase in auxin content during wound-healing of potato tuber tissue. Remarkably, more auxin could be found in the phloem as compared to pith and medulla (16, 42). Also gibberellin-like substances are present in potato tuber slices (95) and a rise in cytokinin content seems to be the consequence of wounding in these tissues (15).

One of the major components of the socalled "ß-inhibitor-complex", originally thought to be responsible for the suppression of sprouting in potatoes (42), is L-abscisic acid (44). This compound or at least a related substance with similar anti-auxin activity increases after wounding in potato tuber tissue. It reaches its maximal concentration when the mitotic activity of the tissue is markedly reduced (101). Abscisic acid may play a crucial role in wound-induced mitotic frequency, since application of the drug inhibits cell divisions considerably (66, 102).

MOLECULAR BASIS OF HORMONE ACTION

One of the more important conclusions emerging from the experiments reported so far points to plant hormones as potential triggers of various wound processes including cell division as a final step. It is evident that there must exist a long chain of interlinked, competitive and synergistic reactions and processes from wounding itself up to mitosis. In no case the whole sequence has been elucidated, but from many systems it is known that the hormone must somehow be recognized by specific receptors (which must not necessarily be localized in the plasmamembrane exclusively, but may be true cyto- or nucleoplasmic receptors as well) in order to induce responsive reactions. After being bound to its specific receptor molecule, the hormone has to be transported to reacting systems (i.e. enzymes, which is possible, or chromatin, which is likely). Inspite of permanent progress in the field of plant hormone research, the reaction of the hormone receptor-complex with the target, i.e. chromatin or one or more of its constituents, is a complete mystery. According to animal systems responsive to steroid hormones (53, 88) one should expect the ultimate turning-on of specific genes with concomitant synthesis of mRNAs and finally formation of distinct proteins. There is no reason to disprove the possibility of a simultaneously occurring turning-off of genes, however. In any case, lack of experimental evidence is balanced by the number of different theories, which altogether try to explain the molecular mechanism of plant hormone action.

The last five years have seen the isolation and partial characterization of some of the hormone-reactive factors from plants. Nuclei of coconut endosperm contain "acceptor proteins" for ß-indole acetic acid (IAA), capable of specifically binding the hormone. The complex of hormone and receptor in turn binds to specific sites on the DNA and - in theory - marks initiation sites for RNA polymerases (6, 84, 85). In soybean tissues a protein has been demonstrated as of plasmamembranes, which specifically reacts with 2,4-dichlorophenoxyacetic acid (2,4-D). After binding of the hormone, the protein together with the bound effector dissociates from the membrane and - in theory - modifies the α-amanitin-sensitive RNA polymerase. This modification is thought to alter transcription of messenger RNAs (39). A somewhat different concept has been proposed for cells of Lens culinaris, which in some way resembles that of steroids in animal systems (5). Steroid hormones usually permeate the plasmamembrane of target cells and are recognized and bound each by a highly affine receptor protein, which after binding changes its conformation. It is not yet known how the hormone-receptor-complex is transported into the nucleus - it might even simply diffuse - where it occupies distinct sites on the chromatin. The transcription of neighbouring sequences is thereby allowed. The mode of action of derepression on a molecular level is completely unknown. May it be histone modification such as acetylation, phosphorylation, thiolation or methylation or modification of RNA polymerases or changes in the non-histone chromosomal proteins, in any case specific messengers are synthesized: molecular response to the hormone signal. In the case of estradiol the hormone-receptor complex firstly induces the synthesis of mRNAs for a "key intermediate" protein, which - possibly as a modifying subunit of the enzyme - stimulates RNA polymerase I and consequently ribosomal RNA synthesis:

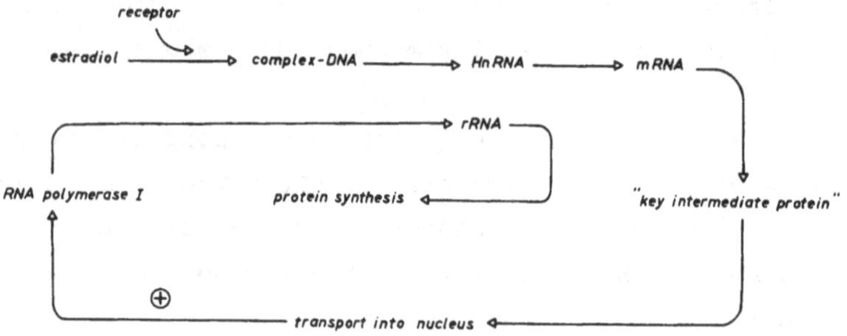

In case of the cells of Lens culinaris the hormone - IAA - is recognized by a receptor protein ("nuclear protein") and in the hormone -receptor-complex switches-on some genes ("early effect"). Among the products of these very genes is a messenger coding for a protein factor , which after its synthesis and transport into the nucleus or nucleolus respectively, modifies nucleolar RNA polymerase Ib. The modified enzyme - in theory - reads out a new set of genes. Result: massive synthesis of ribosomal RNA and activation of protein synthesis ("late effect"": 121, 122).

Taken altogether these data unanimously demonstrate that the hormone-induced synthesis of RNA of whatever kind certainly represents a multiphasic event. In principle the hormone is firstly recognized and bound by specific receptor of target cells. There is as yet no conclusive evidence where the receptors are localized. They may be integral parts of the plasmamembrane (39, 43, 73) or may be components of the nucleoplasm, which move from there to the cytoplasm and vice versa (6). Probably they represent non-histone chromosomal proteins (85). After binding of the hormone the signal is somehow transmitted to the genome. The actual target of the transmitting hormone-receptor-complex is still a matter of controversy in animal as well as plant biology. On one hand the DNA itself seems to react with that complex or parts of it (6, 122), on the other it seems to be more affine for RNA polymerase proteins (39). Whatever the actual target molecule may be, the response to the hormone encircles the synthesis of various RNA species, which may code for intermediate proteins and/ or switch-on the ribosomal protein-synthesizing machinery:

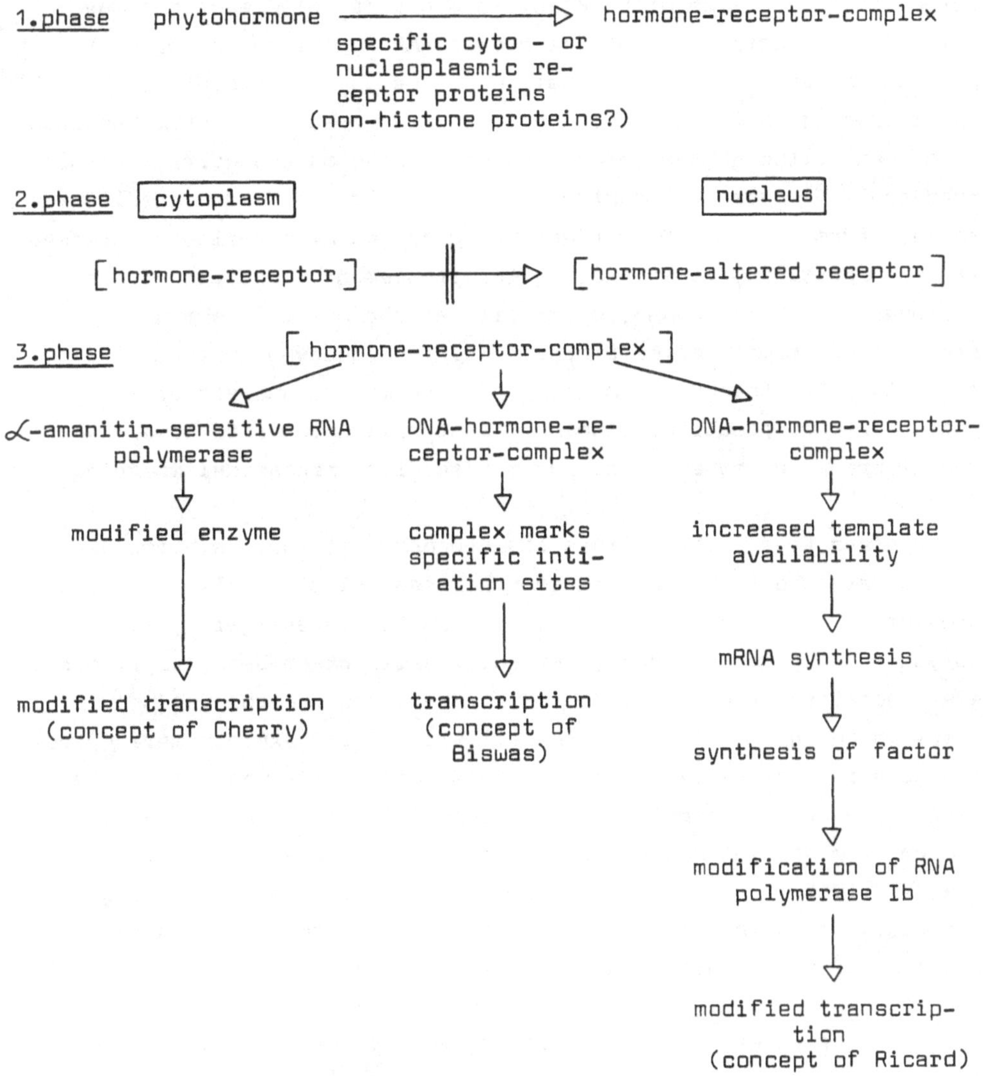

All these models presuppose that the receptors have access to hormones already present. The appearance of hormones in turn is a fairly late event in the chain of reactions, however. In animal systems excitation of parts of the nervous system ultimately leads to the synthesis and excretion of some hormones, which may trigger synthesis and excretion of others and so on. These hormones reach target cells supplied with receptors and secondary reactions may commence,i.e. activation of the cyclic AMP system in the case of proteohormones or complex-formation in the case of steroids. In plant tissues on the contrary, no extensive vascular system has been developed, although the phloem may well serve as channel for hormone transport. However, the synthesis of a particular hormone or its release from inactive form ("activation") has necessarily to preceed its transport. And it is exactly the problem of induction of the hormones in a broad sense, which lies at the heart of our understanding of hormone effects generally. Inadequacy of methods did not allow to solve this problem up to now. So we are unaware of the really primary reactions within a plant cell induced to developmental processes by external parameters, i.e. mechanical wounding.

It is clear from several lines of evidence that various hormones play a key role in wound-induced processes. First of all, they are present in all tuberous plant systems so far investigated, i.e. potato tubers, carrot roots, Jerusalem artichoke tubers, sugar beets, sweet potatoes, red beets, Kohlrabi tubers and others. Secondly, after wounding they increase in "activity" (i.e. exhibit more effectiveness in bio-tests), which does not necessarily mean, that they also increase in concentration. Thirdly, they interfere with a vast number of different metabolic processes, if applied externally (48, 49, 120). Most important, hormones are able to stimulate gene activity, which in our view seems to be one of the more central problems of molecular biology of plant hormones. So the fundamental questions can be posed as follows: What are the effects of hormones on protein formation, chromatin-directed RNA synthesis and the RNA polymerases in wounded plant storage tissues? How do the reactive hormones exert their influence on these systems? What is the molecular basis of the hormon-induced gene activation, presupposed there is such an activation?

GIBBERELLIC ACID AND INDUCTION OF PROTEIN SYNTHESIS

Wounding of resting potato tuber tissue induces the rapid formation of polysomal aggregates as evidenced from electron-microscopic as well as biochemical studies (1, 46; see the contribution of R. Barckhausen, this volume). This has been shown to be true for other sliced storage tissues, too (21, 72, 128; see the chapter of J.H. Cherry, this volume). These polysomes are heavily engaged in protein synthesis and it seems as if the support with messengers is the crucial point in the activation of the ribosomal apparatus. This view is supported by different, albeit insufficient experiments. First of all, polysome formation is totally prevented, if actinomycin D is present during wound-healing. However, this drug has no convincing specificity on mRNA or RNA synthesis in general. It deleteriously acts on membranes and interferes with reactions aside from RNA biosynthesis. So the experiments with this inhibitor and experiments with all other inhibitors indeed are seriously depreciated by the fact, that they rest on the effect of unspecific interactions. On the other hand, ribosomes isolated from resting tissue are vitually inactive in an in-vitro amino-acid incorporating system. These same ribosomes, however, may be stimulated to enhanced synthetic capacity if supplied with artificial messengers (i.e. poly-uridylic acid) or with RNA fractions isolated from polysomes of wounded tissue sedimenting as 9-12S RNA and presumed to be one or more messenger RNAs (Table 6; 47).

The lack of such messengers seems to be the rule for resting cells of other storage organs (i.e. carrot root 72; red beet: 24). Although it should be stressed, that the mere stimulation of an in vitro ribosomal system by endogenous RNAs is no proof whatsoever of the in vivo role of these RNAs, it nevertheless is likely that wounding induces the synthesis of such messengers which allow a

RNA from ribosomes of the intact tuber (50 µg/ml each)	% stimulation of the incorporation of ^{14}C-leucine into TCA-insoluble material by ribosomes from the intact tuber	and ^{14}C-phenylalanine
control	100	100
28S-rRNA	125	100
16S-rRNA	130	120
4S/5S-RNA	100	100
9-12S-RNA	100	100
RNA from ribosomes of aged tissue slices (50 µg/ml each)		
control	100	100
28S-rRNA	120	115
16S-rRNA	130	120
4S/5S-RNA	120	120
9-12S-RNA	165-230	140-200
poly-U	-	600

Table 6: Activities of ribosomal preparations from resting potato tuber tissue and the effect of different RNAs prepared from ribosomes of intact tubers and aged slices respectively.
Ribosomes were isolated and purified and tested in an in-vitro system for their capacity to incorporate labelled aminoacids into TCA-insoluble material. The test mixture consisted of 0.5 mM Tris-HCl, pH=7.6; 0.05 mM KCl; 8 mM $MgCl_2$, 0.01 mM phosphoenolpyruvate, 50 µg pyruvate kinase; 5 mM ATP; 0.5 mM GTP; 1 mg of S-100 protein; 50 µg bovine serum albumin; 0.02 mM 2-mercaptoethanol and 1 µCi ^{14}C-UL-leucine (240 mCi/mM) or 1 µCi ^{14}C-phenylalanine (280 mCi/mM) respectively. The various RNAs were extracted and purified from the ribosomal pellet by LiCl and separated from one another through a linear 17-34% sucrose gradient in 1E-buffer (26 hours at 4°C and 25,000 rpm in an SW 25.1 rotor of the Spinco). The different fractions as detected by OD traces were collected, extensively dialyzed against the incorporation buffer (all proteins and isotopes excluded) and finally lyophilized.
100% of the control correspond to 750 cpm ^{14}C-leucine/mg rRNA respectively. (Kahl, 1971).

substantial rise in protein synthesizing activity of the cell. Clearly there is a need for experiments, which use isolated and highly purified mRNA in a heterologous aminoacid incorporation system and demonstrate the synthesis of proteins, which are specific for the system from which the mRNA originates.

Whatever the reason for induced protein synthesis in wounded tissues, it can dramatically be stimulated by gibberellic acid (GA_3). So an overnight incubation of potato tuber slices with 10^{-7}M GA_3 more than doubles the rate of ^{14}C-leucine incorporation into TCA-insoluble material of polysomes from sucrose density gradients (Fig. 9). So - if one agrees to the messenger hypothesis of the activation of protein synthesis after wounding - GA_3 in some way must accellerate mRNA synthesis.

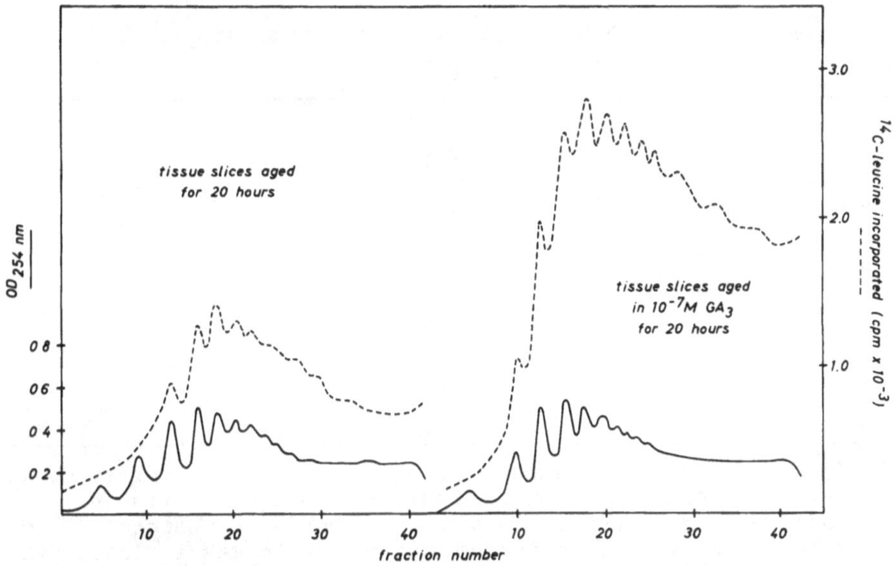

Figure 9. Gibberellic acid-induced stimulation of protein synthesis in aging white potato tuber tissue.
Samples of sliced potato tuber tissue were incubated without (left) and together with 10^{-7}M GA_3 (right) for 20 hours in the dark and at 25°C. Two hours before the end of the incubation period 100 uCi of ^{14}C-labelled L-leucine was added to the incubation mixture. Ribosomes were extracted from the tissue samples essentially as described previously with slight modifications (Kahl 1971) and centrifuged through exponential 10-48% sucrose gradients in 10 mM Tris-HCl,

pH=8,0; 10 mM $MgCl_2$; 1 mMspermidine at 105 000 g and + 4°C for 1.5 hours in the Spinco (L5 65B; SW 40 rotor).

GIBBERELLIC ACID AND THE INDUCTION OF RNA SYNTHESIS

If protein synthesis in wounded potato tissues really is dependent on supply with newly synthesized messengers, then RNA synthesis in general and mRNA synthesis in particular should be stimulated by GA_3. This is actually the case. However, resting cells do not seem to be responsive to the hormone. Even high dosages of GA_3 do not at all influence the poor RNA synthesis of quiescent tissue. Shortly after wounding, however, the tissue develops an increasing capability to respond to exogenously added GA_3. After about 24 hours this reactivity is maximal (Table 7).

Aging time	% stimulation of RNA synthesis by 10^{-7}M GA_3
0	0
6	12
12	35
16	68
18	75
20	78
24	80
36	38
48	15

Table 7: Development of responsiveness of the RNA polymerase system to gibberellic acid after wounding white potato tuber tissue. Tissue slices were incubated in phosphate buffer containing 10^{-7}M GA_3 for the time periods indicated and chromatin was isolated from the tissue essentially as described (Kahl and Wechselberger, 1977). RNA synthesizing capacity was determined in a test mixture consisting of 100 mM Tris-HCl, pH=8.0; 2.5 mM $MgCl_2$; 3.75 mM $MgCl_2$; 2.5 mM dithiothreitol; 2 mM each of GTP, ATP and CTP; 0.037 mM ^{3}H-UTP and chromatin corresponding to 5-8 µg DNA in a final volume of 0.5 ml. RNA polymerase activity is expressed as percentage of the activity of untreated controls. Courtesy of Dr. Bernard Wielgat, Institute of Biochemistry and Biophysics, Polish Academy of Sciences, Warsaw.

It is not clear whether this fact reflects increased uptake of the hormone or - as is suggested tentatively - increased availability of adequate receptors for GA_3. In any event, the hormone induces the formation of all different RNAs. Incubation of tissue slices in 10^{-7} M GA_3 for about 20 hours enhances the incorporation of ^{3}H-uridine into heavy and light ribosomal RNAs,5S- and 4S-RNAs and 9-12S RNAs, which by far exceeds that registered in wounded tissues without any exogenously added hormones (Figure 10).

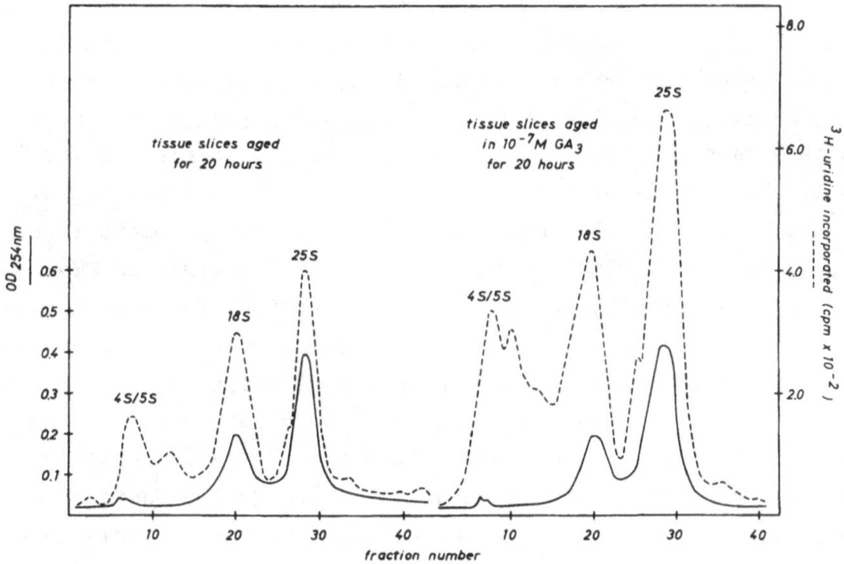

Figure 10. Hormone-induced synthesis of 25S-, 18S-, 9-12S- and 5S/4S-RNA in aging potato tuber tissue slices.
A pulse of 100 µCi ^{3}H-uridine was given to the tissue slices during the last hour of a 20 hours incubation period. Subsequently the ribosomes were isolated and purified. The different RNAs were extracted from these purified ribosomes with phenol and separated on a linear 17-34% sucrose gradient in 10 mM Tris-HCl.pH=8.0; 10 mM $MgCl_2$; 50 mM KCl (24 hours at + 4°C and 25 000 rpm in an SW 25.1 rotor of the Spinco).
Courtesy of Dr. Bernard Wielgat, Institute of Biochemistry and Biophysics, Polish Academy of Sciences, Warsaw, Poland.

GIBBERELLIC ACID INDUCES ACTIVATION OR SYNTHESIS OF RNA-POLYMERASE I AND II.

The hormone-induced changes in RNA synthesis may reflect a whole series of molecular events reaching from the activation of previously inactive genes up to the activation or synthesis of RNA polymerases. Wounding of potato tuber tissue has been shown to enhance Chromatin-bound DNA-dependent RNA polymerase activity, a fact, that has been confirmed for other storage tissues as well (sugar beet slices: 21). The polymerases read out another set of genes in wounded tissues as compared to intact tissues: nearest neighbor frequency analysis revealed changes in A,G and U content of the RNAs of tissue slices (52). In spite of decreased template availability - a fact, that cannot be interpreted as yet - the polymerases are activated and transcribe qualitatively different products in wounded tissues. It is not clear at all, whether the increase in RNA polymerase activity really represents an activation of the enzyme(s) or is due to accellerated synthesis or both. In view of the demonstrated, rapidly occurring activation-inactivation mechanisms, i.e. transfer of ADP or ADP-ribose from NAD^+ to an arginine residue in the α-polypeptide of E.coli RNA polymerase (31) and according to the generally accepted role of initiation or elongation factors in eukaryotic organisms it may be expected, that the increase in enzyme activity after wounding is the consequence of modification processes. These - if occurring - concern both polymerase I and II, since both enzymes show definitely higher activity after wounding potato tuber tissue. In fact a series of different factors seem to play a role in regulating eukaryotic RNA polymerase activity. In some, if not all, prokaryotes the widely known σ- factor temporarily forms a complex with the so-called core enzyme through protein-protein interactions. It is this complex - called holo-enzyme - which recognizes specific promoter sites. After initiation of RNA synthesis σ dissociates from the ternary complex DNA-enzyme-oligonucleotide and may open up a new round of association processes. The current concept of gene regulation in prokaryotes demands a whole set of different of these factors. Such factors have also been shown to be present in various eukaryotic systems (i.e. calf thymus, yeast, slime moulds, coconut endosperm a.o.).

Be it activation or synthesis of the RNA polymerases, in any case the enhanced action of these enzymes ultimately leads to the observed accumulation of different RNAs. Now GA_3 considerably pronounces this effect: if wounded tissue is supplied with the hormone, both polymerase I (copying rDNA, producing mainly or exclusively ribosomal RNAs) and polymerase II (producing various messenger RNAs) are stimulated to greatly enhanced activity. Nucleolar RNA polymerase I, the activity of which is enhanced only 4fold after wounding of potato tuber tissue, is activated about 19fold over the action of the enzyme from resting tissue, if the hormone is present. Nucleoplasmic polymerase II on the other hand is stimulated only 3fold over the enzyme from intact tissue (Table 8; 51, 132).

	Polymerase activity (pMoles UMP incorporated/hour and 100 μg DNA)	
	Polymerase I	Polymerase II
Intact tuber	1.13	0.35
Slices aged for 18 hrs	4.69	0.52
Slices aged for 18 hrs in the presence of 10^{-7}M GA_3	21.50	1.04

Table 8: Relative activities of the two main chromatin-bound DNA-dependent RNA polymerase isozymes in intact and wounded potato tuber tissues and the effect of GA_3.
Chromatin was isolated and purified as described in detail (Kahl and Wechselberger 1977) and the two isoenzyme activities discriminated by use of variable $(NH_4)_2SO_4$-concentrations and α-amanitin in the test (Kahl and Wielgat, 1976).

Again, GA_3 induces the synthesis of qualitatively different RNAs in the potato tissue system, a fact, which can be interpreted by the switching-on of different genes after wounding (Table 9).

	Distribution of label in each 2',3'-mononucleotide (Mole %)			
	CMP	AMP	GMP	UMP
Intact tuber	21.4	26.5	27.0	25.2
Slices aged for 17 hrs	18.4	29.6	28.1	23.9
Slices aged for 17 hrs in the presence of 10^{-7}M GA_3	19.8	30.2	28.1	21.9

Table 9: Nearest neighbor frequency analysis of the RNA produced by chromatin-bound RNA polymerases of intact potato tubers, tissue slices aged for 17 hours in the absence and presence of added gibberellic acid (10^{-7}M). Chromatin was isolated and purified through 1.4 M sucrose gradients and aliquots were incubated with 50 μCi adenosine-5'-α-^{32}P-triphosphate for 40 minutes in the standard assay mixture. The labelled RNAs were isolated and hydrolyzed and the different 2', 3'-mononucleotides separated by high voltage paper electrophoresis. Mole % were calculated from the counts in each peak of the electropherogram (Kahl and Wielgat, 1976).

However, the read-out of rDNA genes obviously is simply enhanced: even in intact tissues rRNA is synthesized, albeit weakly. It is the size of nucleoli that seems to be a fairly reliable indicator for rDNA transcription or pre-ribonucleoprotein particle formation respectively. Whereas the size of nucleoli in the nuclei of intact cells averages to about 25 μ^2, the volume of that part of the chromatin increases as a consequence of wounding (51, 115). GA_3 strengthen this effect (Fig. 11).

The hormone then in some way interferes with the expression of genetic activity.

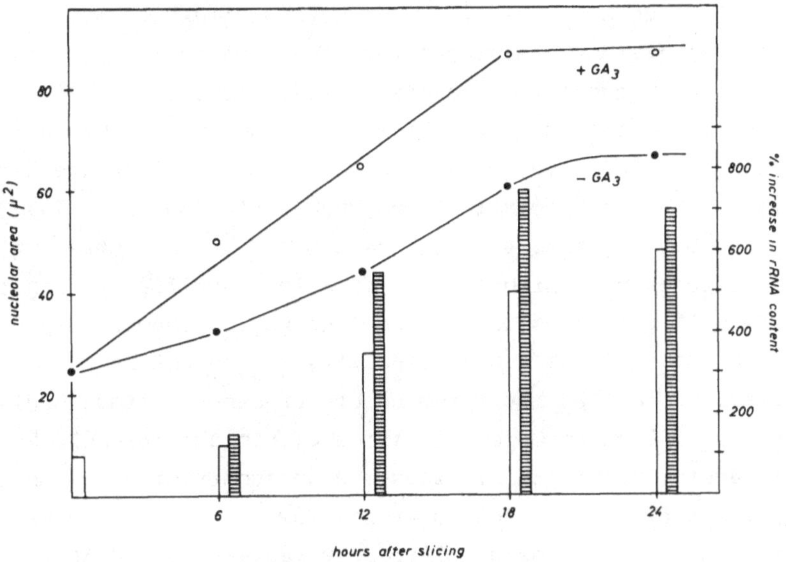

Figure 11. Hormone-induced increase in nucleolar size and rRNA content of isolated ribosomal preparations.
Thin sections of pith tissue from intact potato tubers and from slices after 6,12,18 and 24 hours of aging respectively were taken rectangular to the wound surface, fixed and stained in carmine acetic acid, destained in 5% acetic acid and the nucleolar area calculated from a series of 30 individual measurements. Ribosomes were isolated from ten grams of pith tissue each, purified through sucrose step gradients and dissociated into their subunits by 15 mM EDTA in 10 mM Tris-HCl, pH=8.0 and 15 mM KCl. The different components were separated from one another by centrifugation through 12-30% linear sucrose gradients (17 hours, +4°C, 62000g, SW 25.1 rotor, Beckman L5 65B). The fractions containing the 60S- and 40S-subunits were collected, the RNA extracted with LiCl and the rRNA content determined spectrophotometrically (Kahl 1971; Paulson 1973).

GIBBERELLIC ACID INDUCES CHANGES IN QUANTITY AND QUALITY OF NON-HISTONE CHROMOSOMAL PROTEINS.

The genetic apparatus of eukaryotes, the chromatin, represents a supramolecular complex of DNA, RNA, histones and nonhistone proteins (NHP). Ample evidence now accumulated shows that the regulated expression of specific genes in the differentiated state of eukaryotic cells results from the specific restriction of genetic information in the DNA by chromosomal proteins. Several lines of evidence suggest that the non-histone proteins rather than the basic histones are important in gene regulation generally (57, 62, 91, 110, 117, 123) and specifically in the control of transcription during the cell cycle (89, 116, 118, 119). So the histones suppress most of the template activity of DNA in chromatin in an unspecific way (20, 135), they do not exhibit major tissue differences nor is the number of different histones sufficiently high to allow specific binding to vastly different promoter sites in the eukaryotic genome. A major function of histones is to confer stability to the DNA in chromatin and undoubtedly to act as gross repressors of genes. Although histones are modified at rather specific sites within the molecule by methylations, acetylations, thiolations and phosphorylations - which change the affinity of these proteins towards DNA - they are currently not thought to be the specific gene repressors and derepressors. The selective transcription of DNA seems to be mediated by the non-histone proteins instead. This class of proteins, mostly acidic, is extremely heterogeneous. About 30-40 such proteins can easily be detected on chromatin of higher plants (32, 82, 124) and some 100-200 components my be resolved from gel patterns by appropriate analytical methods (22). The different proteins are subject to cell-specific and cell-stage-specific chemical modifications such as phosphorylation, thiolation, acetylation and possibly methylation.

Phosphorylation and dephosphorylation of such proteins seems to be a dominant factor in gene control: phosphoproteins are highly active in displacing histones from distinct sites of the DNA in chromatin and thereby releasing genes which can be transcribed (90, 111, 123). The interaction of these proteins with DNA appearsto depend on their phosphorylation level (58), so that phosphokinases may be the ultimate determinants of specific gene regulation. This agrees well with the abundant evidence of tissue - and cell-specificity of non-histone phosphoproteins (23, 93). The enzymes responsible for phosphorylation of threonyl- and serylresidues of the NHP as a rule are the actual targets of effectors such as cAMP and the first elements responding to the hormone-signal (59). This seems also to be true for plant systems. In context with a variety of other important observations (Table 10) it may be taken as sure that non-histone chromosomal proteins either as such or modified by phosphorylation act as specific gene activators. The mechanism of reaction, however, still remains obscure.

Table 10

Role of acidic non-histone nuclear proteins in the regulation of gene activity.

1. Active tissues contain more NHP than inactive tissues.
2. Active chromatin contains more NHP than inactive chromatin.
3. NHP restore histone-inhibited DNA-dependent RNA synthesis in vitro.
4. NHP increase the in vitro transcription of inactive chromatin and activate repressed gene loci.
5. NHP interact with DNA in vitro and modify transcription in a way that is characteristic of the tissues of origin.
6. Tissue-specific NHP bind to homologous DNA.
7. NHP possess tissue specificity.
8. NHP regulate cell cycle stage-specific transcription of distinct genes (i.e. histone genes).

Any attempt to explore the role of phytohormones in gene activation of wounded plant tissues then has to take into account the central role of non-histone chromosomal proteins. One major difficulty in interpretation of respective experimental results is the extreme functional heterogeneity of that portion of chromatin proteins. It is a matter of definition, whether or not proteins, which are nucleo - and cytoplasmic at the same time, belong to the class of NHP. Strictly spoken, only proteins being integral parts of the chromatin should be subsummarized in that category. However, unless no definite catalogue will be accepted generally, a number of DNA-dependent enzymes such as DNA-polymerases and RNA-polymerases, DNA-independent enzymes such as poly(A)-polymerases, DNAses and RNAses as "processing" enzymes, proteases, protein-kinases and structural and regulatory proteins (i.e. hormone receptors) all are included.

However, all the enzymes were excluded from the experiments reported here, since these proteins bind less tightly to the chromatin structures and dissociate together with possible cytoplasmic contaminants. These rigid procedures encircle isolation of chromatin, its extensive washing with low salt buffers, a double purification through step gradients of 1.4 M - 1.7 M - 2.0 M - 2.2 M and 1.4 M - 2.0 M linear gradients of sucrose. After dialysis the histones are acid-extracted and the DNA-NHP-complex centrifuged out. The pellet is twice extracted with 0.1% SDS and the solubilized NHP-fraction dialyzed stepwise and lyophilized. Aliquots of the solubilized lyophilisate are electrophoresed in an SDS-acrylamide system.

In virtually inactive chromatin of resting potato tuber tissue a complement of up to 20 distinct non-histone chromosomal proteins can be found. The molecular weight of these compounds ranges from less than 10 000 up to more than 300 000 daltons. It can be taken as sure, that most bands on the gel are composed of more than one compound. Generally the high-molecular weight proteins are more numerous. Wounding of quiescent potato tuber tissue evidently induces the synthesis of NHP, since on a weight basis the mass of these proteins increases about 30 fold. This agrees well with the electrophoretic pattern: more than 29 different bands may be separated from each other (Fig. 12). Since experimental conditions are exactly the same, i.e. mercaptoethanol is present in both preparations to prevent aggregations, it can be concluded that wounding triggers the synthesis of a series of new non-histone proteins. If GA_3 is present during wound-healing, then even more NHP are part of the chromatin complex and up to 35 different bands are visible on the gels. These bands partly correspond to bands also present in wounded tissues without any hormonal treatment. However, at least some of the bands are characteristic for GA_3-treated cells (Fig. 12). GA_3 then triggers extensive synthesis of these proteins (Table 11).

Certainly the simple demonstration, that the hormone in some way induces the synthesis of new NHP is not convincing evidence for the involvement of exactly these proteins in hormone-mediated gene

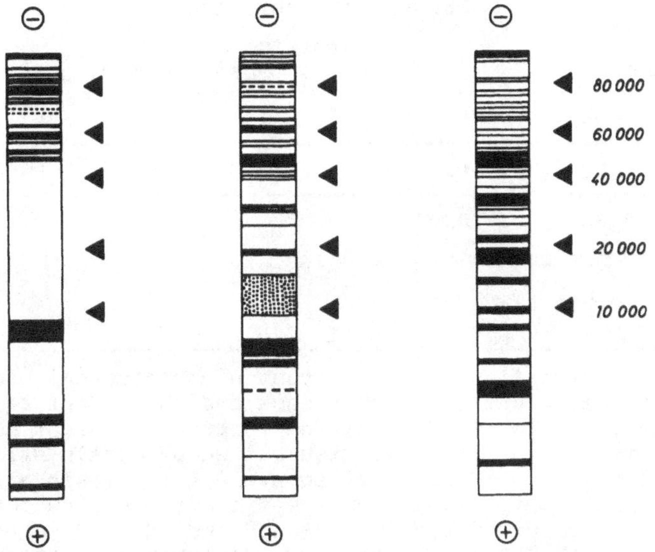

Figure 12. Changes in the pattern of non-histone chromosomal proteins (NHP) in aging white potato tuber tissue and after hormone treatment.
Left: SDS-polyacrylamide gel pattern of NHP from intact tubers.
Middle: NHP from tissue slices aged for 24 hours.
Right: NHP from tissue slices aged for 24 hours in the presence of 10^{-7}M GA_3.
Arrows indicate positions of molecular weight markers. Experimental conditions are outlined in the text.

activation. However, it can be shown that the non-histone chromosomal proteins from hormone-treated tissues are highly active in restoration of histone-inhibited RNA-synthesis in vitro, whereas NHP from resting tissues definitely lack this property. So even fairly high concentrations of NHP from chromatin of resting potato tuber tissue do not release DNA from repression by histones in a heterologous system (Fig. 13A).

Non-histone proteins from wounded tissues on the contrary are potent candidates for counter-acting the inhibitory histones. Some 8 µg histones/µg DNA slow down transcription to about 38%. Addition of only 1 µg of NHP in this case leads to a nearly complete deinhibition.

	Non-histone chromosomal proteins (µg/g fresh weight)	histone proteins
Intact tuber	0.053	1.08
Slices aged for 24 hrs	1.67	0.81
Slices aged for 24 hrs in the presence of 10^{-7}M GA_3	2.20	1.33

Table 11: Variations in the amount of chromosomal proteins after wounding white potato tuber tissue and the effect of 10^{-7}M GA_3. Chromatin was isolated, purified through different sucrose density gradients and the histones dissociated by gentle stirring of the chromatin in 0.3N HCl. After removal of the residual DNA-non histone-protein-complex by high-speed centrifugation the non-histone proteins were extracted in 0.1% SDS and 10 mM 2-mercaptoethanol at 37°C overnight. After gradient dialysis and complete removal of SDS the protein preparation was separated from the DNA and lyophilized.

Higher concentrations seem to reverse this effect to some extent, a phenomenon not yet explainable (Fig. 13B). Non-histone proteins from chromatin of hormone-treated tissues are even more potent activators in so far as they restore histone-inhibition at concentrations as low as 0.2 µg in the same system (Fig. 13C).

At this point should be stressed that the heterologous system used is highly artificial and does not necessarily reflect the situation in situ. However, it readily demonstrates that the non-histone proteins are subject to wound-induced and more pronoucedly hormone-induced changes. These quantitative and qualitative changes doubtless have some stimulating influence on the read-out of histone-inhibited genes. It remains to be elucidated what the chemical background of these changes is and it remains to be seen, whether the proteins are as effective in a homologous system as they are in the heterologous one.

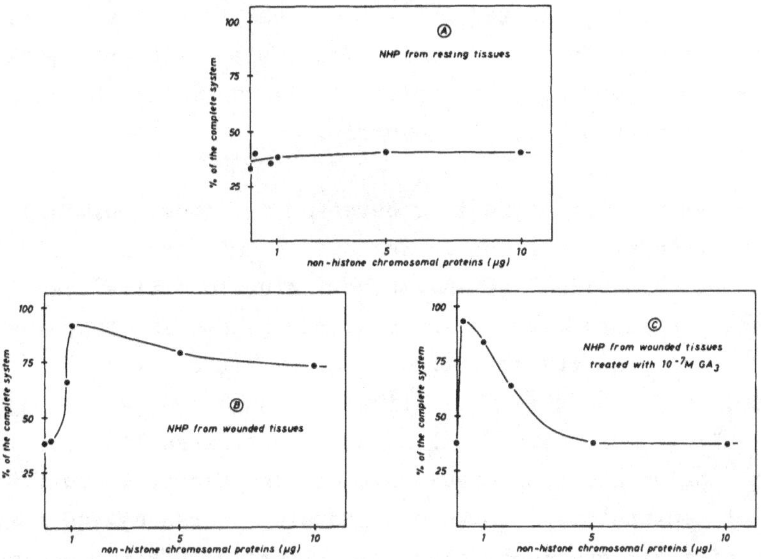

Figure 13. Functional diversity of purified non-histone chromosomal proteins in a heterologous in-vitro transcriptive system. Non-histone chromosomal proteins from resting tissue (A), aged slices (B) and GA_3-treated slices (C) were isolated as outlined in Table 11 and added to 2,5 µg calf thymus DNA complexed with 8 µg of histone protein of the respective tissue. Deinhibition experiments were performed in an incorporation system (52) with rising concentrations of non-histone protein.

PERSPECTIVES

All the experimental results flow into several conclusions. On one hand the wound-induced mitoses are regulated by an interplay of several phytohormones and it is simply superfluous to implicate hypothetical "necro-hormones", although in systems others than the wounded plant storage tissues these might well exist, but have never been demonstrated. In fact, the "lepto-hormone" theory of Haberlandt may have a substantial basis: the "leptom" or phloem seems to be the main transport system of phytohormones and consequently the tissue with the highest concentrations of these effectors. Among these auxins and/or gibberellins evidently are involved in some way in the process of mitosis, whereas cytokinins seem to be preferentially promotive to proliferative growth. The whole series of cytological events at the wound-surface apparently is terminated by abscisic acid or similar inhibitory regulators.

On the other hand we begin to understand how these hormones work on a molecular level. At least in the case of gibberellic acid the hormone is "recognized" by the wound-tissue in a time-specific way, the signal is somehow transferred to the genome and triggers the activation of gene transcription. However, it can't be simple activation, since there is a qualitative change in the RNA population of the cell in wounded as well as hormone-treated tissues: indication for the read-out of previously blocked genes. Concomitantly the number of non-histone chromosomal proteins increases and again, the quality of these compounds alters drastically. After wounding or hormone treatment the NHP are potent gene activators, whereas before they were not. However, it is as yet completely unknown, what the cause and what the effect really is, although in the light of current evidence it may well be the non-histone chromosomal proteins which confer gene-activating specificity.

One of the main, yet unsolved problems lies ahead: what is the primary event after mechanical disturbance of a tissue? Is there any one primary event, which triggers the whole sequence of reactions in a wound-cell? Are there many? We simply feel that one of the first changes upon wounding concerns the osmotic potential of the cell. The link between this change and the sequence of molecular events is still missing and there is enough reason to regard the scheme of proposed events after wounding plant storage tissue as highly hypothetical, albeit discussible (see scheme).

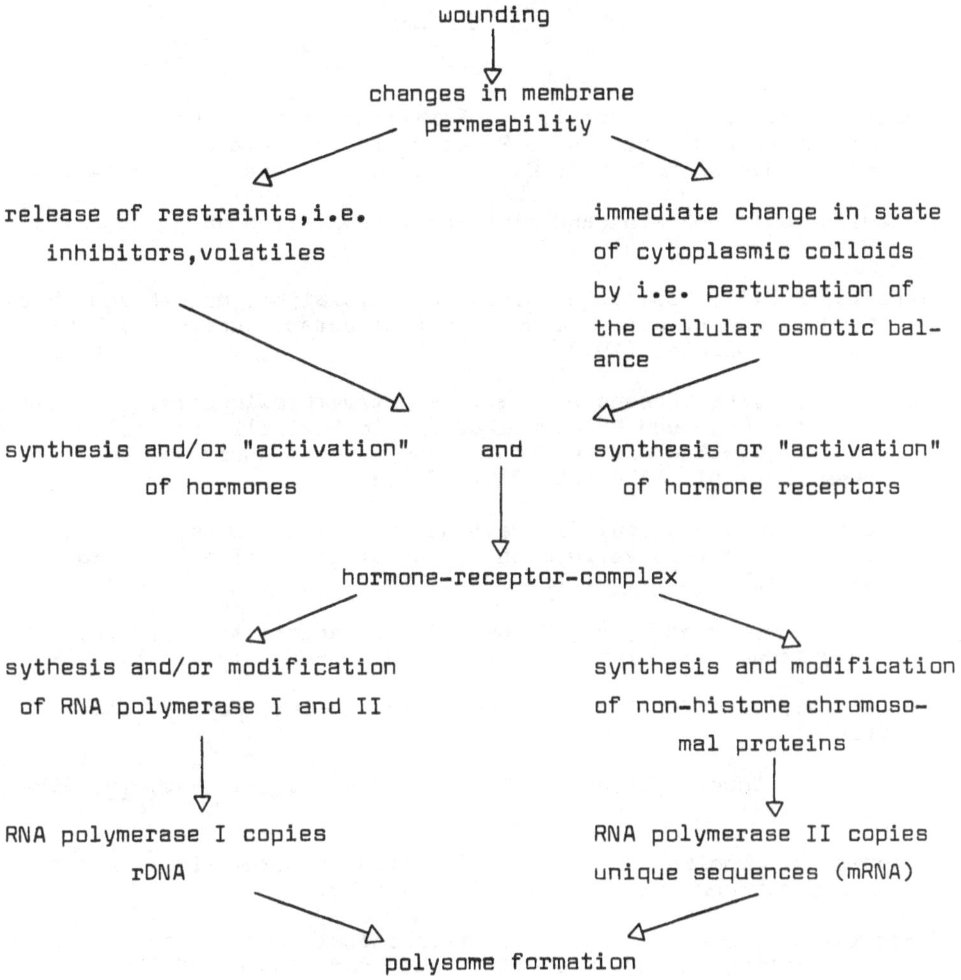

ACKNOWLEDGEMENTS

Part of the work reported in this chapter was supported by the Deutsche Forschungsgemeinschaft (Bad Godesberg) and the Alexander von Humboldt-Foundation. Thanks are due to Miss Heinz, Mrs. Kost and Mrs. Roth for help during preparation of this manuscript and Mr. Wechselberger for expert technical assistance.

References

1 Abdelkader, A.B., Auderset, G.: Formation de polyribosomes dans tranches de tubercules de Pomme de terre (Solanum tuberosum L.) maintenues en "survie". C.R. Acad. Sci. Paris 274, 1311-1314

2 Adamson, D.: Expansion and division in auxintreated plant cells. Can. J. Bot. 40, 719-742 (1962)

3 Anstis, P.J.P., Northcote, D.H.: The initiation, growth and characteristics of a tissue culture from potato tubers. J. Exptl. Bot. 24, 425-441 (1973)

4 Barckhausen, R., Rosenstock, G.: Feinstrukturelle Untersuchungen zum Vorkommen von "Riesenvakuolen" in Nukleoli verletzter Parenchymzellen aus der Knolle von Solanum tuberosum L. Biochem. Physiol. Pfl. 164, 327-337 (1973)

5 Baulieu, E.E., Alberga, A., Raynaud-Jammet, C., Wira, C.R.: New look at the very early steps of oestrogen action in uterus. Nature New Biol. 236, 236-239 (1972)

6 Biswas, B.B., Mondal, H., Mondal, R.K.: Factors that control RNA polymerase from plant cells. Symp.Biol.Hung. 13, 93-111 (1972)

7 Bloch, R.: Wound healing in higher plants. Bot. Rev. 7, 110-146 (1941)

8 Bloch, R.: Wound healing in higher plants II. Bot. Rev. 18, 655-679 (1952)

9 Bonner, J., English J. jr.: Purification of traumatin, a plant wound hormone. Science 86, 352-353 (1937)

10 Broekaert, D., van Parijs, R.: Histological observations and histophotometric DNA measurements in wounded tissues of Pisum sativum L. infected with Agrobacterium tumefaciens. J. Exptl. Bot. 24, 820-827 (1973)

11 Bünning, E.: Entwicklungs- und Bewegungsphysiologie der Pflanze. Springer-Verlag, Berlin-Göttingen-Heidelberg (1953)

12 Bünning, E.: Die Wirkung von Wundreizen. Handb. Pflanzenphysiol. XVII, 119-134 (1959)

13 Burström, H.: Other biogenous growth promoters. Handb. Pflanzenphysiol. XIV, 1156-1161 (1961)

14 Chapman, K.S.R., Trewavas, A., van Loon, L.C.: Regulation of the phosphorylation of chromatin-associated proteins in Lemna and Hordeum. Plant Physiol. 55, 293-296 (1975)

15 Conrad, K., Köhn, B.: Zunahme von Cytokinin und Auxin in verwundetem Speichergewebe von Solanum tuberosum. Phytochem. 14, 325-328 (1975)

16 Coutrez-Geerinck, D., Fries, D.: Respiration et contenu auxinique des parties peripheriques et on centrales de tubercules de Solanum tuberosum. Ann. Physiol. vég. Univ. Bruxelles XIV, 107-150 (1969)

17 Dalessandro, G.: Interaction of auxin, cytokinin, and gibberellin on cell division and xylem differentiation in cultured explants of Jerusalem artichoke. Plant Cell Physiol. 14, 1167-1176 (1973)

18 Das, T.M., Hildebrandt, A.C., Riker, A.J.: Cine-photomicrography of low temperature effects on cytoplasmic streaming, nucleolar activity and mitosis in single tobacco cells in microculture. Amer. J. Bot. 53, 253-259 (1966)

19 Davis, E.A.: Effects of several plant-growth regulators on wound healing of sugar maple. Bot. Gaz. 111, 69-77 (1949)

20 DeLange, R.J., Smith, E.L.: Histones structure and function. Ann. Rev. Biochem. 40, 279-314 (1971)

21 Duda, C.T., Cherry, J.H.: Chromatin- and nuclei-directed ribonucleic acid synthesis in sugar beet root. Plant Physiol. 47, 262-268 (1971)

22 Elgin, S.C.R., Boyd, J.B., Hood, L.E., Wray, W., Wu, F.C.: A prologue to the study of the non-histone chromosomal proteins. Cold Spring Harbor Symp.quant.Biol. 38, 821-833 (1974)

23 Elgin, S.C.R., Weintraub, H.: Chromosomal proteins and chromatin structure. Ann.Rev.Biochem. 44, 725-774 (1975)

24 Ellis, R.J., MacDonald, I.R.: Characterization of amino acid incorporation by subcellular fractions from sterile beet disks. Planta (Berl.) 83, 248-256 (1968)

25 English, J., Bonner, J., Haagen-Smit, A.J.: IV. Structure and synthesis of a traumatin. J.Amer.Chem.Soc. 61, 3434-3436 (1939)

26 Fellenberg, G.: Development Physiology. Fortschr. Bot. 36, 147-163 (1974)

27 Fellenberg, G.: Über die Organbildung von in vitro kultiviertem Knollengewebe von Solanum tuberosu. Zeitschr. f. Bot. 51, 113-141 (1963)

28 Fosket, D.E., Short, K.C.: The role of cytokinin in the regulation of growth, DNA synthesis and cell proliferation in cultured soybean tissues (Glycine max. var. Biloxi). Physiol.Plant. 28, 14-23 (1973)

29 Frank, A.B.: Die Krankheiten der Pflanzen. Verlag Trewendt, Breslau (1880)

30 Gersch, M.: Biologie der Zelle. B.G. Teubner Verlagsgesellschaft, Leipzig (1953)

31 Goff, C.G.: T_4-induced modification of Escherichia coli RNA polymerase in vitro. 3. Int.Symp. on Metabolic Interconversion of Enzymes, Springer Berlin-Heidelberg-New York,p.235-244 (1974)

32 Gregor, D., Reinert, J., Matsumoto, H.: Changes in chromosomal proteins from embryo induced carrot cells. Plant Cell Physiol. 15, 875-881 (1974)

33 Haagen-Smit, Viglierchio, D.R.: Investigation of plant wound hormones. Rec.Trav.chim.Pays-Bas 74, 1197-1206 (1955)

34 Haberlandt, G.: Zur Physiologie der Zellteilung. Sitzungsber. Kgl.Preuß.Akad.Wiss. 1,318-345 (1913)

35 Haberlandt, G.: Wundhormone als Erreger von Zellteilungen. Beitr. Allg.Bot. 2, 1-53 (1921)

36 Haberlandt, G.: Über Zellteilungshormone und ihre Beziehungen zur Wundheilung. Biol.Zentralbl. 42, 145 (1922)

37 Haberlandt, G.: Zur Entwicklungsphysiologie des Periderms. Sitzungsver. preuß.Akad.Wiss.phys.-math.Kl. 23, 317-338 (1928)

38 Hackett, D.P., Thimann, K.V.: The nature of the auxin-induced water uptake by potato tissue. Amer.J.Bot. 39, 553-560 (1952)

39 Hardin, J.W., Cherry, J.H., Morré, D.J., Lembi, C.A.: Enhancement of RNA polymerase activity by a factor released by auxin from plasma membrane. Proc.Nat.Acad.Sci.USA 69, 3146-3150 (1972)

40 Hemberg, T.: Über das Vorkommen wachstumshemmender Stoffe in Kartoffelknollen und die Bildung wachstumsfördernder Stoffe in Wundflächen derselben. Arkiv. Bot. 30B, 1-8 (1943)

41 Hemberg, T.: Weitere Untersuchungen über wachstumsfördernde Stoffe in Wundflächen von Kartoffelknollen. Svensk.bot. Tidskr. 38, 428 (1944)

42 Hemberg, T.: Studies of auxins and growth-inhibiting substances in the potato tuber and their significance with regard to its rest-period. Acta Hort.Berg. 14, 133 (1947)

43 Hertel, R., St.-Thomson, K., Russo, V.E.A.: In-vitro auxin binding to particulate cell fractions from corn coleoptiles. Planta 107, 325-340 (1972)

44 Holst, V.B.: Some properties of inhibitor B from Solanum tuberosum compared to abscisic acid. Physiol. Plant. 24, 392-396 (1971)

45 Johnson, J.M., Jones, L.E.: Behavior of nucleoli and contracting vacuoles in tobacco cells growing in microculture. Amer.J.Bot. 54, 189-198 (1967)

46 Kahl,G.: Synthesis of rRNA, tRNA and other RNA-species concomitant with polyribosome formation in aging potato tuber slices. Z.Naturforsch. 26b, 1058-1064 (1971)

47 Kahl,G.: Activation of protein synthesis in aging potato tuber slices. Z.Naturforsch. 26b, 1064-1067 (1971)

48 Kahl,G.: Genetic and metabolic regulation in differentiating plant storage tissue cells. Bot.Rev. 39, 274-299 (1973)

49 Kahl,G.: Metabolism in plant storage tissue slices. Bot.Rev. 40, 263-314 (1974)

50 Kahl,G., Rosenstock, G., Lange, H.: Die Trennung von Zellteilung und Suberinsynthese in dereprimiertem pflanzlichem Speichergewebe durch Tris-(hydroxymthyl-)aminomethan. Planta 87, 365-371 (1969)

51 Kahl,G., Wielgat,B.: Regulation of transcriptional activity in wounded potato tuber tissues. Physiol.Vég. 14, 725-738 (1976)

52 Kahl,G., Wechselberger, M.: Activation of chromatin-bound DNA-dependent RNA polymerase (E.C.2.7.7.6) in plant storage tissue slices. Z.Naturforsch. 32c, 229-235 (1977)

53 Kalimi, M., Beato, M., Feigelson, P.: Interaction of glucocorticoids with rat liver nuclei. I. Role of the cytosol proteins. Biochemistry 12, 3365-3371 (1973).

54 Kaussmann, B.: Pflanzenanatomie. Gustav Fischer-Verlag, Jena. (1963)

55 Kefford, N. P., Goldacre, P. L.: The changing concept of auxin. Amer.J.Bot. 48, 643-650 (1961)

56 Keller, M.: Die Wirkung synthetischer Traumatinsäure auf verschiedene Pflanzengewebe. Z.Bot. 45, 1-38 (1957)

57 Kleinsmith, L. J., Heidema, J., Carroll, A.: Specific binding of rat liver nuclear proteins to DNA. Nature 226, 1025-1026 (1970)

58 Kleinsmith, L. J.: Specific binding of phosphorylated non-histone chromatin proteins to desoxyribonucleic acid. J.Biol.Chem. 248, 5648-5653 (1973)

59 Kleinsmith, L. J. in: "Chromosomal proteins and their role in the regulation of gene expression", Acad.Press. New York, San Francisco, London, p. 45-57 (1975)

60 Kohlenbach, H. W.: Nucleoli mit "pulsierenden Vakuolen" bei kultivierten Rhoeo-Epidermiszellen. Z.Pflanzenphysiol. 58, 187-190 (1967)

61 Kohlenbach, H. W.: Das durch Kinetin induzierte Nucleolus-, Kern- und Plasmawachstum isolierter Epidermiszellen von Rhoeo spathacea. Z.Pflanzenphysiol. 63, 297-307 (1970)

62 Kostraba, N., Wang, T. Y.: Non-histone proteins and gene activation in regenerating rat liver. Exptl. Cell Res. 80, 291-296 (1973)

63 Krenke, N. P.: Wundkompensation, Transplantation und Chimären bei Pflanzen. Springer-Verlag, Berlin (1933)

64 Küster, E.: Pathalogische Pflanzenanatomie in ihren Grundzügen. Verlag G. Fischer, Jena (1925)

65 Kupila-Ahvenniemi, S., Thermann, E.: DNA synthesis, mitosis and ploidy of dividing cells in early crown gall. Physiol. Plant. 32, 122-127 (1974)

66 Kupila-Ahvenniemi, S., Karjalahti, L.: The role of abscisic acid in sterile and grown gall inoculated wounds. Z.Pflanzenphysiol. 76, 260-269 (1975)

67 Lange, H.: Atmungswege bei vernarbenden und proliferierenden Gewebefragmenten der Kartoffelknolle. Planta 90, 119-132, (1970)

68 Lange, H., Rosenstock, G.: Physiologisch-anatomische Studien zum Problem der Wundheilung. II. Kausalanalytische Untersuchungen zur Theorie des Wundreizes. Beitr. Biol. Pflanzen 39, 383-434 (1963)

69 Lange, H., Rosenstock, G.: Über Alterungserscheinungen bei traumatisch induzierter meristematischer Aktivität. Phytopath.Z. 51, 136-152 (1964)

70 Lange, H., Rosenstock, G.: Kausalanalytische Untersuchungen zum Alterungsvorgang bei der Wundkompensation von Speicherorganen unter besonderer Berücksichtigung des Nährstoffaktors. Phytopath.Z. 52, 188-201 (1965)

71 Lange, H., Kahl, G., Rosenstock, G.: Über den Einfluß hypotonischen Milieus auf den Stoffwechsel dereprimierter pflanzlicher

Gewebe. Physiol.Plant. 23, 80-87 (1970)

72 Leaver, C. J., Key, J. L.: Polyribosome formation and RNA synthesis during aging of carrot-root tissue. Proc.Nat.Acad. Sci. USA 57, 1338-1344 (1967)

73 Lembi, C. A. Morré, D. J., St.-Thomson, K. Hertel, R.: N-1-napthylphthalmic-acid-binding activity of a plasma membrane-rich fraction from maize coleoptiles. Planta 99, 37-45 (1971)

74 Libbenga, K. R., J. G. Torrey: Hormone-induced endoreduplication prior to mitosis in cultured pea root cortex cells. Amer.J. Bot. 60, 293-299 (1973)

75 Lipetz, J.: Wound-healing in higher plants. Int. Rev. Cytol. 27, 1-28 (1970)

76 Loofbourow, J. R., Dwyer, S. C.: Intercellular wound hormones produced by Heteroauxin. Science 88, 191-192 (1938)

77 Loofbourow, J. R., Dwyer, S. C., Morgan, R. S. H.: Intercellular wound hormones from ultraviolet injured cells. Stud.Inst. Diri Thomae II, 137-153 (1938)

78 Loofbourow, J. R., Cook, E.S., Stimson, Sr. M. M.: Chemical nature of proliferation-promoting factors from injured cells. Nature 142, 573-574 (1938)

79 Loofbourow, J. R., Dwyer, S. C.: Production of intercellular hormones. Nature 143, 725-726, (1939)

80 Loofbourow, J. R., Dwyer, S. C., Hart, O. P.: Production of intercellular hormones by mechanical injury. Nature 144, 553-554 (1939)

81 Loon van, L. C., Trewavas, A., Chapman, K. S. R.: Phosphorylation of chromatin-associated proteins in Lemna and Hordeum. Plant Physiol. 55, 288-292 (1975)

82 Matsumoto, H., Hirasawa, E., Takahashi, E.: Fractionation of chromosomal proteins of pea seedlings using procedures which avoid denaturation. Plant Cell Physiol. 17, 955-964 (1976)

83 Mayer, L.: Wachstum und Organbildung an in vitro kultivierten Segmenten von Pelargonium zonale und Cyclamen persicum. Planta 47, 401-446 (1956)

84 Mondal, H., Mandal, R. K., Biswas, B. B.: Factors and rifampicin influencing RNA polymerase isolated from chromatin of eukaryotic cell. BBRC 40, 1194-1200 (1970)

85 Mondal, H., Mandal, R. K., Biswas, B. B.: RNA stimulated by indole acetic acid. Nature New Biol. 240, 111-113 (1972)

86 Nagl, W.: Differential inhibition by actinomycin D and histone f1 of mitosis and endomitosis in Allium carinatum. Z.Pflanzenphysiol. 63, 316-326 (1970)

87 Okazawa, Y., Katsura, N., Tagawa, T.: Effects of auxin and kinetin on the development and differentiation of potato tissue cultured in vitro. Physiol.Plant. 20, 862-869 (1967)

88 O'Malley, B. W., Spelsberg, T. C., Schrader, W. T., Chytil, F., Steggles, A. W.: Mechanisms of interaction of a hormone-receptor complex with the genome of a eukaryotic target cell.

89 Park, W. D., Stein, J. L., Stein, G. S.: Activation of in vitro gene transcription from HeLa S_3 chromatin by S-phase nonhistone chromosomal proteins. Biochemistry 15, 3296-3300 (1976).

90 Patel, G. L, Thomas, T. L.: Some binding parameters of chromatin acidic proteins with high affinity of deoxyribonucleic acid. Proc.Natl.Acad.Sci.USA 70, 2524-2528 (1973)

91 Paul, J., Gilmour, R. S., Affara, N., Birnie, G., Harrison, P., Hell, A., Humphries, S., Windass, J., Young, B.: The globin gene: structure and expression. Cold Spring Harb. Symp. quant. Biol. 38, 885-890 (1973)

92 Pearsall, W. H., Priestley, J. H.: Meristematic tissues and protein iso-electric points. New Phytologist XXII, 185-191 (1923)

93 Platz, R. D., Kish, V. M., Kleinsmith, L. J.: Tissue specificity of non-histone chromatin phosphoproteins. FEBS-Letters 12, 38-40 (1970)

94 Paulson, R. "Isolation, purification and factionation of RNA" in: The ribonucleic acids, ed. Stewart, P. R., Letham, D. S., Springer Berlin-Heidelberg-New York, 243-261 (1973)

95 Rappaport, L., Sachs, M.: Wound-induced gibberellins. Nature 214, 1149-1150 (1967)

96 Rose, R. J., Setterfield, G., Fowke, L. C.: Activation of nucleoli in tuber slices and the function of nucleolar vacuoles. Exp.CellRes. 71, 1-16 (1972)

97 Rosenstock, G.: Physiologische und anatomische Studien zum Problem der Wundheilung. I. Untersuchungen über Zusammenhänge zwischen dem Wassergehalt verletzten Kartoffelparenchyms und dessen Vernarbungsreaktion nebst Bemerkungen zur Theorie des Wundreizes. Beitr.Biol.Pfl. 38, 275-319 (1963)

98 Rosenstock, G., Lange, H., Kahl, G.: Wechselbeziehungen zwischen CO_2-Gaswechsel, Volumen und Oberfläche bei Gewebefragmenten von Speicherparenchym. Z.Pflanzenphysiol. 60, 173-186 (1969)

99 Rosenstock, G., Lange, H., Sponholz, W. R.: Untersuchungen zum Problem traumatogener Zellteilungsinduktion beim Speicherparenchym von Solanum tuberosum L. Beitr.Biol.Pfl. 48 399-412 (1972)

100 Rosenstock, G., Sondermann, C.: Untersuchungen zur Analyse der traumatischen Zellteilungsinduktion. II. Die Beeinflussung der histologischen Wundreaktionen von Sukkulentenblättern durch Wuchsstoffe. Beitr.Biol.Pfl. 50, 431-444 (1974)

101 Rosenstock, G., Fischer, H.: (unpublished results)

102 Rosenstock, G., Röll, G.: Die Wirkung von Abscisin auf die Zellteilung und DNS-Synthese bei verwundetem Speichergewebe von Solanum tuberosum L. 1977 (unveröffentlicht).

103 Schick, R., Klinkowski, M.: Die Kartoffel. VEB Deutscher Landwirtschaftsverlag, Berlin (1961)

104 Schilling, E.: Ein Beitrag zur Physiologie der Verholzung und des Wundreizes. Jb.wiss.Bot. 62, 827-862 (1923)

105 Schneider-Orelli, O.: Versuche über Wundreiz und Wundverschluß an Pflanzenorganen. Cbl.Bakt. 30, 420-429 (1911)

106 Schumacher, W.: Physiologie, in: Lehrbuch der Botanik 28, 260-345 (1962), Gustav Fischer Verlag, Stuttgart

107 Sheldrake, A. R.: The production of hormones in higher plants. Biol.Rev. 48, 509-559 (1973)

108 Söding, H.: Weitere Untersuchungen über die Wuchsstoffregulation der Kambiumtätigkeit. Z.Bot. 36, 113-141 (1940)

109 Söding, H.: Die Auxine. Handb. Pflanzenphysiol. XIV, 450-484 (1961)

110 Spelsberg, T. C., Hnilica, L. S.: Deoxyribonucleoproteins and the tissue-specific restriction of the deoxyribonucleic acid in chromatin. Biochem.J. 120, 435-437 (1970)

111 Spelsberg, T. C., Wilhelm, J. A., Hnilica, L. S.: Nuclear proteins in genetic restriction II. The nonhistone proteins in chromatin. Subcell.Biochemistry 1, 107-145 (1972)

112 Sperti, G. S., Loofbourow, J. R., Dwyer, C. M.: Proliferation-promoting factors from ultra-violet injured cells. Stud.Inst.Divi Thomae I, 163-191 (1937)

113 Sperti, G. S.: Proliferation-promoting substances from cells injured by ultra-violet radiation. Nature 140, 643-644 (1937)

114 Sponholz, W. R.: Veränderungen von Nukleus- und Nukleolusgrößen sowie des Nukleinsäurespiegels in Fragmenten des Knollen-

parenchyms von Solanum tuberosum L. nach Derepression durch Verwundung. Ph. D. Thesis, Univ. Frankfurt/Main (1970)

115 Sponholz, R., Rosenstock, G.: Über die Beziehungen zwischen dem Alter von Speichergewebe und dessen Nukleinsäurespiegel vor und nach Verwundung. Ber.Dtsch.Bot.Ges. 86, 485-497 (1974)

116 Stein, G. S., Farber, J.: Role of nonhistone chromosomal proteins in the restriction of mitotic chromatin template activity. Proc.Nat.Acad.Sci.USA 69, 2918-2921 (1972)

117 Stein, G. S., Spelsberg, T. C., Kleinsmith, L. J.: Nonhistone chromosomal proteins and gene regulation (Nonhistone chromosomal proteins may participate in the specific regulation of gene transcription in eukaryotes). Science 183, 817-824 (1974)

118 Stein, G. S., Park, W., Thrall, C., Mans, R., Stein, J.: Regulation of cell cycle stage-specific transcription of histone genes from chromatin by nonhistone chromosomal proteins. Nature 257, 764-767 (1975)

119 Stein, J. L., Reed, K., Stein, G. S.: Effect of histones and nonhistone chromosomal proteins on the transcription of histone genes from HeLa S_3 cell DNA. Biochemistry 15, 3291-3295 (1976)

120 Steveninck van, R. F. M.: The "washing" or "aging" phenomenon in plant tissues. Ann.Rev.Plant Physiol. 26, 237-258 (1975)

121 Teissere, M., Penon, P., Ricard, J.: Hormonal control of chromatin availability and of the activity of purified RNA polymerases in higher plants. FEBS-Letters 30, 65-70 (1973)

122 Teissere, M., Penon, P., Van Huystee, R. B., Azou, Y., Ricard, J.: Hormonal control of transcription in higher plants. BBA 402, 391-402

123 Teng, C., Teng, C., Allfrey, V. G.: Studies of nuclear acidic proteins (Evidence for their phosphorylation, tissue specificity, selective binding to deoxyribonucleic acid and stimulatory effects on transcription). J.Biol.Chem. 246, 3597-3609 (1971)

124 Towill, L. E., Noodén, L. D.: An electrophoretic analysis of the acid-in-soluble chromosomal proteins from different organs of maize seedlings. Plant Cell Physiol. 16, 1073-1084 (1975)

125 Troll, W.: Allgemeine Botanik. Zell- und Gewebedifferenzierung. p. 643, F. Enke Verlag, Stuttgart (1959)

126 Volk, W.: Die Wirkung von Auxinen auf histologische und biochemische Wundreaktionen beim Speicherparenchym von Solanum tuberosum L. Ph. D. Thesis, Univ. Frankfurt/Main (1977).

127 Vries, H. de: Beiträge zur speziellen Physiologie landwirtschaftlicher Kulturpflanzen. IV. Keimungsgeschichte der Kartoffelknollen. Landw.Jb. 7, 217-249 (1878)

128 Wasilewska, L., Cherry, J. H.: Polyribosome formation and RNA synthesis after breaking the dormancy of sugar beet root. Acta Biochimica Polonica 21, 339-354 (1974)

129 Watanabe, A., Imaseki, H.: Induction of deoxyribonucleic acid synthesis in potato tuber tissue by cutting. Plant.Physiol. 51, 772-776 (1973)

130 Weber, F.: Theorie der Meristembildung. Naturwiss. 12, 289-296 (1924)

131 Wehnelt, B.: Untersuchungen über das Wundhormon der Pflanzen. Jb.Wiss.Bot. 66, 773-813 (1927)

132 Wielgat, B., Kahl, G.: In vivo and in vitro RNA synthesis in potato tuber slices under hormonal control. Proc.9.Int.Conf. Plant Growth Substances,Ed.P.E.Pilet,Lausanne 431-432 (1976)

133 Wiesner, J.: Die Elementarstruktur und das Wachstum der lebenden Substanz. Verlag Hölder,Wien (1892)

134 Wilhelm, A.: Untersuchungen über das Chromogen in Vicia faba. Jb.Wiss.Bot. 72, 203-253 (1930)

135 Wilhelm, J. A., Spelsberg, T. C., Hnilica, L.S.: Nuclear proteins in genetic restriction I. The histones. Subcellular Biochemistry 1, 39-65 (1971)

Subject Index

FSC
www.fsc.org
MIX
Papier | Fördert gute Waldnutzung
FSC® C083411

Kolibri 360 GmbH
Ferdinand-Jühlke-Straße 7
99095 Erfurt, Deutschland
produktsicherheit@kolibri360.de